T0231164

COMPUTATIONAL INTELLIGENCE IN INDUSTRIAL APPLICATION

PROCEEDINGS OF THE 2014 PACIFIC-ASIA WORKSHOP ON COMPUTER SCIENCE IN INDUSTRIAL APPLICATION (CIIA, DECEMBER 8–9, 2014, SINGAPORE)

Computational Intelligence in Industrial Application

Editor

Yanlv Ling
Wuhan Institute of Technology, China

CRC Press
Taylor & Francis Group
Boca Raton   London   New York   Leiden

CRC Press is an imprint of the
Taylor & Francis Group, an **informa** business

A BALKEMA BOOK

CRC Press
Taylor & Francis Group
6000 Broken Sound Parkway NW, Suite 300
Boca Raton, FL 33487-2742

© 2008 by Tianjian Ji and Adrian Bell
CRC Press is an imprint of Taylor & Francis Group, an Informa business

No claim to original U.S. Government works

Printed on acid-free paper
Version Date: 20150622

International Standard Book Number-13: 978-0-415-39774-2 (Paperback) 978-0-415-39773-5 (Hardback)

Visit the Taylor & Francis Web site at
http://www.taylorandfrancis.com

and the CRC Press Web site at
http://www.crcpress.com

Table of contents

Computational Intelligence in Industrial Application – Ling (ed.)
© 2015 Taylor & Francis Group, London, ISBN: 978-1-138-02818-0

Preface

The 2014 Pacific-Asia Workshop on Computational Intelligence in Industrial Application (CIIA 2014) was held on December 8–9, Singapore, and focused on various aspects of advances in these areas. The workshop provided a chance for academics and industry professionals to discuss recent progress, and to share ideas, problems and solutions.

These proceedings include 81 peer-reviewed papers. The topics covered in the book include: (1) Computer Intelligence, (2) Application of Computer Science and Communication, (3) Industrial Engineering, Product Design and Manufacturing, (4) Automation and Control, Information Technology and MEMS.

We would like to acknowledge and give special appreciation to our keynote speakers for their valuable contribution, our delegates for being with us and sharing their experiences, and our invitees for participating in CIIA 2014. We also would like to extend our appreciation to the steering Committee and the International Conference Committee for the devotion of their precious time, advice and hard work to prepare for this conference.

Organizing committee

Keynote Speaker and Honorary Chair

Gerald Schaefer, *Loughborough University, UK*

General Chairs

Wei Deng, *American Applied Sciences Research Institute, USA*
Ming Ma, *Singapore NUS ACM Chapter, Singapore*

Program Chairs

Prawal Sinha, *Department of Mathematics and Statistics, Indian Institute of Technology, Kanpur, India*
Harry Zhang, *SMSSI, Singapore*

Publication Chair

Yanlv Ling, *Wuhan Institute of Technology, China*

The members of Scientific Committees

H.B. Kekre, *MPSTME, NMIMS University, Mumbai, India*
P. Halarnkar, *MPSTME, NMIMS University, Mumbai, India*
Tanuja Sarode, *TSEC, Mumbai University, Mumbai, India*
A. K. Reshamwala, *SVKM's NMIMS University, Mumbai, India*
S. M. Mahajan, *Institute of Computer Science, M.E.T, Bandra, Mumbai, India*
Yanling Yang, *Foreign Languages College, Northeast Dianli University, Jilin, P.R. China*
Han Li, *College of Science, Northeast Dianli University, Jilin, P.R. China*
Ling Zhang, *Beijing Information Technology College, China*
Yutian Chen, *Department of Applied Mathematics, Yanshan University, Qinhuangdao, China*
Daijun Pi, *School of Transportation and Automotive Engineering, Xihua University, China*
Chao Gao, *Kunming University of Science and Technology, China*
Kunqian Wang, *Kunming University of Science and Technology, China*
Zhifeng Liu, *Beijing University of Technology, Beijing, China*
Yonghao Wu, *College Wuhan University of Science and Technology, China*
Xiyan Wang, *Foreign Language Institute, Northeast Dianli University, Jilin, China*

Section 1: Computer intelligence

Computational Intelligence in Industrial Application – Ling (ed.)
© *2015 Taylor & Francis Group, London, ISBN: 978-1-138-02818-0*

Research and implementation of component extraction based on design document

You Qun Shi, Jian Wei Zhou & XiaoYan Guo
School of Computer Science and Technology, Donghua University, Shanghai, China

ABSTRACT: The effective project cases of components are the foundation of realizing component based software development and reuse. At present most research focuses on reuse components retrieval and assembly, however, there is no further specification method of how to extract the components from documentation of the target system, which will give a clear component retrieval target in the component library. The present methods are mostly artificial reading documents and manual retrieval component library. Based on this, this paper puts forward a component extraction method based on design document. Through the agent understanding of design document keywords and semantics, we realize the formation of the target system framework oriented component, and through the human-computer interaction verify the correctness of the system. The test result shows that this method can improve the component extraction efficiency, simplify the process of software reuse, and the extracted component contains the semantics and the relationship between them. With a high degree of accuracy and user satisfaction, it contributes to the retrieval and assembly of the component.

KEYWORDS: Design document; Agent; Component; Component extraction; MAS.

1 INTRODUCTION

Software development based on component (Jyotishman Pathak et al. 2004) is a hot research field of software reuse in recent years, and gradually become one of the important means to improve the efficiency and quality of software reuse. At present, domestic and external have obtained high achievements in the fields of component based software development, but the software reuse process is not smooth, the main problems are as follows: (1) How to determine the components which system needed; (2) Component in what form, for what function and interface; (3) How to establish the mapping relationship, accurate search the matching component in the component library. Existing work focuses more on component retrieval and assemblage, lack of in-depth study of how to extract component.

In early studies, Belady proposed a method for extracting reusable modules, this research focused on object oriented component technology. In general, component extraction methods are divided into two categories: knowledge matching and structural analysis.

1 Knowledge matching using the analysis of the semantic elements in the software system, associate the same or similar semantic entity in procedure. However, this method has the problem: it must be assumed that the system to be analyzed with good semantic features.

2 Extraction method of the component based on structural analysis, using abstract graph, such as class diagram, function call graph and so on to display the relationship between components, and convert it to optimal partition graph structure (NP hard problem). But the operational complexity is high and there are many possible forms of software structure, so this method is the lack of sufficient generality.

In view of this, this paper puts forwards a component retrieval strategy based on design document. First of all, use computer aided software engineering to generate system design document; secondly, convert the design document to XML document with automation tools; thirdly, according to the semantic model of document description specification, use component semantic partition strategy to divide the conceptual component; then put forward the extraction strategy based on multi-Agent system to get the target system architecture, form the topology structure of target system; lastly, use the XML element verification method to verify the correctness of the extraction results. Results show that, component extracted through the extraction strategy contains the semantics and the call relationship between each other, it also has a high accuracy rate and flexible extraction process, it provides a good foundation for component retrieval and assemblage.

2 RELATED TECHNOLOGY

The Target system refers to the application system to be developed, component in the system design document called conceptual component, component collection formed by extracting conceptual components called logical component, component in the component library called physical component.

2.1 *Component*

Component refers to software entity which can be deployed independently in the software system, the interface is specified by contract and always provided by third party. It has a relatively independent function, can be clearly identified, it also has an obvious dependence with context. Reusability is the essential characteristic of software component, it is an important factor to evaluate the quality of components.

General component includes three parts as follows: (1) component entity: including code, service and system these three aspects; (2) component document: including many software resources with reuse value, such as requirement specification, system architecture, design document and test case; (3) requirements for the software component: requirements for the external interface and for the inside of the package (support reuse), it can provide support for technical documents.

2.2 *CASE*

CASE (Computer Aided Software Engineering) (Zhen-gong Cai et al.2011)is a set of methods and tools, including graphic tools, code generators and so on. CASE cover various stages of the software development life cycle, it can accelerate the pace of development and assure software quality. At present the mainstream CASE specification is UML (United Modeling Language) (Ke-Qing He et al. 2001), UML is used for visual modeling the system, it is a standard language for describing, visualization and documentation of products. Use UML to describe the system, formed a standard graph visualization, modeling for system specification, helps to understand the system; on the other hand, the component graph in the CASE is conducive to extract the conceptual component.

UML defines the following types of Graphs: Use Case diagram, Static diagram, Behavior diagram, Interactive diagram, Implementation diagram.

2.3 *Agent and MAS*

Agent is a kind of a computer entity or process, it is able to perceive environmental changes in the specific environment, and it can run autonomously to represent designer or user to achieve a series of goals, it can accomplish the user specified tasks accurately. The main characteristics of Agent in including: autonomy, responsiveness, structure distribution, social.

MAS (Multi-Agent System) (Youqun Shi et al. 2012) is a system composed of a plurality of Agent, is a kind of distributed autonomous system. In MAS, the Agent is independent, may be pre-existed and heterogeneous. Through the collaboration between multiple Agent, MAS can improve the basic ability of each Agent, and according to the interaction of each Agent to further understand the society, therefore the multi Agent system is better than the single Agent system.

3 COMPONENT EXTRACTION

This paper presents a component extraction method based on design document. First of all, using automated tools, switches CASE designed document into XML document; secondly, according UML semantic model, divide the conceptual component by using the component semantic partition strategy; then, using the method for analysis of UML sequence diagram based on the context of constraints to obtain the system structure of the target system.

3.1 *Component extraction process*

Establish a multi-Agent system running on the JADE platform, design Agent with independent function to realize the component extraction function. A specific process is shown in Figure 1, and the steps are as follows:

1 User give a request for reading design document through an interface interaction Agent, interaction Agent encapsulates the request into a corresponding request service ACL message, sent to extraction Agent;
2 Extraction Agent read the design document, generate an XML document, analyze and extract the conceptual component;
3 According to the obtained conceptual component, construction Agent analyzes the XML document and conceptual component, generate a logical collection of components, finally get the system structure;
4 Interaction Agent returns the architecture of logical components to the users.

3.2 *Component description specification*

The component standard description includes four fundamental requests: searchable, measurable, easy to use and supportable (Chris Lamela.2000). In order to achieve these goals, further characterization for implementation technique, function declaration, interface specification, expansibility and data is needed (De-Cheng Zhan.2003).

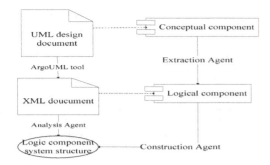

Figure 1. Component extraction process based on MAS.

With the demand of component extraction, in the design document based on the UML, specification description of component mainly involved: functional description, I/O parameter, component types, the interface provided and dependent function. A component description, specification for example, as shown in Table 1.

Table 1. Component description example.

Component name	Shopcart Manage
Function description	Add to Shop cart, Delete from Shop cart, Create Order, Show Shop cart
Input parameters	Id, size, num, price, weight
Component types	Business logic service components
Interface provided	public void Add to Shop cart(String id); public void Delete from Shop cart(String id); public int Num Update(String id, Int num); public void Empty Shopcart(Int num); public int CreateOrder(Int num, Double price, Double weight)
Dependent function	ShopcartShow, NumUpdate

3.3 Generate XML document

ArgoUML is a CASE tool based on Java open environment, it supports 9 graphs in UML. More important, UML diagram can be automatically converted to XML documents in ArgoUML, contains the semantic features of the component: name identification and incoming identification. It can convert graphical semantic to semantic can be identified by computer automatically.

1 name identification
<UML:Package xmi.id = '-34–52–37–127–27024b68: 13deca58cad:-8000:0000000000000AEE'

name = 'ShopcartManage' isSpecification = 'false' isRoot = 'false' isLeaf = 'false' isAbstract = 'false'>
<UML:Class xmi.id='-34–52–37–127–27024b68: 13deca58cad:-8000:0000000000000AAA'
name = 'shopcartManage' visibility = 'public' isSpecification = 'false' isRoot = 'false'

2 incoming identification
In XML document, </UML:StateVertex. incoming> identification reflects dependency relationship between OrderManage and ShopcartManage. The XML fragments are as follows:
name = 'OrderManage' isSpecification = 'false'>
<UML:Transition xmi.idref = '-34–52–37–127–27024b68:13deca58cad:-8000:0000000000000ADE'/>
</UML:StateVertex.incoming>
</UML:SimpleState>
<UML:SimpleState xmi.id = '-34–52–37–127–27024b68:13deca58cad:-8000:0000000000000ADA'
name = 'ShopcartManage' isSpecification = 'false'>

3.4 Component extraction strategy

In order to extract components from XML documents, use the component semantic partition strategy (Horridegem et al. 2004), mainly include: (1) the interaction between classes; (2) service class separated division principle; (3) dependency principle among each class. Combined with the UML semantic model, it is known that: (1) name of component in system must be a certain class name C name in the class diagram, with meaning independent of context or execution way; (2) for the identification r of each component, there must be two certain components in the class diagram to corresponding to it; (3) system may be composed of components, which with formal combination derivation.

The proposed component extraction based on design document is defined as follows:

1 Name: name of two related classes in the UML diagram;
2 t:pre conditions for establishing system structure, in particular, does not need condition is denoted by Empty;
3 p:transfer parameters between components;
4 r:component connection identifier, including incoming, outgoing;
5 Ck: component, the k identifier for the k components;

Six tuple <C1,t,p,r,f,C2> means: when the T condition is satisfied, in sequence diagram, P parameter will be passed from component C1 to C2,with R contact identification between two components, f represents a component connection state, if f is next, means C2 still has the following components, else if f is end, means that C2 is the end.

The component extraction process is shown in figure 2.Mark: (1) name for class collection in component diagram is CnameZ, name for class collection in sequence diagram is CnameX, name for class collection in class diagram is CnameS;(2) component set C=CNameZ∪CnameS∩CnameX, through the component set in component diagram union the component set in sequence diagram ot obtain an intermediate set A, then A intersection with class set in class diagram get the component set C.

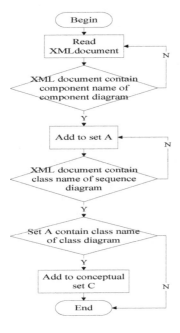

Figure 2. Component extraction process.

3.5 *System structure analysis*

Because the component obtained through the above process is isolated, not fitted with the deployment relationship between components. For this purpose, we propose a context constraints analysis based on MAS to extract the component (James Farrugia.2003), and then build the target system architecture, the process is shown in figure 3, the specific steps are as follows:

1 Mark the components set as C={C1,C2,......, Ck},mark the objects set with name labeled in XML document as X={X1,X2,......,Xn},Xn is an identifier for the N object, mark the system structure set as G, the initial G is empty;

2 Traverse component set C, in set X, if find name identification of identity object Xn is consistent with the name CName of component Ci, exist r connection from component Ci to Cj, public parameter transferred between components is p, then denoted as <Ci,t,p,r,next,Cj>,which means the calling and called relationship exists between component Ci and Cj, parameter flow is from Ci to Cj. Add the six tuple to set G, and remove component Ci from C; if there is no incoming object identifier started from Ci, then remove Ci from C directly.

3 Repeat step (2), until the set C is empty, get the system structure set G;

4 For the elements in the G, construct the topology architecture of the target system.

Traverse set G, if exist element <Ci,t,p,r,next,Cj>, then join sequence pair <Ci,Cj> to the topology sequence set T, until there is no new element can be accessed in the set G. Then traverse the topology sequence set T, for each element in the T, if exist an ordered pair <Ci,Cj>,there is a direct edge Eij between vertices Vi and Vj in the topology. When the set T is traversed end, the topology structure of target system is established.

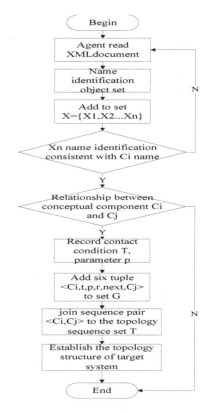

Figure 3. Context constraints analysis process based on MAS.

6

4 APPLICATION EXAMPLE

The target system is a certain electric system, the system including user management, shopping cart management, order generation, favorites management and other functions. The shopping cart system is the core subsystem, provide add, delete, modify the number, empty the shopping cart, billing and other functions. With the help of CASE tools ArgoUml modeling, realize the functions including Shopcart Manage, Order Manage, Favorite Manage, User ManaInfo, Login and so on.

4.1 Extract conceptual component

Use the Argo UML tools to draw UML diagrams, then generate a design document of the target system. As in figure 6, The order Manage and user ItemInfo components are dependent on the shop cart Manage component, favorite Manage component and user ItemInfo component depend on each other, and connected with the user ManaInfo, Login components formed the dependence level.

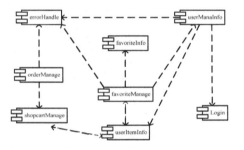

Figure 4.   Component relationship diagram.

Besides considering the component diagram, we pay attention to component attributes and parameters through the class diagram. As shown in figure 6,ShopcartManage class includes Delete from Shop cart and Add to Shop cart interfaces, Add to Shop cart component include customer id and goods id attributes, also include Show Shop cart and Add to Shop cart methods.

Figure 5.   ShopcartManage component class diagram.

Also, check the shopping cart management sequence diagram, get the sequence of every operation.

According to the component semantic partition strategy, extract conceptual components of the shopping system.

Firstly, component diagram class nameset CNameZ={ShopcartManage, OrderManage, FavoriteManage, FavoriteInfo, UserItemInfo, UserManageInfo, Login, ErrorHandle}, class diagram class name set CNameX={ShopcartManage, AddtoShopcart, DeletefromShopcart, CreateOrder}, sequence diagram class name set CNameS= {ShopcartManage, AddtoShopcart, ShowShopcart, DeleteFromShopcart, CreateOrder}.

Then, according to conceptual component set formula C=CNameZ∪CnameS∩CnameX,get C={Shop cart Manage, Add to Shop cart, Delete from Shop cart, Create Order}.Use XML language to describe each conceptual component, including name, Input para, Output pare, Return value and Interface.

4.2 System structure analysis

On this base, using the context constraints analysis method based on MAS to extract the component, form a topological map of the target system, get the target system architecture.

Analyze the XML document of the sequence diagram, the process is as follows:

1 Conceptual component set C={Shopcart Manage, Add to Shopcart, Delete from Shopcart, Create Order}, mark the objects set with name labeled in XML document as X={ShopcartManage, AddtoShopcart, ShowShopcart, DeleteFrom Shopcart, CreateOrder} mark the system structure set as G, the initial G is empty;

2 Traverse conceptual component set C, n set X, Find object X1= ShopcartManage consistent with the conceptual component C1=ShopcartManage, the XML document exists incoming identification from conceptual component C1 to C2,component transfers customerid, goodsid, goodsname, buynum parameters, then records as <C1, empty, (customerid, goodsid, goodsname, buynum),incoming,next, C2>,add this six tuple to set G and remove component C1 from C;

3 Traverse conceptual component set C, in set X, Find object X2=CreateOrder consistent with the conceptual component C2=CreateOrder, the XML document exists incoming identification from conceptual component C2 to C3, component transfers customerid,goodsid,goodsname, buynum parameters, then records as <C2,empty,(customerid, goodsid,-goodsname,buynum),incoming,next,C3>,add this six tuple to set G and remove component C2 from C;

4 Traverse conceptual component set C, in set X, Find objectX3=AddtoShopcart consistent with the

conceptual component C3=AddtoShopcart, the XML document does not exist incoming identification started from conceptual component C3,remove component C3 from C;

5 Traverse conceptual component set C, in set X, Find object X4= Delete From Shopcart consistent with the conceptual component C4= Delete From Shopcart, the XML document exists incoming identification from conceptual component C1 to C4,does not exists incoming identification started from C4,component transfers customerid, goodsid, goodsname, buynum, parameters, then records as <C1,empty,(customerid, goodsid, goodsname, buynum), incoming,end,C4>,add this six tuple to set G and remove component C4 from C;

6 When the set C is empty, the traversal ends. So get the system structure set G={<C1,empty, (customerid, goodsid, goodsname, buynum), incoming, next, C2>,<C2, empty, (customerid, goodsid, goodsname, buynum), incoming, end, C3>,<C1,empty, (customerid, goodsid, goodsname, buynum), incoming,end, C4>}.

After analysis, get the structural map of the target system, the results of simplified is shown in figure 6.

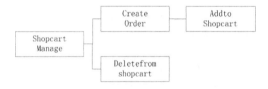

Figure 6.　System structure diagram.

5　RESULTS ANALYSIS

Most of current research results will be directly used for the assembly of retrieval system, with little consideration to the analysis result, whether is consistent with the intention of the designers, even in the presence of inspection, it is often through subjective inspection, so the reuse process is with a certain blindness. In order to ensure the component retrieval and assembly quality, this paper validates the result, in order to improve the efficiency of development.

Via analyze the XML document, validate XML elements which represent the component symbols. In the view of component diagram turn into XML document under ArgoUML tools, the extracted conceptual components are also described in XML language, so the two of them have the common language infrastructure, XML elements is the feature of characterized component. By writing the XML parser to analyze XML document, match the component name string using matcher iteration, then determine whether they are consistent.

Matcher matcher =pattern. matcher (str. substring(i,i+1));
   if (matcher. find( )){
   string = string +str. substring(index,i)+"\n";
   index =i;
   }
   string =string +str. substring(index);
   String[ ] strArray =string. split("\n");

Through the verification of XML elements, get the conceptual component results {Shopcart Manage, Add to Shopcart, Delete from Shopcart, Create Order}, the same as the component name in the component diagram, accord with the XML element validation.

6　CONCLUSION

Component-based software development is the focus of attention of the software reuse. This paper presents component extraction strategy based on design document. The strategy proposed a idea of conceptual component, use Agent for analysis the target system architecture and verify the analysis results.

The focus of the next step is considering how to realize quick positioning from logical component to component retrieval and component assembly, and on this basis to strengthen system validation ability to realize the software reuse efficiencies.

REFERENCES

Chris Lamela. 2000. Breaking Down the Barriers to Software Component Technology. *White Paper, IntellectMarket,Inc.*

Decheng Zhan, Zhongfei Wang,Xiaofei Xu. 2003. Standard description of components based on XML.*Computer Engineering and Applications* 4:89–92.

Horridgen, Knublauch H, Rector A. 2004. A pratical guide to building OWL ontologies using the protégé-OWL plugin and CO-ODE tools edition 1.0. *The Unversity of Manchester and Stanford University* 126~141.

James Farrugia. 2003. Model-theoretic semantics for the web. *The twelfth international conference on World Wide Web(WWW2003).Hungary:* ACM Press 29~38.

Jyotishman Pathak,Doina Caragea, Vasant G Honavar. 2004. Ontology-Extended Component- Based Workflows: A Framework for Constructing Complex Workflows from Semantically Heterogeneous Software Components. *SWDB*: 41~56.

Keqing He,Hong Jiang,Fei He. 2001. Extended UML with role modeling. *China Wuhan University Journal of Natural Science(WUJNS)* 6(1-2):175–182.

Youqun Shi,Cheng Tang,Hengao Wu et al. 2012. Research of Tourism Service Syetem Based on Multi-Agent Negotiation[C].*Gazettez.ICSI2012,* PartI,LNCS7331:pp.583–591.

Zhengong Cai,Xiaohu Yang,Xinyu Wang et al. 2011. A fuzzy formal concept analysis based approach for business component identification. *Journal of Zhejiang University-SCIENCEC(Computer&Electronic)* 12(9):702–720.

Computational Intelligence in Industrial Application – Ling (ed.)
© 2015 Taylor & Francis Group, London, ISBN: 978-1-138-02818-0

Learning to rank experts in heterogeneous academic networks through mutual reinforcement

Zhi Run Liu, He Yan Huang, Xiao Chi Wei & Xian Ling Mao*
School of Computer Science, Beijing Institute of Technology, Beijing, China

ABSTRACT: The problem of evaluating experts is important, and has attracted increasing attention. Various ranking approaches have been proposed and shown to be useful for expert-finding, and most of previous work only focuses on homogeneous networks. Recent works take heterogeneous network structures into consideration. In this paper, we propose a novel co-ranking approach using the mutual reinforcement between papers and venues and apply learning-to-rank technique to rank authors using the rank of papers as features. To evaluate, we collect famous experts of several topics from ArnetMiner. Experiments on a real world dataset collected from ACM Digital Library show that our approach is effective and efficient, and our approach outperforms the three competitors including PageRank, HITS and Co-Rank13 in three metrics. The results of paper rankings and venue rankings also show that our approach is reasonable.

1 INTRODUCTION

Expert finding tasks have been tried to address in different ways in previous research, and it is important for making decisions of appointment and promotion. It has become increasingly important since the late 80's. It is natural to assess the performance of an author by the quality and quantity of the author's papers. Different bibliometric indicators such as g-index (Egghe 2006) and h-index (Hirsch 2005) are proposed, which is a subject of much controversy. To evaluate the quality of papers, citation count metrics are often used. These simple counting metrics have their own advantages that they are convenient to have a single number and easy to interpret. However, these methods do not consider the structure of network available.

Graph-based methods have been widely used for ranking. Many link analysis algorithms like PageRank (Brin & Page 1998) are used to solve the problem (Dom, Eiron, Cozzi, & Zhang 2003, Ma, Guan, & Zhao 2008). Centrality measurements also have been proposed (Baglioni, Geraci, Pellegrini, & Lastres 2012, Zuo, Ehmke, Mennes, Imperati, Castellanos, Sporns, & Milham 2012). For example, a journal can be considered as influential if it is cited by many other journals, especially if those journals are also influential. However, most of these methods are applied on homogeneous networks which contain only one type of entity and relationship.

Recent works begin to consider ranking multiple types of entities in heterogeneous networks. Sun et al. (Sun, Han, Zhao, Yin, Cheng, & Wu 2009) implemented a low-cost clustering method based on a new ranking approach called Authority Ranking. Its main idea is that the rank of an author in heterogeneous networks is determined by its authority and that authors can be clustered by authority. For example, highly ranked authors tend to attend highly ranked venues and highly ranked venues attract highly ranked authors. Another work (Meng & Kennedy 2013) constructs heterogeneous networks by combining citation, authorship and co-authorship together, denoted as **Co-Rank13**. It co-ranks authors and papers simultaneously in heterogeneous networks based on four rules and reinforces the ranking results in a recursive ranking process.

However, previous works never tried to co-rank papers and venues simultaneously using the mutual reinforcement between them. Intuitively, there is a strong interaction between papers and venues, the venue information plays a very important role in evaluating the quality of a paper. And the quality of a venue is mainly determined by the average quality of the papers which published in the venue.

Thus we first propose a novel ranking approach, PV-rank, to co-rank papers and venues simultaneously. Then we regard the ranking results of papers as features to assess authors and apply learning-to-rank technique to optimize the parameters.

* Corresponding author

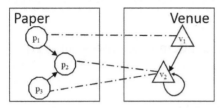

Figure 1. An example of heterogeneous networks.

The heterogeneous networks are comprised of two intra-networks and one inter-network, which are the citation network connecting papers G_P, the citation network connecting venues G_v, and the bipartite network G_{Pv} which ties G_P and G_v. Further details will be given in Sec.3.

A simple example of such heterogeneous networks is shown in Fig.1. We can find that paper p_1 published in venue v_1, p_2 and p_3 published in v_2. And v_1 cite v_2 because p_1 cite p_2. In the intra-networks of papers and venues, there are directed edges representing citations. The undirected edges connecting papers and venues denote the publication relationships.

Experiments on a real world dataset suggest that our approach is more effective and efficient than the state-of-the-art baselines. Specially, for effectiveness, the results of our ranking framework outperform all three competitors in three metrics. In terms of efficiency, the new proposed PV-rank has lower time complexity than *Co-Rank*13 and other co-ranking methods, and the iterative number of our approach is fewer than PageRank.

The contributions of this paper are summarized as follows:

1 A novel efficient approach for co-ranking papers and venues simultaneously in heterogeneous networks is proposed;
2 Learning-to-Rank technique is applied to optimize the performance of author rankings;
3 Different weights of authorship relations among authors and papers are assigned based on the order of the author list.

The rest of this paper starts from reviewing related work (Section 2) of ranking entities in both homogeneous and heterogeneous networks. The novel proposed ranking framework PV-rank is introduced in Section 3 and Section 4. Section 5 is devoted to the experiment and results. Conclusions and further work are provided in Section 6.

2 RELATED WORK

The history of ranking authors and their work can be traced back to the 1960s. Citation information has been long used to evaluate scientific literature.

An important step in bibliometrics was a paper by Garfield (Garfield et al. 1972) in 1972, discussing the methods for ranking journals by Impact Factor. It is the first time that papers were systematically ranked using a large dataset. Gabriel Pinski and Francis Narin proposed several improvements (Pinski & Narin 1976) within a few years, they believed that citations from a more prestigious journal should be given a higher weight.

In 1998, Brin and Page proposed the famous PageRank algorithm (Brin & Page 1998) for ranking web pages. Independently, Kleinberg (Kleinberg 1999) proposed his famous method, the Hyperlink Induced Topic Search (HITS), confirms that ranking can be reinforced through interactions between nodes of various types. Since then, many variants of the PageRank and HITS algorithms were implemented for ranking entities in different applications (Dom, Eiron, Cozzi, & Zhang 2003, Bollen, Rodriquez, & Van de Sompel 2006, Walker, Xie, Yan, & Maslov 2007, Kardan, Omidvar, & Farahmandnia 2011).

Liu et al. (Liu, Bollen, Nelson, & Van de Sompel 2005) compared the rankings of scientists using PageRank with three other rankings using degree, betweeness centrality and closeness centrality. The results confirmed that centrality measures (Baglioni, Geraci, Pellegrini, & Lastres 2012, Zuo, Ehmke, Mennes, Imperati, Castellanos, Sporns, & Milham 2012) are effective with their own advantages. Moreover, topological features have been taken into account (Lou & Tang 2013). It is believed that if a paper or an author is located in a key position in the network, it can be perceived to be important.

However, most of these works focus on homogeneous networks. Zhou, et al. (Zhou, Orshanskiy, Zha, & Giles 2007) proposed a coupled random walk model between citation networks and authorship networks to co-rank authors and documents together. Sun et al. (Sun, Han, Zhao, Yin, Cheng, & Wu 2009) implemented a low-cost clustering method called Authority Ranking. The approach was applied to co-rank authors and venues on the DBLP dataset without citation information and experimental results demonstrated that it is effective.

A recent work (Meng & Kennedy 2013) ranked objects in heterogeneous networks based on a set of rules and reinforced ranking results through a recursive ranking process.

The usage of learning-to-rank methods have been explored for finding experts (Moreira, Calado, & Martins 2011). To our best knowledge, there are not any attempts to co-rank papers and venues through the mutual reinforcement. Using learning-to-rank technique, we can achieve further improvement and lower time complexity.

3 PRELIMINARIES

3.1 Dataset

The dataset used in our experiment is collected from the ACM Digital Library (abbreviated as ACM DL). The ACM DL contains rich citation information as compared to other academic datasets (e.g., DBLP). As shown in Table 1, we totally crawled 98,073 papers and 137,834 authors and 8,755 venues in the domains of databases, information retrieval and data mining. We first crawled the papers from the 10 famous venues, and expanded the dataset based on the references list of each paper. Each paper in ACM DL has the information of authors, publication venue, index terms, references list and cited-by list.

Table 1. Statistics of the dataset.

Number of Papers	98073
Number of Venues	8755
Number of Authors	137834
Number of Citations	1758741
Selected Venues	SIMOD,SIGIR,KDD, WWW,JCDL,CIKM,PODS, WSDM,K-CAP,STOC

Given a topic query, we first select the papers which have the same index term as the query. Then we construct the heterogeneous networks with the selected papers and their publication venues. Our approach has a low requirement for citations that only the citations belong to the selected papers are kept. For example, given the topic "information system", 15,883 papers are selected which are associated with 1,533 venues.

3.2 SVMrank for expert finding

For a give author pairs, the relevance degree can be regarded as a binary value which tells which author ordering is better. SVMrank (Joachims 2006) is a pairwise learning-to-rank algorithm, which builds a ranking model in the form of a linear scoring function for authors, i.e. $f(a) = w^T a$, through the formalism of Support Vector Machines. The idea is to minimize the objective function given below over a set of training queries $\{q_i\}_{i=1}^n$, their associated pairs of authors $(a_u^{(i)}, a_v^{(i)})$ and the corresponding relevance judgment $y_{u,v}^{(i)}$ over each pair of authors.

$$\min \frac{1}{2}\|w\|^2 + C\sum_{i=1}^{n} \sum_{u,v:y_{u,v}^{(i)}} \xi_{u,v}^{(i)}$$

$$s.t. \ w^T(a_u^{(i)} - a_v^{(i)}) \geq 1 - \xi_{u,v}^{(i)}, \ if \ y_{u,v}^{(i)} = 1, \xi_{u,v}^{(i)} \geq 0, i = 1,...,n.$$

The lost function of SVMrank is a hinge loss defined over pairs. The margin term $\frac{1}{2}\|w\|^2$ controls the complexity of the pairwise ranking model w. $\xi_{u,v}^{(i)}$ are the slack variables and C is the coefficient which affects the trade-off between model complexity and the proportion of non-separable samples.

To estimate the expertise of an author towards a given topic query, we use the number of the author's papers in the topic and the scores of the author's top-20 ranked papers in the topic as features.

4 PV-RANK FRAMEWORK

4.1 Data model

Given a topic query t, we denote the heterogeneous network of papers and venues as $G_t = (V, E) = G_P \cup G_V \cup G_{PV} = (V_P \cup V_V, E_P \cup E_V \cup E_{PV})$. G_P and G_V are two intra-networks and G_{PV} is the bipartite network which ties G_P and G_V. Let $P(vi)$ be the set of papers published in venue vi, and $V(pi)$ be the venue where pi published.

$G_P = (V_P, E_P)$ is the unweighted directed network of papers, where V_P is the set of papers belonging to topic t and E_P is the set of edges representing the citations between papers, the weight of edge (p_i, p_j) is calculated using the following formula:

$$w_P(p_i, p_j) = \begin{cases} 1 & p_i \ cite \ p_j \\ 0 & otherwise \end{cases}$$

$G_V = (V_V, E_V)$ is the weighted directed graph of venues, where V_V is the set of venues in topic t and E_V is the set of edges. If there is a paper p_a (published on venue v_i) which has a paper p_b (published on venue v_j) on its reference list, then there is an edge (v_i, v_j) in E_V. The weight of edge (v_i, v_j) in E_V is set as follows.

$$w_V(v_i, v_j) = \sum_{\substack{p_k \in P(v_i) \\ p_l \in P(v_j)}} w_P(p_k, p_l)$$

$G_{PV} = (V_{PV}, E_{PV})$ is the unweighted bipartite graph. $V_{PV} = V_P \cup V_V$ and E_{PV} is the set of edges. If there is a paper pa published on venue v_i, then there is an edge (p_a, v_i) in E_{PV}.

$$w_{PV}(p_i, v_j) = \begin{cases} 1 & V(p_i) = v_j \\ 0 & otherwise \end{cases}$$

In our model, the importance measurements of different entities are as follows:

1 $PR_{paper}(p_i)$: PageRank value of paper p_i;
2 $PR_{venue}(v_k)$: PageRank value of venue v_k;

11

4.2 Mutual reinforcement rules

There are two parameters, $\alpha, \beta \in [0,1]$, that $(1-\alpha)$ and $(1-\beta)$ control the proportion of mutual reinforcement. PV-rank is based on four rules as follows:

- **Rule 1:** The score of a paper is influenced by the scores of the papers which cite the paper.

$$PR_{paper}(p_j) = \sum_i E_P(p_i, p_j) PR_{paper}(p_i) \quad (1)$$

- **Rule 2:** The score of a paper is influenced by the score of the papers which cite the paper and the score of the venue where the paper published.

$$PR_{paper}(p_j) = \alpha \sum_i E_P(p_i, p_j) PR_{paper}(p_i) \\ + (1-\alpha) PR_{venue}(V(p_j)) \quad (2)$$

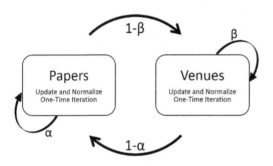

Figure 2. Iterating process between papers and venues.

- **Rule 3:** The score of a venue is influenced by the scores of the venues which cite the venue.

$$PR_{venue}(v_j) = \sum_i E_v(v_i, v_j) PR_{venue}(v_i) \quad (3)$$

- **Rule 4:** The score of venue is influenced by the scores of the venues which refer to it and the average scores of the papers published in it.

$$PR_{venue}(v_j) = \beta \sum_i E_V(v_i, v_j) PR_{venue}(v_i) \\ + (1-\beta) \sum_{p_k \in P(v_j)} \frac{E_{PV}(p_k, v_j) PR_{paper}(p_k)}{|P(v_j)|} \quad (4)$$

Comparing to *Co-Rank13*, PV-Rank is more reasonable and flexible by adding parameters α, β.

4.3 Ranking model

In our approach PV-rank, a papers is ranked through venue and other papers, and a venue is also ranked through papers and other venues. Fig.2 illustrates the iterating process based on above four rules.

Since the venues are composed of papers, it is reasonable to start the process from ranking papers. In the initial phrase, we set the initial PageRank values of all vertexes as 1.

The three main steps of PV-rank are given below:

1. Do one-time iteration in GP based on Rule 1.
2. Update the scores of venues using the updated scores of papers and the scores of venues based on Rule 4, and then normalize the scores of venues and do one-time iteration in GV based on Rule 3.
3. Update the scores of papers using the updated scores of venues and the scores of papers based on Rule 2, and then normalize the scores of papers, and go back to step (1).

Similar to PageRank, there are two ways to control the iterative number. The first way is to set a max iteration number manually, and the second way is to stop iteration when the difference between current iteration and previous iteration is smaller than δ.

$$Diff(t, t-1) = \frac{\sum_{i=1}^{|V|} |PR(i,t) - PR(i,t-1)|}{|V|}$$

4.4 Time complexity

The time complexity of PV-rank is $O(t|E|)$ where t is the iterative number and $|E|$ is the number of the edges. Author information is widely used in previous works (Sun, Han, Zhao, Yin, Cheng, & Wu 2009, Meng & Kennedy 2013), which inevitably increases the number of edges largely. Since the number of venues is much smaller than the number of authors, our approach will have lower time complexity.

5 EXPERIMENTS

All the experiments are conducted in an Intel i5 3.20-GHz computer with 4GB of main memory. The damping factor λ is set to 0.85, and the parameters α and β are set to 0.5.

The evaluation of ranking experts is always a challenge because different approaches have different definitions of importance. ArnetMiner[1] has achieved great success for offering comprehensive search and mining services for academic community.

[1] http://arnetminer.org/

Thus, we collect the top-20 experts list from ArnetMiner as ground truth for each topic query, denoted as *Bench*.

5.1 Evolution metrics

We select three metrics to evaluate the result of author rankings which are Precision@N, binary preference measure(Bperf) (Buckley & Voorhees 2004) and discounted cumulated gain(DCG) (Jarvelin & Kekalainen 2002) respectively.

The first metric is Precision@N,

$$\Pr ecision@N = \frac{\left|S_{GT} \cap S_A\right|}{\left|S_A\right|},$$

where S_{GT} is the ground truth of the query and S_A is the top-N results returned by approach A. Precision@N does not take the ranking order into account.

Our second metric is Bperf. For a topic query, if there are R true experts within M found experts by a method, in which r is a true expert and n is not, then Bpref is defined as follows:

$$Bpref = \frac{1}{|R|} \sum_{r \in R} 1 - \frac{\left|n \ ranked \ higher \ than \ r\right|}{M}$$

The last metric is DCG. There are two formulas of DCG, and we use the binary one, $rel_i \in \{0,1\}$. The DCG accumulated at a particular rank position p is defined as

$$DCG_p = \sum_{i=1}^{p} \frac{2^{rel_i} - 1}{\log_2(i+1)}.$$

Table 2.   Top-10 authors ranked by PV-Rank.

	Data Mining	Information Retrieval
1	**C. Aggarwal(4)**	**W. Bruce Croft(1)**
2	**Jiawei Han(1)**	Ryen W. White
3	**Philip S. Yu(2)**	Chengxiang Zhai
4	**Christos Faloutsos(7)**	**James Allan(12)**
5	Bing Liu	C. Lee Giles
6	**Eamonn Keogh(5)**	**Ellen M. Voorhees(6)**
7	Padhraic Smyth	Jamie Callan
8	Wynne Hsu	**Edward A. Fox(19)**
9	Rakesh Agrawal	Susan T. Dumais
10	Jieping Ye	Zheng Chen

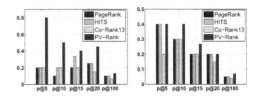

(a) Data Mining   (b) Information Retrieval

Figure 3.   Precision@N for author rankings.

5.2 Result of ranking authors

We select two topic queries, and use one of them to train the ranking model and then evaluate over another one respectively. The parameters are set as $\alpha = \beta = 0.5$. The three competitors are **PageRank**, **HITS** and **Co-Rank13**.

The top-10 experts of domain "*Data Mining*" and "*Information Retrieval*" ranked by PV-Rank are shown in Table 2, where the relevant experts are in boldface. The numbers in parentheses are the rank of the relevant experts in *Bench*.

Table 3 shows the *Bpref*, *DCG20* and *DCG100* to both queries. Fig.3 shows the Precision@N of all four approaches. Results demonstrate that PV-Rank outperforms the competitors in all metrics. For "*Data Mining*", taking DCG_{20} for example, PV-Rank achieves a great improvement of 95.2%, 87.9% and 140.5%, comparing to the three competitors.

The results of "*Information Retrieval*", also outperforms all baselines but the improvements are not as good as the results of "*Data Mining*". PV-Rank achieves an improvement of 43.6%, 43.6% and 88.3%. A possible reason is that the data volume of "*Data Mining*" is smaller than that of "*Information Retrieval*". In our dataset, there are 11915 papers related to "*Information Retrieval*", and only 2411 papers related to "*Data Mining*".

5.3 Result of ranking papers and venues

In the domain of "*Information Retrieval*", SIGIR is the best venue returned by PV-Rank which is obviously one of the best venues. All of the top ranked papers returned by PV-Rank are humanly verified to be important that the least-cited paper has 993 citations. For the topic "*Data Mining*", the top ranked venues are KDD, WWW, SIGMOD, CIKM and PODS.

5.4 Parameter effect

We ran PV-Rank with various setting of α and β, the *Bpref*, DCG_{20} and DCG_{100} of different settings are shown in Fig.4. Also, Fig.5 shows the iterative

13

Table 3. *Bpref, DCG$_{20}$ and DCG$_{100}$* for author rankings.

	Data Mining				Information Retrieval			
	PageRank	HITS	Co-Rank13	**PV-Rank**	PageRank	HITS	Co-Rank13	**PV-Rank**
Bpref	0.8170	0.8230	0.7529	**0.8677**	0.7120	0.7120	0.6700	**0.7600**
DCG$_{20}$	2.8777	2.9892	2.3355	**5.6173**	2.1127	2.1127	1.6108	**3.0330**
DCG$_{100}$	4.0894	4.2231	3.3445	**6.7213**	2.3658	2.3658	1.8837	**3.7934**

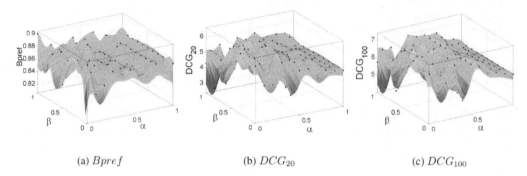

(a) $Bpref$　　　　　　(b) DCG_{20}　　　　　　(c) DCG_{100}

Figure 4. Bpref, DCG_{20} and DCG_{100} under various settings of α and β.

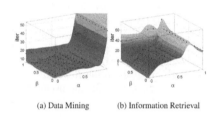

(a) Data Mining　　(b) Information Retrieval

Figure 5. Iterative number until convergence under various settings of α and β.

number until convergence of PV-Rank under different settings of α and β.

We arrived at following conclusions: (1) Using the mutual reinforcement can improve the performance of author ranking; (2) For smaller α and β (more mutual reinforcement), the approach converges faster.

6   CONCLUSIONS

This paper proposes a novel fast ranking approach, PV-Rank, to co-rank papers and venues with lower time complexity. PV-Rank employs the mutual reinforcement relationships among papers and publication venues to rank papers and venues simultaneously. Moreover, this paper first applies learning-to-rank technique in the context of expert finding regarding the rank of papers as features. Using the benchmark collected from ArnetMiner for authors, we show

through experiments that PV-Rank outperforms the state-of-the-art competitors including PageRank, HITS and Co-Rank13 in all three metrics (i.e. Precision@N, Bpref and DCG).

In the future, we will enrich the set of topic queries and consider more features to achieve better performance.

ACKNOWLEDGMENTS

The work described in this paper was fully supported by the Key Project of Chinese National Programs for Fundamental Research and Development (973 Program, No. 2013CB329605) and the Fund of Beijing Institute of Technology for Fundamental Research (No. 3070012211413).

REFERENCES

Baglioni, M., F. Geraci, M. Pellegrini, & E. Lastres (2012). Fast exact computation of betweenness centrality in social networks. In Proceedings of the 2012 International Conference on Advances in Social Networks Analysis and Mining (ASONAM 2012), pp. 450–456. IEEE Computer Society.

Bollen, J., M. A. Rodriguez, & H. Van de Sompel (2006). Journal status. Scientometrics 69(3), 669–687.

Brin, S. & L. Page (1998). The anatomy of a large-scale hypertextual web search engine. Computer networks and ISDN systems 30(1), 107–117.

Buckley, C. & E. M. Voorhees (2004). Retrieval evaluation with incomplete information. In Proceedings of the 27th annual international ACM SIGIR conference on Research and development in information retrieval, pp. 25–32. ACM.

Dom, B., I. Eiron, A. Cozzi, & Y. Zhang (2003). Graph-based ranking algorithms for e-mail expertise analysis. In Proceedings of the 8th ACM SIGMOD workshop on Research issues in data mining and knowledge discovery, pp. 42–48. ACM.

Egghe, L. (2006). Theory and practise of the g-index. Scientometrics 69(1), 131–152.

Garfield, E. et al. (1972). Citation analysis as a tool in journal evaluation. American Association for the Advancement of Science.

Hirsch, J. E. (2005). An index to quantify an individual's scientific research output. Proceedings of the National academy of Sciences of the United States of America 102(46), 16569–16572.

Jarvelin, K. & J. Kekalainen (2002). Cumulated gain-based evaluation of ir techniques. ACM Transactions on Information Systems (TOIS) 20(4), 422–446.

Joachims, T. (2006). Training linear svms in linear time. In Proceedings of the 12th ACM SIGKDD international conference on Knowledge discovery and data mining, pp. 217–226. ACM.

Kardan, A., A. Omidvar, & F. Farahmandnia (2011). Expert finding on social network with link analysis approach. In Electrical Engineering (ICEE), 2011 19th Iranian Conference on, pp. 1–6. IEEE.

Kleinberg, J. M. (1999). Authoritative sources in a hyper-linked environment. Journal of the ACM (JACM) 46(5), 604–632.

Liu, X., J. Bollen, M. L. Nelson, & H. Van de Sompel (2005). Co-authorship networks in the digital library research community. Information processing & management 41(6), 1462–1480.

Lou, T. & J. Tang (2013). Mining structural hole spanners through information diffusion in social networks. In Proceedings of the 22nd international conference on World Wide Web, pp. 825–836. International World Wide Web Conferences Steering Committee.

Ma, N., J. Guan, & Y. Zhao (2008). Bringing pagerank to the citation analysis. Information Processing & Management 44(2), 800–810.

Meng, Q. & P. J. Kennedy (2013). Discovering influential authors in heterogeneous academic networks by a co-ranking method. In Proceedings of the 22nd ACM international conference on Conference on information & knowledge management, pp. 1029–1036. ACM.

Moreira, C., P. Calado, & B. Martins (2011). Learning to rank for expert search in digital libraries of academic publications. In Progress in Artificial Intelligence, pp. 431–445. Springer.

Pinski, G. & F. Narin (1976). Citation influence for journal aggregates of scientific publications: Theory, with application to the literature of physics. Information Processing & Management 12(5), 297–312.

Sun, Y., J. Han, P. Zhao, Z. Yin, H. Cheng, & T. Wu (2009). Rankclus: integrating clustering with ranking for heterogeneous information network analysis. In Proceedings of the 12th International Conference on Extending Database Technology: Advances in Database Technology, pp. 565–576. ACM.

Walker, D., H. Xie, K.-K. Yan, & S. Maslov (2007). Ranking scientific publications using a model of network traffic. Journal of Statistical Mechanics: Theory and Experiment 2007(06), P06010.

Zhou, D., S. A. Orshanskiy, H. Zha, & C. L. Giles (2007). Coranking authors and documents in a heterogeneous network. In Data Mining, 2007. ICDM 2007. Seventh IEEE International Conference on, pp. 739–744. IEEE.

Zuo, X.-N., R. Ehmke, M. Mennes, D. Imperati, F. X. Castellanos, O. Sporns, & M. P. Milham (2012). Network centrality in the human functional connectome. Cerebral Cortex 22(8), 1862–1875.

Research on security defense model of oil & gas gathering and transferring SCADA based on factor neural network

P. Liang, X.D. Cao, Y. Qin, W.W. Zhang & Q.H. Hu
Electrical & Information School, Southwest Petroleum University, Chengdu, Sichuan, China

ABSTRACT: This paper applies the method of framework and rules combined to describe the analytical factors neuron and proposes the security defense model of oil & gas gathering and transferring SCADA based on factor neural network. Each analytical factor neuron is a knowledge element which is on the basis of frame representation. And the factors are used to compose the structure of the neuron. The neuron uses the related factors in the frame structure's slot to construct the relationship chain of factor neural network, which can form a network system. The network system takes the framework as whole system's nodes and uses relation-type factors as the relationship chains. Then the model can accomplish protection of host security of oil & gas gathering and transferring SCADA at a system level. This paper provides a new idea for the means of oil & gas gathering and transferring SCADA security defense.

KEYWORDS: Neural network; Security defense; Oil & gas gathering and transferring; SCADA; Method of framework and rules combined.

1 INTRODUCTION

SCADA (Supervisory Control and Data Acquisition) system [1] has been widely applied in the field of oil & gas gathering and transferring [2], especially in large scale oil & gas gathering pipe network. In the past, the SCADA system has been in a closed environment with a relatively low safety risk. In recent years, for the reason of economic, political and military purpose, the hostile countries and organizations frequently attack the SCADA system of energy [3], as an important constituent of national major infrastructure and energy development strategy, face serious security challenges. The existing method of security defense applied in oil & gas gathering and transferring SCADA system is usually by means of installing industrial firewall. However its ability of protecting host is very limited and cannot form security defense at a system level.

In order to improve the level of host security in distributed oil & gas gathering and transferring SCADA system and then to increase the defense capability of the whole SCADA system, the paper combines the theory of factor neural network [4-5] with the method of framework and rules combined for describing the analytical factors neuron, then proposes the security defense model of oil & gas gathering and transferring SCADA based on factor neural network. Compared with the traditional way to install firewall as a method of defense, the model not only aims at protection of the individual host, but also can form the protection for the SCADA's hosts at a system level.

2 FACTOR NEURAL NETWORK THEORY

2.1 *Introduction to the factor neural network theory*

Factor neural network theory is one of the basic theory about the intelligent engineering field. It's based on the factors representation of knowledge. The theory take factor neuron and factor neural network as the formal framework, which can realize the storage and use of knowledge for completing the engineering simulation of intelligent behavior.

2.2 *Factor and factor state space*

The factor is a primitive describing objects, and it is a kind of cognition and expression. Besides, it is abstracted from the process of cognition. People use it to cognize and describe objects, such as properties of objects, all kinds of conditions in the process of reasoning, etc. Factor $f \in F$ can be regarded as a mapping, acts on certain objects $u \in U$, and gets a certain state $f(u)$. f: $D(f) \rightarrow X(f)$, where $X(f) = \{f(u)|u \in U\}$ is the state space of f [4-5].

2.3 *Analytical factor neuron*

FN (Factor Neuron) is the basic unit of knowledge expression and information processing [5].

The analytical factor neuron based on the analytical model is a kind of neurons for knowledge representation and information processing. It use functions, rules and prototype to express and deal with the knowledge and information. It's also capable of presenting and applying knowledge which is a simulation of the patterns of human mind.

From the perspective of automaton, an analytical factor neuron is like a miniature automaton. It has a set of transition rules and states expressed by factors. When external input triggers automaton, automaton does the corresponding operation according to perceived information and accomplishes the conversion of state. At last, it outputs the response according to its response functions.

For purpose of describing an object, the analytical factors is constructed on basis of the framework. It uses factors to compose the structure of neuron and to organize and encapsulate the description, explanation and process. The neuron uses the related factors in the frame structure' slot to construct the relationship chain of factor neural network, which can form a network system. The network system take the frameworks as whole system's nodes and use relation-type factors as the relationship chains.

3 THE METHOD OF FRAMEWORK AND RULES COMBINED

3.1 *Framework representation*

A frame [6] is composed of a frame name and a group of slots for describing all aspects of the specific attributes. Each slot has a slot name, and its value describe the integral attributes of the things represented by frame. Slot can be further divided into a number of facets. Each facet has one or more of the facet value. Each facet value can also be a value of statement of a concept. These can be chosen based on specific issues and specific needs. The general structure of a framework is as follows:

<frame name>
<$slot_1$><slot $value_1$> <facet $name_{11}$>
    <facet $value_{111}$, ...>
    <facet $name_{12}$><facet $value_{121}$, ...>

    <facet $name_{1m}$><facet $value_{1m1}$, ...>
<$slot_2$><slot $value_2$>|<facet $name_{21}$>
    <facet $value_{211}$, ...>
    <facet $name_{22}$><facet $value_{221}$, ...>

    <facet $name_{2n}$><facet $value_{2n1}$, ...>

<$slot_k$><slot $value_k$>|<facet $name_{k1}$>
    <facet $value_{k11}$, ...>
    <facet $name_{k2}$><facet $value_{k21}$, ...>

    <facet $name_{ki}$><facet $value_{ki1}$, ...>

Besides, facet values and slot values can be another framework. Then it forms a frame network. The framework representation has the advantages of inheritance and natural, and it is appropriate to represent structural knowledge. But it is not good at representing processing knowledge.

3.2 *Rule representation*

The rule representation [7], also known as production representation, is usually used to indicate a causal relationship between knowledge and basic form: IFATHEN B. Rule presentation has the advantage of natural, effective, it is suitable to represent structural knowledge.

3.3 *Method of framework and rules combined*

In the preceding sections, we know the production method is suitable for the processing knowledge, not good at representing structural knowledge. And framework is suitable for structural knowledge and not good at representing processing knowledge .Thus, the framework and production can be combined, called as method of framework and rules combined [8-9] to achieve mutually beneficial results.

The method of framework and rules combined is: When knowledge is represented by framework, rules can be embedded in the framework. Moreover, in the process of realization, framework can call rules and rules can also call framework. Rules and framework are integrated [10]. The storage structure of framework and rules is as follow:

Table 1. Storage structure of frame.

Name	Type	Meaning
Frame name	CString	The name of frame
Slot	Class	Slot
Relation	Class	Relation type
Rule	Class	Reasoning

Table 2. Storage structure of rules.

Name	Type	Meaning
Rule ID	Int	Rule number
Subject frame	CString	Name of Subject frame
Condition	Class	Rule condition
Conclusion	CString	Reasoning result
Belief	float	Rule's confidence

The architecture of storage is easy for framework to find corresponding class of rules, and rules find the framework. It offers an interface for the framework and the rules.

As the slot value or facet value in the framework can be another framework, this can make the connection between framework, and a framework can be found by another framework. In a certain environment, several objects will inevitably have some common properties, these common properties can be extracted to construct an upper framework. And then the unique properties of each kind of object can extracted to constitute a number of respective lower frames. Through those relations, such as ISA, Part of, subclass, instance of [6], etc., inheritance of lower frame to the upper frames' properties can be realized.

Taking a certain Execution neuron as an example, this paper uses the method of framework and rules combined to describe the process of dealing with Stuxnet virus:

Frame: < Neurons Information >
Neuron categories: Execution neuron
Location: 192.168.0.8
Objects controlled by host: Siemens PLC (step7, SIMATICWinCC)
Framework name: <Rule 1 >
Premise: Behavior1: automatically create a shortcut
Behavior2: replace s7otbxsx. DLL
Conclusion: < Stuxnet attack >
Information output action: Output information to the Statistics neuron (<Neurons Information >, Attack information, Solution Information)
Framework name: < Stuxnet attack >
Information: attack categories: Stuxnet attack
Attack description: Try to attack SIMATIC WinCC, replace the Step7 s7otbxsx, and ultimately achieve the goal of control PLC
Danger Level: severe
Solution: processing action (Delete created shortcuts, End All the Isass.exe process whose parent process is not winlogon.exe, Delete derived files)

When Execution neuron perceives suspicious behavior1 and behavior2. , framework of Rule 1 will be matched and make a conclusion of Stuxnet Attack. In the framework of Stuxnet attack, it makes a brief description of the attack and implement the processing action. At last, through the information output action in Rule1, the Execution neuron will send the neuron information, attack information, and solution information to the Statistics neuron.

4 ESTABLISHMENT OF THE DEFENSE MODEL

According to topology structure of SCADA of certain oil & gas filed, combining with the neural network, we established the security defense model of oil & gas gathering and transferring SCADA based on

factor neural network. The model has the distributed character and consists of various analytical factor neurons. Each neuron is a knowledge element based on method of framework and rules combined.

As what is shown in figure 1, defense model includes four types of analytical factor neurons: execution neuron, factors knowledge base neuron, statistics neuron, management neuron, and dispatch and assembly neuron.

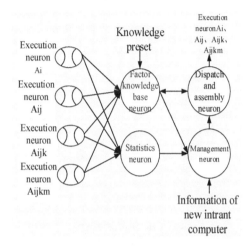

Figure 1. The security defense model.

Execution neuron: as shown in figure2, they are deployed on industrial computer, engineer station, and operator station in SCADA system at all levels. They capture and judge the program behavior factors of the running program. Compared with the internal factor knowledge base, they can judge whether the program is a malicious program and take corresponding measures. The factor knowledge base, including the legal factor knowledge base and malicious factor knowledge base is part of the knowledge stored in the knowledge base neuron.

Factor knowledge base neuron: including legal knowledge base and malicious factor knowledge base, it is rule base storing program factors. Its means of knowledge acquisition comes from the preset mode and legal factors judged in the process of system operation. The legal knowledge base is constitute of those legal program behavior factors judged by execution neurons deployed on all hosts of oil & gas gathering and transferring. Malicious factor knowledge base, consisting of malicious program behavior factors, is knowledge base of attack recognition.

Statistics neuron: it totals the configuration information of hosts and the attack information the hosts suffered from. Attack information includes: detailed location of attacked hosts, types of attack, and the origin of attack.

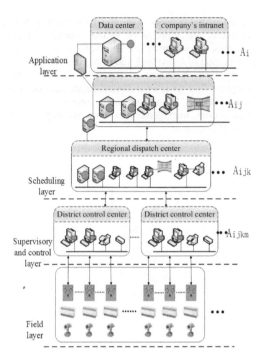

Figure 2. Topology structure of SCADA.

Management neuron: It receives the update information sent by Factor knowledge base neuron and configuration information of hosts sent by the statistics neuron. It also receives the new intrant computers' information from outside interface. It aims at the registration, cancellation and the discovery of neurons and other services. Besides, the management neuron assign assembly tasks to existing kindred execution neurons and new execution neurons on new intrant hosts. Those tasks will be sent to dispatch and assembly neuron for execution. The new intrant computers' information mentioned above consist of the information of the hosts' location and configuration information of hosts' software.

If management neuron receives the update information sent by Factor knowledge base neuron and configuration information of hosts sent by the statistics neuron at the same time, then it will sends the dispatch information of update of knowledge base in existing kindred execution neurons to dispatch and assembly neuron. However, If only information coming from statistics neuron is received, the dispatch information will not be delivered. If the new intrant computers' information from outside interface is received, then management neuron will deliver an instruction to dispatch and assembly neuron. The instruction aims at letting the dispatch and assembly neuron assembles a new corresponding execution neuron and deploy it on the new intrant computer.

Dispatch and assembly neuron: it receives the task information from management neuron, get update knowledge from the factor knowledge base neuron, and is responsible for the knowledge update of existing kindred execution neurons and assembly of new execution neurons. And then it will assemble a new corresponding execution neuron and deploy it on the new intrant computer.

When Dispatch and assembly neuron receive the update instructions to specific execution neurons' knowledge base from management neuron, then it will send the knowledge acquisition instructions to factor knowledge base neuron. After the instructions are received by factor knowledge base neuron, the corresponding knowledge in the factor knowledge base neuron will be sent back to dispatch and assembly neuron and be dispatched to corresponding execution.

If the instructions, assembling a new execution neuron from, from management neuron are received, then Dispatch and assembly neuron will still assemble a new execution neuron before sending the knowledge acquisition instructions to factor knowledge base neuron.

To sum up, the knowledge's storage in all kinds of analytical factor neurons (knowledge elements) are based on the method of framework and rules combined. With the inheritance between slots and the connection of relation-type factors, we can realize the organic combination between various neurons and construct an SCADA security defense model with distributed feature.

5 CONCLUSIONS

According to SCADA topology structure of a certain oil & gas field, this paper integrates factor neural network theory and establish the security defense model of oil & gas gathering and transferring SCADA based on factor neural network. The model consists of various analytical factor neurons. Rules representation and frame representation are combined to from a method of framework and rules combined. The method is applied to the construction of the analytical factor neuron and make the factor neurons become knowledge elements. They can not only express procedural knowledge, but also the structural knowledge. By using the related factors in the frame structure' slot to construct the relationship chain of factor neural network, the model can let these various types of factor neuron combine as an organic network system. Therefore, the model provides a new thought for the means of oil & gas gathering and transferring SCADA security defense.

REFERENCES

Huazhong, Wang. 2012. *SCADA and its application.* Beijing: Publishing House of Electronics Industry.
Qian, Wang & Minghui, Jiang. 2014. Exploration of oil & gas transmission pipeline based on SCADA system.

Journal of China petroleum and chemical standard and quality (3):43–43, 31.

Yong, Peng. 2012. Industrial control system cybersecurity research. *Journal of Tsinghua university (natural science edition)* 52(10):1396–1408.

Zengliang, Liu. 1994. *Factor neural network theory and its application*. Guizhou: Guizhou Science and Technology Publishing House.

Zengliang, Liu & Youcai, Liu. 1992. *Factor neural network theory and implementation strategy study*. Beijing: Beijing Normal University Press.

Guangju, Zhu. 2007. Research and application on knowledge representation with product ion and framework base on domain Ontology. Yunnan normal university.

Yue, Jiuang. 2007. Research on knowledge representation with frame-rule and reference base on fuzzy logic. *Yunnan normal university (natural science edition)* S2:171–174.

Wei, Fu. 2002. Expression approach research on expert knowledge based on frame network structure. *Journal of computer application* 22(1):3–6.

Chi, Ma & Baoguo, Wu. 2009. Forestation expert system based on the production and framework knowledge representation. *Journal of agriculture network information* (5):22–24.

Cengyang, Li & Zhongsheng, Cao. 2000. Knowledge base system based on the frame and middle ware model. *Journal of mini-micro system* 21(12):1297–1300.

Research and implementation of K-Means algorithm based on Hama

Haiyan Wei & Hongliang Chen
Personnel Department, Northwestern Polytechnical University, Xi'an, China

Li You
School of Computer Science and Technology, Northwestern Polytechnical University, Xi'an, China

ABSTRACT: With the rapid development of cloud computing, the Internet of Things technology and social networking, the massive data computing has become an important issue in the field of data mining. Traditional stand-alone data mining algorithms can hardly meet the production demand on processing speed. Though lots of data mining algorithms have been implemented on Hadoop platform, it supports iteration only with multiple external MapReduce job chain. This method not only needs the developers to actively interact the execution process, but also leads to massive unnecessary duplication cost. Hama platform is based on the implementation MapReduce and BSP (Bulk Synchronous Parallel). It uses the transportation of messages in BSP to accelerate the speed of the iterative process, and doesn't need multiple MapReduce jobs, the cost of HDFS persistence and data serialization stay unchanged. So we, in this paper, design and implement the classical data algorithm K-Means in Hama platform, to pave the way for further complex data mining algorithm.

KEYWORDS: Hadoop; MapReduce; BSP; Hama; K-means.

1 INTRODUCTION

With the rapid development of cloud computing, the Internet of Things, and social networking technology, the massive data processing and massive data data mining has become an important issue. Traditional serial data mining algorithms are often only able to handle some toy data, but when referred to massive data, its execution speed will reduce or even unacceptable, which is the exact tough challenge for current data mining.

Clustering is one of the important unsupervised machine learning methods in data mining. In the cases of massive data, existing clustering algorithm's time complexity and space complexity encountered a bottleneck, which demands the study of the field of parallelizing clustering algorithms.

MapReduce [1], a parallel programming model designed for processing very large scale of distributed data sets proposed by Google, is now becoming a core computing model of cloud computing. The Apache's Hadoop [2] is an open source implementation of MapReduce, and has good scalability and fault tolerance in complex data-intensive computing, such as multi-dimensional scientific simulation, machine learning and data mining. Currently there are a variety of data mining algorithms have been implemented on Hadoop platform.

Hadoop can be deployed on inexpensive computers, which can save a considerable amount of the system implementation costs. While large enterprise systems need to handle the increasing amount of data, some of the antiquated method's implementation is very complex, therefore the cost will be drastically increased and won't meet the needs of massive data processing. Hadoop platform has an outstanding performance in the field of distributed data processing and has developed rapidly maturing in security and stability. But Hadoop is not a panacea, it can only be called multiple times by an external chain of MapReduce jobs to support an iterative and interactive data processing, which not only requires developers to actively intervene in the implementation process, but also inevitably will introduce a lot of unnecessary of duplication cost [3]. As is for an outside job chain call, each iteration inevitably will produce overheads like job initiation (including job distribution, initialization input division, division of tasks, etc.), the immutable network data transmission, iterative intermediate results the HDFS persistence and so on. At this under circumstances Hadoop's HDFS cannot play to its strengths and the researchers urgently need a platform focused on solving large-scale computing, Apache Hama project so started.

Hama [4] is a framework built on top of the Hadoop distributed parallel computing model and based on MapReduce and BSP [5] (Bulk Synchronous Parallel block synchronous parallel), mainly composed by

Zookeeper, HBase, HDFS. The greatest advantage of using BSP computing technology is accelerating the speed of iteration iterative process, because the iterative process may need to pass data of the message several times before the final output, but do not need to start multiple MapReduce jobs avoiding HDFS persistence and constant data serialization etc. overheads .Though Hama is a distributed computing framework based on BSP and Hadoop, it does not rigidly adhere to restrict distributed file system, but also capable of combine various advantages of BSP, MapReduce, HDFS, etc., aiming at establishing a set number of programming frameworks, specializing in large-scale computing.

This article below provides an overview of the Hadoop and Hama platform and discusses details of parallel K-Means algorithm design and implementation based on Hama platform.

1.1 Hadoop platform

Google proposed a distributed file system Google FileSystem [6] and distributed computing framework MapReduce in order to meet the growing data storage needs, as well as rapid analysis of large-scale data. Apache Foundation achieves a corresponding open source version: Hadoop Distributed File System (HDFS), Hadoop MapReduce. Together, these two form the core parts of the Hadoop Project's. Therefore Hadoop is not just one distributed file system, but should be considered as an open source framework providing distributed computing on clusters.

1.2 Hama platform

Pregel, a distributed programming framework introduced by Google focuses on graph computing, is another sharp weapon for distributed computing after MapReduce. It's primarily designed to facilitate the realization of a variety of graph algorithms, such as graph traversal (BFS), the shortest path (SSSP), PageRank calculation in the case of large-scale graph data. As is said, Google has 20% of the computing tasks been completed by the Pregel. Unfortunately Pregrel not open source, so the Apache Foundation launched Hama, an open source version of Hama.

Block synchronous parallel computing model (Bulk Synchronous Parallel Computing Model), referred to as the BSP model, was proposed by Bill McColl, Oxford University and Viliant, Harvard University, which aims to establish a parallel and scalable theoretical model that does not rely on specific architecture and builds a bridge between the software and hardware in computing. BSP model is both a parallel architecture model and a parallel programming model that can accurately analyze and predict the performance of a parallel program execution [7].

In BSP model, each job is composed of a series of parallel super step (Supersteps), each of which constitutes a Phase Parallel. Each super step is divided into ordered three-step stages:

- Local calculate phases: processors complete computing tasks in parallel.
- Global communication phase: parallel tasks complete interactive data using a message passing mechanism before the end of the super-step.
- Barrier synchronization stages: synchronization wait all interactive parallel tasks within the same super-step to complete, only after this, the entire parallel step can move down into the next round in parallel.

BSP model run with super-step as a unit, which greatly simplifies the programmers to write and debug parallel programs. At the same time, it actually separates the communication and synchronization, which not only makes the parallel program structure more clearly, but also helps to avoid deadlocks. The vast majority of the architecture can achieve BSP model effectively, including HAMA, therefore, parallel algorithms can be achieved by programming the target machine.

2 IMPLEMENTATION

2.1 Standalone K-Means algorithm

K-Means algorithm input k as parameter, clustering dataset which has n objects into k parts, achieving high similarity within a cluster, and low degree of similarity between clusters. The cluster similarity is measured by the objects' similarity in the cluster, which can be regarded as the cluster centroid or center of gravity.

K-Means algorithm processing flow is as follows. First of all, randomly select k objects, each object represents an initial cluster mean or center. For each of the remaining objects, assign to the most similar clusters in accordance with the mean distance between each cluster. The new mean value is then calculated for each cluster afterwards. This process is repeated until the criterion function converges. Typically, the squared deviation criterion is defined as:

$$\sum_{i=1}^{k} \sum_{p \in C_i} (p - m_i)^2)$$

Wherein, E is the square deviation of all objects in the data set; p is a point in the dataset, said a given object; m_i is the mean of the cluster C_i. In other words, for each object in each cluster, the cluster computing the sum of the objects to their distance square from the center. Here is K-Means algorithm process overview.

Algorithm 1: K-Means

Input:
- k: cluster number
- D: dataset contains n nodes

Output: k clusters containing correspond nodes
Method:
1. Randomly select k objects from the dataset as initial cluster center

repeat
2. Assign each object to its most similar cluster according to average distance;
3. Update cluster average distance, computing objects' distance in each cluster **until** difference criterion is met

2.2 Parallel K-Means algorithm on a single Hadoop platform

Since calculating the distance between data object with the initial cluster centers and calculate new cluster centers are independent from each other, so it can be achieved in MapReduce model. Algorithm process is as follows:

Algorithm 2: K-Means on Hadoop

1. In the main function, randomly select k initial clustering center and save them in the global Configure object.
2. In map function, read data object fromHDFS, compute the distance between each data and k initial clustering, and assign object into the nearest initial cluster. Print results (initial k as the clusteringcenter, object as the value), and deliver them to the reducer.
3. Reducer accepts a key/value pair clustered in the initial clustering center, all objects belonging to the same cluster will be delivered to the same reducer.
4. Loop step (2) and (3) until the sum of distance in very k cluster meets the predefined criterion.

2.3 K-Means algorithm for Hama

Algorithm 2 requires iteration, composed by a series of MapReduce job chaining. After each MapReduce job is completed, the output Reducer will automatically be persisted to the storage medium (default is HDFS) eventually. Then, when the next time the job performed MapReduce process, it needs to read data from the HDFS. This approach relies on a series of operations to achieve MapReduce iteration, which greatly increases the computational cost. Hama relies on BSP model, using message exchange of the data calculation process. To ensure that the overall message is delivered in an orderly manner, after asynchronous parallel computing, there need gate barriers to be inserted into tasks within the same super-step sends message and wait synchronization. Tasks inserted barrier gate into need to wait until all tasks within the super-step to finish for interactive message data through, and then all the obstacles task inserted blocking gate into inside this super task can be re-awakened and use the received collection data message to continue running the next super-step.

General steps for programming in Hama:

1 Implements the setup function, initialize the task and set the number of the specified task peer.
2 Implements bsp function, concrete the iteration code, pass the computing results to master a task as is for map tasks.
3 Implements the cleanup function, conclude all middle results of each peer, output results according to the final result form as is for reduce task.
4 Set the BSP Job object and set the parameters for distributed tasks up.

Below shows K-Means algorithm in parallel on Hama

Algorithm 3: Parallel algorithm on Hama

Input:
    k: Number of clusters
    D: Dataset containing n objects

Output: k clustered groups
Method:

1. In setup function, randomly select k objects as initial centers and store them into parameters (centers []), set up the maximum iteration times.
2. In bsp function: doing the following operations (step 3-16)
3. Loop (true)
4. Peer.ReadNext(key, value) //read all data in
5. id= get NearestCenter (key) //compute the distance between the nodes and center of the cluster and return the nearest node id into new Center Array and increase the counter summation Count[id] by one.
6. Assign Centers Internal (new Center Array[], summation Count[], key)
7. Iterate all BSPPeer, send all local cluster center to other BSPPeer, in which Center Message is the message type.
8. for (String peer Name : peer. Get All Peer Names()).
9. peer.send(peer Name, new Center Message (i,summation Count[i], new Center Array[i]));
10. Peer. sync(); //process synchronize operation
11. After synchronization, every BSPPeer receive messages coming from all others and update its new center node centers[].
12. Compute the deviation between the new and old center, according to the following formula:

$$\sum_{i=0}^{m} | oldCenters[i][j] - newCenter[i][j] |$$

13. If the deviation is acceptable, then return else update the old Center[i] to new Center[i];
14. peer. Reopen Input (); //point back to the initial data and proceed the next iteration.
15. Write the clustering results into output file. Key means cluster id(start from 0) and value is the data object
16. Recalculate assignments and write ()

3 EXPERIMENTS AND RESULTS

The experiments are executed on a six-machine cluster. Each machine runs the Ubuntu Linux (version 10.04), has a memory size of 16G, disk storage of 160G and four Intel Xeon E5502 CPUs with the frequency of 1.87GHz. Used Hadoop version is hadoop-1.0.3, Hama version hama-0.6.0. We test and evaluate the algorithm on different kinds of datasets [8].

Experimental results are shown below in Table 1:

Name	N	k	Max Iteration	Tasks	Job Execution Time (Hama)	Job Execution Time (Hadoop)
Turkiye Student Evaluation	5820	10	10	2	35.172s	116.342s
Gesture Phase Segmentation	9900	10	10	5	45.328s	212.391s
NYSK	10421	10	10	10	70.591s	293.092s
Amazon Access Samples	30000	10	10	20	115.068s	483.648s

Table 1. Experiment results.

These experiments show that Hama as an iterative aimed platform is more efficient than Hadoop when coming to these iterative algorithms.

4 CONCLUSION

In this paper, we analyzed the design and implementation of the classical algorithm in parallel architecture like MapReduce and especially Hama for its highly efficient iteration over tasks.

Then, we test the efficiency on real Hama and Hadoop cluster separately and verified the assumption of Hama's advantage on iterative jobs. We hope to introduce a more efficient way of which are appropriate for Hama and to pave the way for further complex data mining algorithm on it.

REFERENCES

Dean J, Ghemaw at S. MapReduce: Simplified data processing on large clusters/ / Proceedings of the Conference on Operating System Design and Implementation (OSDI. 04). San Francisco, USA, 2004: 137–150.

Hadoop: http://hadoop.apache.org/

PanWei, LiZhanhuai, Chen Qun. Evaluating Large Graph Processing in MapReduce Based on Message Passing [J].CHINESE JOURNA L OF COMPUTERS, 2011, 34(10).

http://hama.apache.org/

Leslie G. Valiant, A bridging model for parallel computation, Communications of the ACM,Volume 33 Issue 8, Aug. 1990.

Ghemawat S, Gobioff H, Leung S T. The Google file system, F, 2003[C]. ACM.

D.Borthakur.The hadoop distributed file system Architec -ture and design. Apache Software Foundation.2007.

https://archive.ics.uci.edu/ml/datasets.html?format= &task=clu&att=&numAtt=&type=&sort=nameUp &view=table

Computational Intelligence in Industrial Application – Ling (ed.)
© 2015 Taylor & Francis Group, London, ISBN: 978-1-138-02818-0

Education in the age of big data

YanLing Yang
Foreign Languages College, Northeast Dianli University, Jilin, P.R. China

Han Li
College of Science, Northeast Dianli University, Jilin, P.R. China

ABSTRACT: Big data puts us into an age featured with data, data collection and analysis, and the flipped classroom and MOOC based on big data leads to great changes in the educational field. In this paper, first, the definition and features of big data are introduced, and then the application of big data is discussed. Finally, characteristics of education are analyzed, that is, we should realize that educational resources become generative, teaching individualized, and the role of teachers repositioned in an age of big data.

KEYWORDS: Big data; Educational data mining; Learning analytics; Flipped classroom; MOOC.

1 INTRODUCTION

Big data, as the hot keywords, have a significant impact on education, just as on all walks of life. We can feel the effect deeply when the flipped classroom and MOOC develop. Big data puts us into an age featured with data, data collection and analysis, and the flipped classroom and MOOC based on big data leads to great change in the educational field. In the age of big data, educational resources become generative, teaching individualized, the role of teachers repositioned.

2 AGE OF BIG DATA

2.1 *Definition of big data*

Big data have three features: volume, velocity and variety according to Gartner company in 2012. First, learners registered in a course may reach thousands of people, even up to hundreds of thousands of people, so sample size is quite big. Second, big data can be produced by learners in the course of learning, which happen immediately, rather than spend much time collecting after that. Third, the types of big data are various.

2.2 *Characteristics of big data*

2.2.1 *Massiveness*
Big data makes it possible that information which cannot be got, calculated, storied and analyzed in the past becomes data-orienting. With the support of internet, telecommunications and satellite communication technology, rapid-spread PCs, tablet computers, smart phones and other devices are recording people's daily life in the form of data and storing them in the applicable database: Mobile telecom carrier has individual's whereabouts; credit card companies and online payment platform record individual's shopping, traveling and payment capacity; Social Networking Services collect and store all the information concerned with users' social relationships and personal preference. The idea and technology of big data give people much more opportunities to get information in many fields and levels than ever.

2.2.2 *Potentiality*
"Data" in big data and "figure" have different meanings. Background and comment data behind a figure interpret it all-around. For example, if a student gets 90marks, 90 is a "figure". Only when we combine it with metadata behind it such as learning capacity, learning attitude, intelligence level, family environment and social relations, we can interpret the true meaning of the "figure". In the present age when data have been produced massively, a small part of value about data can be known, while most are not explored. In the past, data were considered to be useless and discarded once they were used. Nowadays, there is no limit to data. It is quite important whether we can bring now value to data and realize the successful transformation from support by figures to by data in order to draw out the potential value of data.

2.3 *Application of big data in education*

Technology Office of Ministry of Education in U.S.A. released 《Improvement of Teaching and Learning by Educational Data Mining and Learning Analytics: Problem Description》 (hereafter referred to as 《Description》) in 2012. Educational data mining means that data generated in the course of teaching and learning are collected and analyzed with statistics, machine learning, as well as the data mining technology and developing methods. It tests learning theory and guides educational practice. Learning analytics means data collected in educational management and service are analyzed by applying technology of informatics,sociology,statistics,psychology,machine learning and data mining. Application program created in learning analytics have an effect on educational practice.

2.3.1 *Educational data mining*

The early educational data mining refers to mine data of web's log. Nowadays, interactive learning method and tools (intelligent tutoring system, simulation, games) bring new opportunities for quantifying students' behavioral data. Online learning system which is more integrated, modularized and complicated gives us more types of data, which include variables used in data mining algorithm. Educational data mining can discover schema and rules in data, establish prediction model so as that we can rediscover and predict students' learning. For instance, the evaluation on online course was mainly conducted through questionnaires at the end of the course. People developed online course evaluation method, which analyzed Students' learning logs on the online course, classified students, predicted students' achievements and attained students' performance and satisfaction on the course with the combination of demographic characteristics data and questionnaires at the end of the course.

2.3.2 *Learning analytics*

Learning analytics is an emerging field of study, with the aim of providing references for decision-making in education system with data analysis. The key point of it is to analyze the big data applied in educational field. This research method came from commercial field. Merchants mined and analyzed the data about consumers' activities to grasp the trends in consumption. For example, Taobao can deduce its users' preference to certain types of products according to the products users have browsed or bought. Sina microblog, a social networking site, can recommend what its users may well be interested in based on what they focus on. Learning analytics aims at predicting and judging statistically through data extraction, classification and analysis.

The analysis of data related to students takes students as the main body, takes aim at designing perfect teaching methods, concentrates on giving students high-quality, individualized learning experience and evaluates whether learning plans can help improve students' learning effectively. The results of data analysis are of great significance for both teaching parties in educating applications. For educators and researchers, learning analytics is very important to understand the relations among students, online texts and courseware. For students, mobile software and online platform are developed to analyze data about students in order to provide students learning support system that accords with their learning needs and improve their learning marks and efficiency.

Just as what the chief economist Hal Varian in Google said: what we need is the ability to extract knowledge from data that are widely available. The purpose of data collection is to extract useful knowledge from data according to needs and apply it to specific aspects. "Research Hotspot of Educational Informationization and Word Frequency Analysis of Development Tendency" and "Contrast Analysis System of Effects of Subject Construction" in South China Normal University and "Transactional Analysis of Micro-Blog Platform" in Anyang Teachers College are good attempts at big data analysis and mining. Innovation of big data will develop alone the lines of data – big data – analysis and mining – discovery and prediction. In the course of data creation, big data are bound to influence educational innovation and provide the limitless possibilities for informative teaching.

2.4 *Educational reform in an age of big data*

2.4.1 *Flipped classroom*

Flipped classroom is a new teaching model. Students watch teaching videos recorded by the teacher at home or after school, while they can complete their exercises and communicate with their teacher and classmates in class. This practice is contrary to traditional one that the teacher gives lecture in class and students finish their homework after class. Flipped classroom is to pass on knowledge after class and put knowledge internalization in class. It, first, was conceived by two chemistry teachers- Jonathan Bergman and Aaron Sams from Woodland Park in U.S.A. In 2007, they videotaped their lectures and put on the internet for the students who missed a scheduled class. This pioneering teaching practice is being paid increasing attention. Flipped classroom highlights the importance of big data in promoting informative teaching transformation.

Flipped classroom has different characteristics from traditional teaching methods. Table1 shows the contrast between flipped classroom method and traditional teaching method.

Table 1. Contrast between flipped classroom method and traditional teaching method.

	Flipped Classroom	Traditional classroom
Teacher	Guider, promoter	knowledge transmitter, organizer
Student	Active learner	Passive receiver
Teaching Model	Pre-class learning + classroom inquiry	Classroom lectures + homework
Teaching Contents	Exploration of knowledge	Transmission of knowledge
Teaching Mode	Combination between face-to-face instruction and web-based instruction	Face-to-face instruction
Evaluation Mode	Multi-dimension and multi-mode assessment	Traditional assessment

1 Change of roles

Flipped classroom turns teachers from knowledge transmitter to guider ,promoter and producer of teaching videos. In flipped classroom, knowledge transmission has been carried out through teaching videos before class and in class the teacher spends most of time on communication with students. The teacher does not only prepare teaching materials, but also take part in the discussion about difficult points with students both in and out of class.

In the flipped classroom, students change from passive receivers to active learners. Before class, they learn the knowledge through teaching videos, choose what, where and when to learn, furthermore they can communicate with their classmates and teacher on qq, We Chat or blog. In class, they have more opportunity to join the various teaching activities and discussion in order to internalize the knowledge. They do not absorb knowledge passively, but construct actively.

2 Chang of teaching model and contents

There is a reverse between learning time and teaching time. The teaching contents in class are now shown in the teaching videos for students to learn independently before class. In class the teacher do not spend most of time in giving a lecture, but organize various activities in which students can have more discussion, communication, role-playing, presentation and the like. In this way, students have more chance to analyze, explore and solve problems actively, and then have the knowledge internalized.

3 Change of teaching and evaluation mode

Knowledge is mainly transformed through the face to face instruction given by teachers, textbooks and other paper materials in the traditional

way. Besides these modes, multi-media teaching equipments, web-based teaching platforms and other electronic resources are applied in the flipped classroom. Diversification of teaching resources makes evaluation change from traditionally paper testing to Multi-dimension and multi-mode assessment possible.

2.4.2 MOOC

The success of flipped classroom, reconstruction of education with videos, leads to the development of MOOCs (Massive Open Online Course) in higher education. In 2012, MOOCs were rapidly spreading. The leading "three carriages" were Coursera, Udacity in Stanford University and edX co-founded by MIT(Massachusetts Institute of Technology) and Harvard University. In 2013, Tsinghua University and Peking University joined in edX. In Tsinghua University, 4 courses were chosen to be online and open to the world in the early stage. In Peking University, 14 courses which cover many subjects of Arts and Science were online in Septembers, 2013. With the popularity of MOOC in advanced education, it is quite possible that students who complete the credits in MOOCs will reach the level of normal education.

Massive participation manifests the "openness" of MOOC. The learners from all over the world can freely take the courses they are interested in because there is no limit to the number of registration in MOOC. In 2011, for example, the free online course, 《Introduction to Artificial Intelligence》, offered by Thrun was registered by 160 thousand students from more than 190 countries and 20 thousand students completed it. In 2013, the registered users in Coursera, the largest platform of MOOC, have reached more than 5 million people, over 450 courses offered, more than 90 universities joined in, which include Fudan University and Shanghai Jiaotong University.

With massive visits, global cooperation and dynamic information resources, MOOCs inevitably produce complex big data. These large numbers of real-time data are recorded in MOOC platform. If they learn in MOOC, learners do not only watch videos and answer questions, but also communicate with the teacher and other learners on social networks, forums, blogs and so on. In this way, they leave long and rich trace of data, and then the big data are analyzed and processed with technology to know how and what students learn and find what results in the learners' success.

3 CHRACTERICS OF EDUCATION IN AN AGE OF BIG DATA

3.1 Generativity of teaching resources

With the change of educational subjects, the providers of learning resources have changed. MOOCs indicate the important trend of education in an age of big data- generativity of teaching resources, which reflects the diversity of learning. The designers and developers of resources have changed from authorities to diverse providers. Traditional education cannot meet the needs of learning in social learning network that change rapidly. Moreover, cloud resources are very popular in an age of big data because massive amounts of information resources are stored in "cloud". Generally speaking, no matter what you want to look for, texts, videos, audios and animation, if you input key words, you can find it rapidly and easily. As for image resources which are hardly found, we can find them with "image perception", which means you should form representation in your mind according to learning needs, give it a keyword and form an image and input the keyword into search bar to find your image resources.

3.2 Individuation of teaching

<<National Medium- and Long-Term Plan for Education Reform and Development (2010–2020) >>issued by Ministry of Education mentions: student's individual differences should be paid attention to and each student's potentiality should be developed. Individualized education environment should be constructed. Students should be guided according to their pace and condition of learning, which requires to assess students' ability, interest, talent and style of learning. In the past, we could not understand the student's individual differences owing to the lack of method of getting and analyzing information data. We could not create the supportive learning environment which helps learners promote their academic achievement. Nowadays, these problems can be improved and solved with the ideas and technologies which can process data.

In an age of big data, students can leave massive digital fragments in course selection, online learning, interaction and feedback, activity involved with the social networks and use of campus cards. When the data are integrated, mined and analyzed, students' behavioral model can be reflected. "Individualized Curriculum Evaluation Based on Degree Compass" in Austin Peay State University, "Pre-warning System of Students' Dropout Based on Key Factor Analysis" in University of Phoenix and "Pre-warning System of Students with Financial Difficulties" in East China Normal University are good attempts and practices that big data are applied to individualized education.

3.3 Reposition of teachers' role

The traditional teaching is mainly empirical. The teacher chooses the teaching contents and methods according to his experience. In as age of big data, however, the teacher takes not only the role of "teacher" in a traditional meaning, but also the role of analyst. Data collection and learning analytics, can free teachers from the traditional teaching. The teacher's teaching decisions are not based on experiences, but on data analysis.

4 CONCLUSIONS

The emergence of flipped classroom and MOOC shows that data collection and learning analytics in an age of big data have had a great influence on education and informative teaching innovation age has come. In this paper, first, the definition and features of big data are introduced, and then the application of big data is discussed. Finally, characteristics of education are analyzed, that is, we should realize that educational resources become generative, teaching individualized, the role of teachers repositioned in an age of big data.

REFERENCES

Aragon, S.R. & Johnson, E.S. 2008. Factors influencing completion and noncompletion of community college online courses. *American Journal of Distance Education* 22: 146–458.
Bergendahl, C. & Tibell, L. 2005. Boosting complex learning by strategic assessment and course design. *Journal of Chemical Education* 82: 645–651.
Bienkowski, I. & Feng, M. & Means, B. 2012. Enhancing teaching and learning through educational data mining and learning analytics: an issue brief. *Washington, D.C. Office of Educational Technology,* 9–10.
http://en.wikipedia.org/wiki/Big-data.
Hung, J. & Hsu, H. C. & Ricc, K. 2012. Integrating data mining in program evaluation of k-12 online education. *Educational Technology & Society* 15(3): 27–41.
Jiyuan, Li. 2013. The idea of MOOC. *China Education Network* 4:3 9–41.
Kai, Liu. & Jia, Zhong. & Haitao, Gao. & Jun, Yue. 2013. Analysis of the influence of network to college students' ideological and political quality. *Journal of Northeast Dianli University* 28(3): 75–78.
Mullen, G.E. & Tallent – Runnels, M.K. 2006. Student outcomes and perceptions of instructors' demands and support in online and traditional classrooms. *The internet and Higher Education* 9:257.
Saettler. P.L. 1990. *The evolution of american educational technology*. Englewood, CO: Libraries Unlimited.
S. Papert. 1980. *Mindstorms*. New York: Basic Books.
Viktor, Mayer-Schönberger. & Kenneth, Cukier. 2013. *Big data*. Hangzhou: Zhejiang People's Publishing House.
Yufan, Fu. 2013. "Pathfinder" of China online education. *China Education Network* 4: 26–28.

Computational Intelligence in Industrial Application – Ling (ed.)
© 2015 Taylor & Francis Group, London, ISBN: 978-1-138-02818-0

Research on FPN-based security defense model of oil and gas SCADA network

W.W. Zhang, X.D. Cao, Q.C. Hu & P. Liang
College of Electrical & Information, Southwest Petroleum University, Chengdu, Sichuan, China

Y. Qin
College of Mechatronic Engineering, Southwest Petroleum University, Chengdu, Sichuan, China

ABSTRACT: With a focus on SCADA system security status, mainly malicious behavior as defense factors, based on factor space theory and combined fuzzy Petri network, this paper proposes a Fuzzy Petri Network-based security defense model for SCADA. It is done by monitoring the malicious behavior point and fuzzy quantified to get the weights of malicious behavior. We proposed a fuzzy Petri reasoning method in the input weights, the transition threshold, output reliability multi-constraints, then used part of malicious program behavior in SCADA to verify the feasibility in the SCADA security defense system.

KEYWORDS: Fuzzy Petri nets; SCADA; Security defense; Factor.

1 INTRODUCTION

SCADA (Supervisory Control and Data Acquisition), data acquisition and monitoring system are the important components of the industrial control system. Through the HMI interactive and data acquisition and transmission system, SCADA real-time monitoring to the scene of field equipment, so as to realize the function such as field device control, parameter adjustment and measurement [1]. Due to the SCADA system from birth to now, is considered a closed system, almost no consideration a vulnerability. The application of EtherNet, Fieldbus, OPC technology makes the industrial equipment interface more open and enterprise information in SCADA system is no longer isolated. SCADA system is vulnerable to hackers.

This paper, according to the theory of factors space and based on the active defense oriented to the host and server, puts forward an industrial control SCADA system security defense model based on Factor Neural Network[2] (Factor Neural Network-based SCADA Security And Defense Model, referred to "FSDM"). Combined with synchronization, concurrent ability of Petri net [3], and proposes the knowledge representation method of fuzzy Petri net for FSDM, presents a fuzzy reasoning Petri net model [4]. Knowledge representation method of fuzzy Petri net provides an effective fuzzy knowledge representation and reasoning tool, the knowledge model has the parallel reasoning ability and the model structure is clear, is a model of the system to describe the system structure and the dynamic variation of the system state.

2 MODEL STRUCTURE

Combined with oil and gas gathering and transferring enterprise SCADA system network topology, as shown in Figure 1, deployed the neurons in the host and server of the application layer, scheduling layer and monitoring layer. In the absence of host and server, the field layer is mostly sensors, PLC, RTU. The field devices work instructions are issued by the upper computer, just to protect the safety of the host and server, so the field layer doesn't have to deploy defenses neurons. Mutual awareness between neurons, decompose and elaborate the task [5], then execute the subtasks.

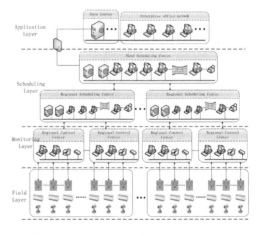

Figure 1. SCADA system topological structure.

According to the distributed SCADA system architecture, FSDM model is shown in Figure 2. Execution neurons contain Ai, Aij, Aijk, Aijkm. Among them, the Aijkm⊂ Aijk⊂Aij⊂Ai(i, j, k, m = 1, 2,3...) and have a certain range of motion. Ai deployed in the host and server of the application layer, Aij deployed in the host and server of head scheduling layers, Aijk deployed in the host and server of the regional scheduling center layer, Aijkm deployed on the host of monitoring layer. FSDM model of the execution task of security defense is a dynamic programming problem, the calculation process is commonly backward recursive and according to the dynamic process to find the optimal path. First, determine the initial input factor, transition factor and output factors of the system, then, according to the grades of fuzzy classification, and integrate all the results according to the rules in the knowledge base. Finally, trigger the transitions of the fuzzy Petri net [6] to get the actual output of the system.

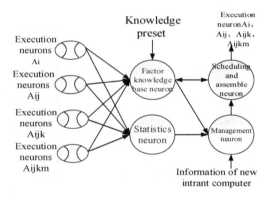

Figure 2. FSDM overall model structure.

3 THE CONSTRUCTION OF FUZZY PETRI NETS

3.1 *Fuzzy Petri net knowledge representation of model*

Factor is a kind of cognition and expression category, it is abstract from the process of cognition, is the foundation of knowledge representation and reasoning. The researchers use it to recognize and describe things. The analytical factors, neural network is used to construct the defense system, using the fuzzy Petri net can define a six dimensional tuple [7] FPN= (P, T, F, I, O, f), stated below:

P= {p1, p2,...,pn},P is a limited set of place node, means the factor set;

T= {T1, T2,...,Tn},T is a limited set of transition node, it means the implementation of established rules;

F is a limited set of proposition, means the flow between the place node and transition node, expressed as a fuzzy set which meet the condition: F⊆(P×T) ∪(T×P);

I:P×T→ [0,1],I is a fuzzy relation of P to T, represents the connection status and weights of place node to transition node and meets the condition:0≤I(pi,Tj)≤1,i=1,2,...,n,j=1,2,...m;

O: T×P→ [0,1],O is a fuzzy relation of T to P, represents the connection status and credibility of transition node to place node, meets the condition: 0≤I(Ti,pj)≤1,i=1,2,...,m,j=1,2,...n;

f is defined as a real value of [0,1] in the transition node T, means the trigger threshold value of transition node[8].

3.2 *Fuzzification of connection relationship*

FSDM model, for example, the malicious programs have more than one program behavioral factors. By using the hook technique to monitor malware [9], the malicious program behavioral factors are divided into four hazard levels: low, medium, high, highest. They represent different hazard level of malicious behavioral factors. Table 1 lists some malicious behavioral factor, and the malicious level in the hierarchy.

Table 1. Behavioral factors hazard rating (part).

Behavioral factor name	Behavior identity	Hazard rating
Terminate system process	TSP	medium
Set windows hook	SWH	higher
Create startup service	CSS	medium
Delete service	DS	higher
Modify system startup	MSS	medium
Traverse disk	TD	higher
Alter file attribute	AFA	medium
Replace Registry File	RRF	highest

Fuzzy set is completely characterized by membership functions, different membership functions determine the different membership degree, the size of the membership degree can reflect the essential characteristics of things rightly and intuitively. Based on the normal form membership function distribution, the malicious behavioral factors which are listed in Table I can be fuzzified. The rank of low, medium, higher and highest respectively quantify for four grade X = {1, 2, 3, 4}, then the malicious behavioral factors of the fuzzy set in Table 2 can be obtained.

Table 2. Hazard rating fuzzy membership degree.

Grade	Behavioral factors identity							
	TSP	SWH	CSS	DS	MSS	TD	AFA	RRF
1	0.2	0.1	0.3	0	0.1	0.1	0.2	0
2	0.9	0.5	0.9	0.3	0.9	0.4	1.0	0.3
3	0.2	1.0	0.4	1.0	0.3	0.9	0.2	0.5
4	0	0.5	0.1	0.3	0	0.4	0	1.0

Combined with the hazard membership degree of malicious behavior factors in Table II, using the weighted average method [10] can calculate the weights of malicious behavioral factors ,is also the weights of place node to transition node.

$$U=\sum U(u_i)\ u_i/\sum U(u_i) \qquad (1)$$

Take the "Terminate system process (TSP)" as an example, by the formula (1), the P to T input weight I $(TSP)=\left(\dfrac{0.2\times1+0.9\times2+0.2\times3+0\times4}{0.2+0.9+0.2+0}\right)/(1+2+3+4)=0.2$, then get the weight of "Terminate system process". By the same, the weight of other malicious behavioral factors can be obtained. When the weight is greater than the trigger threshold, the system can enter the next level place node.

3.3 FPN-based SCADA security defense system design

The SCADA security defense system is to imitate human experts to solve specific problems, to be able to use observation, calculation, judgment, reasoning, and selection and a series of smart way of thinking to analyze the problem. Fuzzy Petri net makes the system have certain soft features. Fuzzy Petri net is clearer to describe the defense system under the concurrent attack. The fuzzy production rules for malicious behavioral factors reasoning have the following three basic forms. The fuzzy Petri nets models are shown in Figure 3.

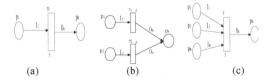

(a)   (b)   (c)

Figure 3. Fuzzy production rule basic model of FPN.

Form (a): Using formula (1) to calculate the weight of place node to transition node, only the input weight greater than transition node trigger threshold f, has the following rule: if pi then pk (output credibility On=Ii), namely pi→ pk.

Form (b): Calculated the weight, only the input weight greater than transition node trigger threshold f, has the following rule: if pi and pj and pm then pk (output credibility Ok=(max(Om, On)), namely pi∪pj→ pk.

Form (c): Calculated the weight, only the input weight greater than transition node trigger threshold f, has the following rule: if pi or pj then pk (output credibility On=(Ii+Ij+Im)/3), namely pi∩pj∩pm→ pk.

Using the above three basic model, combining with the SCADA security defense system, transform FSDM into FPN model, as shown in Figure 4. When the malicious behavioral factors place node input weights are greater than the transition threshold value, continue to reasoning, otherwise stop.

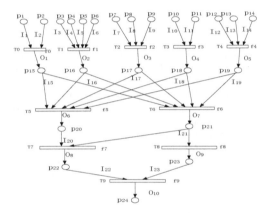

Figure 4. FPN-based SCADA security defense system.

The FPN adopts the hierarchical structure, reflects the structure of the distributed SCADA system. In Figure 4, p1- p14 is malicious behavioral factors;p16- p19 are implicit place nodes, as the sensory information processing units; p20 is statistical place node; p21 is factors represent place node; p22 is management place node; p23 is scheduling place node; p24 to task end place node; I1- I23 are the weights of place nodes to transition nodes; O1- O10 are credibility of transition nodes to place nodes, also the credibility of the neuron output; f0- f9 are transition nodes threshold, which trigger the next neuron threshold. According to the description of defense behavioral factors, to set up the SCADA security defense rules table (part), as shown in Table 3 (part).

Table 3. Defense rules description (part).

Rule number	Rule description
R1	If TSP and DS then stop closing process or restart the process, set honeypot.(Credibility O=0.9)
R2	If SWH then start backup program and alarm(Credibility O=0.85)
R3	If MSS then prevent to modify the related and reported to terminal (Credibility O=0.8)
R4	If CSS then monitoring its behaviors, get the parameters (Credibility O=0.7)
R5	If RRF then close modify permission, access denied and alarm(Credibility O=0.95)

4 RELATIONSHIP EXPERTS OF FPN-BASED DEFENSE MISSION

4.1 *Relationship between defense mission and task*

The relationship between defense mission and task is the relationship between the overall mission and specific tasks, which is the relationship between an abstract class and its instantiated task. "Task" sets are affiliated to "Mission" set, Mission ={ Task1, Task2,...Taskn},if $\forall$ Taski $\in$ Mission (i = 1,2,...,n), then SpecifyOf(Taski, Mission) means that the specific defense task "taski" is an instance of the overall defense mission. The overall defense missions centered on the global defense. According to behavioral factor rules to decompose the mission into tasks, then decompose tasks into several operations, the task decomposition tree shown in Figure 5. The decomposition of mission to tasks is a process of abstract concept specification essentially. A mission can be decomposed into multiple specific tasks.

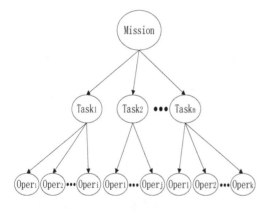

Figure 5. Mission decomposition tree.

4.2 *Relationship between defense task and operation*

A defense task includes defense operations longitudinally. After the specification of the defense motion to defense task, also need to decompose specific defense tasks. The defense task can be formalized operationally set: Taski= {Oper1, Oper2,...,Opern}. If $\forall$ Operi$\in$Taski (i=1,2,... n), the AffiliationTo (Operi Taski) means that defense action Operi affiliated to Taski, namely defense operations Operi are the subtasks of defense Taski.

4.3 *Relationship between defense operation*

The relationship between defense operations is horizontal "brother" relationship, defined as BrotherOf (Operi, Operj), means that Operi and Operj are peer relations. The operation execution needs a certain sequential relation, including the continuous and intermittent relation, and equal, overlapping, during the period, start and end relation in concurrent events [11]. To the mission, the execution of operations performed as logical relationship, including the logic and, or, select, exclusive or relation. Combined above relation, more logical relationship of defense operation can be obtained. The BrotherOf ={Sequence-And, Sequence-Cho, Parallel-And, Parallel-Or, Cir, XOR},It means a set of relationships.

If Operation=(Oper1,Oper2),then the formalization below:

1 Sequence-And: if Sequence-And (Oper1, Oper2), then T.endOper1<T.startOper2, Taski= Oper1& Oper2. It means Oper1 completion time before the start time of Oper2, requires Oper1 and Oper2 be successfully implemented, then the Mission completed.
2 Sequence-Cho: if Sequence-Cho (Oper1, Oper2), then T.endOper1<T.startOper2,Taski=Oper1 → (choose) Oper2, signified to be executed in sequence. The first implementation of Oper1, if successfully implemented, the Mission succeed, Oper2 is not executed. If the execution of Oper1 is failed, Oper2 is executed, the Mission success depends on Oper2.
3 Parallel-And: if Parallel-And (Oper1,Oper2), then TimeOper1 $\wedge$ TimeOper2 $\neq$ 0 and Taski =Oper1&Oper2. Oper1 and Oper2 have overlapping in execution time. Both Oper1 and Oper2 performed successfully, the taski success.
4 Parallel-Or: if Parallel-Or(Oper1,Oper2), thenTimeOper1 $\wedge$ TimeOper2 $\neq$ 0, Taski=Oper1|Oper2, means Oper1 and Oper2 have overlapping in execution time, one of them executed successfully, the Taski succeed.
5 Cir: Cir (Oper1,Oper2) means executed Oper1 and Oper2 circularly, it equals to Cir(Oper2, Oper1).

5 CONCLUSIONS

This paper analyzes the present status of SCADA system security, and abstracts the proposed FSDM model furtherly. Using the fuzzy Petri net theory to the SCADA security defense system modeling, build the fuzzy reasoning Petri net of security defense. The methods to calculate the transition node input weights and the credibility of rules are introduced. The tasks and relationships in the fuzzy Petri net are described. But the transition threshold value of fuzzy Petri net largely depends on the expert's experience, how to set up the fuzzy Petri net transition node threshold value is crucial to improve the system of defense, so expansion of the knowledge base and transition node threshold value optimization are currently studying.

ACKNOWLEDGMENT

This work is financially supported by the National Natural Science Foundation of China (No.61175122).

REFERENCES

Shuai Zhang. 2012. Industrial Control System Security Risk Analysis. *Journal Information Security and Communications Privacy*, Vol 34(2012),p.15–19.

Zengliang LIU. 1994. Factor Neural Network Theory and Application. Guiyang: Guizhou Science and Technology Press, 17–36.

Claude Girault, Rudiger Valk, Wang Sheng-yuan (translated).2005. Petri Nets for Systems Engineering. Beijing: Publishing House of Electronics Industry.

Li-xin Jia, Jun-yi Xue, Feng Ru. 2003. Formal Reasoning Algorithm and Application of Fuzzy Petri Nets. *Journal of XI 'AN Jiaotong University*, Vol 37 (2003),p.1263–1266.

Qian-yi Zhan, Qiang Sun, Yu-sen Zhan.2012. For Solving the Dynamic Task Cooperation Alliance Model. *Journal of Frontiers of Computer Science and Technology*,Vol.6(2012),p.1098–1108.

Sheng-jun Wei, Chang-zhen Hu, Ming-qian Sun.2007. A Method of Dynamic Knowledge Representation and Reasoning Based on Fuzzy Petri Nets. *Science & Technology Review*,Vol 25(2007),p 13–17.

Qiao Peng, Li-feng Guo, Jie Ma.2012. Study of Nuclear Power Plant Expert System Based on Fuzzy Petri Net Reasoning. *Journal of Atomic Energy Science and Technology*,Vol 46(2012),p.356–360.

Jun-ren Pan, Dao-wu Pei.2009. Fuzzy Reasoning Algorithm Based on Fuzzy Petri Nets. *Journal of Zhejiang Science Technology University*, Vol 26(2009),p.874–879.

Xin-juan Yan, Min-sheng Tan, Ming-e Lv. 2012. Based on the Analysis of Active Defense Technology Research. *Journal of Computer Secuerity*, Vol 10(2012),p38–39.

Xie-dong Cao.2003. The Fuzzy Information Processing and Application. Beijing: Science Press.

Kai Cheng, Jun-hui Che, Hong-liang Zhang. 2012. Formal Description of Operational Task and Its Process Expression. *Journal of Command Control and Simulation*,Vol 34(2012),p15–19.

Investment Project Management Software System (IPMSS): Ideas and practice

H. Xie
China Railway Resources Group Investment Ltd., Beijing, China

Q. Li
China Railway Resources Group Co., Ltd., Beijing, China

ABSTRACT: Investment Project Management Software System (IPMSS), which is developed based on the ASP.net technology, consists of four modules: Information database, office decision-making, project management and instant messaging. The IPMSS is characterized by informatization, standardization, multi-target, expandability and instantaneity. Aimed at the existence of network information security risks, the IPMSS takes physical security, virtual private networks, user permission settings and network security management methods to deal with them. Mobile platform operation and intelligence of online data processing are thought to be the further development direction of the IPMSS.

1 INTRODUCTION

As a specialized company of mining venture capital management and mineral resource development, China Railway Resources Group Investment Limited (CRRI) receives thousands of project information each year. Only those projects that meet the company's criteria of investment will be placed into the "project pool". CRRI chooses the projects worth of investing from the pool to carry out further due diligence and business negotiations. Finally, a few projects selected enter into the ultimate substantive cooperation. However, it seems to be an endless process from getting project information to investment, management and quit the project.

During these processes, the company inputs a lot of staff, money and time on the storage of vast amounts of information, the identification and control of investment risks, the supervision of series of related managing activities, the collective decision-making of vita issues, as well as the report to the company's top managers who want to know the project updates. That's why the company needs a fast and efficient platform to disclose project information, publish project progress, share project resources, report and track investment risks and help managers to make decisions anytime and anywhere.

Although there are a lot of similar softwares and systems on the market, they are mostly based on traditional single project management (Sun 2013), such as Microsoft Project Manager, DotProject, ZenTao PMS software; or special aspects in project management like schedule management, cost management, risk management, information management and process

supervision. For example, HilionEAS project evaluation and economic analysis software, NavalPlan schedule management software, PHProjekt project information and document sharing software, Jinniu cost control software and Hongda office affairs management software. Therefore, an investment project management system including office decision-making, multi-project management, mass data storage and sharing, and real-time online management seems particularly important to satisfy the management requirements of each phase, i.e. investment project approval, management, supervision and quit.

2 IPMSS MODULES DESIGN

The IPMSS was designed for diversified investment project information, based on which the database was set up. The real-time operation, update and release on this database are according to the company's quality management system, investment system, processes, and forms in order to achieve the purpose of the investment project risk control and operational management.

Table 1 has listed the four main modules of the IPMSS: information database modules, office decision-making modules, project management modules and instant messaging modules.

2.1 *Information database module*

Project information of different regions, different types, different exploration and development phases, and different properties of investment were classified, numbered and property assigned to implement database. Operations on the database principally include data classifying, numbering, inputting, storing, modifying, updating, searching and sharing. This module of information database is the basis of the IPMSS.

Table 1. Main modules and contents of the IPMSS.

Modules	Work flows	Functions
Information Database (DB)	Information Classification Input Database	Storing Updating Searching Downloading
Office Decision-making (OA)	Filtering Due Diligence Reviewing Approval	Decision-making Nodes controlling
Project Management (PM)	Design Construction Change Completion	Systematism Standardization Programming Charting
Instant Messaging (IM)	Forum Message Calendar	Updates releasing Nodes warning Online chatting

2.2 Office decision-making module

It's now linked with the company's current Office Automation (OA) systems. The OA system aimed to release information and to implement the functions of signing and issuing documents, reviewing contracts, approving projects, controlling project nodes, arranging staff and large sums of money, and decision-making on vita change.

2.3 Project management module

Filling fixed forms or generating online data as the input to complete the project design, cost budgeting, construction organization, schedule control, change management and project completion and so on.

2.4 Instant messaging module

Using the online network or linked with mobile phones and other mobile devices to complete project nodes tracking and early warning, online chatting forums, e-mail client, calendar, to-do list of projects, voting systems and other functions.

The main operators of the IPMSS are from all levels of the company managers including the president,

vice presidents in charge, function department heads, project directors, project members, system engineers and network administrators. In addition, group managers, investors and potential buyers often learn about project-related information and updates according to the IPMSS. Above crowd of people constitute the IPMSS stakeholder network (Figure.1).

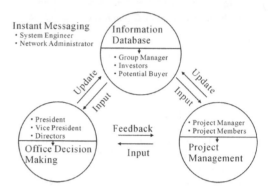

Figure 1. The operation mechanism of the IPMSS. Arrows point to the output.

Internal modules of the IPMSS are not isolated subsystems but related to each other. Information database is the basis of project management and decision-making. Only projects of great value will be approved and a cross-functional project team of investment, management and marketing will be established as a result. So office decision-making processes are throughout the whole investment activities. Undoubtedly, any management information and decision-making behaviors makes updates of previous database.

The group managers, investors and potential buyers outside the company visit the information database to master the basic project information and historical data according to their privileges. So the information database is like a window of the company to the outside and also a project marketing platform. Daily maintenance of the IPMSS is operated by both the system engineers and network administrators.

3 IPMSS CHARACTERISTICS

3.1 Informatization

The IPMSS based on modern communications, networking, database technology gathers various related elements of investment projects into the information database. Using this database the company managers and all kinds of engineering and technical personnel can accurately grasp the project information timely and systematically use the accumulated information resources of the company to increase productivity and employee skills(Zhu & Luo 2003). Managers can take advantage of the gradual accumulation information of project management information to provide the basis for decision making. For example, establishing a financial statements system to assist the analysis of project data and to real-time forecast project costs in real time. Based on the network connection, the sign and issue of internal and external files can be transferred electronically and in informatization to achieve the low-carbon paperless office management and the trace management. More importantly, the user of the IPMSS can work at home, in cities or anywhere in the world at any time to know the project progress.

3.2 Standardization

A key item of the company's risk management on the venture capital projects is to develop a set of risk management system. Based on standardized processes of operation, strict node behavior limits were done during the access, due diligence, review, approval and operation of investment projects. However, in reality, the executive layers often consider too much about the short-term interests and unintentionally forget the established company standards. Therefore, the IPMSS characterized by the managing manner of systematism, standardization, programming and charting may help the managers overcome short-term behavior and speculation and hold the direction of investment decisions constantly. This system provides not only the daily office decision-making processes but also the project management tools. The person in charge of investment project uses the basic theory, knowledge and methods of project management, in accordance with the company's quality management system to manage and control the schedule and various resources of the project and finally ensure the implementation of investment objectives. The IPMSS provides a class of non-approval standardized forms such as project site visit schedule, due diligence report, meeting minutes, monthly report format, etc. These standardized forms can be documented as the company's process assets.

3.3 Multiple-target

This is a significant feature of the IPMSS, which has three main targets. Firstly, managers at all levels make decisions and control risks based on the standard approval and management processes of the investment project. Secondly, the project managers carry out the projects as required and give feedback of management performance timely. Thirdly, the company needs to manage and output the mass project information to attract all levels of management, investors and potential buyers to understand the projects. In this way, it opens up the channels of project approval,

investment, management and quit. So the IPMSS is actual an OA (office decision-making)-PM (project management)-DB (information database) system.

3.4 Expandability

The architecture of the IPMSS meets future expansion and upgrades and reserves docking interfaces with other systems.

3.5 Instantaneity

Network online and interaction with the mobile phone to implement instant communication is a major feature of Web-based application. On the one hand, the managers can grasp the project information dynamically, allocate the resources rationally and make decisions timely and effectively. On the other hand, nodes controlling and risk warning designed based on the online early warning mechanism of objective management will avoid information silos and significantly improve the investment risk prevention and control capabilities.

4 IPMSS PROGRAM DESIGN AND FUNCTION IMPLEMENTATION

4.1 Analysis

IPMSS is a fast and efficient investment management platform based on the ASP.net technology combined with OA-PM-DB-IM modules to configure the responsibilities of company leadership and staff (Figure 2).

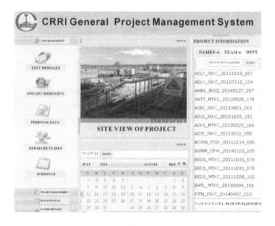

Figure 2. The system interface of IPMSS.

4.2 Algorithm design and programming

Part of the program used some classical algorithms of the DoNet C # to improve operational efficiency, such as the recursive method and the greedy algorithm. Using recursion method to get all the parent department numbered Code of sub-sectors and return a collection; while the greedy algorithm for memory allocation were processed as the first, the best and the worst adaptation.

4.3 Program description

4.3.1 Hardware environment

IBM server +500 G hard drive.

4.3.2 Software environment

Window Server2008 (Server OS) + SQL Server2008 (Database) + DoNet C # (Programming language) + DoNet FrameWork4.0 (Assembly frame) + IIS6.0 (Server platform).

4.3.3 Clien

Ordinary clients, namely personal PC or laptop.

4.3.4 Network configuration

100M fiber optic line, enterprise-class routing.

5 IPMSS SECURITY

Undoubtedly, the most important thing for a network information platform is security (Li & Tian 2008). In order to protect the hardware, software and data from accidental or malicious reasons of destruction, change and leakage and to ensure that the system runs properly, continuously and reliably, a series of measures have been taken as follows.

5.1 Physical security

Physical isolation was carried out between the project data storage and external information supply. That is, data processing of internal and external using a separate server, host, hard disk and data transmission path without cross, and physically cut off input and output functions of the two separated parts to ensure safer supervision and management of the function of external data and information supply.

5.2 Virtual Private Network (VPN)

VPN is using data encryption and access control technology based on a public data network to interconnect between two or more trusted internal network. VPN is usually built by a router or firewall of encryption functions to achieve trusted data transfer in the public channel.

5.3 User permission setting

IPMSS strictly controls the access using dynamic password generation and verification. Moreover, for company staff not involved in the project team, they can just see the title, the person in charge and the status of the projects but cannot enter in to view details. All documents stored have been encrypted and strictly searching using and downloading of the documents are strictly controlled. This means that outside people can not see the bookmarks but only to see part of the project attachment the company has been shared. Even more, network administrator is only responsible for adding or deleting visitors but forbidden to see the project information.

5.4 Network security management

Changes on the key elements of the system are supported by trace management. The operation trace of upload, update, modification and delete will be retained which include the information of logging in IP display, logging in accounts, traces time point, traces of information extraction records, etc. In addition, IPMSS provides online short messaging service (send text messages to mobile phones, or text messages or emails within platform) and gradually replaces the existing Tencent QQ and other social networking tools to enhance information security within the platform.

Currently, login in IOS operating system is under research now and Android environment will be developed. In addtion, the IPMSS will make project management develop from the procedures and forms to the intelligent operation and finally implement the online dynamic data production and management.

ACKNOWLEDGMENTS

H. Xie thanks Mr. X. Wang for his technical support and also thanks Dr. Lv from SUNTRANS for his improving of the English expression. This project was financially supported by the China Railway Resources Group Co., Ltd. Major Project Plan (No. 2013-Major-02).

REFERENCES

Sun, H. 2013.Design and implementation of investment project management system, *Technological Development of Enterprise* 32(7):97–99.

Zhu, D. &Luo, J. 2003. Design and Development of Venture Capital Investment Project Management System Based on B/S Structure, *Journal of North China Institute of Water Conservancy and Hydroelectric Power* 24(4):58–60.

Li, C. & Tian, D.2008. Study on Enterprise Investment Item Information Management System by Integrate Method of C/S Mode and B/S Mode. *Modern Electronics Technique* 18:84–86.

Computational Intelligence in Industrial Application – Ling (ed.)
© *2015 Taylor & Francis Group, London, ISBN: 978-1-138-02818-0*

The pedestrian detection algorithm with history tracking based on the vehicle vision

B. Zhao, D. Pang & M. Wang

Key Laboratory of IOT Terminal Pivotal Technology, Harbin Institute of Technology, Shenzhen Graduate School, Shenzhen, Guangdong, China

ABSTRACT: In order to improve the accuracy and performance of the pedestrian detection system based on the vehicle vision, this paper proposes a history tracking method with a few improvements. The pedestrian detection algorithm is based on Adaboost algorithm using Haar-like features. The whole image is scanned periodically every N (e.g. 10) frames to identify the number and position of people. And the results of each frame are stored in the history tracking module to make the system focus on specific regions when processing the next frame. Moreover, if pedestrain targets are missed in one or some continous frame, the Kalman filter is used to evaluate the most possible position at the current frame based on the preceding trace information. In the single frame detection, different sizes of scanning windows according to their position are used to reduce calculation. The comparative experiments showed that pedestrians can be tracked more effectively.

KEYWORDS: pedestrian detection; history tracking; Kalman filter.

1 INTRODUCTION

The current mainstream research of pedestrian detection is to learn from the statistical point of view, extracting features from a large number of training samples and creating a human model, then the pedestrian detection process is considered as a pattern classification problem. The advantage is to focus on learning from samples of different variations of the human body, with a robust feature and a reasonable choice of training samples. Combined with the structure of reasonable classification algorithm, this method can overcome many adverse conditions, such as the pedestrians diversity, the scenes diversity, and the lighting conditions diversity.

But the problem is that due to the complexity of the algorithm and lack of temporal relation, the detection of the single-frame pedestrian image is time-consuming, so this method may not be suitable for real-time detection of multiple frames. This paper provides an optimization method for the pedestrian detection based on multi-frame images. The algorithm reduces the computational complexity and time-consuming while improving the detection accuracy, and can be used for the purpose of real-time pedestrian detection.

2 RELATED WORK

2.1 *Feature selection and calculation*

In the process of pedestrian detection, an analysis to determine whether the target region is or not candidate pedestrian region in an image is necessary. Therefore, multiple features can be used in pedestrian modeling, which should make it clear to distinguish if there are pedestrians in the regions. In this paper, we use Adaboost algorithm to make pedestrian modeling, in which a lot of simple features should be extracted from pedestrian samples. Haar-like feature is first presented by Papageorgiou (Barron, J.L. et al. 1992) (Papageorgiou, C. et al. 1998) and applied to facial feature. On this basis, Viola and Jones (Mohan, A. et al. 2001) (Viola, P. & Jones, M. 2001) presented three types and four kind of forms, as shown in Figure 1(Guo, Lie et al. 2008).

Haar-like features consist of two or three rectangles. Their function is mainly edge detection and linear feature. The value of features can be calculated by the gray-level value integration of rectangle regions.

$$feature = \sum_{i=1}^{N} w_i \, \mathrm{Re}\, cSum(r_i) \qquad (1)$$

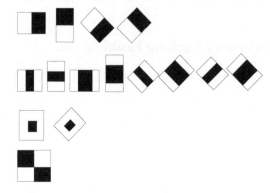

Figure 1. Haar feature.

In formula (1), $W_i \in R$ is the weight of rectangle; $RecSum(r_i)$ is the gray-level value integration of rectangle r_i; N is the quantity of rectangle (Lienhart, R. & Maydr, J. 2002).

2.2 *Adaboost algorithm*

Adaboost is presented by Freund and Schapire in 1997. It is a kind of self-adapting mode. The basic purpose of this algorithm is turn weak learning algorithm into strong learning algorithm. Cascading a lot of strong classifier into a final cascade classifier, and finish the searching of images. In view of the characteristic of Adaboost algorithm, we choose discrete Adaboost algorithm to get a cascade classifier (Li, Wenbo & Wang, Liyan 2009). In this way lot of regions which doesn't contain pedestrians can be excluded. The structure is shown in Figure 2.

The process of training strong classifier is shown as following (Viola, P. & Jones, M. 2004)

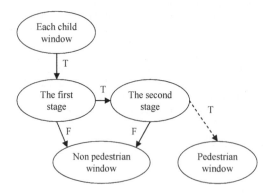

Figure 2. Detection principle of Adaboost cascade classifier.

1 Given N training samples (x_1,y_1), (x_2,y_2), $\cdots$, (x_N,y_N), where x_i feature vector, $y_i = \{0,1\}$ corresponds to the non-pedestrians and pedestrians samples.

2 Initialization weights: when the sample is non-pedestrian, $w_{1,i} = k/2$; when the sample is pedestrian, $w_{1,i} = l/2$. k and l are respectively the number of samples for non-pedestrians and pedestrians.

For each t=1, 2, $\cdots$, T(T is the number of training) proceed as follows:

1 Normalized weights
$w_{t,i} \leftarrow w_{t,i}/\sum_j w_{t,i}$

2 For each feature j, training for the corresponding weak classifiers
When $p_j f_j(x) < p_j \theta_j$, $h_j(x) = 1$; others, $h_j(x) = 0$. Where, p_j indicates the direction of the inequality; $f_j(x)$ is feature value; θ_j is threshold.

3 calculation error
$\varepsilon_j = \sum_i w_i |h_j(x)-y_i|$, And selecting a minimum error ε_t simple classifier $h_t(x)$ added to the strong classifier.

4 update sample weights
$w_{t+1,i} = w_{t,i}(\beta_t)(1-e_i)$, If x_i is the i sample correctly classified, then $e_i = 0$; contrary $e_i = 1$, $\beta_t = \varepsilon_t/(1-\varepsilon_s)$.

3 Finally, following the formation of a strong classifier:
When $\sum_t \xi_t h(x) >= \sum_t \xi_t/2$, R(x)=1; others, R(x)=0. Where, $\xi_t = \lg(1/\beta_t)$.

From the result of training, this algorithm is used to adjust the weight of the training samples, and strengthen the training of error classifying samples. At last, a strong classifier will be presented by the combination of weak classifiers.

2.3 *Detection and tracking*

2.3.1 *Single frame detection method*

The resolution of detection image is 640×480. When detecting one image, the distance between target pedestrians and the automobile are different, this lead to different sizes in one image. For detection on different sizes, we choose scaling the detector instead of scaling the image itself. On all scales, features can be find in same way. Traditional scanning method for single frame image is fixed the size of image, scanning the image with the minimum detection window(training classifier use the size of pedestrian), when finish one image extend the detection window(extend 1.25 times one time),and scanning the image again, until the detection window is the same size as the image.

However, in this paper, we divided every single frame image into four regions, upper part, above middle part, below middle part and bottom part. And use different proportion scanning windows to matching every regions one by one:

1 If lower edge's coordinate of scanning window is less than or equal to the lower edge of upper part, choosing the minimum window to matching;
2 If lower edge's coordinate of scanning window is less than or equal to the lower edge of above middle part, choosing the smaller window to matching;
3 If lower edge's coordinate of scanning window is more than the lower edge of above middle part and also less than or equal to the lower edge of below middle part choosing the bigger window to matching;
4 If lower edge's coordinate of scanning window is more than the lower edge of below middle part, choosing the bigger window to matching;

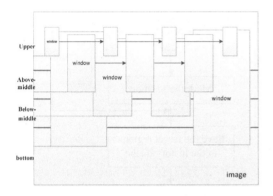

Figure 3. Single frame pedestrian detection with windows in different sizes.

2.3.2 Pedestrian tracking between adjacent frames

In 2.3.1 we have mentioned pedestrian detection and recognition in moving regions of a single frame image (Guo, Lie et al. 2008). On the basis of the extracted feature confirming the target is pedestrian or not and also confirm the target pedestrian exist in image sequences or not. In this way, we can get the coordinate of pedestrians. As a general rule, after confirming the target pedestrian, this pedestrian doesn't disappear at once, it is continuously present. This means the location of a pedestrian is continuous, its features is related to each other. Therefore, for multiple frames we choose a method that set a region of interest(ROI), and use the method which used in single frame image detection to the ROI, this also can reduce a lot of tracking time. Considering the height will not change so much when pedestrian is walking, but the width will change obviously because of the walking. In this case, we get the size of the scanning rectangle from previous frame, and extend the rectangle for several times as region of interest, and then detect this ROI for the location of the pedestrian.

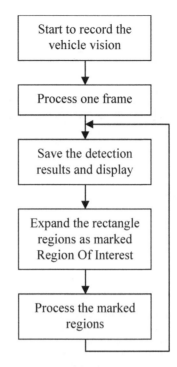

Figure 4. Pedestrian tracking flow chart using regions of interest based on the previous frame.

Pedestrian tracking between adjacent frames not only improve the speed of single frame detection but also optimize the images between adjacent frames. In fact, the distance between pedestrian and automobile is growing closer, so the scanning rectangle should be larger and larger, this exactly right this method.

3 METHOD

In this system, a single input frame image is used to detect the pedestrian's quantity and location, and predict the location of pedestrian in next frame by the image information between consecutive frames.

3.1 System flow

Pedestrian detection module is used for detecting pedestrians' quantity and position and then save the result in history tracking module. History tracking module realizes the purpose to predict the pedestrian information of next frame for pedestrian detection module according to the combination of current frame's information and previous frame's information.

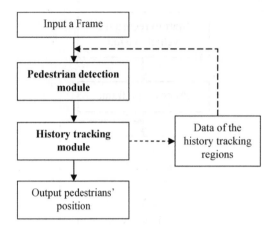

Figure 5. Top-level system flow chart of pedestrian detection with history tracking module.

3.2 History tracking module

History tracking module realize the purpose of target tracking by means of the target pedestrian's information provided by pedestrian detection module, and record the location of the closest pedestrian. At the same time a new algorithm is presented for preventing the problem of shade from other objects. The new algorithm is to improve the security of the system, which record the location of pedestrian who is probably sheltered by other objects, as the basis for next frame detection. In this module, lots of images which are adjacent frames are connected to each other, greatly reducing the complexity of the system for processing.

In history tracking module, we present a new kind of tracking method which is called prediction for preventing shelter. This method combine Kalman filtering and the method of extend scanning rectangle as region of interest to make the tracking process more accurately, and narrowing the detection time. The flow chat of history tracking module is shown in Figure 6.

Firstly, target location R_{n-1} of previous frame is provided by pedestrian detection module. On the basis of the method presented in 2.3.2 extend the scanning rectangle a certain number of times as region of interest, and then detect this ROI for detection result A_n. Secondly, make a judgment that in this detection result A_n the detection target of previous frame exist or not. If it exist(e=1), mark the location of pedestrian in current frame and save result in history tracking module for detection in next frame, $R_n=A_n$. If it doesn't exist (e=0), we present prediction for preventing shelter. On the basis of record results in history tracking module, take use of Kalman filtering to predict the location of pedestrian in current frame, and save the result in history tracking module, $R_n=K_n$. As shown in formula (2).

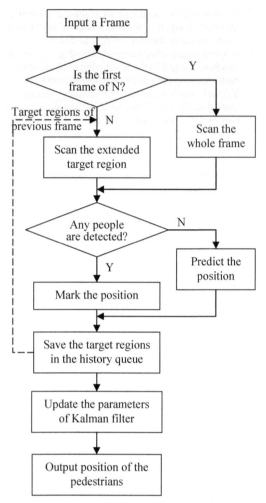

Figure 6. Detailed system flow chart of pedestrian detection with history tracking module.

$$Rn = \begin{cases} A_n, & e = 1 \\ K_n, & e = 0 \end{cases} \qquad (2)$$

However, what called for special attention is that on account of new pedestrian appeared sporadically, history tracking module does not save every frame of the target information, it has a storage limitation. When reaching this limitation, the system will empty the history tracking module. At this time, the system will detect the whole image of next frame, and give the new initial target pedestrian's information and save it in history tracking module, then return to the loop mentioned above.

Kalman filtering (Guo, Lie et al. 2012) (Li, Juan et al. 2009) (Chang, Haoli & Shi, Zhongke 2006) is a minimum variance estimation algorithm to a dynamic system state sequences. The system predict the next state for optimal estimation on the basis of previous state sequences, the forecast has no bias, stability and optimal characteristics. The state equation and observation equation are (Welch, G. & Bishop, G. 2001)

$$X_k = \Phi_{k-1} X_{k-1} + W_{k-1} \tag{3}$$

$$Z_k = H_k X_k + V_k \tag{4}$$

In these formulas, X_k is an n-dimension state vector at time k; Z_k is a m-dimension observation vector at time k; n-dimension matrix Φ_{k-1} is state transition matrix; m×n-order matrix H_k is a observation matrix; W_{k-1} and V_k are two zero-mean Gaussian white noise sequence follow normal distribution, variance matrix are Q_k and R_k.

The process of kalman filtering can be expressed by prediction equation

$$\hat{X}_k(-) = \Phi_{k-1} \hat{X}_{k-1}(+) \tag{5}$$

and filter recurrence equation

$$\hat{X}_k(+) = \hat{X}_{k-1}(-) + K_k[Z_k - H_k \hat{X}_k(-)] \tag{6}$$

Among them

$$K_k = P_k(-)H_k^T / [H_k P_k(-)H_k^T + R_k] \tag{7}$$

is filter gain;

$$P_k(-) = \Phi_{k-1} P_{k-1}(+)\Phi_{k-1}^T + Q_{k-1} \tag{8}$$

is variance matrix to predict error;

$$P_k(+) = (I - K_k H_k)P_k(-) \tag{9}$$

is filter error variance matrix.

4 RESULTS AND ANALYSIS

4.1 *Overlapping pedestrian can be handled by history tracking*

Figure 7.   Frame 5.

Figure 8.   Frame 6.

Figure 9a.   Frame 10 (history   Figure 9b.   Frame 10 (ROI). tracking).

Figure 10a.   Frame 12 (his-   Figure 10b.   Frame 12 (ROI). tory tracking).

When two pedestrians become close and overlap, tracing is likely to miss targets for one or two frames. With the ROI tracing, pedestrian tracing gets a miss while pedestrians can be predicted and traced by history tracking. Adjacent pedestrians can be detected correctly.

5 CONCLUSION

By contrast of history tracking and simple ROI (region of interest), the results obtained showed that history tracking module can improve the tracing performance, preventing from target missing and reducing time consumed. So the driver could be noticed more effectively.

Furthermore, we will consider adding an external environment module, which covers the major elements affecting road safety. The environment module indicates the potential pedestrians more accurately and reduces unnecessary calculation. In this case the safety condition is determined by driving speed, environment factor, the nearest pedestrian's information and so on.

REFERENCES

Barron, J.L. et al. 1992. Performance of Optical Flow Techniques. IEEE Conference on Computer Vision and Pattern Recognition.

Chang, Haoli & Shi, Zhongke 2006. A Method of moving pedestrian detection and tracking based on monocular vision technology. Journal of Traffic and Transportation Engineering 6(2): 56–58.

Guo, Lie et al. 2008. Pedestrian Detection Method Based on Adaboost Algorithm. Computer Engineering 34(3): 202–204.

Guo, Lie et al. 2012. Pedestrian Detection and Tracking Based on Automotive Vision. Journal of Southwest Jiaotong University 47(1): 21–23.

Li, Juan et al. 2009. Study on Pedestrian Tacking Based on Kalman Filter. Journal of Transportation Systems Engineering and Information Technology 9(6): 148–153.

Li, Wenbo & Wang, Liyan 2009. An approach of vehicle detection based on Adaboost algorithm. Journal of Changchun University of Science and Technology: Natural Science Edition 32(2): 292–295.

Lienhart, R. & Maydr, J. 2002. An Extended Set of Haar-like Features for Rapid object Detection. IEEE ICIP: 900–903.

Mohan, A. et al. 2001. Example-based object detection in images by components. IEEE Transactions on Pattern Analysis and Machine Intelligence (4).

Papageorgiou, C. et al. 1998. A general framework for object detection. Proceedings of the 6th International Conference on Computer Vision.

Viola, P. & Jones, M. 2001. Rapid object detection using a boosted cascade of simple features. Proceedings of IEEE Conference on Computer Vision and Pattern Recognition.

Viola, P. & Jones, M. 2004. Robust real-time objectdetectionI. International Journal of Computer Vision 57(2): 137–154.

Welch, G. & Bishop, G. 2001. An Introduction to the Kalman Filter. Proceedings of the 28th annual conference on Computer graphics and interactive techniques. Los Angeles: ACM Press, Addison—Wesley.

Computational Intelligence in Industrial Application – Ling (ed.)
© *2015 Taylor & Francis Group, London, ISBN: 978-1-138-02818-0*

Exploration on teaching reform of competency-based "JSP web development"

Xi Zhu Zhang
Tianjin Bohai Vocational Technical College, Tianijin, China

ABSTRACT: Problems in the teaching process of improving ability of higher vocational college students in JSP dynamic web development were analyzed. A course reform that takes occupational activities as the guidance, quality as the basis, students as the main body, projects as the carrier, and practical training as the means and highlights ability goal was put forward; course design and development process were introduced, and problems found during the process of implementation were analyzed.

KEYWORDS: JSP, Competence-based, Project development.

1 INTRODUCTION

High-skilled talent cultivation is the goal of vocational colleges. With rapid development of e-commerce industry, advertising, online exhibitions, consultation, negotiation, online ordering, online payment, electronic accounts and other functions are presented, bringing more market demand for web development talents. Many higher vocational colleges set up the course of "JSP Web Site Development", aiming to improve employment situation and help the students adapt to market demand. However, there is a certain gap between goal setting of the course in higher vocational level and the practical need of society, and students fail to meet the demand of enterprises. This paper explores teaching reform and implementation plan of "JSP Web Site Development", introduces the competency and project course-based design and development, to contribute certain experience to the improvement of educational quality for the majority of software technology in higher vocational education level.

2 PROBLEMS EXISTING IN JSP WEB DEVELOPMENT TEACHING

Cultivation of technical application, skill and operation-oriented talents are the main goal of higher vocational education. However, inappropriate teaching mode, content and methods are adopted in colleges, and practices of traditional colleges and universities are introduced directly into vocational education, ignoring the role and positioning of vocational education and forming "compressed biscuit" of course of regular colleges and universities. "Duck-stuffing" type of teaching is also adopted in

"JSP Web Development", thinking that "knowledge is competence". In fact, competence is quite different from knowledge, and their difference can only be eliminated through "practice", as knowledge can be imparted, but competence can only be obtained in practice. Moreover, the content of teaching fails to follow closely with technology development, and the gap between ability of students and demand of employing units is large.

Currently, teaching materials on JSP dynamic web development highlight basic knowledge and skills. The content is lagging behind and fail to meet market demand for talents in JSP dynamic web development. In the mode of teaching, though project teaching is integrated into the course, the projects only serve for certain knowledge points, without accurate requirements on skills of students, and there is no scale project case to run through the entire course. Separation of teaching and training, lack of workplace atmosphere and defective university-enterprise cooperation distinctly make the knowledge learned fail to meet employers' demand. To cope with insufficient learning interest and initiative of students in practice, and flinch or escape in difficult code debugging, effective and thorough adjustment, of course teaching should be made, to help students have the ability to handle affairs and interpersonal relationship, apply knowledge and possess advanced skills in this specialty.

3 SIGNIFICANCE OF THE COMPETENCY AND PROJECT COURSE-BASED COURSE REFORM

According to the "employment-oriented and service-targeted" goal of vocational education, besides learning the specialized knowledge, students should be cultivated with noble ethics, excellent quality, proficient skills, solid expertise and ability of sustainable development after they step into the society. Teachers should take occupational activities as the guidance, quality as the basis, students as the main body, projects as the carrier, and practical training as the means, highlight ability goal, and integrate knowledge, theory and practice as a whole in a course. In the process of delivering all courses in higher vocational education, it is necessary to highlight the cultivation of "students' self learning ability", to ensure their ability of sustainable development in the society. Clarifying the goal of "JSP Dynamic Web Development" is the core issue in improving the quality of high-skilled talents cultivation. Teachers should select the content of courses according to the requirements of industry and enterprise development and practical post for knowledge, ability and quality, and help students know what to learn, what to do and how to do before graduation, so as to lay a good foundation for their sustainable development.

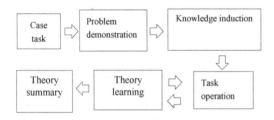

Figure 1. Competence-based teaching mode.

4 COMPETENCE-BASED COURSE DESIGN AND DEVELOPMENT

4.1 *Post analysis*

According to analysis of website information on recruiting JSP web developers, the posts include Web background development, website development and maintenance, background development of malls, backstage development of a portal website, front-end development of the Web, and other management information.

4.2 *Course analysis*

This course, as a professional core course for students majored in software technique, is set up in the first semester of sophomore year after "Technology of Web Page Making", "Java Realization of Object-oriented Programming", "SQL Server Database Application", and its subsequent courses include "JavaEE Open source Framework Technology".

Professional core abilities required by enterprises include skills in communication, cooperation, self learning, problem solving, information processing, creation and innovation, digital application, and foreign language application. JSP website developers should be familiar with HTML, JavaScript, and CSS static webpage development technology, master the core technology of JSP and Servlet, etc. in Java web development, proficient in using MVC model design and developing Web applications.

JSP web developers are expected to be all-rounder, who should not only master various development technologies and web project development process, but also have a certain ability of programming, program debugging, teamwork and communication skills, independent learning and innovation in practice. Work tasks and professional skills of JSP web development are shown in Table 1.

Table 1. Work tasks and professional skills of JSP web development.

Task	Professional skills
MyEclipse9.0 installation and application	Using IDE environment
Tomcat installation and configuration	Integrating MyEclipse and Tomcat
Basic Web page design	Language XHTML and CSS styles
Front-end page development	JavaScript and Ajax technology
JSP syntax and command	Basic elements, compiling directives and action elements of JSP
JSP server object application	Using JSP common built-in objects
Javabeans coding	Developing JavaBean components
Using JavaBean	Invoking JavaBean component in JSP page
Servlet programming	Compiling Servlet, deploying Servlet (configuration of web. xml file)
Using Servlet	Servlet requests and responses in JSP
Application of MVC design model	Application of three components: Model, View and Controller
Constructing Struts2 framework	Importing Struts1 class Lib, filter configuration
Model component development	Connecting database, defining POJO class
Development of Action functional module	Compiling Action class
Configuration file preparation	Configuring struts-config.xml file

Training contents:

1 JSP environment construction.
2 Basis of JSP development.
3 JSP+JavaBean development (Model 1).
4 JSP+JavaBean+Servlet development (Model 2).
5 JSP comprehensive application and development.

4.3 *Objective design*

According to work task and requirements for professional ability, the teaching objectives of this course include:

Table 2. Teaching objectives of "JSP Website Development".

Skills	Knowledge	Quality
1. Installing and configuring the development environment 2. Employing integrated development tools 3. Employing JSP directives and built-in objects 4. Using JavaBean 5. Using Servlet 6. Using Struts framework	1. Mastering basic knowledge of Web development 2. Mastering basic knowledge of JSP programming 3. Mastering theory and knowledge of Model1 development model 4. Mastering theory and knowledge of Model2 development model 5. Mastering theory and knowledge of Struts framework	1. Dedication 2. Study hard 3. Bearing hardships and standing hard work 4. Scientific and rigorous manner 5. Solidarity and collaboration 6. Honesty and trustworthiness

4.4 *Learning-situated context*

In the process of teaching, teachers should complete teaching unit according to the ability object, set up project of reasonable scale to meet the requirements of skill training. In course exploitation of JSP website development, teachers should employ Web application system development of student information management system, according to requirements of software engineering and actual development process, the learning context should be designed as follows:

1 Learning context 1: analysis and design of student information management system, database design
2 Learning context 2: construction of JSP development environment
3 Learning context 3: based on JSP-JavaBean student information management system
4 Learning context 4: based on JSP-JavaBean-Servlet student information management system
5 Learning context 5: Struts-based student information management system

4.5 *Teaching means and methods*

Teachers should focus on cultivation of students' professional ability and adopt various teaching means such as project teaching, situational teaching, scene teaching, case teaching, task-driven teaching, discussion-based teaching, to improve the students' interest in learning basic knowledge and skills and enhance their enthusiasm and initiative of learning.

Teachers should lay emphasis on the students and focus on "teaching, learning and doing" integration and interaction, positively inspire and induce students' creativity and cultivate their creativity while promoting students' learning of the basic specialized knowledge and professional skills.

4.6 *Assessment*

Periodic assessment should be adopted, while process assessment and objective assessment should be integrated; Diversified assessment should be adopted to evaluate students' achievement comprehensively, by taking their performance on homework, results in ordinary exams, practical learning experience and basic skills into consideration. However, emphasis shall be laid on students' practical ability.

Assessment of three aspects should be involved in each project: ability, knowledge and quality, based on the achievement submitted, project summary, internal evaluation of groups, and project defense, etc.

5 PROBLEMS EXISTING IN THE PROCESS OF IMPLEMENTATION

The course of JSP website development requires much for theoretical and practical competence, while some teachers just commute between school and school, they know little about market competition, and become unused to strict management; in addition, their impetus in educational reform, sense of responsibility and crisis for overall development of the whole school, practical experience of the application, as well as practice and leading edge knowledge are insufficient. Therefore, teachers should be encouraged to get more experience in factories and actively participate in project development, to have the level of education and improved teaching.

Practical teaching environment and facilities cannot meet the needs of project teaching. Integrated teaching requires more for teaching environment and hardware supporting, thus facilities construction should be promoted.

The process of teaching can be divided into "teaching" and "learning", and should be student-oriented. Students of higher vocational level often cannot be absorbed in project training from beginning to end, but always ended in anticlimax. Therefore, in the process of teaching and learning, teachers should keep helping students overcome fatigue period in project development and encouraging them to persevere.

6 CONCLUSION

Teachers should simulate actual operation in delivering the competence-based course of JSP website development, so as to cultivate the students' practical ability (comprehensive ability to solve practical problems). Moreover, case introduction, problem motivation, case analysis and other teaching methods should be used comprehensively, to realize put the knowledge learned into practice, and improve the practical ability of higher vocational students. The practice has proven that the course reform of competence-based JSP website development can improve the teaching quality and promote the improvement of students' vocational capability. Some problems have been found in the reform and exploration, and reform needs to be optimized further, to accumulate experience for reform of other courses.

REFERENCES

Zhanjun Hu,Xin Zhang, Jianrong Dong, Wenjie Chen. The implementation of the project teaching scheme based on CDIO model[J],Chinese Vocational and Technical Education; 2009,24:55–58.

Shuowang Wang. CDIO :The classical model of Massachusetts Institute of Technology engineering education American - Based on the interpretation of CDIO Curriculum[J],Journal of Higher Education in Science & Technology;2009.

Guiying Du. Vocational Colleges "Webpage design" course teaching design[J], Theory and Practice of Education;2010,30(9):57–59.

Computational Intelligence in Industrial Application – Ling (ed.)
© 2015 Taylor & Francis Group, London, ISBN: 978-1-138-02818-0

Classification tag sorting algorithm based on web group behavior

Dong Gao & Shu Hang Guo
School of information, Central University of Finance and Economics, Beijing, China

ABSTRACT: During Big Data Time, there are massive web resources urgently waiting for classification annotation. This paper aims to research music tag sorting problems of classification annotation, providing a classified tag recommendation algorithm based on tag utility drifting. Firstly, the paper creatively put forward the concept of tag utility and defined its calculation method. Secondly, it proposed a fuzzy comprehensive evaluation system of music tags utility and the definition of its measurement metric. Thirdly, the study discussed tag utility drifting and established a mathematical model. Then, the research raised an improved classification annotation algorithm with tag utility drifting and dynamic measurement. Finally, the algorithm was applied to verify and determine the recommended order of tag utility in a label cluster time series, as well as complete the performance analysis.

1 INTRODUCTION

Massive network resources not only can be refined and developed into a useful classification system tools, in the actual implementation, data is ubiquitous and can be expanded of use. "Crowd labeling algorithm," as a typical example, making big data active, is to give data vigor. "Crowd labeling algorithm" is to collect data through online crowd behavior, extract and mark product-oriented label, via the crowd label clustering, and analyzes the behavior of crowds and trends of interest (Yin 2012, Song & Zhang 2011).

Actual application analysis shows that the current labels belong to plane classification system Stolpe & Morik (2011). Its main function is to achieve multi-dimensional browsing and searching of the label. This method has not yet reflected the sequence of tag utility, and then brings the label system of casualness, chaotic attribution and other issues. In addition, tag utility is a time-efficient and dynamic. In order to achieve an orderly labeling system and dynamically adapt users' access needs, This paper creatively proposes a tag utility function, and then a tag utility metric algorithm and its mathematical model are established, which are based on time measurement and fuzzy comprehensive evaluation.

2 RELATED CONCEPT

Definition 1 Tag: A tag is a term or keyword assigned to an item. This kind of metadata depicts specific information and allows it to be discovered again via searching or browsing. Tagging, known as the symbol of Web2.0, is very popular in websites Jun-zhi & Ning (2012).

Definition 2 Tag utility: It refers to the knowledge with meaningful and significant reference or the decision-making value, by obtaining, extraction, processing, and mining label data.

Definition 3 Dynamic measurement: Due to the dynamic and the timely nature of tag data, the measure of tag utility must consider and take into account these features Tong-guang & Shi-tong (2014). So dynamic measurement adds this time element to the mathematical model, in order to achieve a more realistic and comprehensive information extraction and knowledge discovery of tag utility.

Definition 4 Utility drifting: While dynamic measurement considers the timeliness nature of label, utility drifting measures its dynamic feature. By calculating the label stability, heat, popularity, driftance and other indicators, synthetically reflecting the dynamic changes in tag utility.

3 FUZZY EVALUATION MODEL OF TAG UTILITY

3.1 *The bottom measurement of indicators*

In order to streamline the model, thus making several key assumptions.

The number of clicking people equals to visiting people.

The number of clicks equals to visits.

This paper considers there exists significantly difference between the user's click behavior (click) and access behavior (visit). For example, a user clicked on a label but found no favorite music under the label, and therefore did not produce access behavior. In order to simplify the model and achieve statistical convenience, this paper makes two assumptions above. Taking into account the access behavior is more important than the click behavior, thus the "visitors" is attributed to "visit" module and "clicks" attributed to "click" module, which has a higher weight.

Total duration of songs (A_1):

$$\sum_{m \in TAG(\rho)} m \cdot time$$

"m" is a song name. If "m" belongs to the current label, then accumulating the total length of the current label.

Total number of songs (A_2):

$$\sum_{m \in TAG(\rho)} N$$

Update rate (A3):

$$\sum_{m \in TAG(\rho)t_0}^{t_0 + \Delta t} N$$

"m" is a song name. If "m" belongs to the current label, then calculating the total number of new songs under the current label from t_0 to $t_0 + \Delta t$.

Total number of clicks (B_1):

$$\sum_{\tau \in TAG(\rho)} N, N = \omega_1 p + \omega_2 d + \omega_3 f + \omega_4 s + \omega_5 g + \omega_6 b$$

"τ" is the records of access. If "τ" belongs to the current label, then accumulating the total number of clicks of the current label. The total number of clicks is a weighted average number, including play (p), download (d), favorite (f), share (s), good (g), and bad (b).

Average clicks per unit time (B_2):

$$\sum_{\tau \in TAG(\rho)} N \Big/ T_n$$

"τ" is the records of access. "T_n" is the total open time of the current label. If "τ" belongs to the current label, then accumulating the total number of clicks.

Average clicking users per unit time (B_3):

$$\sum_{\tau \in TAG(\rho)} N_{people} \Big/ T_n$$

Total duration of access (C_1):

$$\sum_{\tau \in TAG(\rho)} \tau \cdot period$$

"τ" is the records of access. If "τ" belongs to the current label, then accumulating the total duration of access.

The number of users accessing (C_2):

$$\sum_{\tau \in TAG(\rho)} N_{people}$$

"τ" is the records of access. If "τ" belongs to the current label, then accumulating the number of users accessing.

Average access duration per unit time (C_3):

$$\sum_{\tau \in TAG(\rho)} \tau \cdot period \Big/ T_n$$

"τ" is the records of access. "T_n" is the total open time of current label. If "τ" belongs to the current label, then accumulating average access duration per unit time.

3.2 Establishing fuzzy evaluation model of tag utility

In fact, label's meaning is vague, since there is not the only accurate norm if a song can be attributed to a label. So it is difficult to describe it by probability theory and classical mathematics (Xiong-fei et al. 2010).

Fuzzy math can effectively solve the problem of the ambiguity of the music label and its attribution values (Symeonidis et al. 2010). Although the description of the tag is a vague concept, membership functions can be used to measure the subordinate degree of the label attributions (Yi-Zhang et al. 2012). Inputting and normalizing music tag attribute behavioral data, so we can get the membership values, and finally output tag utility via a fuzzy evaluation model of tag utility.

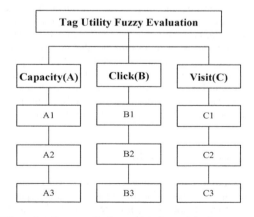

Figure 1. Fuzzy evaluation model of tag utility factor.

This paper makes use of fuzzy comprehensive evaluation methods to establish a music tag utility model. Considering massive factors associated with the music tag utility, we adopt two-level fuzzy synthesis evaluation to measure tag utility. The first layer of indicators consist of the capacity, click and duration three elements. The second layer is a kind of refinement and concretization of the first layer, and they are all indicators can be computed based on the access record. For $Tag(\rho)$, we build the indicator system as follows.

Set $U=\{A,B,C\}$, $A=\{A_1,A_2,A_3\}$, $B=\{B_1,B_2,B_3\}$, $C=\{C_1,C_2,C_3\}$.

4 TIMELINESS AND DRIFTING OF TAG UTILITY

4.1 Dynamic measurement of tag utility

Due to the dynamic and timely nature of tag data, the measure of tag utility must consider and take into account these features. Dynamic measurement adds a time element to the math model, for obtaining a more realistic and comprehensive information extraction and knowledge discovery of tag utility. For $Tag(\rho)$, $U_{\rho 1}, U_{\rho 2},..., U_{\rho t}$ means the tag utility of $Tag(\rho)$ in the period $i(i=1,2,...,t)$.

$U_{\rho 1}, U_{\rho 2},..., U_{\rho t}$ can be viewed as a time series of tag utility of $Tag(\rho)$.

4.2 Utility drifting measurement

4.2.1 Stability (S)
Stability indicates the degree of stability or variation of tag utility in a continuous time period. Here, we refer to the definition of time series stability in econometric models, viewing the tag utility series $U_{\rho 1}, U_{\rho 2},..., U_{\rho t}$ as a time series with stochastic error (disturbance).

Referring to the concept of stationary time series in econometric models, we can easily define the stability of tag utility. A stationary tag utility must meet these three prerequisites.

$$E(U_{\rho t})=\mu, t=1,2,\cdots$$

$$Var(U_{\rho t})=E(U_{\rho t}-\mu)^2$$
$$=\sigma^2, t=1,2,\cdots$$

$$Cov(U_{\rho t},U_{\rho (t+k)})=E[(U_{\rho t}-\mu)(U_{\rho (t+k)}-\mu)]=r_k$$

$t=1,2,...;k\neq 0$. If the tag utility series of $Tag(\rho)$ $U_{\rho 1}, U_{\rho 2},..., U_{\rho t}$ satisfies the above three conditions above, then we consider it is stationary.

In particular examples, we can use Unit Root Test in *Eviews* to test the stability of tag utility.

4.2.2 Heat (H)
Heat means the click condition of tag utility in one or more periods.

$$H=\sum_{i=1}^{t}U_{\rho i}\cdot r_B=\sum_{i=1}^{t}B_2\cdot(r_{V1},r_{V2},r_{V3},r_{V4},r_{V5}) \qquad (1)$$

4.2.3 Popularity (P)
Popularity represents the visit condition of tag utility in one or more periods.

$$P=\sum_{i=1}^{t}U_{\rho i}\cdot r_C=\sum_{i=1}^{t}B_3\cdot(r_{V1},r_{V2},r_{V3},r_{V4},r_{V5}) \qquad (2)$$

4.2.4 Driftance (D)
Driftance indicates the degree of variation of tag utility in a continuous time period.

If $U_{\rho 1},U_{\rho 2},...,U_{\rho t}$ meets the prerequisites of Stability, then $D = 0$; otherwise, $D \neq 0$, which has the following two situations.

$D > 0$. It means there is a positive utility drifting, indicating that tag utility increases.

$D < 0$. It means there is a negative utility drifting, indicating that tag utility decreases.

We define the *Driftance* as follows.

$$D=\left.\sum_{i=1}^{t}\overline{U_i}\middle/\overline{U_0}\cdot t\right.-1 \qquad (3)$$

5 THE ALGORITHM OF TAG UTILITY FUZZY EVALUATION

5.1 Building up the membership function

Definition 5 Fuzzy set A in domain U, is a mapping from U to [0,1], we use $\mu_A: U\rightarrow[0,1]$ to represent A.

For $u\in U$, $\mu_A(u)$ is called membership (u to A),indicating the degree u belongs to A, whose maximum is 1 and the minimum is 0. μ_A is the membership function of A.

Definition 6 According to the characteristics of tag attributes, the membership function is constructed as follows (Wei et al. 2010).

$$\mu_A(x)=\begin{cases} 0 & x\leq a-b \\ \dfrac{1}{2}+\dfrac{1}{2}\sin\dfrac{\pi}{b}[x-(a-\dfrac{b}{2})] & a-b<x<a+b \\ 1 & a+b<x \end{cases}$$

5.2 The algorithm steps and flow chart

Firstly, establishing the first level factor set of music label, known as factor sets: $U=\{U_1,U_2,...,U_k\}$, among them, $U_i=\{U_{i1},U_{i2},...,U_{in}\}$, n is the number of second level factor in factor set i.

Secondly, establishing an evaluation set. Assuming that the evaluation set has m comments: $V=\{V_1, V_2,..., V_m\}$.

Thirdly, Calculating all layers' weight vector by using triangular fuzzy number, which are represented as: $A=(a_1,a_2,...,a_k)$, $A_i=(a_{i1},a_{i2},a_{in})$, each weight vector's sum of the weights equals 1.

According to the membership function depicted in *Definition 6*, we can easily get a single factor fuzzy evaluation set (first factor set i, secondary factor j):

$$\tilde{R}_{ij} = (r_{ij1}, r_{ij2}, \cdots, r_{ijm}).$$

The comprehensive evaluation matrix of factor set i is as follows.

$$\tilde{R}_i = \begin{bmatrix} r_{i11} & \cdots & r_{i1m} \\ \vdots & \cdots & \vdots \\ r_{in1} & \cdots & r_{inm} \end{bmatrix}$$

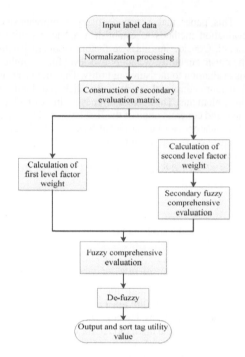

Figure 2. Algorithm description of tag utility model.

Then we can achieve the fuzzy comprehensive evaluation result of factor i:

$$B_i = A_i \circ_T \tilde{R}_i = (a_{i1}, a_{i2}, \cdots, a_{in_i}) \circ_T \begin{bmatrix} r_{11}^{(i)} & r_{12}^{(i)} & r_{13}^{(i)} \\ r_{21}^{(i)} & r_{22}^{(i)} & r_{23}^{(i)} \\ \vdots & \vdots & \vdots \\ r_{n1}^{(i)} & r_{n2}^{(i)} & r_{n3}^{(i)} \end{bmatrix}$$

$$= (b_{i1}, b_{i2}, \cdots, b_{im})$$

Similarly, we calculate final fuzzy comprehensive evaluation result as:

$$B_i = A_i \circ_T \tilde{R}_i = (b_{i1}, b_{i2}, \cdots, b_{im}).$$

Algorithm description are as follows.

6 EXPERIMENTS AND RESULTS

6.1 Data source

In order to verify the tag utility model, this paper randomly generates experimental data within a reasonable range, referring to *the Baidu Music label system*. In order to assess the performance of the tag utility and utility drifting model, the article also considers the law of the basic data, constructing multiple sets of time series data to simulate real situations. The experiment of tag utility fuzzy comprehensive evaluation. The experimental source partly refers to music tag data to *Baidu music*. For example, tag name, total duration of songs, and total number of songs. Other data randomly generate within a reasonable range.

Part of tag data statistics are listed in Table 1.

Among these underlying tag data defined in section 3.1, A_1=Total duration of songs[hour], A_2=Total number of songs, A_3=Update rate[per month], B_1=Total number of clicks per unit time[per day], B_2=Average clicks per unit time[per day], B_3=Average clicking users per unit time[per day], C_1=Total duration of access[hour], C_2=The number of users accessing, C_3=Average access duration per unit time[hours per day].

In this experiment, the article divides tag utility into five degrees, $V=\{$Very Low (V_1), Low (V_2), Common (V_3), High (V_4), Very High $(V_5)\}$.

Giving the comments $V = \{$Very Low, Low, Common, High, Very High$\}$ values of $\{-1, 0, 1, 2, 3\}$, then multiplying $B=(b_1,b_2,...,b_m)$ by $V=(-1, 0, 1, 2, 3)^T$ to get a tag value from -1 to 3.

Then, We set the first level factor weight set as $U=(0.25,0.35,0.4)$, second level factor weight set as $A=(0.3,0.5,0.2)$, $B=(0.3,0,4,0.3)$, $C=(0.2,0.5,0.3)$.

Table 1. Part of underlying data of tag.

Tag Name	Duration(A_1) hours	Click(B_1) times	Access(C_1) hours
World Cup	9.7	34842	8250.8
Oldie	58.4	71173	24244.4
Travel	54.4	68959	15068.9
Japan	57.5	54853	18750.7
Europe	65.9	96377	43114.9
Light Music	64.1	40635	33722.9
Chinese Style	32.4	62674	13474.1
Network Song	57.8	30143	36245.8
Teana Music	15.4	77614	9181.7
Love	67.5	32958	62441.9
Madden	27.4	55536	9773.0
Nursery Song	64.6	72917	50474.0
The 1990's	13.6	90182	10583.4
Classical	60.6	96337	26796.4
Popular	68.2	59249	41647.0
Rock	67.3	23436	11323.7
Military Song	53.9	24635	7998.2
Hip-Hop	47.0	33176	27164.9
Country Song	49.9	22476	40001.6
Famous	34.9	33176	32836.2
After 70s	41.7	85665	9040.4
After 80s	48.5	32885	29667.2
After 90s	67.3	83286	35180.6
Graduate	16.4	20709	1811.9
Romantic	67.8	31917	27352.7

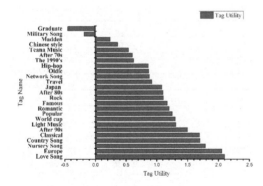

Figure 3. Tag utility values and their rankings.

Tag utility values and their rankings are as follows.

Calculation results are basically consistent with the actual situation, but more in line with people's cognition. For example, the tag "World cup" has a very low capacity, total number of clicks and total duration of access are not significant compared to other labels. However, with higher average click and average access indicators, ultimately, it obtains a front ranking. This fully shows the fuzzy comprehensive evaluation system is a scientific and practical model.

Then, let us consider the utility drifting in this model.

6.2 The experiment of tag utility drifting

On the basis of results in the previous experiment, referring to the law of the underlying data, within reasonable ranges, constructing time series of tag utility.

During this experiment, we use *Eviews* to test the Stability of tag utility time series. If it meets the premise of Stability, then we think this tag utility is stationary, which means the Driftance of this tag is 0. The values of Heat, Popularity, and Driftance can be calculated by their formulas. They estimate the click, visit, and drift situation of tag utility, respectively.

Calculating each tag's Stability, Heat, Popularity, Driftance and then sorting them. In order to simplify the process, the calculated Heat and Popularity are set at time t_0.

We chose 10 tags to draw their tag utility drifting curves, then analyze them.

Tags are sorted by their Heat, Popularity, and Driftance respectively, and we get the following table.

Table 2. Part of tag utility drifting data.

Tag Name	Period		
	t_0	t_5	t_9
World Cup	1.302	1.464	1.644
Oldie	0.872	0.886	0.908
Travel	0.926	0.925	0.908
Japan	1.083	1.168	1.207
Europe	2.055	2.183	2.144
Light Music	1.313	1.311	1.315
Chinese Style	0.365	0.460	0.391
Network Song	0.885	1.086	0.883
Teana Music	0.542	0.684	0.637
Love	2.100	2.651	2.708
Madden	0.242	0.250	0.343
Nursery Song	1.787	1.785	2.563
The 1990's	0.628	0.516	0.292
Classical	1.692	1.301	1.363
Popular	1.257	1.457	1.782
Rock	1.110	1.488	1.575
Military Song	−0.183	−0.176	−0.232
Hip-Hop	0.865	0.685	0.475
Country Song	1.699	1.495	1.276
Famous	1.171	1.335	1.503
After 70s	0.611	0.502	0.435
After 80s	1.103	0.848	0.889
After 90s	1.498	1.684	1.750
Graduate	−0.449	−0.448	−0.644
Romantic	1.202	1.818	1.708

Table 3. Tags' heat ranking.

Ranking	Heat (H)	Tag Name
1	1.985	Classical
2	1.950	After 90s
3	1.888	World cup
4	1.788	Famous
5	1.712	Love Song
6	1.666	The 1990's
7	1.660	Teana Music
8	1.492	Europe
9	1.329	Oldie
10	1.180	Nursery Song
11	1.071	After 80s
12	0.946	Network Song
13	0.865	Country Song
14	0.810	Light Music
15	0.741	Popular
16	0.691	Romantic
17	0.592	Japan
18	0.587	Madden
19	0.519	Chinese Style
20	0.476	Travel
21	0.226	After 70s
22	0.142	Hip-hop
23	−0.296	Rock
24	−0.725	Military Song
25	−0.890	Graduate

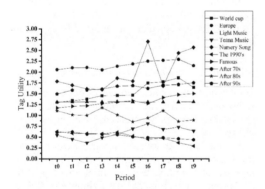

Figure 4. Tag utility drifting curve.

Table 4. Tags' popularity ranking.

Ranking	Popularity(P)	Tag Name
1	2.228	World cup
2	2.119	Europe
3	1.953	Love Song
4	1.946	Country Song
5	1.717	Nursery Song
6	1.430	Rock
7	1.212	Classical
8	0.959	Famous
9	0.858	Light Music
10	0.813	Travel
11	0.789	Romantic
12	0.691	The 1990's
13	0.620	Popular
14	0.592	Japan
15	0.560	Hip-hop
16	0.433	After 80s
17	0.283	After 90s
18	0.222	Network Song
19	0.193	After 70s
20	0.132	Teana Music
21	0.095	Graduate
22	0.088	Madden
23	0.058	Chinese Style
24	−0.597	Oldie
25	−0.928	Military Song

Combining tag utility drift curves and Table 3–5, we can find that the curves of "World cup", "Teana Music" and "Famous" have evident positive drifting, which means the tag utility increases. Besides, while the curves of "Light Music", "Nursery Song" and "After 80s" have a certain level of fluctuation, it can be considered as a time series with a stochastic error by referring to the definition of stationary time series. In addition, the curves of "The 1990's" and "After

Table 5. Tags' driftance ranking.

Ranking	Driftance(D)	Tag Name
1	0.270	Rock
2	0.232	Popular
3	0.222	Love Song
4	0.204	World cup
5	0.199	Chinese Style
6	0.152	Madden
7	0.143	Network Song
8	0.138	Famous
9	0.127	Military Song
10	0.105	After 90s
11	0.103	Teana Music
12	0.074	Japan
13	0.054	Europe
14	0.000	Oldie
15	0.000	Travel
16	0.000	Light Music
17	0.000	Nursery Song
18	0.000	Classical
19	0.000	After 80s
20	0.000	Graduate
21	0.000	Romantic
22	−0.110	Country Song
23	−0.152	After 70s
24	−0.217	The 1990's
25	−0.219	Hip-hop

70s" have a decreasing tendency. We can well find the direction and degree of tag utility by calculating its Driftance. Finally, we can clearly and dynamically measure the tag utility by observing the ranking in Table 3–5.

7 SUMMARY

This paper constructs a music tag hierarchy system, a tag utility metric model, and proposes a kind of tag utility drifting measurement model. The constructed music tag system clearly shows the hierarchical relationship between the label and its measurable properties.

The purpose of this study is to measure the tag utility and its drifting from the underlying data of the music label. In order to achieve this goal, firstly builds a hierarchical music label system. Taking Baidu music for case analysis and research, to build a multi-perspective and hierarchical structure. Building the tag utility model of capacity, click and visit three aspects. We chose the fuzzy comprehensive evaluation model to measure the tag utility and use the weights constructed of triangular fuzzy numbers, rather than the traditional *AHP*, to calculate the weight of each layer index weight, which makes the results more in line with people's awareness. Finally, the experiments show that the proposed fuzzy metric of tag utility achieves good results. Proposed tag utility drifting measurement model effectively reflects the dynamic and the timeliness nature of labels. In the process of improving the model, innovatively proposing four indicators measuring a tag utility: Stability, Heat, Popularity, and Driftance, adding the time series factors to reflect the dynamic drifting of tag utility.

REFERENCES

Jun-zhi, JIA. & Ning, ZHANG. 2012. Application function analysis of social labels. Information studies: Theory & Application 35(11):112–116.

Song, Y. & Zhang, L. 2011. Automatic tag recommendation algorithms for social recommender systems. ACM Transactions on the Web 5 (1): 4–39.

Stolpe, M. & Morik, K. 2011. Learning from label proportions by optimizing cluster model selection. Proc of ECMLPKDD. Berlin: Heidelberg.

Symeonidis, P. & Nanopoulos, A. 2010. A unified framework for providing recommendations in social tagging systems based on ternary semantic analysis. IEEE Trans. on Knowledge and Data Engineering 22(2):179–192.

Tong-guang, NI. & Shi-tong, WANG. 2014. Transfer support vector machine for learning from data with uncertain labels. Control and Decision 29.

Wei, ZHENG. 2010. A Multi Label Classification Algorithm Based on Random Walk Model. Chinese Journal Of Computers 2010,33(8):1419–1425.

Xiong-fei, LI. et al. 2010. Rough set model based on the labelled tree. Journal on Communications 31(6):35–42.

Yin, ZHANG. 2012. Autonomy Oriented Personalized Tag Recommendation. Acta Electronica Sinica 40(12):2354–2358.

Yi-Zhang, JIANG. et al. 2012. Mamdani-Larsen Type Transfer Learning Fuzzy System. J ACTA AUTOMATICA SINICA 38(9): 1393–1409.

Computational Intelligence in Industrial Application – Ling (ed.)
© 2015 Taylor & Francis Group, London, ISBN: 978-1-138-02818-0

Integration of autonomous, cooperative and exploratory learning in modern college education

Y.H. Zhang, X.H. Su & H.L. Zhang
School of Computer Science and Technology, Harbin Institute of Technology, Harbin, Heilongjiang, China

ABSTRACT: Compared to the learnings of traditional teaching, autonomous, cooperative and exploratory learning are three useful learning methods for college students. Only, mastering one learning method is not enough. In order to improve students' abilities such as innovation and cooperation, an integration of multiple learning methods in the teaching system is proposed in this paper. Firstly, we introduce the characteristics of three learning methods. And then features of modern knowledge are pointed out. We discussed relationships among autonomous, cooperative and exploratory learning further. Finally, we illustrated how to integrate the above three learning methods in college education.

1 INTRODUCTION

The rapid advances in the economy and technology have changed the way of students study today. In current years, research of effective learning methods in college education has been aroused extensive attention (Raymond, 2012; Riek, 2013). On the one hand, most people think cooperative learning is a useful skill that should be mastered by a university student (Steve et al. 2012). On the other hand, autonomous learning is highlighted by some researchers (Gwen et al. 2014; Douglas Elliot, 2013). Moreover, some people argue that exploratory learning should be paid more attention to by the students (Foote Stephanie et al. 2012; Seri & Gal, 2014). In fact, autonomous, cooperative and exploratory learning are three effective learning methods which are different from each other, but have correlative dependence. Only applying one of them is not enough in undergraduate phase. For that reason, it's necessary to explore difference and relationship among themselves. To meet the specific learning needs of students, how to integrate the three learning methods effectively in a college education is very important, especially for improving the undergraduate learning efficiency.

2 CHARACTERISTICS OF THREE LEARNING METHODS

2.1 Autonomous learning

Autonomous learning is a type of highly self-conscious, self-directed and personalized learning. Autonomous learning is flexible. An investigation shows that most excellent students master the skill of learning knowledge autonomously. During the study process, they usually have a clear aim and choose their learning content with a suitable learning method. They excel in solving problem and overcoming difficulties.

Autonomous learning is the core competence required in lifelong learning, it is not only important for the students' academic achievements, but also benefits improving their learning motivations. Particularly important is autonomous learning increases individual autonomy.

2.2 Cooperative learning

Cooperative learning is a form of autonomous learning (AL) in which small groups of students work together on an issue. This method provides opportunities to develop social and communication skills and group thinking. The aim of cooperative learning is to achieve higher levels of cognitive learning using personal knowledge within a collaborative team.

2.3 Exploratory learning

Exploratory learning, proposed by the scientist of the university of Chicago, is the production of education modernization of the USA in the 50s of the last century. For the students, learning process is similar to the explore work of scientists. This kind of learning is open, and not limited to, class hour and classroom practice. What the students study is not fixed at text books or subjects. Raising questions and solving the problems are very important in exploratory learning. So openness and critical thinking are the main characters of exploratory learning.

3 CHARACTERISTICS OF THREE LEARNING METHODS

In modern society, the knowledge has characteristics of constructiveness, situatedness, complexity and tacitness.

3.1 *Constructiveness*

Knowledge building includes individual knowledge building and collaborative knowledge building, which was proposed by Carl Bereiter and Marlene Scardamalia. Task of knowledge building is to create or modify public knowledge. The built knowledge can be used by the people.

3.2 *Sociality*

What the students learn in the university should be similar to the situations that they will meet probably off campus in the future.

3.3 *Complexity*

When listing facts use either the style tag List signs or the style tag List numbers.

3.4 *Equations*

The knowledge structure has the characteristic of openness. We can't master the knowledge with an isolated way.

4 RELATIONSHIPS AMONG THE THREE LEARNING METHODS

It was pointed out that modern learning includes three aspects, namely content, motivation and interaction (Illeris Knud, 2010). The content means what knowledge should be studied. The motivation is the power source of learning, and it plays an important and necessary role in the learning process. The motivation affects persistence and effectiveness of learning also. The study usually happens in a social situation, it needs interpersonal communication. Therefore, Autonomous learning, exploratory learning and cooperative learning correspond to, content, motivation and interaction.

Group work has been recognized as an effective way of improving students' learning experiences. Cooperative learning helps to promote work-related skills, develop a deeper level of understanding of complex tasks and manage time-consuming tasks.

However, cooperative learning is based on the autonomous learning. As any individual of a group, if he wants to contribute to a rise in knowledge of the team, he must learn more and more knowledge autonomously as he obtained knowledge from his group member. The more knowledge he learns autonomously, the more chances he will have to exchange information and cooperate with others, which will promote autonomous learning further.

Exploratory learning is based on the autonomous learning also. But different from cooperative learning, interest is highlighted in exploratory learning. More in-depth autonomous learning may change with exploratory learning. So exploratory learning is a higher level of study. When exploratory learning is extended to a certain degree, an effective learning approach that is called cooperative exploratory learning is needed. In process of cooperative exploratory learning, every member executes a sub-exploratory task, and they cooperate with each other in a group. Cooperative exploratory learning integrates the features of cooperative learning, exploratory learning and autonomous learning.

5 INTEGRATION OF LEARNING METHODS IN COLLEGE EDUCATION

College education comprehends the training of many kinds of ability. In undergraduate phase, only master one learning approach for a student is not enough. They should obtain new knowledge through various channels and improve their abilities such as innovation, cooperation and so on. Therefore, integration of multiple learning methods in modern college is proposed which is shown as figure 1.

The teaching activities include theory and practice. And these activities are executed in class or outside class. For the in-class activities, anywhere cooperative discussion and cooperative project belong to the cooperative learning training, which will provide the student's ability of Knowledge fusion and interpersonal communication. The other two in-class activities are lecture teaching and lab work, which are traditional teaching methods and implanted in fixed room.

Anywhere outclass reading is a useful theory, learning approach, and it can widen student's thoughts. This reading work can be done outside class. In order to increase the student's ability of exploring an unknown world without limits, the anywhere exploratory experiment is essential. This type of experiment is based on the interest of the student. Any fantastic idea is reasonable and encouraged.

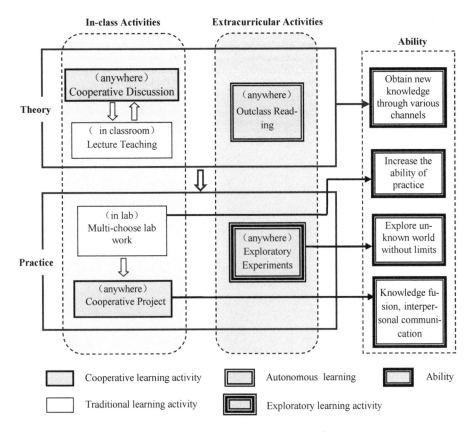

Figure 1. Active, cooperative and exploratory learning in modern college education.

6 CONCLUSIONS

In today's higher education institutions, students should not to be passive recipients of knowledge from the lectern. Therefore, educators should integrate multiple learning methods training in the teaching activities, and create opportunities for students to master various learning methods. Autonomous, cooperative and exploratory learning are three useful learning methods for college students. Relationships among the three methods are mutual independence, but complementing each other.

ACKNOWLEDGEMENTS

The authors would like to thank the funds of "985 Project" construction project on Undergraduate Teaching, our research was also financially supported by the Computer Education Association of the National College of China (Grant No. ER2013012).

REFERENCES

Raymond, S. 2012. An evaluation of a group project designed to reduce free-riding and promote active learning. *Assessment & Evaluation in Higher Education.* 37, 285–292.

Riek, L. D. 2013. Embodied computation: An active-learning approach to mobile robotics education. *IEEE Transactions on Education*, 56, 67–72.

Steve, H., Emma, M., Liz, B. & Andrew, J. G. 2012. Multi-touch tables and collaborative learning. *British Journal of Educational Technology*, 43, 1041–1054.

Gwen, N., Amber, D. H., Katerina B. C. & Wim G. 2014. Essential knowledge for academic performance: Educating in the virtual world to promote active learning. *Teaching and Teacher Education.*37, 217–234.

Douglas, Elliot P. 2013. Student construction of knowledge in an active learning classroom. *Research in Engineering Education Symposium, REES2013. Putrajaya, Malaysia. 4-6 July, 2013.*

Foote, Stephanie, M., Harrison, David, S., Ritchie, C. M. & Dyer, A. 2012. Exploratory learning through critical inquiry: Survey of critical inquiry programs at mid-sized U.S. Universities. *IADIS International*

Conference on Cognition and Exploratory Learning in Digital Age, Madrid, Spain. 19-21, October, 2012.

Seri, O. & Gal, Y. 2014. Visualizing expert solutions in exploratory learning environments. *International Conference on Intelligent User Interfaces, Proceedings IUI, Haifa, Israel. 24-27, February, 2014.*

Chilton, M. A. & Bloodgood, J. M. 2008. The dimensions of tacit and explicit knowledge: a description and measure. *International Journal of Knowledge Management*, 4, 75–91.

Illeris Knud. 2010. *How we learn.* Beijing: Educational Science Publishing Hous.

Computational Intelligence in Industrial Application – Ling (ed.)
© 2015 Taylor & Francis Group, London, ISBN: 978-1-138-02818-0

Identifying optimum time interval in a sequence database

A.K. Reshamwala
MPSTME, Computer Engineering Department, SVKM's NMIMS University, Mumbai, India

S.M. Mahajan
Institute of Computer Science, M.E.T, Bandra, Mumbai, India

ABSTRACT: Mining educational data is an emerging discipline, concerned with developing methods for exploring the unique information from educational settings, and using those methods to learn students' behavior. *"Take Action Before It's Too Late"*. Imagine being able to change a student's path if you were aware of risk factors early on. With students' browsing patterns or sequences of web click patterns, it makes it easy for the faculty to see immediately who's not participating, missing deadlines, and/or performing below average. In this paper, we propose a time interval sequential pattern mining algorithm using a priori property. The algorithm not only detects an optimum time for a particular event, but also identifies where these events can occur in the future with the optimum time interval detected. Experimental results of a real time database of Blackboard Learning Management System (LMS) of undergraduate students confirm the efficiency of the algorithm.

1 INTRODUCTION

The tremendous volume and diversity of real-world data embedded in huge databases clearly overwhelm traditional manual methods of data analysis, such as spreadsheets and ad-hoc queries. An urgent need exists for a new generation of techniques and tools with the ability to intelligently and automatically assist users in analyzing mountains of stored data for nuggets of useful knowledge. During the last decades, the potential of analytics and data mining methodologies that extract useful and actionable information from large datasets has transformed one field of scientific inquiry after another. Indeed, a data mining method can extract from raw data patterns of interest to the application domain. While these patterns are useful as the starting point of an analytic process, the challenge here is to navigate and explore these patterns in order to come up with a meaningful analysis: an interpretation or a model that can explain the patterns and be used to exploit them as useful insights for decision making. In recent years, the sophistication and ease of use of tools for analyzing data make it possible for an increasing range of researchers to apply data mining methodology.

Sequential data are omnipresent. Sequential pattern mining is the mining of frequently occurring ordered events or to discover frequent subsequences as patterns in a sequence Database. Sequential Pattern Mining finds interesting sequential patterns among the large database. It finds out frequent subsequences as patterns from a sequence database. An example of sequential pattern is "Customers who buy a Smart phone are likely to buy a Scratch guard followed by phone insurance within a day". Such patterns have been used to implement efficient systems that can recommend based on previously observed patterns, help in making predictions, improve usability of systems, detect events, and in general help in making strategic decisions. Many approaches have been proposed to extract information, and mining sequential patterns are one of the most important ones as introduced by Agrawal &Srikant (1995); Chen et al. (2002); Han&Kamber (2001). It is firstly proposed by Agrawal&Srikant (1995) in the shopping basket data analysis.

However, from these discovered sequential patterns discussed above, the time gaps between successive patterns cannot be determined. For example, smart phone (*one hour*) scratch guard (*one day*) phone insurance and smart phone (*one day*) scratch guard (*two days*) phone insurance. It is a clearly distinct customer behavior where some customers buy scratch guard one hour after they bought smart phone and others may buy the next day. To study the customer behavior is very important for the enterprise to provide the right products at the right time to the right customers. In current competitive and fast-changed business environment, becoming your customers' trusted advisor will reduce customer churn and build customer loyalty. As the saying, "you do business with people you trust". The first step to this is to predict client future behavior early. For example, to learn which offer works more effectively to which customer and when? To address these time gaps based sequential patterns, in this paper, we have proposed an algorithm to predict the optimum time interval when the precious customers will buy a particular product. In our research we have used a real time dataset of Blackboard Learning Management System (LMS) of undergraduate Students.

2 RELATED WORK

Sequential pattern mining was first introduced by Agrawal & Srikant (1995), which can be described as follows. A sequence database is formed by a set of data sequences. Each data sequence includes a series of transactions, ordered by transaction times. This research aims to find all the subsequences whose frequencies exceed the minimum support threshold in the database. Following Agrawal& Srikant (1995) in mid-1990, one after another many scholars provided more efficient algorithms (Han et al., 2000; Pei et al., 2001; Srikant & Agrawal, 1996; Zaki, 2001) which differ in representing and applying the searching technique. Another important research in this category includes finding frequent episodes in event sequences(Mannila, Toivonen, & Inkeri Verkamo, 1997); finding similar patterns in a time-series data base (Agrawal, Faloutsos, & Swami, 1993; Faloutsos, Ranganathan, & Manolopoulos, 1994; Li, Yu, & Castelli,1996; Stejic, Takama, & Hirota, 2003); finding cyclic patterns in transaction database (Han, Dong, & Yin, 1999; Han, Gong, & Yin, 1998; Ma &Hellerstein, 2001); finding traversalpatterns in a web log (Chen, Park, & Yu, 1998; Cooley,Mobasher, & Srivastava, 1999; Pei, Han, Mortazavi-Asl, & Zhu, 2000); finding sequential alarm patterns in atelecommunication database (Wu, Peng, & Chen, 2001).

In Agudo-Peregrinaet al. (2012), presents three different extant interaction typologies in e-learning and analyzes the relation of their components to students' academic performance. The three different classifications are based on the agents involved in the learning process, the frequency of use and the participatory mode, respectively. Petrushyna et al. (2011), distinguishes different phases of the self-regulated learning process and aim to identify them in learners' activities. Also they, attempt to recognize patterns of their behavior and consequently their roles in a community, analyzing the success or failure of a community. A large amount of educational content is available as lecture videos, which is recorded by teachers as they proceed through a course. Students watch these videos in different ways. They rewind, skip forward, and watch some scenes repeatedly. Ullrichet al. (2013), investigates what can be learned by analyzing such viewing patterns. Fortenbacher etal. (2013), focus is on the LeMo system architecture, user path analysis by algorithms of sequential pattern mining, and visualization

of the learners' activities. Rojas et al. (2012), proposed a tool to make use of the students' logs, learning analytics and visualization techniques for providing monitoring and awareness mechanisms for leveraging the detected problems and thus improving the learning and assessment processes. Govindarajan et al. (2013), discusses a method to continuously capture data from students' learning interactions. Then, it analyzes and clusters the data based on their individual performances in terms of accuracy, efficiency and quality by employing Particle Swarm Optimization (PSO) algorithm. Corredor & Gesa in (2012), designed a frame work considering multimodal interaction mechanisms by means of using different communicative channels (visual, auditory and speech).

The preceding discussion shows that past research does not address the intervals between successive events. To address these time gaps based sequential patterns, Chen et al. (2003) has proposed a an algorithm considering the time gap between events, called time-interval sequential patterns, which reveals not only the order of events, but also the time intervals between successive events. Chen et al. (2003) developed algorithms to find sequential patterns using two approaches: Candidate generation and Pattern growth based. By assuming the partition of a time interval as fixed, developed two efficient algorithms I-Apriori and I-PrefixSpan. The first algorithm is based on the conventional Apriori algorithm, while the second one is based on the PrefixSpan algorithm. But no theory in previous work is talking about "providing the right products at the right time to the right customers".

Therefore, this work proposes a new approach that clearly involves the time intervals between events. Accordingly, not only the orders, but also the time intervals of items can be determined. We believe that mining sequential patterns with time intervals can yield more meaningful patterns and provide more valuable information.

3 SEQUENCE DATASET

Data acquired from the transaction dataset may be not sequential. A sequence is an ordered list of items as in Agrawal & Srikant (1995). In our research we have used a real time dataset of Blackboard Learning Management System (LMS) of undergraduate Students of SVKM's NMIMS, Mukesh Patel School of Technology Management and Engineering. The dataset is about 14,00,000 with 2725 students, 1140 different web clicks events on LMS enrolled in a course at the university for the period of 3 months, i.e. from January 2014 to March 2014.To predict student's behavior on Blackboard Learning Management System, we need to transform the *Event Web Page Clicks*, into unique identifier for each unique web click event. Table 1 shows the result of finding the unique identifier for each *Event Web Page Clicks*.

Table 1. LMS event details.

Event-Id	Event Web Page Clicks
20	/webapps/Assignments
105	/webapps/cmsmain/webui/courses
110	/webapps/Experiments
120	/webapps/Lecture
136	/webapps/Tutorial
175	Announcements
354	Department Academic Calendar
373	Discussion Board
760	mobile. view. announcements

These unique web page click events are used to form the sequences of web page navigation of students on the LMS.

Table 2. Students web page sequences on LMS.

Sequence_id	Student_ID	Event Sequences
1	71115120026	(<0> (760) <10> (105 20 373) <14> (354) <14> (110) <15> (20 120 20))
2	71106130052	(<10> (110 105 136) <12> (110) <13> (175))

From the sequences generated, in the Table 2, it can be seen that Sequence_id 1 on 10th day has click events (105 20 373) all the three web click events are occurring on the same day with same SessionId. Thus the sequences are generated with respect to timestamp and the SessionId.

4 I-APRIORI_EVENT ALGORITHM

As discussed in earlier section, time interval sequential pattern provides more valuable information than a conventional sequential pattern. The major drawback of the algorithm by Chen (2003) is that the observation is drawn on a set of events and hence cannot predict an optimum time for a particular event's future occurrence. The theory discusses for a general prediction of future sequence of events.

4.1 *Definition*

Consider the sequence database as shown in Table 3, the sequences are of the format $< (e_1,t_1) \dots (e_k,t_k)>$, where e_j is an event-set and t_j stands for the time at which e_j occurs, $1 \leq j \leq n$, and $t_{j-1} \leq t_j$, for $2 \leq j \leq n$.

In the sequence, if event-set occur at the same time, they are ordered alphabetically. The time interval values can be calculated as $ti_j = | t_{j+1} - t_j |$, where $j = 1,2,\dots,n-1$. The *length* of a sequence is the number of event sets in the sequence. A sequence of length k is called a k-sequence. We assume that the time interval has already been partitioned into a set of fixed time intervals before executing the algorithm. For example, consider the time interval to be partitioned into four partitions with time difference of three units. Hence the time interval partitions sets can be written as TI- 4 {I0; I1; I2; I3} where, I0: t = 0, I1: 0 < t ≤ 3, I2: 3 < t ≤ 6, I3: 6 < t ≤ ∞.

Table 3. Sequence dataset.

Sid	Sequence
10	<(a,1),(b,4),(a,5),(b,9),(e,29),(a,30)>
20	<(d,1),(a,2),(d,24),(a,25),(a,27),(b,30)>
30	<(b,1),(a,11),(e,18),(b,28),(a,30)>
40	<(f,1), (b,5),(c,19),(b,23),(a,27),(b,30)>
50	<(a,4),(b,5),(a,6),(d,10),(b,14),(a,15),(e,18),(b,22), (a,27),(b,30)>
60	<(a,0),(b,5),(e,15),(a,15),(e,18),(b,22),(b,30)>
70	<(j,2),(a,17),(h,17),(b,21),(a,22)>
80	<(c,3),(i,10),(b,14),(f,18),(b,24),(a,27)>
90	<(h,4),(a,10),(b,21),(a,21),(b,24),(a,27),(b,30)>
100	<(g,0),(a,0),(b,3),(e,3),(b,4),(a,5),(b,14),(a,15), (e,18),(b,22),(a,27),(b,30)>

Figure 1, depicts the working of the algorithm to find an optimum time interval for the future occurrence of a given event; "a" in the sequence dataset of Table 3. The minimum support for the working of the algorithm is taken to be 30%.

Sequence	Support
<a>	1.0

Sequence	Support
<a I1 a>	0.2
<a I2 a>	0.4
<a I3 a>	0.7

Optimum Time interval: I3 (6 < t ≤ ∞)
1- length: <a>
2- length : <a I2 a>, <a I3 a>
3- length : <a I3 a I3 a>
    With Sequence-ID: 1, 4, 8
    and 9a

Sequence	Support
<a I2 a I3 a>	0.2
<a I3 a I2 a>	0.1
<a I3 a I3 a>	0.4

Figure 1. I-Apriori_Event algorithm.

On applying the algorithm on the dataset of Table 3, as shown in Figure 1, we get L_1 as input event "a" with minimum support greater than or equal to minimum support of 30%. Then the following candidate time-interval sequences exist in C_2: (a I1 a); (a I2 a) and (a I3 a) with support of 20%, 40% and 70% respectively. In summary, C_2 can be generated as L_1 X TI X L_1; where X denotes join. L_2 will be the C_2 satisfying the minimum support threshold. Next, similarly generate C_k and L_k. Calculate the frequent time interval in the L_k. From Figure 1, I-Apriori_Event algorithm not only predicts the optimum time interval: *MaxCount* (*t*); for the event "a" but also identifies the sequences in which it can occur in future. The detailed algorithm is as follows:

Algorithm 1:

Input

Event Sequence Database S, Minimum Support *min_sup*, Event *e* and Time Interval set TI

Output

The complete set of time interval sequential patterns with sequences-Id *s* and optimum time interval *t*.

C_1 = find event *e* patterns in S.

$L_1 = \{c \in C_1 | \left(\dfrac{c.count}{|S|} \right) \geq min\_sup\}$

for (k=2; L_k-1 ≠ $\emptyset$; k++) {
    C_k=new candidates generated from L_{k-1}
    for each $p_1 \in L_{k-1}$ {
        for each $p_2 \in L_{k-1}$ {
        If (k=2) {
            for each *i* $\in$ TI {
            c= p_1* *i*$ p_2$;
            add c to C_k;
        }
}
}
}
if (k>2)
    Build the time interval candidate tree from C_k;
    for each sequence s $\in$ S
    Traverse the time interval candidate tree and accumulate the supports;
$L_k = \{c \in C_k | \left(\dfrac{c.count}{|S|} \right) \geq min\_sup\}$
t=Occurrence time of events in L_k
}
MaxCount (*t*) in L_k
return UL_k

---

5 RESULTS AND DISCUSSIONS

For the experimental evaluation and performance study of the I-Apriori_Event algorithm, we have implemented I-Apriori_Event as well as Apriori based algorithm: I-Apriori algorithm by Chen (2003) in Sun Java language and tested on an Intel Core Duo Processor, 2.10 GHz with 4GB main memory under Windows 7 operating system.

Consider the data set discussed in section 3, the raw data of LMS Blackboard for three months i.e. 90 days. Hence, by taking time interval of 90 days, we get values of 30, 15 and 7 days partitions. The detailed partitions are:

a. For time interval of 30 days in three months' time (90 days), we get three time interval partitioned sets as TI- 3 {I0; I1; I2} where, I0: t = 0, I1: 0 < t ≤ 30, I2: 30 < t ≤ 90.
b. For time interval of 15 days in three months' time (90 days), we get six time interval partitioned sets as, TI- 6 {I0; I1; I2; I3; I4; I5} where, I0: t = 0, I1: 0 < t ≤ 15, I2: 15 < t ≤ 30, I3: 30 < t ≤ 45, I4: 45 < t ≤ 60, I5: 60 < t ≤ 90.
c. For time interval of 7 days in three months' time (90 days), we get a thirteen time interval partitioned sets as, TI- 13 {I0; I1; I2; I3; I4; I5; I6; I7; I8; I9; I10; I11; I12} where, I0: t = 0, I1: 0 < t ≤ 7, I2: 7 < t ≤ 14, I3: 14 < t ≤ 28, I4: 28 < t ≤ 35, I5: 35 < t ≤ 42, I6: 42 < t ≤ 49, I7: 49 < t ≤ 56, I8: 56 < t ≤ 63, I9: 63 < t ≤ 70, I10: 70 < t ≤ 77, I11: 77 < t ≤ 84, I12: 84 < t ≤ 90.

The experimental evaluation and performance study is done by considering the minimum support of 82%. The execution time is measured in seconds (s). The memory usage is measured in megabytes (mb).

Table 4. Performance and memory usage.

| | LMS at min supp 82% | | | |
| | Performance | | Memory Usage | |
Time Interval	I-Apriori	I-Apriori-Event	I-Apriori	I-Apriori-Event
TI-3	2519	186	955.75	323.73
TI-6	980	249	700.13	291.53
TI-13	263	243	609.42	340.57

The algorithm I-Apriori was executed for all events, whereas I-Apriori_Event algorithm was executed for event with a support count of 99%.

From the Table 4, we observe that, execution of I-Apriori results in a linear decrease in the execution time and memory usage to increase in the partitions of the time interval, whereas execution of I-Apriori_Event results in a linear increase in execution time as well as memory usage with the increase in partitions of time interval.

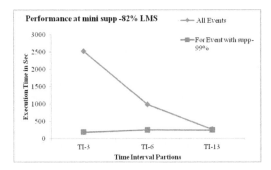

Figure 2.   Performance at minimum support of 82%.

From the Figure 2, we observe that approximately same time is taken at TI-13 and as the partitions in time intervals are reduced the execution time varies for both the algorithms.

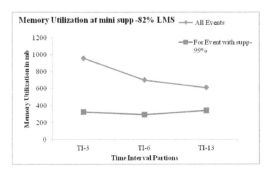

Figure 3.   Memory utilization at minimum support.

Figure 3 plots the readings of memory usage of a priori based time interval sequential pattern mining algorithm from Table 4. We observe that there is a linear increase in memory usage with increase in partitions of time interval on executing the I-Apriori_ Event algorithm. With few events satisfying minimum support threshold the algorithm utilizes less memory than the algorithm with maximum events satisfying minimum support threshold.

The preceding discussion shows the execution or runtime efficiency and the memory utilization efficiency. Now, we discuss about the optimum time interval prediction for a given event in the sequence dataset by implementing I-Apriori_Event algorithms which predicts an optimum time interval for the future occurrence of a given event. Thus, a solution to, "predict the optimum time for future occurrences of a given event by the customers/ students".

Table 5.   Event description- LMS.

Events	Description	Support Count
89	Upload/Download	98.5%
21	Attendance	88%
293	Evaluation	80%

Let us consider the events with their support count as in Table 5. The optimum time intervals are detected on these events by implementing the I-Apriori_Event algorithm as shown in Table 6.

As observed from Table 6, for all the time interval partitions such as TI-3, TI-6 and TI-13, an optimum time interval is discovered. For event "89" with 98.5% and event "21" with 88% of support count, at minimum support of 82% and 65%, respectively, the optimum time interval detected for future occurrence was zero time units or on the same day. Similarly, for event "293" with 80% support count, at minimum support of 50% the optimum time interval detected for future occurrence was between 0 and 7 time units or within 7 days. On execution of the I-Apriori_Event algorithm, the algorithm not only predicts an optimum time for a particular event, but also identifies with the help of support count set where these events can occur in future with optimum time interval detected.

Table 6.   Optimum time interval by i-apriori_event algorithm.

Optimal Time Interval for LMS			
Events	Support	Time Interval	Optimum Time Interval (t)
Event "89" with 98.5% support	82%	TI-3	I0 (t = 0)
		TI-6	I0 (t = 0)
		TI-13	I0 (t = 0)
Event "21" with 88% support	65%	TI-3	I1 (0 < t ≤ 30)
		TI-6	I0 (t = 0)
		TI-13	I0 (t = 0)
Event "293" with 80% support	50%	TI-3	I1 (0 < t ≤ 30)
		TI-6	I1 (0 < t ≤ 15)
		TI-13	I1 (0 < t ≤ 7)

6   CONCLUSION

This work presents a new approach that clearly involves the time intervals between events. The I-Apriori_Event algorithm predicts an optimum time interval for the future occurrence of a given event. Thus, a solution to, "predict the optimum time for future occurrences of a given event by the customers/ students". This study can also be applied to predict customer behavior for the enterprise to provide the right products at the right time to the right customers.

Experimental results ofI-Apriori_Event indicate that there is a linear increase in execution time as well as memory usage to increase in partitions of time interval. Thus, with a few events satisfying minimum support threshold the algorithm will utilize less memory than the algorithm with maximum events satisfying minimum support threshold. The algorithm not only detects an optimum time for a particular event, but also identifies with the help of support count set where these events can occur in future with optimum time interval detected. Thus, in a learning management system this algorithm can be applied, *"to identify for a given event, who's regular and who's not"*.

REFERENCES

Agrawal R.& Srikant R.1995. Mining sequential patterns. *In Proc. Int. Conf. Data Engineering*: 3–14.

Agrawal R., Faloutsos C. & Swami A. 1993. Efficient similarity search in sequence databases. *Proceedings of Conference on Foundations of Data Organization and Algorithms*: 69–84.

Agudo-Peregrina A.F., Hernandez-Garcia A.& Iglesias-Pradas S. 2012. Predicting academic performance with learning analytics in virtual learning environments: A comparative study of three interaction classifications, *in 2012 International Symposium on Computers in Education (SIIE)*:1–6.

Chen M. S., Park J. S. & Yu, P. S.1998. Efficient data mining for path traversal patterns. *IEEE Transactions on Knowledge and Data Engineering*, 10(2): 209–221.

Chen Y. L., Chiang M. C. & Ko M. T. 2003. Discovering time-interval sequential patterns in sequence databases. *Expert Syst. Applicat.*, vol. 25, no. 3: 343–354.

Chen. Y. L., Chen S. S. & Hsu P. Y. 2002. Mining hybrid sequential patterns and sequential rules. *Inf. Syst.*, vol. 27, no. 5: 345–362.

Cooley R., Mobasher B.& Srivastava J. 1999. Data preparation for mining World Wide Web browsing patterns. *Journal of Knowledge and Information Systems*, 1(1): 5–32.

Corredor C.M.& Gesa R.F. 2012. Framework for Intervention and Assistance in University Students with Dyslexia, *IEEE 12th International Conference on Advanced Learning Technologies (ICALT)*: 342–343.

Faloutsos C., Ranganathan M. & Manolopoulos Y. 1994. Fast subsequence matching in time-series databases. *Proceedings of the 1994 ACM SIGMOD International Conference on Management of Data*: 419–429.

Fortenbacher A., Beuster L., Elkina M., Kappe L., Merceron A., Pursian A., Schwarzrock S.& Wenzlaff B. 2013. LeMo: A learning analytics application focussing on user path analysis and interactive visualization. *IEEE 7th International Conference on Intelligent Data Acquisition and Advanced Computing Systems (IDAACS)* vol.02: 748–753.

Govindarajan K., Somasundaram T.S., Kumar V.S. & Kinshuk. 2013. Continuous Clustering in Big Data Learning Analytics. *IEEE Fifth International Conference on Technology for Education (T4E)*: 61–64.

Han J. & Kamber M. 2001. Data Mining: Concepts and Techniques, *New York*: Academic.

Han J., Dong G., & Yin Y. 1999. Efficient mining of partial periodic patterns in time series database. *Proceedings of 1999 International Conference on Data Engineering*: 106–115.

Han J., Gong W. & Yin Y. 1998. Mining segment-wise periodic patterns in time-related databases. *Proceedings of 1998 International Conference on Knowledge Discovery and Data Mining*: 214–218.

Li C., Yu P. S. & Castelli V. 1996. Hierarchyscan: A hierarchical similarity search algorithm for databases of long sequences. *Proceedings of the 12th International Conference on Data Engineering*: 546–553.

Ma S., & Hellerstein J. L. 2001. Mining partially periodic event patterns with unknown periods. *Proceedings of the 17th International Conference on Data Engineering*: 205–214.

Mannila H., Toivonen H. & Inkeri Verkamo, A. 1997. Discovery of frequent episodes in event sequences. *Data Mining and Knowledge Discovery*, 1(3): 259–289.

Pei J., Han J., Mortazavi-Asl B. & Zhu. H. 2000. Mining access patterns efficiently from web logs. *Proceedings of 2000 Pacific-Asia Conference on Knowledge Discovery and Data Mining*: 396–407.

Pei J., Han J., Mortazavi-Asl, B.& Zhu H. 2000. Mining access patterns efficiently from web logs. *Proceedings of 2000 Pacific-Asia Conference on Knowledge Discovery and Data Mining*: 396–407.

Pei J., Han J., Pinto H., Chen Q., Dayal U. & Hsu M.C. 2001. PrefixSpan: Mining sequential patterns efficiently by prefix-projected pattern growth. *Proceedings of 2001 International Conference on Data Engineering*:215–224.

Petrushyna Z., Kravcik M.& Klamma R. 2011. Learning Analytics for Communities of Lifelong Learners: A Forum Case. *11th IEEE International Conference on Advanced Learning Technologies (ICALT)*: 609–610.

Rojas I.G. & Garcia R.M.C. 2012. Towards Efficient Provision of Feedback Supported by Learning Analytics. *IEEE 12th International Conference on Advanced Learning Technologies (ICALT)*: 599–603.

Srikant R.& Agrawal R. 1996. Mining sequential patterns: Generalizations and performance improvements. *Proceedings of the 5th International Conference on Extending Database Technology*: 3–17.

Stejic Z., Takama Y., & Hirota K. 2003. Genetic algorithm-based relevance feedback for image retrieval using local similarity patterns. *Information Processing and Management*, 39(1): 1–23.

Ullrich C., Ruimin Shen& Weikai Xie. 2013. Analyzing Student Viewing Patterns in Lecture Videos. *IEEE 13th International Conference on Advanced Learning Technologies (ICALT)*:115–117.

Wu P.H., Peng W.C. & Chen M.S. 2001. Mining sequential alarm patterns in a telecommunication database. *Proceedings of Workshop on Databases in Telecommunications* (VLDB 2001): 37–51.

Zaki M. J. 2001. SPADE: An efficient algorithm for mining frequent sequences. *Machine Learning Journal*, 42(1/2):31–60.

Computational Intelligence in Industrial Application – Ling (ed.)
© 2015 Taylor & Francis Group, London, ISBN: 978-1-138-02818-0

Introduction of uncertainty in complex event processing

Na Mao & Jie Tan
Institute of Automation, Chinese Academy of Sciences, Beijing, China

ABSTRACT: In recent years, sensor-kind network devices have been deployed in many applications, which can generate a lot of raw data continually. However, the raw data is of a high level of uncertainty because of the inherent unreliability of sensor readings. Meanwhile, the rules used to model the real world are simplified so that they do not capture the complexity of application scenarios and introduce the uncertainty. Consequently it is of great value to model uncertainty in complex event processing. In this paper, we survey the uncertainty in CEP to help researchers in understanding different approaches and mechanisms discussed so far, according to the classification of uncertainty for CEP from uncertainty in events to in rules. In particular, we propose a general framework to deal with uncertainty in complex event processing and look into the future about the research work.

KEYWORDS: Uncertainty; Complex event processing; Uncertainty in events; Uncertainty in rules.

1 INTRODUCTION

In recent years, sensor-kind network devices have been deployed in many applications, such as environment monitoring [Broda et al. 2009], fraud detection [Schultz-Møller et al. 2009], RFID-based inventory management systems [Wang & Liu 2005], weather monitoring [Kurose et al. 2006]. These devices can generate a lot of raw data continually. Following the processing and analysis the raw data streams, continuously arriving events could be transformed into complex event streams which are rigidly matched against complex event pattern and outputted. We call this processing course Complex Event Processing, or CEP [Luckham 2001]. Several famous event processing systems for complex event matching have been proposed[Agrawal et al. 2008].

Because of the inherent unreliability of sensor readings, the raw data is of high level of uncertainty. The data is often incomplete, imprecise and even misleading. We call the uncertain raw data uncertain events. Since uncertain event streams widely exist in reality, how to generate complex event streams which can be trusted in is becoming an urgent problem.

However, there is other aspect of uncertainty in complex event processing besides the inaccurate information provided by event source. In complex event processing systems, complex events or composite events are derived from rules of primitive ones. When we use the virtual world to describe the physical world, it is simplified [Cugola & Margara 2012]. That is to say, the rule presented above are a simplified version of rules expected to exist in the

real world, which are not suitable to capture the complexity of the aforementioned application scenarios. So it is necessary for us to capture the derivation uncertainty in complex event processing systems for higher efficiency.

In this paper, we discuss the classification of uncertainty for complex event processing and propose a general framework for dealing with the uncertainty. At the same time, we introduce the methods used in each aspect respectively, and use it to provide an extensive review of the state of the art in the area.

The rest of the paper is organized as follows. Section 2 presents a background work on complex event processing and several CEP engines characterized with the data model and rule definition language. A general framework for dealing with uncertainty in complex event processing systems is described in Section 3, where we discuss the two kinds of uncertainty in CEP respectively. Section 4 shows the future work in this area. Finally, the conclusion is highlighted in Section 5.

2 BACKGROUND

2.1 *Complex event processing*

Complex Event Processing, or CEP, is an event processing technique that analyzes multiple events with the goal of identifying meaningful complex event within both large collections of events and event streams. A complex event is a composite of primitive events. Indeed, the CEP system relies on the ability to specify composite events through event patterns or rules that match incoming event notifications on the basis of their contents and on some ordering relationships of them. The general architecture of the CEP system is shown in Figure 1.

At the peripheral of the system are the sources and the sinks. The former observe primitive events and report about them, while the latter receive composite event notifications and react to them. The task of identifying so called composite events from primitive ones is referred as Complex Event Processing (CEP). It relies on an Engine which operates with a set of rules conceived and deployed by rule managers.

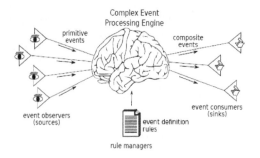

Figure 1.   The high-level view of a CEP system.

The kernel of complex event processing is described with four steps as followed:

i. Extracting primitive events from large volume data;
ii. Creating meaningful events with operators using event correlation or event aggregation detected according to specific rules;
iii. Extracting the time, causal, hierarchical and other semantic relationships by processing primitive or composite events;
iv. Sending responses to the applications to guarantee successful delivery of events to the subscribers.

2.2   CEP engines

Although complex event processing is a relatively new area of research, several CEP engines have been developed in the last few years. Each of them has its own data model, its own rule definition language, its own processing algorithm and implementation.

In practice, the data model a system adopts significantly affects the structure of the rule definition language it uses. At the same time, the languages interpret messages flowing into the system as notifications of events occurred in the observed world at a specific time, and they define how to generate composite events from primitive ones.

Table 1.   Data model, rule definition language and support for uncertainty of several CEP engines.

Name	Nature of Items	Language	Support for Uncertainty
Cayuga	Events	Cayuga Event Language (CEL)	No
Amit	Events	Java	No
Sase	Events	SQL-like	No
Sase+	Events	SQL-like	No
Coral8 CEP	Data	Continuous Computation Language	No
Oracle CEP	Data	Java	No
Esper	Data	Event Processing Language (EPL)	No
Tibco BE	Events	UML	No
IBM System S	Data	Spade, Semantic Solver	No

Table 1 focuses on the data model, the rule definition language and the support for data uncertainty of several CEP engines. Cayuga [Brenna et al. 2007], Amit [Adi & Etzion 2004], Sase [Wu et al. 2006], Sase+ [Gyllstrom et al. 2008], Coral8 CEP. [Coral8 2014], Oracle CEP [Oracle 2014], Esper [Esper 2014], Tibco BE [Tibco 2014] and IBM System S [Wu et al. 2007] are given in Table 1. As shown in Table 1, existing CEP engines mostly have no support for uncertainty.

3   UNCERTAINTY IN COMPLEX EVENT PROCESSING

CEP rules are conceived to describe and model some aspects of interests of the real world. As for any modeling approach, accuracy is crucial. At the same time, our ability to capture a phenomenon is often affected by some form of uncertainty. Ignoring it may lead to incomplete, inaccurate, or even incorrect decisions concerning the phenomenon itself. To achieve an effective management, we propose a general framework of dealing with uncertainty in CEP. The following three aspects must be considered [Cugola et al. 2014]:

Identification of the sources of uncertainty. This requires a careful analysis of the phenomenon to model. We consider the environment where CEP is deployed in general, the way we capture the phenomenon, as well as, the domain knowledge, in order to identify the potential sources of uncertainty.

Modeling of uncertainty. Once the sources of uncertainty have been identified they must be incorporated into the CEP system. The modeling phase aims at providing a sound mathematical foundation to represent uncertainty, so the CEP engine can be aware of it, and manipulate it consistently.

Propagation of uncertainty across the system. An uncertainty-aware CEP engine should produce a result characterized by an appropriate degree of uncertainty, consistent with the identified sources of uncertainty and the models adopted to represent them.

The framework proposed above accords with the general process of handling of events, so it can be widely used to extend the uncertainty on existing CEP engines.

As it was said in the introduction, we classify the uncertainty in complex event processing into the two aspects as followed [Artikis et al. 2012]:

Uncertainty in events, i.e., the uncertainty deriving from an incorrect observation of the phenomena under analysis. This means to admit that the notifications entering the CEP engine can be characterized by a certain degree of uncertainty.

Uncertainty in rules, i.e, the uncertainty deriving from incomplete or erroneous assumptions about the environment in which the system operates. This means to admit that the CEP engine has only a partial knowledge about the system under observation, and consequently the CEP rules can not consider all the factors that may cause the composite events they are in charge of detecting.

We discuss the uncertainty in complex event processing with the two aspects respectively in this section and give a review of the method used in the area.

3.1 *Uncertainty in events*

In this section we focus on uncertainty that affects event notifications, showing the method about how to model it and propagate it during processing.

3.1.1 *Modeling uncertainty in events*
For each event e, we considers two forms of uncertainty:

the uncertainty regarding the content of e;

the uncertainty regarding the occurrence of e;

The first form of uncertainty comes from the fact that the measures of real world are affected by some degree of uncertainty, which derive from the inaccuracy, imprecision, and noise.

To solve the problem, the author in [Cugola et al. 2014] considers the value of each attribute $Attr_i$ of event e as a sample from a random variable $X_i' = X_i + \varepsilon_i$, where X_i is the real, and ε_i is the measurement error. For ε_i, we assume that it depends on the noise at sources and the probability distribution function is known. [Wasserkrug et al. 2008] describes a probabilistic event model based on rules, which define the probability space of interest at each time point. The author extends deterministic rule definitions to handle uncertainty, and develops a sampling algorithm that heuristically assesses materialized event probabilities.

In some applications, the value of the related attribute is difficult to measure. [Cao et al. 2013] has proposed a event model introducing fuzzy concept. It uses fuzzy ontology to create query model to support event query of uncertainty and fuzzy.

The second form of uncertainty derives from that event occurrence times are often unknown or imprecise in the real-world applications. And the events from various sources can not be merged into a single stream with a total or partial order. [Zhang et al. 2010] proposed a temporal model that assigns a time interval to each event to represent all of its possible occurrence times.

3.1.2 *Propagating uncertainty in events*
After the modeling phase presented above, we can translate the event streams into probabilistic event streams. Here we discuss the main idea of approaches to show how uncertainty in primitive events propagates to composite events.

Recently, some work on detecting complex events in probabilistic event stream based on NFA(Nondeterministic Finite Automaton) has been proposed. The event detection is started at the "start" state of the NFA and it enters the next state with a probability when an event is scanned. When the "end" state of the NFA is entered, the detection enters the accepting state which means that the required event is detected.

In the work of [Xu et al. 2010], a data structure called Chain Instance Queues (CIQ) is used to detect complex events satisfying query requirements with single scanning probabilistic stream. Conditional Probability Indexing-Tree (CPI-Tree) is defined to store the conditional probabilities of a Bayesian network to improve the performance. In another paper [Kawashima et al. 2010], an optimized method is proposed to not only calculate the probability of outputs of compound events but also obtain the value of confidence of the complex pattern given by a user against uncertain raw input data stream generated by distrustful network devices. This method is based on an existing stream processing engine SASE+, and extends its evaluation model NFA[b] automaton to a new type of automaton in order to manage the runtime against probabilistic stream. In the work of [Shen et al. 2008], a query language is designed which allows users to express Kleene closure patterns when detecting probabilistic events. This method uses a new data structure AIG to detect sequence patterns over probabilistic data streams. With the benefit of lineage, the probability of an output event can be directly calculated without considering the query plan. In [Govindasamy & Thambidurai 2012.], a Multi-layer Event Filtering (MEF) approach is proposed to filter and correlate events by matching the large number of incoming events and rules. It includes Estimated Semantic Matching (ESM), NFA Matching and optional filtering based on domain specific Key Indicators (KI).

Besides NFA, there are other CEP methods to deal with uncertainty based on the data structure of trees. [Li et al. 2010] used a tree-based CEP method and optimized the algorithm by event grouping. [Zhang et al. 2014] proposed an improved method to not only process volumes of real-time event stream effectively but also calculate the probability of outputs of compound events. It introduces a new type of matching tree to process the historical probabilistic event. What's more, in the work of [Cugola et al. 2014], the author presents the model called CEP2U to computes the probability that each constraint is satisfied based on the probability distribution function of attribute's measurement error. It is designed to support the commonly offered operators.

3.2 Uncertainty in rules

In the previous section we focused on the uncertainty associated with event notifications, under the assumption that the rules that detect composite events from primitive ones were definite and "certain". Here we relax this assumption and discuss how to deal with uncertainty deriving from rules. Mostly, we describe the uncertainty using Bayesian Networks(BNs).

3.2.1 Bayesian networks

Before introducing the details of dealing with uncertainty in rules, we provide an intuitive description of Bayesian networks. A BN is the most widespread paradigm for the representation and efficient inference of a probability space defined over a set of random variables [Cugola et al. 2014]. A BN has two main components: qualitative and quantitative. The qualitative component of the network is a graph, $G_{BN} = (V_{BN}, E_{BN})$, where V_{BN} is a set of vertices which describe the random variables and E_{BN} is a set of edges that describe the direct probabilistic dependencies between the random variables. For example, an edge from node N_1 to node N_2 represents a causal dependency between N_1 and N_2, i.e., N_1 causes N_2. Unlike the static triggering graph that represents event types and their relationships, this graph represents event instances. The quantitative component is a Conditional Probability Table (CPT), which defines the probabilities of each random variable based on the probabilistic dependencies.

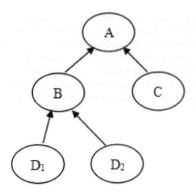

Figure 2. Bayesian network graph example.

For example, Figure 2 contains a depiction of the qualitative (graph) part of a Bayesian network which defines five events: A, B, C, D_1, and D_2, depicted as vertices. The edges indicate direct probabilistic dependencies. Table 2 contains the quantitative relationship between A to B and C. For each possible value of B and C the table provides the conditional probability of each value of A. For example, Pr(A=true | B=false, C=true) = 0.8. Such a table exists for each vertex in the Bayesian network. For vertices which have no incoming edges, the table contains the prior probabilities.

Table 2. CPT example.

	A=true	A=false
B=true, C=true	0.3	0.7
B=false, C=true	0.8	0.2
B=true, C=false	0.05	0.95
B=false, C=false	0.1	0.9

3.2.2 Modeling uncertainty in rules through Bayesian networks

In the real-world applications, the need of considering uncertainty not only in events but also in rules plays an important role to model the reality effectively.

As an example taken in [Cugola et al. 2014], a Tunnel Ventilation System (TVS) malfunctioning event can be detected through the contemporary observation of high temperature and low oxygen concentration. It implicitly assumes a closed world in which only the three entities of oxygen, temperature, and malfunctions exist, and there are no additional factors that influence their occurrence and their causal dependencies. However, reality is a lot of other factors may actually influence the TVS malfunctioning, such as the happening of traffic jam. Some of these factors could be impossible to observe and measure, or it could be too complex to precisely monitor them. In general every modeling activity throws away those details that are considered marginally relevant, but in doing so it trades simplicity for precision, exposing to the risk that these omissions could result in incomplete models, possibly leading to wrong deductions. The ability to model the level of uncertainty present in rules is precisely what we offer to avoid this risk.

To the best of our knowledge, the first model proposed for dealing with uncertainty in CEP is described in [Wasserkrug et al. 2005]. This model has been extended in [Wasserkrug et al. 2008], where the authors introduce a general framework for CEP in presence of uncertainty. It captures the uncertainty in events and in rules, and adopts Bayesian Networks to model both of them. Further details about the framework are presented in [Wasserkrug et al. 2012].

In particular, [Cugola et al. 2014] models uncertainty in rules through Bayesian networks using the next steps:

i. Translating the rule R into a corresponding BN;
ii. Enriching such BN to include additional factors that may influence the occurrence of events by rule manager or other domain experts;
iii. Evaluating the updated BN to compute the probability of occurrence of composite events;
iv. Propagating the computed value, which is integrated with the results obtained by considering the uncertainty in events, to the composite events generated by the rule R.

4 FUTURE WORK

From the discussion in Section 3, we claim that the theory of probability is used in dealing with uncertainty in events and Bayesian Networks are built to model uncertainty in rules. However, the epistemic uncertainty[Lakshmi & Kurian 2013] can be represented in many ways including probabilistic theory, fuzzy set [Ryoo & Aggarwal 2008], Bayesian analysis, soft computing technique [Lee & Smith 1995], rule-based classification technique [Qin et al. 2009], etc. So we can think about applying other methods to uncertainty for complex event processing.

Uncertainty is the main obstacle of modeling event processing in the real world, so designing a CEP engine to satisfy the uncertainty is very promising.

Future work also includes investigating further sources of uncertainty and the application to a real case study. Furthermore, learning mechanisms to automatically generate rules from historical data analysis can be considered [Margara 2013]. This could also help domain experts in determining some critical parameters related to uncertainty (e.g., tune the minimum probability for considering events based on past occurrences).

5 CONCLUSION

The need for processing a lot of event streams in a timely manner is becoming more and more common in several domains, from environment monitoring to finance. While uncertain event streams widely exist in reality, it is of great value to generate the trusted complex event streams. This need was answered by different researchers, each bringing its own work in dealing with the uncertainty in complex event processing.

In this paper, we surveyed the uncertainty for complex event processing from uncertainty in events to uncertainty in rules. We also proposed a general framework for dealing with uncertainty in CEP. Meanwhile, we focused on the classification of uncertainty for CEP with different approaches and mechanisms discussed so far. Researchers are going on this area to find an efficient mechanism to handle uncertainty in CEP.

ACKNOWLEDGMENTS

This research work is sponsored by the National Natural Science Foundation of China Grants U1201251 and the National Science & Technology Support Plan of China (125) Grants 2012BAF11B04.

REFERENCES

Adi, A. & Etzion, O. 2004. Amit - the situation manager. *The VLDB Journal* 13(2): 177–203.

Agrawal, J., Diao, Y., Gyllstrom, D. & Immerman, N. 2008. Efficient Pattern Matching over Event Streams. *ACM SIGMOD*. Vancouver: Canada.

Artikis, A., Etzion, O., Feldman, O. & Fournier, F. 2012. Event processing under uncertainty. *Proceedings of the annual ACM International Conference on Distributed Event-Based Systems*. Berlin: Germany.

Brenna, L., Demers, A., Gehrke, J., Hong, M., Ossher, J., Panda, B., Riedewald, M., Thatte, M., & White, W. 2007. Cayuga: a high-performance event processing engine. *Proceedings of the 2007 ACM SIGMOD international conference on Management of data*: 1100–1102. New York: USA.

Broda, K., Clark, K., Miller, R. & Russo, A. 2009. Sage: A logical agent-based environment monitoring and control system. *Ambient Intelligence volume 5859*: 112–117.

Cao, K., Wang, Y., Li, R. & Wang F. 2013. A distributed context-aware complex event processing method for Internet of Things. *Journal of Computer Research and Development* 50(6): 1163–1176.

Cugola, G. & Margara, A. 2012. Processing flows of information: From data stream to complex event processing. *ACM Computing Surveys* 44(3): 15:1–15:62. New York: USA.

Cugola, G., Margara, A., Matteucci, M. & Tamburrelli, G. 2014. Introducing uncertainty in complex event processing: Model, implementation, and validation. *Computing.* Springer Vienna.

Coral8. 2014. http://www.aleri.com/WebHelp/coral8_docume ntation.htm. Visited Sep. 2014.

Esper. 2014. http://www.espertech.com/. Visited Sep. 2014.

Govindasamy, V. & Thambidurai, P. 2012. An efficient and generic filtering approach for uncertain complex event processing.*Proceedings of International Conference on Data Mining and Computer Engineering* (ICDMCE '12): 211–216. Bangkok: Thailand.

Gyllstrom, D., Agrawal, J., Diao, Y., & Immerman, N. 2008. On supporting kleene closure over event streams. *Proceedings of the IEEE 24th International Conference on Data Engineering*: 1391–1393. Washington:USA.

Kawashima, H., Kitagawa, H. & Li, X. 2010. Complex event processing over uncertain data streams. *Proceedings of the Fifth International Conference on P2P, Parallel, Grid, Cloud and Internet Computing*: 521–526.

Kurose, J., Lyons, E., McLaughlin, D., Pepyne, D., Philips, B., Westbrook, D. & Zink, M. 2006. An end-user-responsive sensor network architecture for hazardous weather detection, prediction and response. *Proceedings of the Second Asian Internet Engineering Conference volume 4311* :1–15. Pathumthani: Thailand.

Lakshmi, S.D. & Kurian, M. 2013. Handing of uncertainty- A survey. *International Journal of Scientific and Research Publications, Volume 3.*

Lee, M.A. & Smith, M.H. 1995. Handling Uncertainty in Finance Applications Using Soft Computing. *Proceedings of 3rd International Symposium Uncertainty Modeling and Analysis.*

Li, Y., Wang, J. & Feng, L. 2010. Accelerating sequence event detection through condensed composition. *Proceedings of the 5th International Conference on Ubiquitous Information Technologies & Applications.* Sanya: China.

Luckham, D.C. 2001. The Power of Events: An Introduction to Complex Event Processing in Distributed Enterprise Systems. Boston: USA.

Margara, A., Cugola, G. & Tamburrelli, G. 2013. Towards automated rule learning for complex event processing. Technical Report.

Oracle. 2014. http://www.oracle.com/technologies/soa/complex-event- processing.html. Visited Sep. 2014.

Prasad, G. & Murahari reddy, D. 2013. Effective event derivation of uncertain events in rule-based systems. *Proceedings of International Conference on Computer, Control and Cognitive Sciences.* Bengaluru.

Qin, B., Xia, Y., Prabhakar, S. & Tu, Y. 2009. A Rule-Based Classification Algorithm for Uncertain Data. *Proceedings of the 25th International Conference on Data Engineering (ICDE2009)*: 1633–1640. Shanghai: China.

Ryoo, M.S. & Aggarwal, J.K. 2008. Human Activities: Handling Uncertainties Using Fuzzy Time Intervals. *Proceedings of the 19th conference of the International Association for Pattern Recognition(ICPR 2008)*: 1–5. Florida: USA.

Schultz-Møller, N.P., Migliavacca, M. & Pietzuch, P. 2009. Distributed complex event processing with query rewriting. *Proceedings of the Third ACM International Conference on Distributed Event-Based Systems (DEBS '09).* New York: USA.

Shen, Z., Kawashima, H. & Kitagawa, H. 2008. Probabilistic event stream processing with lineage. *Proceedings of the Data Engineering Workshop.*

Tibco. 2014. http://www.tibco.com/software/complex-event-processing/businessevents/default.jsp. Visited Sep. 2014.

Tran, T., Peng, L., Li, B., Diao, Y. & Liu, A. 2010. PODS: A New Model and Processing Algorithms for Uncertain Data Streams. *ACM SIGMOD.* Indianapolis: USA.

Wang, F. & Liu, P. 2005. Temporal management of RFID data. *Proceedings of the 31st international conference on Very large data bases (VLDB '05):* 1128–1139. Trondheim: Norway.

Wang, Y.H., Cao, K. & Zhang, X.M. 2013. Complex event processing over distributed probabilistic event streams. *Computer and Mathematics with Applications* 66: 1808–1821.

Wasserkrug, S., Gal, A. & Etzion, O. 2005. A model for reasoning with uncertain rules in event composition systems. *Proceedings of the 21st Annual Conference on Uncertainty in Artificial Intelligence*: 599–606.

Wasserkrug, S., Gal, A., Etzion, O. & Turchin, Y. 2008. Complex event processing over uncertain data. *Proceedings of the second international conference on Distributed event-based systems (DEBS '08)*: 253–264. New York: USA.

Wasserkrug, S., Gal, A., Etzion, O. & Turchin, Y. 2012. Efficient processing of uncertain events in rule-based systems. *IEEE Transactions on Knowledge and Data Engineering* 24(1): 45–58.

Wu, E., Diao, Y., & Rizvi, S. 2006. High-performance complex event processing over streams. *Proceedings of the 2006 ACM SIGMOD international conference on Management of data*: 407–418. New York: USA.

Wu, K.L., Hildrum, K.W., Fan, W., Yu, P.S., Aggarwal, C.C., George, D.A., Gedik, B., Bouillet, E., Gu, X., Luo, G., & Wang, H. 2007. Challenges and experience in prototyping a multi-modal stream analytic and monitoring application on system s. *Proceedings of the 33rd international conference on Very large data bases*: 1185–1196.

Xu, C., Lin, S. & Lei, W. 2010. Complex event detection in probabilistic stream. *Proceedings of the 12th International Asia-Pacific Web Conference*: 361–363.

Zhang, X., Wang, Y., Zhang, X. & Zhu, X. 2014. Complex Event Processing Over Uncertain Event Stream. *Proceedings of the 2nd International Conference on Communications, Signal Processing, and Systems*: 647–656.

Zhang, H., Diao, Y. & Immerman, N. 2010. Recognizing Patterns in Streams with Imprecise Timestamps. *Proceedings of 36th International Conference on Very Large Data Bases (VLDB '10).* Singapore.

Zhang, X.L., Wang, Y. & Zhang, X.M. 2014. Complex event processing over distributed uncertain event streams. *Proceedings of International Conference on Computer Science and Service System(CSSS2014)*: 357–361.

Computational Intelligence in Industrial Application – Ling (ed.)
© 2015 Taylor & Francis Group, London, ISBN: 978-1-138-02818-0

QGA-SVM study on temperature compensation of liquid ammonia volumetric flowmeter

T. Lin

School of Automation, Chongqing University, Chongqing, China
School of Applied Electronics, Chongqing College of Electronic Engineering, Chongqing, China

P. Wu

School of Automation, Chongqing University, Chongqing, China
Chongqing Chuanyi Automation Co., Ltd. Chongqing, China

F.M.Gao & Y. Yu

School of Biomedical Engineering, Xinxiang Medical University, Xinxiang, Henan, China

L.H. Wang

School of Applied Electronics, Chongqing College of Electronic Engineering, Chongqing, China

ABSTRACT: This paper introduces a Support Vector Machine (SVM) inverse regression compensation method to improve temperature compensated precision of volumetric flowmeter, for measuring the liquid ammonia quality in the coal chemical industry. The SVM kernel function parameter can be optimized by Quantum Genetic Algorithm (QGA), global search performance for optimal solutions. The experimental results indicate that this method is more accurate than the traditional quadratic expression method.

KEYWORDS: QGA; SVM; Volumetric Flowmeter; Temperature Compensation.

1   INTRODUCTION

Ammonium hydroxide is one of an important cooling medium for many devices such as gas collecting tubes in the process of the coal chemical technology, and preparation of ammonia hydroxide of liquid ammonia is an important part of the coal chemical process. With the development of fine coal chemical industry, the requirement for device temperature control performance become more and more critical, so the request to the control and measurement accuracy of liquid ammonia keeps higher and higher.

Volumetric flow meters such as turbine flowmeter and vortex flow meter are used extensively in theflow measurement to volatile and high saturated vapor pressure liquid like liquid ammonia, ethylene, etc. But the given flow valuedirectlyon the volumetric flow meteris merelythe volume flow value,and the conversion of mass flow and volume flow isoftenperformed according to the following relation in order to satisfy customer demands for mass flow measurements [Fu, H. et al., 2013], [Ji, G., 2012].

$$q_m = \rho_f q_v \qquad (1)$$

Where q_m ___ the mass flow, kg / h;
ρ_f ___ the fluid density in working condition, kg / m^3.

While liquid ammonia density is a function of temperature, that is $\rho = f(t)$. Because the nonlinear relationshipbetween density and temperature of liquid ammonia, the fitting coefficients and constants need vary constantly for different temperature sections. With the improvement of intelligent of volumetric flowmeters such as turbine flowmeter and vortex flowmeter, the traditional method using quadratic density-temperature expression is often not effective, especially when the temperature span is relatively large. It is necessary to developa nonlinear regression method to satisfythe high precision requirements for liquid ammonia flow control and measurement.

It is possible to use SVM (Support Vector Machine) to establish a regression inverse model of temperature compensation to liquid ammonia, which can reduce temperature sensitivity during the translation of mass flow and volume flow and improve stability and accuracy of volumetric flow meter.In view of the nonlinearity and complexity of regression system, the trial and error is usually needed for kernel function parameters matching, and it is exactly these results that are at risk in the SVM training process[Fu, H. et al., 2013], [Xu, Z.B., 2002].With the aid of powerful global searching of QGA (Quantum Genetic Algorithm), the SVM kernel function parameter σ can be optimized, which can modify the parameters of the density-temperature regression model, and thus the measuring precision is improved.

2 ESTABLISHMENT OF SVM REGRESSION MODEL OF TEMPERATURE COMPENSATION

2.1 SVM regression model

The SVM regression method is different from multivariate regression analysis one, whose constructors that have function non-object parameters to be eliminated needn't be established, can geta theoretical optimal solution using convex quadratic optimization problem transformation through estimation and prediction of the small sample on the basis of VC dimension theory coming from statistic learning theory and structural risk minimization. Sampling group points $\{(x_i, y_i)\}_{i=1}^{N^+}$ in the input analytical space X can be mapped to become training group points $(\varphi(x_i), y_i)$ in high dimensional Hilbert space F by SVM kernel function algorithm, and the training set $D = \{\varphi(x_i), y_i)\}_{i=1}^{N^+}$ which has been mapped is regressed by constructinga linear discriminant function in Hilbert space F. Thus the regression inverse model has better generalization ability, and the dimension disaster is avoided, which means that algorithm complexity is unrelated to sample dimension[Liu, J.H., 2010].

Set sample set to $\{(x_i, y_i)\}_{i=1}^{N^+}$, where $x_i \in R^d$ is input vector, y_i is corresponding expected value. A dual problem model of constrained convex quadratic optimization is defined as

$$\arg \max_{\alpha} \omega(\alpha) = \sum_{i=1}^{N} \alpha_i - \frac{1}{2} \sum_{i=1}^{N} \sum_{j=1}^{N} y_i y_j \alpha_i \alpha_j K(x_i, x_j)$$

$$s.t. \sum_{i=1}^{N} \alpha_i y_i = 0; \ 0 \le \alpha_i \le C, \ i = 1, 2, \cdots, N \quad (2)$$

α_i is the Largrang multiplier and $K(x_i, x_j)$ is the kernel function.

Let $\alpha^* = (\alpha_1^*, \alpha_2^*, ..., \alpha_{N^+}^*)$ be the solution vector of (2), in which usually only part of the solutions are not zero. The sample input of nonzero solutions x_i serves as support vectors, which determined decision boundary. The data fusion based on SVM was done to establish the fitting relationships between input x and output y, that is

$$y(x) = \omega^T x + b = \sum_{i=1}^{s} \alpha_i K(x, x_i) + b \quad (3)$$

In (2), x_i is the support vector; s is the number of the support vectors; b is the SVM offset; ω is the weight coefficient whose number is similar to support vector number. A Gaussian radial basis function which meets Mercer condition was chosen to be Kernel function. That is

$$K(x, x_i) = \exp\left(-\frac{\|x - x_i\|}{2\sigma^2}\right) \quad (4)$$

σ is the kernel function parameter. The forecast accuracy of SVM would be improved by regulating σ properly.

2.2 Preparation of the data sample

The number of the overall sample pairs (x_i, y_i) $(i = 1,2,... N)$ is $N = N_p + N_t$, and N_p is the number of training samples (N_p accounted for 1/2~2/3 of the overall sample number N), N_t is the number of test samples. The $N_p = 26$ training samples and $N_t = 25$ testing samples were randomly selected from liquid ammonia density temperature relation table ($-20°C\sim30°C$).

The relation data of liquid ammonia density temperaturesare in Table 1.

Table 1. Liquid ammonia density temperature relation table.

NO	Temperature	Density
	°C	kg/m³
1	−20	665.005
2	−19	663.721
3	−18	662.433
4	−17	661.141
5	−16	659.846
6	−15	658.546
7	−14	657.243
8	−13	655.936
9	−12	654.625
10	−11	653.31
11	−10	651.991
12	−9	650.668
13	−8	649.341
14	−7	648.009
15	−6	646.673
16	−5	645.333
17	−4	643.989
18	−3	642.64
20	−1	639.929
21	0	638.567

(continued)

NO	Temperature	Density
	℃	kg/m^3
22	1	637.2
23	2	635.828
24	3	634.451
25	4	633.07
26	5	631.684
27	6	630.293
28	7	628.897
29	8	627.496
30	9	626.089
31	10	624.678
32	11	623.261
33	12	621.838
34	13	620.411
35	14	618.978
36	15	617.539
37	16	616.094
38	17	614.644
39	18	613.188
40	19	611.726
41	20	610.258
42	21	608.784
43	22	607.303
44	23	605.817
45	24	604.324
46	25	602.824
47	26	601.318
48	27	599.805
49	28	598.285
50	29	596.759
51	30	595.225

3 QGA OPTIMIZATION

QGA is a kind of genetic algorithm based on principles of quantum computing, in which one chromosome can express the superposition of multiple states because the quantum state vector is used in epigenetic code and the chromosome evolution is achieved by the use of quantum logic gate, and ultimately find out the optimal solution suitable for the objective [Wu, P. et al., 2014], [Zhang, X.H. et al., 2012].

3.3 Quantum bits code

In QGA, one gene can be stored and expressed by quantum bits. The quantum bitis different than-theGAbit in that the former can be in two quantum superposition states at the same time. This express method, which is the superposition of multiple states, makes QGA has better individual (and chromosome)

diversity that traditional GA, and can effectively restrain premature convergence especially when solving these complex problems.

$$|\varphi\rangle = \alpha|0\rangle \beta|1\rangle \qquad (5)$$

(α, β) is the probability amplitude, and satisfies the following condition

$$|\alpha|^2 + |\beta|^2 = 1 \qquad (6)$$

$|0\rangle$ and $|1\rangle$ represent spin up state and spin down state respectively in (5). A quantum bit includes the state $|0\rangle$ and state $|1\rangle$ at the same time, and the chromosome using quantum bit code will converge to a single state when $|\alpha|^2$ or $|\beta|^2$ tends to 0 or 1. The binary code form is usually used in the QGA, whose gene, including m parameters using multiple quantum bits code represented as shown below

$$q_j^t = \begin{bmatrix} \alpha_{11}^t & \cdots & \alpha_{1k}^t \\ \beta_{11}^t & \cdots & \beta_{1k}^t \end{bmatrix} \begin{matrix} \alpha_{21}^t & \cdots & \alpha_{2k}^t \\ \beta_{21}^t & \cdots & \beta_{2k}^t \end{matrix} \cdots \begin{matrix} \alpha_{m1}^t & \cdots & \alpha_{mk}^t \\ \beta_{m1}^t & \cdots & \beta_{mk}^t \end{matrix} \qquad (7)$$

Where, q_j^t is the chromosome of the jth individual in the tth generation; m is the gene number of the chromosome; k is the quantum bit number of each gene.

During initialization, the probability amplitude (α, β) of quantum bits code is initialized to $(1/\sqrt{2}, 1/\sqrt{2})$ to make all possible states of chromosomes on each individual in one population equally probable.

3.4 Quantum rotation door update strategy

The quantum door is the actuating mechanism of QGA evolution, which is chosenaccording to specific problems. Here, the quantum rotation door is applied for adjustment in QGA evolution, whose matrix expression is as follows.

$$G(\theta_i) = \begin{bmatrix} \cos(\theta_i) & -\sin(\theta_i) \\ \sin(\theta_i) & \cos(\theta_i) \end{bmatrix} \qquad (8)$$

The process of regulation and updating of the quantum rotation door is

$$\begin{bmatrix} \alpha_i' \\ \beta_i' \end{bmatrix} = G(\theta_i) \begin{bmatrix} \alpha_i \\ \beta_i \end{bmatrix} = \begin{bmatrix} \cos(\theta_i) & -\sin(\theta_i) \\ \sin(\theta_i) & \cos(\theta_i) \end{bmatrix} \begin{bmatrix} \alpha_i \\ \beta_i \end{bmatrix} \qquad (9)$$

$(\alpha_i, \beta_i)^{\mathrm{T}}$ and $(\alpha_i', \beta_i')^{\mathrm{T}}$ is the probability amplitude before and after update respectively to the ith quantum bit rotation door on a chromosome; θ_i is the rotation angle.

According to (8) and (9), α_i' and β_i' can be shown as

$$\begin{cases} \alpha_i' = \alpha_i \cos(\theta_i) - \beta_i \sin(\theta_i) \\ \beta_i' = \alpha_i \sin(\theta_i) + \beta_i \cos(\theta_i) \end{cases} \tag{10}$$

So after the transform, $\left|\alpha_i'\right|^2 + \left|\beta_i'\right|^2$ and $\left|\alpha_i\right|^2 + \left|\beta_i\right|^2$ equal in value, remain 1.

The selection strategy of rotation angle θ_i is shown in table 2. x_i is the ith bit on the current chromosome; $best_i$ is the ith bit on the current optimal chromosome; $f_F(\bullet)$ is fitness function; $s(\alpha_i, \beta_i)$ is the direction of rotation angle; $\Delta\theta_i$ is the rotation angle; the rotation direction and rotational angle size of rotation angle θ_i based on the selection strategy.

Table 2. Selection strategy of rotation angle θ_i

x_i	$best_i$	$f_F(x_i) <$ $f_F(best_i)$	$\Delta\theta_i$	$s(\alpha_i, \beta_i)$			
				$\alpha_i\beta_i>0$	$\alpha_i\beta_i<0$	$\alpha_i=0$	$\beta_i=0$
0	0	F	0	0	0	0	0
0	0	T	0	0	0	0	0
0	1	F	0.01π	$+1$	-1	0	±1
0	1	T	0.01π	-1	$+1$	±1	0
1	0	F	0.01π	-1	$+1$	±1	0
1	0	T	0.01π	$+1$	-1	0	±1
1	1	F	0	0	0	0	0
1	1	T	0	0	0	0	0

3.5 The optimization algorithm design of SVM kernel function parameter by QGA

Trained SVM with training samples is tested by MSETD which represent the standard deviation of mean square error between the density calibration values and the predicted values of testing samples, in order to reduce parameter selection reliance on testing samples. The experiments show these learning parameters in SVM, including the boundary of Lagrange multiplier C, the condition parameter of convex quadratic optimization $\lambda1$ and ε-neighborhood parameter around solutions ε, have no obvious effect on the output results, but the kernel function parameter σ that have the largest influence on the output

results is often difficult to identify only by trial and error. Taking MSETD as fitness function, the kernel function parameter σ is optimized by virtue of QGA global search performance for optimal solutions and then the proper offset b and weight coefficient ω are found, so that output results are optimal or suboptimal to meet the precision and accuracy of system measurement [LI, X.J. et al., 2012]. The fitness function can be expressed as

$$f_F = MSETD \tag{11}$$

Inaddition, the object founction is

$$f_{Obj} = \min(f_F) = \min(MSETD) \tag{12}$$

QGA algorithmprocesses are as follows[Wu, P. et al., 2014]:

Step 1 Initialize population $P(t_0)$, n Chromosomes encoded by quantum bits are randomly generated;

Step 2 Measure each individual in initial population $P(t_0)$ and get the corresponding solution $S_p(t_0)$;

Step 3 Calculate the fitness value of each solution and compare them;

Step 4 Record those best individuals and their fitness values;

Step 5 The algorithm exits if the condition is met, otherwise it continues normally;

Step 6 Measureeach individual in population $P(t)$ and get the corresponding solution $S_p(t)$;

Step 7 Calculatethe fitness value of each solution and compare them;

Step 8 Adjust those individuals according to the quantum rotation gate $G(\theta_t)$to generate the next population $P(t+1)$;

Step 9 Record those best individuals and their fitness values;

Step 10 Add 1 to the iterations t and return to *Step* 5.

3.6 The operating result of QGA

The population size of the QGA and GA both equal to 30, their maximum number of generations are 50, the binary length of each individual is 20; In addition, their crossover probability is 0.7, their mutation probability is 0.01. The corresponding SVM configuration parameters are set as follows: the kernel function is RBF function, regularization parameter C is 500 and the non-sensitive ε is 0.001.

After the program runs, the optimal solution of kernel function parameter σ with QGA is 2.305, the corresponding fitness function value is 9.0606; the optimal solution of kernel function parameter σ with GA is 2.231, the corresponding fitness function value is 9.0621. As can be seen fromthecomparison between the QGA evolution curve and the GA evolution curve in fig. 1 and their optimum solutions of each generation in Fig. 2. QGAis superior to GA in the convergence speed and the immaturity convergence.

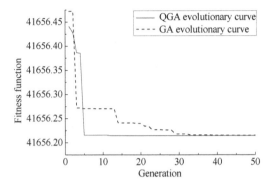

Figure 1.   Evolutionary processes of QGA & GA.

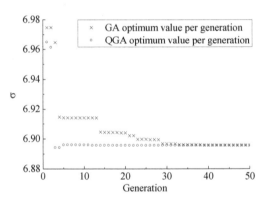

Figure 2.   The optimum solutions of each generation.

According to the results of QGA operation, MSETD, namely the standard deviation of mean square error between the density calibration values and the predicted values of testing samples, is 41656.2148 when kernel function parameter σ obtained by QGA equals6.8956.

4   CONCLUSIONS

With the traditional quadratic expression method, consider the process as in the following equation:

$$\rho = \rho_d[1 + \mu_1(t - t_d) \times 10^{-2} + \mu_2(t - t_d)^2 \times 10^{-6}] \quad (13)$$

Here, t——the temperature of liquid ammonia, ℃;
t_d——the reference temperature of liquid ammonia, ℃;
ρ_d—— the density of liquid ammonia that correspond to t_d, kg / m^3;
μ_1—— the linear compensation coefficient of liquid ammonia, 10^{-2} °C ;
μ_2—— the quadratic compensation coefficient of liquid ammonia, 10^{-6} °C^{-2}.

When $t_d = 5$°C ,then $\rho_d = 631.684 kg / m^3$, the two endpoint temperature values –20°C and 30°C and the density data are substituted in (13) using binary equation groups to obtain $\mu_1 = -0.2209$ and $\mu_2 = -3.9741$, and the relation between the temperature t and the density ρ can be established so long as μ_1 and μ_2 are substituted in (13) again.

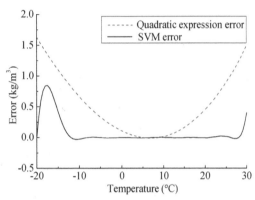

Figure 3.   Errors comparison.

The error curves of liquid ammonia density by the quadratic expression method and the SVM regression method are shown in fig. 3. By comparing their error curves, the temperature compensation error of the SVM regression method significantly less than the SVM regression method.

ACKNOWLEDGMENTS

This research was partially supported by the Natural Science Foundation of China under the contract number 61305147 and the Natural Science Foundation of Chongqing under the contract number cstc2012jjA10129.

REFERENCES

Fu, H. & Shi, D.D. 2013. Study on gas emission prediction model based on IGA-LSSVM. *China Safety Science Journal* 23(10):51–55.

Ji, G. 2012. *The flow measurement instrument application skills.* Beijing: Chemical Industry Press.

Xu, Z.B. & Nie, Z.K. 2002. Almost sure convergence of genetic algorithms: a martingale approach. *Chinese Journal of Computers* 25(8):785–793.

Wu, P. & Lin, T. 2014. Research on identification modeling of sheathed thermocouple sensor based on hybrid QGA-SVM. *Chinese Journal of Scientific Instrument* 35(2): 343–349.

Zhang, X.H. & Li, Z.Y. 2012. Application of vortex flow-meter and mass flowmeter in liquid ammonia measurement. *Soda Industry* 2012(8):34–36.

LI, X.J. & RAO, F. 2012. Rough set-based feature weighted kernels for support vector machine. *Journal of computational and theoretical nanoscience* 9(12): 2250–2254.

Liu, J.H. 2010. *Intelligent sensor system (second edition).* Xi'an: Xidian University Press,

Gao, F.M. & Lin, T. 2013. Application of SVM optimization based on GA in electronic sphygmomanometer data fusion. *International Journal of Computer Science Issues* 10(1-1):1–6.

Computational Intelligence in Industrial Application – Ling (ed.)
© 2015 Taylor & Francis Group, London, ISBN: 978-1-138-02818-0

A NetFlow-based traffic and direction analysis software

Yan Chen
Softwares Section, School of Software Outsourcing, Tianjin College of Commerce, Tianjin, China

Ya Jing Niu
School of Economics and Management, Tianjin ChenJian University, Tianjin, China

ABSTRACT: As more and more terminals are connected to the network according to the requirements of the user market, and the terminal market is being saturated, it is urgent for the telecom operators to establish an operable and manageable broadband network of high performance, high reliability and high security. This network is expected to provide more and better services than other operators and bear the telecom-level loads of high-end customers, in order to forge its own brand in broadband network. By deploying the traffic and direction analysis software system (hereinafter called the System), we promptly understand the loads of CTT (a large telecom operator in China) in A Province and the bandwidth usage of important services, accurately measure the traffic and service types of international and domestic operators or over the backbone network and MAN (analyze the traffic within and out of the Province, outbound traffic, etc.), properly evaluate and plan network extension and upgrading, carry out timely and accurate analysis of the traffic of the services over the network, better analyze the user services in different traffic directions. It is helpful to optimize the network traffic and direction, maintain relations with large customers, control costs, support market expansion, and improve the return on the network investment.

KEYWORDS: Communications, Networking, Traffic, Software, Analysis, Optimization.

With rapid advancement in the Internet, the telecom operators are entering a new stage in terms of the IP network infrastructure and the complexity of the network structure. After the Internet infrastructure is constructed, the provincial operator needs to optimize the Internet structure. Our work targets the telecom operator in A Province. The company has to follow the construction of the backbone data network with the optimization of the routes over the provincial network. Besides, it has to devise and implement policies for traffic control in order to improve user experience in video and game applications. To achieve this, we must first obtain accurate data about the traffic and its directions by properly analyzing the Internet traffic. Thus, it is technically urgent for CCT in A Province to carry out timely and accurate analysis of the services' traffic over the network and the traffic direction. By developing the System and deploying it at the traffic collection sites, we collect the IP addresses of CTT's backbone routers. Moreover, we enable the operator to analyze the traffic and the direction according to the IP address classification of China Telecom, China Unicom and peering, offering enormous support for route analysis, route adjustment, careful route management and traffic distribution.

1 POLICY FORMULATION

The software development is centered on the processes of NetFlow traffic collection, IP address collection, NetFlow data processing and data query. The flow chart of our project implementation is presented as Figure 1:

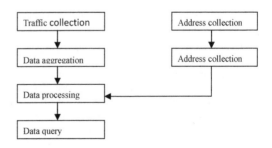

Figure 1.   Flow chart of the devised policy.

1.1 *NetFlow traffic collection*

In order to analyze the NetFlow traffic, we need to first collect it at the corresponding ports. The System achieves this by using the NetFlow data collected by the IP Integrated Network Management System of CTT.

The timing programs are developed to regularly download the collected NetFlow data into the System's traffic collection server for aggregation. [1]

1.2 *IP address classification*

The source and destination IP addresses of the NetFlow traffic has to be analyzed before we classify the traffic directions. Therefore, it is highly important to properly collect and classify IP addresses. There are many approaches to collecting the IP addresses. For example, the addresses can be collected via the Internet or the existing IP address database (e.g. the IP data of the pure network). These methods are flawed, because their collected data is either incomplete or inaccurate. [2] In order to accurately collect and classify the IP addresses, we decide to use the data from the BGP routing lists of routes in CTT's backbone network. Then the IP address classification table is obtained by sorting out the data using the self-developed programs.

1.3 *NetFlow data processing*

After aggregation, the collected NetFlow data will be loaded into the database to form the basic data table. In this process, format conversion will be performed to make the data readable to the System.

1.4 *NetFlow data query*

A set of windows is developed to enable conditional query and form query lists. Thus, the data can be analyzed in different ways, and the data sheet can be output in the EXCEL format.

2 POLICY IMPLEMENTATION

2.1 *Establishment of the hardware platform and installation of the database*

1 Overview of the Requirements on the System's Hardware Configurations

Two database application servers that support hot backup are in place so that in the case of master server breakdown, the backup server can work as a substitute. The backup function is available from the backup software of the backup server provider. The data of the database application server is stored in the disk arrays. Optical fibers are employed to connect the database application servers and disk arrays. Two optical fiber switches are set up separately. The free MYSQL (official free version) is adopted as the database system to reduce the development costs. [3]

The disk array is equipped with two control cards which are connected to the pair of optical fiber switches using optical fibers. Backup of the database files are made and stored in the disk arrays of different physical mediums to ensure database recovery in the case of disk array malfunction. The capacity of the disk array should suffice to back up the data in the database for at least 5 years.

Two NetFlow application servers for data sensing and collection that support hot backup are in place so that in the case of one server breaking down or undergoing repairs, the other server can work real-time to ensure that NetFlow can collect data completely.

Two NetFlow servers for data report and analysis that support hot backup need to be provided.

One access-layer switch is available that has 24 optical fiber ports and dual power supplies.

Dual GE optical fiber ports are used to connect the System to the provincial backbone network via the access-layer switch.

2 Details of the Requirements on the Hardware Configurations

The database application server: CPU of 4–8 cores, memory of 32-64G; hard disk: RAID5 backup, speed of 10,000 rotations; power supply: redundant 220V AC power. The dual GE optical ports are connected with the disk array optical switches, and the dual GE electrical ports are connected with the upstream switches of the NetFlow application servers.

The disk arrays: 12 1T hard disks with a speed of 10,000 rotations and RAID0+1+HOT backup; power supply: redundant 220V AC power; each of the control cards use 2-4 or more GE optical fibers to connect with the disk array optical switches; the capacity of the disk array should suffice to back up the data in the database for at least 5 years.

The NetFlow application server for data sensing and collection: 2 CPUs of multiple cores, memory of 64G; hard disk: RAID5 backup, speed of 10,000 rotations; power supply: redundant 220V AC power; the dual GE electrical ports are connected with the upstream switches.

The NetFlow server for data report and analysis: 4 CPUs of multiple cores, memory of 64G; hard disk: RAID0+1+HOT backup, speed of 10,000 rotations; power supply: redundant 220V AC power; the dual GE electrical ports are connected with the upstream switches.

The hardware firewall: Huawei E1000e, database application servers and disk arrays should be put behind the shield of the hardware firewall.

The disk array optical switch: full-Gigabit optical port Switch.

The access-layer switch: Huawei S9312 Gigabit Switch.

1 Traffic Collection and Aggregation

A program is developed to automatically download the originally collected data files of NetFlow from the IP Integrated Network Management via FTP. Data in the files will be aggregated into a new file.

2 IP Address Collection and Classification

Collecting IP addresses is essential to the System. Inaccurately collected IP addresses may bias the results of the analysis. The collected IP addresses come from the network of the Province or other sources (China Unicom, China Telecom, and CTT outside the Province).

1 Collection and classification of IP addresses from the network of the Province:

The IP addresses of CTT in the Province are classified according to cities and uses. Address tables of all cities, NDC and IDC address tables are obtained.

2 Collection and classification of IP addresses from other sources:

AS numbers of all operators, service providers and organizations are collected via the operator headquarters or other provincial companies. BGP routing lists of backbone routes are collected. Then, a program is developed to associate IP addresses with the same AS numbers with the corresponding operators to form the classification tables of other IP addresses. [4]

The large segments of addresses usually contain many small segments of addresses because the data in the BGP routing lists are highly complex. Moreover, the large and small segments of addresses may belong to different institutions. Thus, the program is difficult to process and requires skills.

The IP addresses are extremely variable and need to be updated regularly.

Table 1. Classification table of AS numbers.

Autonomous Domain	Description of the Peer	Natur
4538	Cernet	Peering
9808	China Mobile	Peering
9308	21Vianet	Peering
17429	Bgctv	Peering
9306	Nowec	Peering
9389	Hebei.com.cn	Peering
17430	Hebei.com.cn	Peering
7497	China Science and Technology Network	Peering
17964	Teletron	Peering Extra ports are added for charging fees on Nov.25,2009 and Teletron released some telecom routes
9800	China Unicom	Peering Since its merger with China Netcom, China Unicom has adjusted this part of the traffic and it is no longer the peering exit.
4859	China Unicom customers	Connected via China Unicom
9298	China Unicom customers	Connected via China Unicom
9391	China Unicom customers	Connected via China Unicom
9484	China Unicom customers	Connected via China Unicom (Inner Mongolia Mobile, now connected to China Telecom)
9789	China Unicom customers	Connected via China Unicom
9814	China Unicom customers	Connected via China Unicom
9934	China Unicom customers	Connected via China Unicom

3 Database Establishment and Table Creation

The desired database and tables are created according to the projects and contents to be analyzed. The System creates the traffic database, basic data tables, IP address tables of operators, IP address tables of branches of all cities, IP address tables of NDC servers, port classification tables, and operator name tables.

After the tables are created, the collected basic data (e.g. IP address classification tables) is input into the corresponding tables. In order to improve input efficiency, LOAD is adopted to directly load the files into the database tables. [5]

The structures and properties of the tables are specified. Examples of the designed tables are presented as Table 3, 4, 5, 6, 7 and 8:

Table 2.   BGP router table.

*>i	1.51.16.0/20	222.39.160.2	2000	0	9394	4538i
*>i		222.39.160.1	2000	0	9394	4538i
*		10.100.100.11	1000	0	9394	4538i
*		10.100.100.11	1000	0	9394	4538i
*>i	1.51.144.0/20	222.39.160.2	2000	0	9394	4538i
*>i		222.39.160.1	2000	0	9394	4538i
*		10.100.100.11	1000	0	9394	4538i
*		10.100.100.11	1000	0	9394	4538i
*>i	1.51.168.0/21	222.39.160.2	2000	0	9394	4538i
*>i		222.39.160.1	2000	0	9394	4538i
*		10.100.100.11	1000	0	9394	4538i
*		10.100.100.11	1000	0	9394	4538i

Table 3.   T_U_OperatorIPAddress (IP address tables of operators).

No.	Column	Type	Length	NULL	Default	PK	FK Table	Description
1	ID	int	4			Yes		
2	IPAddress	varchar	50					
3	MASK	varchar	50					
4	OperatorNameID	int	4					Associate with the operator name tables
Total			108					

Table 4.   T_U_OperatorName (operator name tables).

No.	Column	Type	Length	NULL	Default	PK	FK Table	Description
1	ID	int	4					
2	OperatorName	varchar	50					
Total			54					

Table 5.   T_U_NDCIPAddress (IP address tables of NDC servers).

No.	Column	Type	Length	NULL	Default	PK	FK Table	Description
1	ID	int	4			Yes		
2	IPAddress	varchar	50					
3	MASK	varchar	50					
4	NDCNameID	int	4					Associate with NDC name tables
Total			108					

Table 6.   T_U_NDCName (NDC name tables).

No.	Column	Type	Length	NULL	Default	PK	FK Table	Description
1	ID	int	4					
2	NDCName	varchar	50					
Total			54					

Table 7. T_U_CityIPAddress (IP address tables of cities).

No.	Column	Type	Length	NULL	Default	PK	FK Table	Description
1	ID	int	4			Yes		
2	IPAddress	varchar	50					
3	MASK	varchar	50					
4	CityNameID	int	4					Associate with city name tables
Total			108					

Table 8. T_U_CityName(city name tables).

No.	Column	Type	Length	NULL	Default	PK	FK Table	Description
1	ID	int	4					
2	CityName	varchar	50					
Total			54					

4 Input of the Traffic to the Database and Data Processing

The NetFlow traffic data is loaded into the corresponding basic tables of the database five minutes after data aggregation (Table 9). LOAD is used to load the files directly for higher efficiency. The format of some original traffic data is different from the actual data representation. For example, the time of the data stream is represented by the time stamp, but the IP address is expressed using the decimal scale. Thus, format conversion has to be performed before loading the data. There are many approaches to data conversion. For instances, we can add the conversion function to the LOAD statement, or convert the data format by using the triggers or creating new tables. The use of triggers is preferable.

In order to improve the query efficiency after the data is loaded, we add to the basic table the fields of affiliation of source and destination addresses, regions and ports. We use the source and destination addresses, devices' IP addresses and port numbers in the original data to search for the corresponding address and port tables. Then we can identify the affiliation of the source and destination addresses, regions and exits, update the column data accordingly, and obtain the details of the traffic. (Table 10)

2.3 Test of the proposed system

After the System is developed, the NetFlow collection sites are set up at the exits of the provincial and backbone networks of the core routers at nodes 1 and 2 and also at the corresponding ports with which NDC is connected. The collection ratio is set at 1,000:1. The NetFlow data collected by the IP Integrated Network Management is processed and loaded into the database by the System. We carry out analysis of the traffic at the exits of the backbone and provincial networks and the traffic of NDC at busy hours on Nov.26.

Table 9. Basic tables of the traffic (truncated fields).

Date	Time	InPort	OutPort	LiuBao	Flow	ProtocolSource	ProtocolDest
2013-9-1	20	24	21	1	95	161	27286
2013-9-1	20	22	0	2597	3409570	9002	9002
2013-9-1	21	22	0	1	103	161	61839
2013-9-1	21	22	0	1	95	161	9137
2013-9-1	22	24	12	1	54	1457	80
2013-9-1	22	22	0	1	104	161	56042
......	......	......	......	......	......	......	......

The inbound traffic of branches of CTT in each city largely consist of the inflow of the backbone network, provincial network and NDC, i.e. the traffic shown in the three traffic analysis tables above. The sum of the three items of the traffic at busy hours approximates to the traffic at ports of branches in all cities at busy hours. The inbound traffic of the System at busy hours on Nov. 26, Nov. 29 and Sept. 6 is compared with the traffic of the ports at busy hours from the IP Integrated Network Management. The results are shown in Tables 11, 12 and 13.

The analysis results above reveal that there is a deviation of 3.87% between the total traffic of the System and the IP Integrated Network Management. For branches of CTT, the errors are mostly below 5% and seldom in the acceptable range 5%-10%. This shows that the results of the proposed System are accurate and reliable.

Table 10. Flow tables ofi in busy traffic at the entry and exit of the backbone on Nov.26, 2013.

ChinaTeletron (GB)	ChinaUnicom (GB)	ChinaMobile (GB)	Outside the province(GB)	Peering (GB)	Inside the province(GB)	International and other(GB)	Total (GB)
15.88	15.58	110.81	2.36	76.29	976.10	59.36	1256.38
2.86	4.07	36.75	3.51	20.04	279.70	13.19	360.12
54.97	58.23	179.41	4.57	122.24	1316.22	99.53	1835.17
3.61	4.28	38.34	0.64	21.93	279.99	13.73	362.51
51.06	51.39	250.29	33.25	181.41	1567.64	200.72	2335.77
13.68	16.53	76.14	3.23	46.95	602.93	33.71	793.17
22.15	60.90	242.01	4.11	153.13	2093.34	119.44	2695.08
29.82	69.02	319.35	19.02	239.26	2380.66	224.40	3281.53
4.18	19.19	75.07	3.42	57.35	743.24	47.22	949.68
9.41	34.30	172.73	6.91	131.18	1339.37	204.88	1898.77
0.68	0.35	6.36	0.77	11.11	65.65	0.17	85.09
0.42	0.28	0.06	0.08	0.09	0.19	0.18	1.31
208.72	334.11	1507.31	81.88	1060.99	11645.04	1016.53	15854.58

Table 11. Comparison of traffic from the proposed on Nov.26, 2013.

Names of Branches	Data of the System					Errors
	Traffic at the entry and exit of the backbone (GB)	Traffic at the entry and exit of the provincial network (GB)	Traffic at the entry of NDC (GB)	Sum of the three items (GB)	Traffic at the ports for the IP Integrated Network Management (GB)	
Branch 1	1256.38	1225.16	442.498	2924.04	3,000.45	2.55%
Branch 2	360.12	286.17	142.901	789.19	837.07	5.72%
Branch 3	1835.17	1406.51	460.782	3702.46	3,819.99	3.08%
Branch 4	362.51	329.84	104.802	797.16	789.78	-0.93%
Branch 5	362.51	329.84	104.802	797.16	789.78	-0.93%
Branch 6	793.17	698.83	277.928	1769.93	1,838.08	3.71%
Branch 7	2695.08	1726.88	760.469	5182.43	5,295.49	2.14%
Branch 8	2695.08	1726.88	760.469	5182.43	5,295.49	2.14%
Branch 9	949.68	541.01	307.331	1798.02	1,964.80	8.49%
Branch 10	1898.77	1808.72	329.84	4037.33	4,266.57	5.37%
Total	15854.58	12831.40	4148.1	32747.5702	33,806.91	3.13%

Table 12. Comparison of traffic from the proposed on Nov.29, 2013.

| Names of Branches | Data of the System | | | | | Errors |
	Traffic at the entry and exit of the backbone (GB)	Traffic at the entry and exit of the provincial network (GB)	Traffic at the entry of NDC (GB)	Sum of the three items (GB)	Traffic at the ports for the IP Integrated Network Management (GB)	
Branch 1	1084.62	1132.90	284.38	2501.90	2,596.36	3.64%
Branch 2	328.63	292.49	109.74	730.86	780.08	6.31%
Branch 3	1550.92	1312.87	326.25	3190.05	3,366.39	5.24%
Branch 4	288.54	274.17	66.38	629.09	655.55	4.04%
Branch 5	1896.97	1813.96	336.81	4047.74	4,178.81	3.14%
Branch 6	706.31	665.48	176.37	1548.17	1,614.29	4.10%
Branch 7	2236.89	1668.39	561.97	4467.25	4,693.67	4.82%
Branch 8	2814.70	3048.18	708.19	6571.06	6,696.56	1.87%
Branch 9	844.45	597.02	216.39	1657.87	1,775.56	6.63%
Branch 10	1666.25	2069.80	313.75	4049.79	4,148.61	2.38%
Total	13418.30	12875.26	3100.22	29393.77	30505.86765	3.65%

Table 13. Comparison of traffic from the proposed on Sept.6, 2013.

| Names of Branches | Data of the System | | | | | Errors |
	Traffic at the entry and exit of the backbone (GB)	Traffic at the entry and exit of the provincial network (GB)	Traffic at the entry of NDC (GB)	Sum of the three items (GB)	Traffic at the ports for the IP Integrated Network Management (GB)	
Branch 1	1185.83	970.96	256.02	2412.81	2,555.83	5.60%
Branch 2	388.06	244.05	105.75	737.86	791.22	6.74%
Branch 3	1704.99	1139.84	398.36	3243.19	3,383.62	4.15%
Branch 4	329.66	250.60	77.48	657.75	676.83	2.82%
Branch 5	2170.57	1536.33	387.18	4094.08	4,268.31	4.08%
Branch 6	729.20	545.60	186.00	1460.80	1,587.42	7.98%
Branch 7	2345.12	1663.20	551.64	4559.95	4,742.41	3.85%
Branch 8	2974.95	3054.37	743.36	6772.69	7,031.75	3.68%
Branch 9	907.99	636.55	281.84	1826.38	1,937.31	5.73%
Branch 10	1749.00	1965.16	306.32	4020.48	4,279.36	6.05%
Total	14485.36	12006.67	3293.95	29785.99	31254.06011	4.70%

3 CONCLUSIONS

By deploying the System, CTT can obtain accurate data of the traffic and its directions. This is helpful to carry out proper analysis of the Internet traffic in the Province, understand the loads of the network and the bandwidth usage of important services, and accurately identify the quantities and types of the traffic over the backbone and provincial networks. Moreover, it enables CTT to statistically analyze the network and the services over the network.

The proposed System provides a powerful tool for optimizing the network and carefully managing the routes.

REFERENCES

Bao hzTie, Liu Shu-fen. Network Fault Management Formal Description Based on Communication Sequential Processes (CSP) [J]. Journal of Jilin University Engineering and Technology Edition, 2007, 37(1).

Xuo Yian-xun. Research and Application of Traffic Identification Technology on IP Network [D]. Chongqing University, 2007.

Liang Rui. Research and Implementation of Network Traffic Identification Model [D]. University of Electronic Science and Technology, 2010.

Wang Lin. Research on Classification of Intelligent Applications for High-Speed Network [D]. Jinan University, 2008.

Ka Hung HUI and OnChing YUE, Mobile Technologies Centre (MobiTeC), The Chinese University of Hong Kong, Analysis of a Distributed Denial- of- Service Attack.

Computational Intelligence in Industrial Application – Ling (ed.)
© 2015 Taylor & Francis Group, London, ISBN: 978-1-138-02818-0

Teaching method investigation in international education

M.Y. Xiao & Y.F. Nie
Department of Applied Mathematics, Northwestern Polytechnical University, China

G. Crouk
Mathematics Department, University of Alabama Huntsville, Huntsville, AL, USA

ABSTRACT: Higher international education becomes the trend with the step of economic globalization, especially in China, where more international students come with different cultures and different backgrounds of basic knowledge, and then the appropriate teaching methods with innovations are mandatory to fit the case. In this work, innovation in mathematical course teaching is first proposed and discussed based on experience in Northwestern Polytechnical University in the aspects of teaching content, teaching form and evaluation form. The teaching practice shows that low theory, large practical content combined with directed study and various evaluation form can obtain a positive and competitive efficiency in different fields of study such as motivation, creativity, practical ability and team cooperation capability etc.

KEYWORDS: Teaching method; International program; Education reform; Teaching innovation.

1 INTRODUCTION

1.1 Overview of current educational systems

The needs of today's students in industrial applications (a solid background in mathematics and excellent problem solving skills), combined with an increasingly competitive market and the demand for lifelong learning, presents a host of challenges and opportunities to higher education institutions, particularly universities. However, the traditional teaching methods like theory content, indoctrinating teaching, mono-evaluation form, etc. are still applied in normal teaching processes. This leads to the high score low capability phenomenon with which we are very familiar. It's even harder for most of them to continue academic careers in the future. Most of graduates who go to work in the industrial sectors find that their education did not provide the practical knowledge and capability that they really need. In order to ameliorate this relationship, teaching innovation becomes imperative to achieving success in the case occasioned by dynamics of globalization (Ibrahim D. et al. 2012). Innovations usually come with original ideas. However, ideas don't have to come randomly. That is why academic environment can conduce to the expansion of innovation with support of the whole education process. The European Union accepted a strategic program based on a strong support of innovations with the aim to create with the EU Innovation Union scope (EP Commission, 6.10.2011). The Chinese Government is also try its best to improve education systems such as recruitment of foreign students to study in china, but also invitation of academic staff to come to work in Chinese universities. It is necessary to modernize our educational systems at all levels. Top quality is more important than ever before. We need more universities that serve at the global level, teaching a broader range of skills and bringing in the best talents from abroad (Dana et al. 2013). As a mathematical teacher with five years' study experience in EU. I am concentrating on international mathematical course teaching reform. In the following sections, NPU's education system is first introduced and teaching reform and innovation with mathematical course is then discussed.

1.2 Background introduction on improvement of traditional educational systems

In recent years, more and more foreign students are coming to study in China. So also in my university (Northwestern Polytechnical University, NPU). In order to improve the current education situation and foster more high-end talent, an experimental platform for exploring new teaching systems has been set up at NPU China. One international class has been established with mixed-nationals, consisting of about 40% of Chinese student, and 60% Of non-Chinese students. Four major departments currently participate in the program. Classes are taught in English with English Textbooks. All non-Chinese students are enrolled automatically in the international program.

Chinese students are allowed to enter the program and take the mixed-nationals classes if they so choose. This gives the Chinese students very valuable experience in innovational teaching methods, exposure to foreign cultures, and opportunities to improve their English language skills.

In order to improve our traditional education system, the following three aspects should be considered:

- Attract abroad foreign students to study in Chinese universities. This enhances knowledge, education, and business exchange among other countries;
- Attract famous foreign professors to teach in our Chinese universities. This is an excellent way to improve teaching quality and knowledge exchange. Chinese nationals with educational and/or teaching experience abroad enhance diversity and ideas for improving teaching methods;
- Send domestic teachers to learn and interchange teaching methods in abroad universities.

With the support of the above advantages, what have we done to innovate the teaching reform? We will use the description of our mathematics courses teaching in the mixed-nationals classes as a particular example.

2 EDUCATION REFORM AND INNOVATION IN MATHEMATICS COURSES OF NPU

After three years' teaching tests in five international mathematical courses. The closed loop proposed is covered in five steps during the teaching process to get the positive performance seen in figure 1.

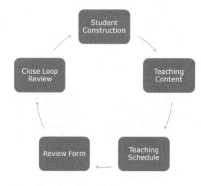

Figure 1. Closed cycle teaching process.

2.1 Student construction

Before exploring a new teaching method, a global view of the character of mixed-national classes is a first necessary step. In the current situation, the great difference of educational background/mathematical basis poses a big problem. The entry level foreign students do not have as strong a mathematical background as the native students in China. They need more computational exercises and applications. Most of them at the undergraduate level are not ready for abstract mathematical thought. Furthermore, they have a complete different way of thinking. In this case, we must offer many remedial mathematics classes so that the average student will have a chance to succeed. Without remedial assistance the material is too difficult for students to finish the course. Even though today's young people are very visually oriented due to their immersion in computer and smart-phone technologies. Thus, the first way to improve the traditional teaching method is to build the remedial course. Then, the teaching system has to be improved and connected in teaching content, teaching form and evaluation form.

Based on the placement of University of Alabama in Huntsville, USA and with the help of lecturer Mrs. Gena Crook. We also made up a similar placement and remedial form.

2.2 Methods proposed to remedial mathematics

2.2.1 Remedial algebra classes
In order to get a good teaching performance, students are required to take a mathematics placement test upon registering for the university. This placement test reveals their current level of algebraic, computational, and Calculus skills. The test score determines the class at which the student should begin mathematical study. For example, a score of 1 places the student in a remedial algebra class; a score of 3 places the student in beginning calculus shown in table 1.

Table 1. Placement table.

Math test score	Level	Equivalent math course
80–100	3	Calculus A
60–79	2	Basic theory
≤ 59	1	Pre-algebra/geometric

2.2.2 Guided problem solving work sessions
In addition to lecture classes, separate "laboratory" sessions are required for the students to receive credit for the course. These "laboratory" sessions are taught by mathematics graduate students as part of their scholarships for graduate study at the university. Students work in groups on problem sets. The graduate student teacher guides their progress, aggressively answers questions, offers hints and suggestions for solving problems, and generally encourages the students.

2.2.3 Free tutoring

Tutoring is offered in two ways. The first is in a designated room in the mathematics department. This tutoring laboratory is staffed all day long by mathematics graduate students. Undergraduate students taking entry level classes may come to the tutoring laboratory at any time and work with the graduate teachers. They can ask questions concerning mathematical concepts and ask for guidance on solving homework assignments. Students are allowed to sit in the laboratory for as long as they like and work on mathematics assignments. Some students choose to work in the tutoring laboratory between classes and during lunch. The second type of tutoring is available in the "Student Success Center." We are offering some version of this concept to our undergraduate students. The hope is that if students can get assistance when their mathematical background is weak, then they have a better chance to succeed and actually graduate with a degree. At the Student Success Center, students make an appointment to study one-on-one with a fellow undergraduate student tutor. These tutors are top-performing undergraduate students, usually tutoring subjects in the major area of study, and are paid a small stipend by the university. This is good for the student tutors as well since they are getting excellent work experience.

2.2.4 Computer problem solving- software programs available online

Many of the textbooks used in our mathematics classes have coordinated problem solving software. Students purchase access on the World Wide Web for a reasonable fee. The instructor can create assignments to accompany the course work and set due dates, all online. Students work their exercises on the computer just as they would on paper. The software automatically gives feedback as to whether the answers are correct or incorrect. Learning tools are also available in the software: step-by-step example solutions, video lectures on the topic, similar exercises for more practice, etc.

We find these online problem solving to be especially suited to today's students who typically have spent a significant number of hours in their childhood doing computer "gaming." The online problem solving software that accompanies our textbooks used are called Web Assign, offered by Advanced Instructional Systems, Inc.; and MyMathLab, offered by Pearson Education, Inc.

2.3 Teaching content

Less emphasis on theory will allow time to teach computational techniques, to teach practical applications of mathematical concepts in real world situations, and to explore the use of technology tools in problem solving. Practical applications and experience with technological tools such as Matlab, Mathematica, Console and Lingo, etc. are extremely valuable to students when they graduate and enter the work force, especially in the industrial and computational fields. Another way to improve course content is to choose textbooks that contain practical application examples, diagrams, graphs, projects utilizing technological tools.

Supplemental materials on the World Wide Web including free video lecture series from prestigious universities around the world are an incredible resource to enhance course content. A very important aspect of changing course content and delivery would be to incorporate a way for students to provide feedback. An evaluation of the instruction method in the form of a survey taken by the students at the end of the course is invaluable for determining which strategies work and which do not. It gives the students the opportunity to tell the instructor what most helped them learn.

2.4 Teaching form/schedule

In order to improve student's innovation, various teaching methods are recommended for teaching students. A similar view is proposed in (Shahida Sajjad, report). In fact, lecture is the direct approach to transfer the knowledge to students. Group discussion, individual presentation and assignments would be the difference. For each activity, we could assign a specific rate of time to make the class more active and positive, seen in Figure 2.

Figure 2.   Various teaching methods.

- Lecture method rated as a best teaching method. It provides all knowledge related to the topic, time saving. Students attentively listen to the lecture and take notes etc.;
- Group discussion is rated as a second best way to give students a chance to listen to other's opinions and express own opinion. They don't have to rely on note learning. It can develop creativity among students;
- Individual presentation can induce students to thoroughly understand the project and increase confidence among students;

- Assignments are a very important step during teaching. They can enhance the ability of students to do research on any relative problem, Assignments facilitate active learning and consolidation of what is learned during lecture time.

2.5 Evaluation/review form

The traditional method of evaluating an individual student performance in a Chinese university is the final examination, and nothing else. In order to better evaluate the studying performance, based on various teaching forms, periodic testing over smaller amounts of material during the course proposed can give most students a better opportunity to perform well. It also increases student motivation to study throughout the session rather than cramming at the end of the course for an exam. Study on a continuous basis such as this enhances learning and retention. For example in an entry level Calculus course, 20% of the course grade is based on class participation and graded homework assignments. 50% is based on periodic chapter tests. The comprehensive final examination counts only 30% of the overall course grade. The following reasonable view is given to different percentages:

- A 20% class participation grade motivates regular attendance. Students are also motivated to prepare for each class with questions about previous assignments. It gives the teacher the opportunity to review concepts and to utilize small group work in class. It is usually easy for students to maintain a high class participation grade. This has the effect of lowering the stress placed on exam scores;
- A 50% part for periodic chapter tests which cover small amounts of related material. Students can concentrate on this material and better assimilate it and relate it to previous concepts. Students usually perform better on these shorter examinations simply because there is less material to study and the stress level is lower;
- A 30% weight for the comprehensive final examination is large enough to ensure that students must have a high level comprehension of the material in order to pass the course. A performance of 70% is in line with widely recognized passing requirements.

3 CONCLUSIONS AND PERSPECTIVES

3.1 Conclusions

An overview of the current situation in higher education has been presented. Possibilities for improving teaching methods have been proposed and have been successfully implemented in the international courses at Northwestern Polytechnical University, Xi'an China.

3.2 Perspectives

One possible improvement for the future is to reduce the gap between foreign students and local students ans to make teaching methods more interesting.

The first strategy is humanistic care built up not only to local students, but also foreign students. In fact, the foreign students have many obstacles to overcome at the beginning. Everything is new for them. They are eager to have a friendly and warm atmosphere during class periods. The importance is just like the Mrs Androulla Vassiliou, European Commissioner for Education, Culture said "Whenever I think about the people who have most inspired or motivated me during my life, I return, without fail, to my student days. I can still remember in particular the professor who inspired me in International and European affairs. His positive influence on me then continues to this day" (Report to the European commission. 2013).

The second one is to give more possible opportunities for students to work in teams and to communicate with classmates. For effective discussion the students should have prior knowledge and information about the topic to be discussed. For example, group projects are a great way to train students to work with other people. This is absolutely essential in today's industrial work fields. It would be beneficial even for teachers to work together and share teaching methods and ideas.

ACKNOWLEDGMENTS

This work is supported by the Higher Education Innovation Foundation of Shaanxi Province (Grant No. 13BY12) and Teaching Reform Foundation of Northwestern Polytechnical University, Xi'an in China (Grant No. 2014JGY14).

REFERENCES

Ibrahim Danjuma & Amran Rasli. 2012. Imperative of service innovation and service quality for customer satisfaction: Perspective on higher education. *Procedia-Social and Behavioral Science* 40: 347–352.
EP COMMISSION: Unie inovací. Sdělení EP, 6.10.2011. Innovation Union.
http://ec.europa.eu/enterprise/policies/innovation/policy/innovation-union/communication/iu_cs.pdf
Dana Likeschová M.A. & Alena Tichá. 2013. Multicultural education, creativity and innovation at universities in the Czech Republic. *Procedia-Social and Behavioral Sciences* 93: 349–355.
Shahida Sajjad. Report: Effective teaching methods at higher education level.
Report to the European commission. 2013. High level group on the modernisation of higher education.

Research and application of database course teaching methods

HuiTing Wu

College of Information Engineering, Chutian College Huazhong Agricultural University, Wuhan, China

ABSTRACT: Database course is mainly systematically introducing the database principle knowledge and the application process of the principle combined with the specific database management system software. It can make students choose the correct development platform in the process of database software development and design databases reasonably through study. Thereby the overall quality of software development is improved. According to the characteristics of the database course teaching contents and the principle of linking theory with practice, the teaching method which uses the style of question guide, task scenario, multi-answer question and case is put forward to preliminary study on classroom teaching of database course.

1 THE PRESENT TEACHING SITUATUON

Database course is a professional basic course of computer specialty and the preferred course which is involved in information processing in many professional. The course requires students to master the basic principles, concepts of database system and application technology of the general database system. It helps the students lay a solid foundation for future work. Now colleges want to transform to the applied universities. It is necessary to pay attention to the cultivation of students' ability in order to bring up applied talents. A database course mainly cultivates the students' ability of the following aspects: data abstraction and modeling ability; analysis and design ability; data system development ability; database configuration and management ability; innovation ability. In teaching practice, teachers form the general process of database theory teaching: the establishment of speculative knowledge-- the explanation of speculative knowledge. The concepts and definitions in the teaching material are often directly showed in a single text. It not only causes that students' enthusiasm is not high to accept new knowledge, and the learning concept thinking is not smooth, and the use is not flexible, but also can't improve the comprehensive ability of students. It is concluded that it mainly exists the following problems in the teaching of database course:

Problem1: The difficult point of database principle is that it is very abstract. Students tend to feel inadequate, especially when studying about the chapters of the relational model and normal form.

Problem2: The knowledge is scattered. Students do not know why to learn one knowledge point and what exactly is the use of the knowledge learning.

Problem3: The teaching methods and means are single. Students are in passive learning if using the traditional cramming method of teaching.

Problem4: Theory is divorced from practice. It makes students appear many problems in the practical application process that the theory and practice operation of internal computer database course are not unified.

2 THE DESIGN OF THE TEACHING PROCESS

2.1 *General instructions*

Teachers must constantly improve and enrich their teaching process and their teaching methods when analyzing the difficulties for students to learn the basic principles of database. In order to make the teaching more close to students and make students more easily understand and grasp the abstract concept, we can try to use the following methods of theory teaching.

2.2 *The method of question guide*

The method of question guide is that the abstract concept is decomposed to set the decomposition of each link to visualize the problem in order to introduce, understand and consolidate the learned concept by means of the process of solving the setting problems. The relation normalization theory teaching is used as an example to try to use the method of question guide to preliminary research to the basic theory of teaching.

2.2.1 Leading to new knowledge with the help of a familiar case

Normal form is a knowledge point which is difficult to understand in the database. How can we motivate students to learn and accept such an abstract and very important concept? Teachers can start from the example of real life to make students realize the necessity of the introduction a new concept. For example, teachers can let the student associate the teaching management systems which they often use. Students can be asked if we want to store the student information, course information and student achievement, how we can store. A student relational scheme is given specifically in the following Student(sno, sname, sex, age, dept, dname, cno, cname, credit, grade). Students can be asked to think about whether the relationship mode is reasonable. Then the teacher takes the students one by one to analyze the existing problem of the Student relational scheme: Data redundancy. For example, when a student takes many elective courses, student's information (sname, sex, age, Dept) will be stored for many times repeatedly. It is data redundancy that data is repeatedly stored in the database; Update anomalies: If a student changes department, the department information will be modified one by one. If there is a little carelessness, some of the records will be missed. Insertion anomalies: If a department just set up and don't have any students, director of the department and the department information can't be inserted. It is insertion anomalies that the data which should be inserted cannot be inserted. Deletion anomalies: If all of the students in a department graduate, we will delete the department information of students and the director of the department and the department information will be also deleted at the same time. It is deletion anomalies that the data which should not be deleted is deleted. What are the causes of these problems? The reason is that all kinds of data exist in one relationship. How to solve? Decompose relational scheme. Decomposition basis: Normal form. New concepts of 1NF, 2NF, 3NF, BCNF are eventually led to through the layers of questions. Students experience the practical application of new concepts in life when booting from these problems. The enthusiasm of students is enhanced through setting such links.

2.2.2 Consolidating new knowledge with the help of the problem analysis

When students learn a new knowledge, it does not mean that they can understand and master accurately. The keywords of new knowledge's connotation, extension and concept are needed to strengthen in the teaching. And students' errors on conceptual understanding are needed to remove. So it is necessary to lead the students to analyze the new knowledge. For example, students often will confuse 2NF, 3NF, BCNF in the concept of normal form. In order to make students master the concept accurately, keywords in the concepts and students' error prone points are needed to analyze the chart type and practice through true or false items. And it will make the concept in the minds of students more clear.

2.2.3 Developing new knowledge with the help of the migration problem

It has been pointed out that the normal form of the database is the specification which is needed to meet in the database design. The database which meets these specifications is concise and well-structured. In the meanwhile, the abnormal operation of insertion, deletion and update will not occur. So can the students master it? The answer is negative. It will make the students know about the new concept to consolidate and develop after the migration of application. So, students need to be allowed to design a teaching management system after learning the relational datum theory. Then it leads to the design knowledge database. When students design according to the six steps of database design, problems of optimizing data model will be encountered in the transformation from E-R model to the relational model. It will link up the knowledge of normal form which is learned before. It makes students know that if I learn the knowledge point, where can be used and how to use. This not only learns new knowledge, but also consolidates old knowledge.

2.3 The method of task scenario

Index and view are the objects of the database. Although it is not very difficult, students can't always understand their use. When introducing this kind of concept, the method of task scenarios is the best. If we let the students to remember in what kind of circumstances it needs to use, the effect of both can be understood naturally. For example, you can use the following scenario to lead to the view of knowledge: In the teaching management system, the storage data of students list contain student ID, name, gender, date of birth, family address, home phone and other information. But the home address, home phone and date of birth are confidential data. Not everyone can see this data. Only certain people can see the columns of data. But anyone can see the name and gender information. In order to solve this problem, please put forward a kind of solution.

In order to solve the above problems, view can be used. If we want to restrict the data which the user can use, the view can be used. For example, the user can access to some data to query and modify, but the rest of the table or database is not visible. Students can be allowed to design a small task next. Task 1 understands view of the creation and use. Task 2 is using the view to query the student information about the computer science department. Task 3 is deleting the record of ID 001 in the view. The final task is analyzing tasks and giving the solution through the use of T-SQL statement. It is easier for students to accept than directly introducing the view definition.

2.4 The method of multi-answer question

Multi-answer question is conducive to strengthen the training of students' thinking. The practice of multi-answer question in the teaching is beneficial to develop students' thinking, improve students' comprehensive use knowledge skills and guide the students to grasp the flexible connection between the knowledge. For example, the student gender field in the table is required to use only "male" or "female" two values. Then the two different methods of the constraint and trigger can be used to achieve. Firstly a constraint is realized and the result is demonstrated for students. Then a trigger is realized and the result is demonstrated. Finally, the two together is realized, and then the result is demonstrated. Students will find that not only constraint's statement is simpler than a trigger, but also constraint's priority is higher than a trigger. Through this comparison, we can lead the students to the conclusion: When the treatment is related to the integrity of the data content, the constraint is simpler. But when the function which the constraint supports can not meet the functional requirements of application, the trigger is extremely useful. After comparing and summarizing, students can have a more profound understanding and memory in the knowledge point.

2.5 The method of case teaching

Case teaching is very suitable for database course teaching. It introduces the basic principle and method of the database through the analysis of information system typical case which is based on a database. For example, in the teaching process, teaching management database has always been as the case to run through the whole teaching process in the database. When introducing the relational model, relational algebra and SQL language, basic principle and basic operation are introduced based on the case. In the introduction of database design, teaching

management system is finally designed through the demand analysis, concept structure design, logic structure design and model refinement based on the case. In the teaching, multiple cases are not advocated, because the explanation of one case can make students proficient. It can make the students have a more thorough understanding of the database design process through the analysis and the design of the case. It also helps students lay a solid foundation for the future design work.

3 COMBINATION OF INSIDE AND OUTSIDE CLASS

The teacher grasps knowledge goal, ability goal, the difficulty and the knowledge frame in the classroom. It requires students to extend the practice of class content outside the classroom. Extended practice can be divided into two types: One is to consolidate the classroom teaching such as the case which carries out teaching management system from beginning to end. The teacher should be targeted to lay out some exercises to let the students in the practice independently outside of the classroom. So that the students can not only consolidate the knowledge, but also find and solve problems timely to achieve the good consolidated effect. Another extension is that the teacher puts database cases which are related to enterprise demand and some design topic into platform with the help of an auxiliary teaching platform. Students can do extended training with the help of platform at any time. They can also discuss and communicate with classmates and teachers in the platform. So it achieves the purpose of learning in practice.

4 CONCLUSION

We try to introduce the method of question guide, task scenario, multi-answer question and case on the teaching of database. And the knowledge is consolidated and extended through the combination of inside and outside class. Practice shows that using these methods can significantly improve the quality of teaching in the database class. The students understand the basic principle and method imperceptibly through the teacher's guide and master data operation in a relaxed and enjoyable experience. In a word, teacher use the different ways flexibly for different knowledge. Through the teaching of elaborate design, it can make the students learn more positive, think smoother, understand more accurately, apply more flexible.

REFERENCES

Chris Bunch,Navraj Chohan,Chandra Krintz.2010.Key-Value Data stores Comparison in App Scale. UCSB Tech Report,17,54~58.

Mary Dugan. 2011."Database of the Week": Successfully Promoting Business Databases to Faculty. Journal of Business & Finance Librarianship, 16,159~166.

Sabah Currim, Sudha Ram, Alexandra Durcikova, Faiz Currim.2014. Using a knowledge learning framework to predict errors in database design.Information Systems,40, 11~31.

Baha Sen, Emine Ucar, Cosmina Ivan. 2012. Evaluating the achievements of computer engineering department of distance education students with data mining methods. Procedia Technology, 1,262~267.

Computational Intelligence in Industrial Application – Ling (ed.)
© 2015 Taylor & Francis Group, London, ISBN: 978-1-138-02818-0

Application of multimedia and network in college English translation teaching

XiYan Wang & TianHao Wang
Foreign Language Institute, Northeast Dianli University, Jilin, China

ABSTRACT: Translation is an important language skill, but it is not emphasized in the college English teaching. Based on the network and multimedia, the translation teaching in college English could be reformed to cultivate the qualified translators in the non-English majors.

KEYWORDS: Multimedia, Network, College English translation teaching.

English is the most widely used language in the international communication nowadays. As the globalization develops, translation is the important media between China and the outside world. Many companies require graduates who are good at the professional skills as well as translation in English. The college English course should pay more attention to the translation teaching assisted by the multimedia and the Internet.

1 CURRENT STATUS OF COLLEGE ENGLISH TRANSLATION TEACHING

As the teaching reforms in college English is developing in recent years, the teaching strategies and the skills of college English teachers improve a lot. However, the translation teaching is relatively weak, which is discussed in the following details.

1.1 *The minor emphasis on the translation teaching*

There are many disadvantages in the traditional translation teaching. Many teachers and students consider the translation ability as the result of the improvement in reading, writing, listening and speaking. The translation teaching doesn't need to take some special attention with the introduction of the relevant theories and the skills. The content of translation teaching is mainly on the phrase translation and some grammatical rules. These ideas of the teachers and students should be updated.

There are five classes for each unit in college English teaching plan in the most colleges. One is for the viewing, listening and speaking course. The other four classes are for the comprehensive reading and some exercises. There is no special course for translation. The translation teaching is always mixed with the translation exercises. There is no special class for the translation teaching and relevant exercises. Translation is subordinate to the comprehensive reading classes.

The proportion of translation in College English Tests Band 4 and Band 6 is much less than listening and reading, which results in the fact that students and teachers pay less attention on translation. Some students consider the translation exercises as redundant when they prepare for these exams.

1.2 *The improper teaching materials*

The available teaching materials in the market are various. However, the main content of these books is about the western culture. We can hardly find any passage about the Chinese traditional culture. This phenomenon leads to the less input of Chinese culture in English. Many teachers and students feel helpless when they come to the translation about the history, culture, economy, social development of China.

The exercises of translation in the teaching materials are designed to test the key words, key phrases and the important sentence patterns. There is no reference to the theory and skills in translation. Without the proper direction of the translation theory and skills, students could hardly adapt themselves to the translation part in CET 4. [1]

The appropriate introduction of translation theory is necessary in the teaching materials. After learning the theories in translation students could apply these theories into the different situations when they become translators sometimes after graduation. They should have a system of translation skills in mind to cope with the complex environment. In a word, the reform of the teaching materials is necessary.

2 THE APPLICATION OF MULTIMEDIA IN COLLEGE ENGLISH TRANSLATION

Applying the network and multimedia in the college English translation classes is a new educational device. It can construct a student-centered curriculum environment. The high efficiency of the network and multimedia could make the translation classes active. The students could be vigorous in class. The application of multimedia in college English translation could be done in the following aspects:

2.1 Making the courseware with multimedia

The courseware disks are mostly ready in the teaching materials. According to the different levels of students and the equipment, these disks are usually lack of the systemic introduction of translation. Therefore, it is necessary to make the proper courseware with the multimedia by teachers. This courseware should be designed according to the teaching plans, the English levels of different classes, and the teaching content. The purpose of making the courseware is to cultivate the autonomic learning ability of the students and to ensure the proper exercises and the improvement in translation. [2]

In these courseware, the pictures, the sound, the animation and the videos should be made fully use of to stimulate the students' brains and to make the abstract knowledge into concrete and interesting. The content of the courseware should be focused on the introduction of the translation theories and the details of translation skills. The Chinese traditional culture and the common sense in the western society should also be explained in details.

This new courseware is convenient in two aspects. On the one hand, this new teaching form could save a lot of teaching time and is flexible in teaching the students of different levels. The boring translation texts are designed in the modernized forms. All this could improve the translation ability of the students in such a relaxing environment. On the other hand, the teachers could design many chained addresses in the courseware. These chained addresses are chosen by the students according to their different interests and levels of English. Under the direction of the teacher the students could do the translation task in a lively language environment.

2.2 More information to the class by the multimedia and network

Firstly, interest is the best teacher always. The motivation of the learners contains the attitude, which is the key in their study. For the students from the non-English majors, their interests in English become less when they find the traditional English classes are boring. They would feel nervous when they find translation difficult in front of the teachers. Adding more information to the class by multimedia could arouse this interest in the study.

Secondly, the time of English classes is so limited that students can't finish enough exercises in class. Teachers could put the unfinished tasks or the homework on the blogs or bbs by network. Students could also revise the teaching content by network in their free time.

2.3 More cultural input

The second language education should put more emphasis on culture. By making use of the multimedia teachers could show the students the background information about the history, the geography, the life style, the literature, and the traditions which can hardly be discussed about clearly in class. There are many cultural connotations in a language. Understanding the nonverbal information in a conversation or film is also very important. Learning the background information is helpful for the students to cultivate the intercultural communication competence.

The details of culture are difficult to be described clearly in the traditional English classes. However, it is different in the multimedia assisted English classes. Pictures and videos could be used in class to show the concept vividly. Students understand the complex details easily and hardly to forget.

3 MATTERS THAT NEED ATTENTION

First, teachers should adapt themselves to the new teaching ideas towards translation. Translation should be emphasized in college English teaching. It is a useful skill when students work in the society. Translation is as important as other abilities in English, such as listening, speaking, reading and writing.

Second, in the courseware, the size of the letters should be clear to the students. The time of the videos and animations should be set by the teacher in advance. [3] Film editing is a technical task for most teachers. English teachers should improve themselves in relevant field.

Third, the translation teaching should emphasize the autonomic learning by the students. As the technology grows rapidly, students will come for the latest information when doing research or surfing online. As a translator in the IT times, the traditional ways to collect information in the familiar fields are not efficient. Students should learn how to collect information in the network, how to use the E-dictionary, how to maintain the translation software and websites.

4 CONCLUSION

The application of multimedia and network in college English translation teaching is relatively new in the teaching reforms. Translation is an important part in English teaching. To cultivate the language ability and the autonomic learning is the purpose of college English teaching. Because of the limited space here, the author discussed the current status of the translation teaching as well as the application of multimedia in the translation classes in this thesis. More researches should be done in the detailed teaching strategies and other aspects.

REFERENCES

Chen Fen, *Enlightenment in College English Translation Teaching Based on the New Items in the Latest CET-4*, [J] English Square, 2014.8.

Huang Hong, *A Probe into New Pattern of College English Translation Teaching by Multimedia and Internet Techniques*, [J] Journal of Changsha University, 2006.5.

Liu Jian, *A Probe into the Application of College English Translation Teaching by Multimedia*, [J] Journal of Kaifeng Institute of Education, 2013.9.

Computational Intelligence in Industrial Application – Ling (ed.)
© 2015 Taylor & Francis Group, London, ISBN: 978-1-138-02818-0

Application of multiple intelligences theory to the individualized teaching of college English in the multimedia and network environment

XiYan Wang & TianHao Wang
Foreign Language Institute, Northeast Dianli University, Jilin, China

ABSTRACT: Multimedia and the network have been widely used in the college English education in recent years. There are some advantages and limitations in the multimedia assisted college English class. According to the Theory of Multiple Intelligences, there are more effective ways to improve the teaching effect as well as the students' English level.

KEYWORDS: Individualized teaching, Multimedia and network, College English.

Nowadays, when quality education is advocated, individualized teaching is one of the topics which aroused great interest both in the educational theory and its practice. Based on the educational reforms in recent years, College English is the course which could be taught in the multimedia and network environment in most universities. However, the traditional teaching strategies are still taking up too much time in class, which causes many students lack of interest and curiosity. According to the Multiple Intelligences Theory, teachers could apply many effective teaching strategies with the multimedia and network to improve the teaching effect.

1 THE MULTIPLE INTELLIGENCES THEORY

This theory was first put forward in the book *Frames of Mind: the Theory of Multiple Intelligences* in 1983. As a psychologist, Howard Gardner of Harvard has identified eight distinct intelligences. They are verbal – linguistic intelligence, logical – mathematical intelligence, musical – rhythmic intelligence, visual – spatial intelligence, bodily – kinesthetic intelligence, interpersonal intelligence, intrapersonal intelligence and naturalistic intelligence. This theory could give a hint to the phenomenon that students possess different kinds of minds and therefore learn, remember, perform, and understand in different ways.

According to this theory, we are all able to know the world through language, logical-mathematical analysis, spatial representation, logical thinking, and the use of the body to solve problems or to make things, an understanding of other individuals, and an understanding of ourselves. Where individuals differ is in the strength of these intelligences. In the ways in which such intelligences are invoked and combined to carry out different tasks, solve diverse problems, and progress in various domains.

Gardner says that these differences challenge an educational system that assumes that everyone can learn the same materials in the same way and that a uniform, universal measure suffices to test students learning. Indeed, as currently constituted, our educational system is heavily biased toward linguistic modes of instruction and assessment and, to a somewhat lesser degree, toward logical-quantitative modes as well. Gardner argues that a contrasting set of assumptions is more likely to be educationally effective. Students learn in ways that are identifiably distinctive. The broad spectrum of students – and perhaps the society as a whole – would be better served if disciplines could be presented in a number of ways and learning could be assessed through a variety of means. [1]

At first, it may seem impossible to teach in all learning styles. However, as we move into using a mix of media or multimedia, it becomes easier. As we understand learning styles, it becomes apparent why multimedia appeals to learners and why a mix of media is more effective. It satisfies the many types of learning preferences that one person may embody or that class embodies. A variety of decisions must be made when choosing media that is appropriate to learning styles, for example, visual media, printed words, sound media, motion, color, realia, instructional setting, learner characteristics, reading ability and categories of learning outcomes.

2 THE STATUS OF COLLEGE ENGLISH TEACHING IN THE MULTIMEDIA AND NETWORK ENVIRONMENT

The multimedia and network is very important in the modernized instruction. Making use of the advanced technology in college English teaching, the teacher could improve the instructional efficiency as well as the students' English level. However, every coin has two sides. The multimedia assisted English teaching is no exception. The status implies that there are some limitations as well.

2.1 Some advantages of the multimedia assisted English class

Firstly, the multimedia assisted English class requires the English teachers should learn some basic knowledge about the computer science. In making the multimedia courseware, teachers should grasp some skills to show their teaching characteristics. Some colleges hold the specialized lectures for this special purpose. Some young teachers learn by themselves and finally do well in it.

Secondly, the multimedia assisted English class requires the innovative ideas of the teachers. Teachers should think a lot about how to attract the students' attention by this new technology. They must make the class vivid and lively. To arouse the students' interest and curiosity, teachers should sum up the experience and explore as many new ideas as possible.

Thirdly, the application of multimedia improves the teaching effect of the college English class by creating a harmonious atmosphere. It is a big problem for many teachers that how to create an active atmosphere in a class. Teachers could make use of the video, the audio, the text and the animation to transfer the teaching directions. This teaching strategy is easier to understand and students are often active in such a relaxed environment. Some multimedia could be shared by all teachers and students, which is an important aspect. The courseware could be saved for long and be learned by other teachers. The best resources online offers the students, many choices to learn English. They could choose what to learn according to their own interests. [2]

2.2 Some limitations of the multimedia-assisted English class

Firstly, there are some teachers who independent on the multimedia while teaching. They use the screen as the blackboard to show the PowerPoint courseware, one page by another until the class is over. Students feel rather tired and boring in class. They don't know the emphasis and the difficulties of the lesson. Students are passive learners in class.

Secondly, there are less communication between the students and teachers. Teachers pay more attention on the design of the courseware and neglect the activities with the students. In this way, students pay much attention on the screen rather than the teacher. Students are not the center of the class, nor is the teacher.

Thirdly, all the classes are taught with the same courseware. Ignoring the differences of the students, many teachers use the disks with the books as a uniform. Some classes may have a good response to the content in the disks. Some classes may not follow the disks because of their limited English performance. Teachers should design the individualized courseware by the multimedia based on the different degrees of the students' English level.

3 APPLICAITON OF MULTIPLE INTELLIGENCES THEORY TO THE COLLEGE ENGLISH TEACHING IN THE MULTIMEDIA AND NETWORK ENVIRONMENT

Personalized teaching is a relatively new educational idea applied in college English in the university. Teachers should take the individual differences of the students into consideration. Different needs of the students are met by different teaching strategies. In the students' eyes, the teacher is not the center of the class anymore. The purpose of the teaching is to let the students choose the learning materials and the teaching speed according to their needs and characteristics. Therefore, individualized teaching could promote the students in the overall development model without losing their own characteristics.

3.1 The diversified teaching models

In Howard Gardner's opinion, the purpose of education is not the instruction of knowledge, but to explore potential ability and to direct these intelligences. The first step is to make the students realize their own advantages. Then it comes to how making use of these advantages to strengthen their beloved intelligences. The teaching models of the college English could be the oral instruction, self-learning with the help of multimedia and network and other independent learning models. The teachers could choose the proper models according to the need of the teaching activities. The Powerpoint courseware, the pictures, the cards, the role play, the drama play are some effective and interesting activities in class.

3.2 The differences of the instructional objectives

The instructional objectives are the start and the end of the teaching activities which play an important role in the instructional evaluation. Under the direction of the teaching program, teachers could set the different and flexible instructional objectives and teaching strategies according to the mood, interest, and mind state of the students of different English levels. There should be a uniform standard for all the students to reach after learning one unit or passage. There should also be some space for the better students to develop in some further directions.

3.3 The personalized development opportunities

Every student should have his own intelligence advantages and his learning style. Teachers should know their interests and intelligences and offer them the individualized opportunities to develop their personalities and potentials. Teachers should allow the different ways and strategies to solve the problem. Students are encouraged to express their different views. After being questioned, they could think about it from other aspects on their own. This could help them to be deep in thought and further in development. [3]

3.4 The individualized evaluation

According to the theory, the system of evaluations should base on the individual. It emphasizes the diversified inspects and evaluation criteria. The exam of a course should be reformed. As to the evaluation criteria, the learning intelligence on this course is not enough. Their development during learning this course is as important as the score on the final exam. The change in their attitude and emotion in the teaching activities should be paid close attention to. Everybody has his advantages and disadvantages. The cognitive way is unique of each person. Teachers should help students to transfer their strong or advanced intelligences to the weak ones in order to promote their all-round development.

4 CONCLUSION

The application of the Multiple Intelligences Theory could offer a new perspective in the college English teaching reform. The realization of the theory should be in the individualized teaching strategies in order to offer all the students of different intelligences the coequal education. Because of the limited space here, the author discussed the current status of the college English teaching as well as the application of the Multiple Intelligences Theory to the actual teaching models in this thesis. More researches should be done in other innovative teaching strategies and other aspects.

REFERENCES

Gardner H, Frames of Mind: the Theory of Multiple Intelligences [M]. New York: Basic Books, 1983.

Li Hongmei, On Pros and Cons of the Multimedia Technique in College English Teaching and the Settlement,[J] Journal of An' shun College, 2007.9.

Wang Honglei, A Probe into the Application of Individualized College English Teaching According to the Theory of Multiple Intelligences, [J]Contemporary Educational Science, 2009(03).

Computational Intelligence in Industrial Application – Ling (ed.)
© 2015 Taylor & Francis Group, London, ISBN: 978-1-138-02818-0

Indirect speech acts applied to the multimedia-assisted college English listening teaching

XiYan Wang & TianHao Wang
Foreign Language Institute, Northeast Dianli University, Jilin, China

ABSTRACT: Indirect speech acts are frequently used in verbal communication. The interpretation of them is of great importance in order to meet the demands of the development of students' language skills and communicative competence. This paper, with a focus on the analysis of the functions of indirect speech acts, is intended to emphasize the importance of the application of functional strategies of indirect speech acts into the students' listening activities, and its greatest effects on improving students' oral communication abilities, thus providing some hints to foreign language teaching.

KEYWORDS: Indirect speech acts, Multimedia, College English listening teaching.

1 INTRODUCTION

The listening and speaking ability is very important in English study. College English teaching should not only meet the needs of the social development, but also the requirement of the students. All the college students in China want to pass the CET 4 and CET 6, however, listening is a big obstacle for them. As English teachers, we should put more emphasis on cultivating the listening ability in order to improve their pragmatic ability.

Pragmatics is a subfield of linguistics and semiotics which studies the ways in which context contributes to meaning. Pragmatics encompasses speech act theory, conversational implicature, talk in interaction and other approaches. Speech Act Theory, advanced by J. Austin, promoted the development of Ordinary Language Philosophy to further study, while the Indirect Speech Act Theory founded by J. Searle, who perfected the Speech Act Theory. It expounded the meaning of language from the perspective of use and communication, and provides a unique way to explain the essence of illocutionary force. This theory could be applied in the listening teaching to improve students' listening ability.

2 THE INDIRECT SPEECH ACT THEORY

Speech Act Theory was put forward by J. L. Austin, a philosopher in Oxford, with a book named *How to Do Things With Words* in 1962. It is the first time that he introduced the Speech Act Theory in this book. He introduced the notions of locutionary act, the illocutionary act and the perlocutionary act. According to

Austin, the idea of an illocutionary act can be captured by emphasizing that "by saying something, we do something", such as the action of nominating, sentencing or promising could be performed by the utterance of the sentence itself. [1]

John Searle, a famous American philosopher and linguist, published a book named *Speech Act* in 1969. He introduced the notion of indirect speech act as follows: "In indirect speech acts the speaker communicates to the hearer more than he actually says by way of relying on their mutually shared background information, both linguistic and nonlinguistic, together with the general powers of rationality and inference on the part of the hearer." He set up the following classification of illocutionary speech acts, the assertives, the directives, the commissives, the expressives and the declarations. [2]

If there is a direct match between a sentence type and an illocutionary force, we have a direct speech act. If there is no direct relationship between a sentence type and an illocutionary force, we are faced with an indirect speech act. Thus, when an explicit performative is used to make a request, it functions as a direct speech act; the same is the case when an imperative is employed. By comparison, when an interrogative is used to make a request, we have an indirect speech act.

Indirect speech acts are usually considered to be more polite than their direct counterparts. For example, refusing sometimes may be hard to utter. If it is said in the indirect speed act, the refusing is more polite and kind to the friends. The use of indirect speech act is based on the conversation context, therefore, understanding it right or not is important for the speaker and the hearer.

3 INDIRECT SPEECH ACT THEORY AND THE LISTENING COMPREHENSION

Listening comprehension is a complicated information processing capability. To improve the students' listening ability, teachers should help them to solve the biggest obstacle, that is, the indirect speech. There are many exercises that are concerned with the indirect speech. In order to understand this phenomenon well, teachers should classify, analyze and explain the exercises as clearly as they can.

The following is the 16th from the CET 4 in June, 2013.

M: Frankly, Mary is not what I'd called easy-going.
W: I see. People in our neighborhood find it hard to believe she's my twin sister.
Q: What does the woman imply?
A) Mary is not so easygoing as her.
B) Mary and she have a lot in common.
C) She finds it hard to get along with Mary.
D) She does not believe what her neighbors said.

In this dialogue, the man is talking about his opinion of Mary. The woman agrees with him. The literal meaning of the woman is to declare that she and her twin sister are different in character. However, the illocutionary act is to describe her sister's weirdness. Therefore, the right answer to the question is A.

Although it is not complicated, teachers should explain the examples of indirect speech acts in class. Students would get some clues about how to deal with the similar questions in the listening comprehension. The students should also observe the life around to apply the theory and understand it deeply.

4 APPLICATION OF INDIRECT SPEECH ACT THEORY TO THE LISTENING CLASSES

Influenced by the structuralism since 1960s, the foreign language teaching in China puts too much emphasis on the forms of the language, while neglect the communicative competence. In the increasingly more and more communication between the countries, to cultivate the communicative competence is necessary, which is also the purpose of the college English teaching. The communicative competence includes the language ability as well as the pragmatic ability. How to communicate properly with the different people in different situations needs much concern. Because of the different languages and cultures, how to understand each other and how to use the indirect speech act is very important. In the daily life, people often misunderstand each other because of misinterpreting the indirect speeches, which result in the failure of the communication. Therefore, teachers should make a brief introduction of the Indirect Speech Act Theory to the students in class. Students could understand how to use it properly. Here are some suggestions as followed:

4.1 The diversified language forms in teaching

In the traditional English teaching, the language forms and the communicative function of the language are connected, which is wrong in the actual use of language. For example, students would consider using the imperative sentences to order. They could consider using questions to raise a question. To put as much emphasis on the forms of language is right. The key is to combine the forms with the functions. In fact, according to the Indirect Speech Act Theory, one language form could express the different language functions while one communicative function could be expressed by more language forms. If the teacher doesn't pay much attention on the language forms in the daily education, students would have limited language input which is mostly composed of the written language. Therefore, teachers should use the diversified language in the lectures. Students would perform better in the listening comprehension.

4.2 More emphasis on the cultural differences and context communication

The teacher should not only explain the literal meaning, but also the background information about different cultures, different pragmatic functions and the contexts. The language form should follow the communicative purposes, while the communication is going on in the social and cultural environment. It is necessary for the students to know more about the history, the culture, the tradition, the customs, and the lifestyles and so on. Understanding this background information well is the first step to communicate properly.

Context is the key factor for the speaker and the hearer to understand each other without mistakes. In the listening and speaking lessons, teachers should offer more opportunities to the students to communicate or role play in some contexts. Students could use the indirect speech to express themselves and understand others. Their listening and speaking abilities could be improved. [3]

4.3 To cultivate the inferential capability of the students

The inferential capability is important because it is the key to understand each other in communication. The inferential capability refers to the illocutionary meaning of the speaker. Students should focus on the social, psychological, linguistic levels to make a conclusion about the real intention of the speaker. This could help students to communicate successfully.

5   CONCLUSION

Indirect speech acts play an important role in communication. It is significant in directing the listening and speaking in college English teaching. Therefore, teachers should introduce more about the theory and emphasize on cultivating the communicative competence while teaching. It also gives more tasks to the editing of the teaching materials, the instructional objectives as well as the curriculum provision. More emphasis on the communicative competence cultivating, more fluently the communication will be when students graduate from college.

REFERENCES

Austin J.L. *How to Do Things with Words?* [M], Cambridge: Harford University Press, 1962.

Searle, John R. *Indirect Speech Acts* [M], New York: Academic Press, 1975.

Qiao Mu, *A Probe into the Application of the Indirect Speech Act Theory to the College English Teaching* , [J]Science and Technology Information, 2007(33).

Computational Intelligence in Industrial Application – Ling (ed.)
© 2015 Taylor & Francis Group, London, ISBN: 978-1-138-02818-0

A GIS simulation system development with offline GPS data generation

ZhongLi Wang, YiFeng Hua & BaiGen Cai
School of Electronic and Information Engineering, Beijing Jiaotong University, Beijing, China
Beijing Engineering Research Center of EMC and GNSS Technology for Rail Transportation, Beijing , China

ZhenHui Huang, GuiGuo Wang & TianBai Zhang
CNR Tangshan Railway Vehicle Corporation, Tangshan, China

ABSTRACT: Based on OLE Customer Control (OCX) provided by MapX, which is extensively used in Geographic Information System (GIS) development, an offline GPS data generation method and a GIS simulation system are presented in this paper, which includes map displaying, the movement simulation of the vehicle, and information exchange functions are analyzed in simulation software. The proposed method of offline GPS data generation is used for the movement simulation of a vehicle, the initial position, velocity, direction of the tram, and the frequency of data generation are considered. The GIS simulation system can be applied to other scenarios for receiving real-time GPS data and more than one vehicle movement can be simulated. Some sample interactions are provided in the simulation system, it makes the operation more flexible. It has been proved that the simulation system presented in this paper is practical and cost effective.

KEYWORDS: Geographic Information System (GIS); GPS; MapX; GIS Simulation System.

1 INTRODUCTION

Currently, the GPS/GIS integrated system has been widely used in the transportation system, military, civilian, etc. The global positioning system (GPS) is a high technology to achieve global position and navigation, which has the virtues of globality, all-weather, rapid-positioning, high-accuracy, providing continuous information of real-time three dimensional position, three-dimensional velocity and time[1]. The geographic information system (GIS) is a spatial information system. Based on the GIS database and GNSS information, many applications can be realized [2].

MapX OCX with powerful map analysis functions is provided by MapInfo Corporation, and has been widely used for developing GIS system. Map features are easily embedded in the software using MapX control, which enhance the ability of spatial analysis[3-4]. Current studies are mainly used for secondary development of electronic map software in vehicle motion simulation, introducing the development of an electronic map based on MapX methods. In this paper, general GPS simulation data are applying for the development of MapX electronic map software. The use of the early stage of the GPS data has been always in route data, which make the required data gotten by one-time simulating come true. Providing a generic method of building the simulation stage, enhances the universality and extensibility of the GIS display simulation system. The current problems such as poor portability, simple functions are solved either. This system gets the simulation by setting different initial status of the vehicle. Compared with using a GPS simulator to simulate the vehicle movement, the new method costs less with simple operation

This method can be used in the transportation sector data of the vehicle motion simulation in different lines.

For most of current GIS system, only limited interaction is provided, such as zooming in or out, and movement. This paper puts forward the solution to tracking the status of vehicles in real-time, combined the state of vehicle's location with monitoring, which providing the idea and implementation method for designing the vehicle monitoring center.

In this paper, we take Beijing six-ring line data as an example and focus on the composition and implementation of the simulation system, simulation capabilities and interactive features has been achieved.

2 SYSTEM ARCHITECTURE

The system includes: the module of offline GPS data generator, map display, interactive interface of GIS system, are shown in Figure 1.

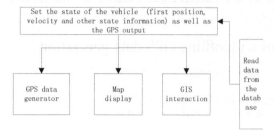

Figure 1. The composition of simulation system.

2.1 Implemented features of the simulation system

The simulation system has the following features:

1 The output of the receiver can be performed in accordance with the statement of the receiver output parameters
2 The initial speed, vehicle's location and direction of the vertical row can be set
3 It can be configured for serial output port
4 It shows the movement of the vehicle in real time, while having a display function of the multi-car movement
5 It has an interaction function between vehicles
6 It can be used in different areas of the transportation, which has better adaptability.

2.2 Function definitions of the offline GPS data generator

The GPS data generator should have the following characteristics:

1 The output parameters of GPS can be configured such as output frequency (simulation time), the format of the required output of GPS statement.
2 Statements can be output. The output of the software is configured through the serial port, while the software interface has also been shown more intuitive.
3 The vehicle running state can be set, including the direction of the line, the initial location, speed and so on.

3 THE ALGORITHM OF OFFLINE GPS DATA GENERATION

According to existing data base of Beijing's sixth rings, which is collected from the real scene before, the initial speed of the vehicle, the direction of the driving, the output frequency of GPS, and the output data format are configured first, then calculating

the position of the vehicle in real-time, outputting corresponding GPS data message, and displays the location of the vehicle on the map.

The process of offline GPS data generation is shown in Figure 2.

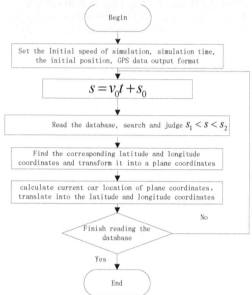

Figure 2. The process of offline GPS data generation.

1 Latitude and longitude are transformed into plane coordinates[5]

Assuming that the target geographical coordinates (λ, φ), longitude is λ, latitude is φ, geographical coordinates of the origin of Cartesian coordinates are (λ_0, φ_0).

According to simplified Gauss - Kruger projection model:

$$\begin{cases} x = S(\varphi) - S(\varphi_0) \\ y = \dfrac{\alpha \cos \varphi}{\sqrt{1 - e^2 \sin^2 \varphi}} \cdot \dfrac{\Delta \lambda}{\rho} \end{cases} \quad (1)$$

Where,

$$S(\varphi) = \alpha(1 - e^2) \cdot (\frac{A}{\rho}\varphi - \frac{B}{2}\sin 2\varphi + \frac{C}{4}\sin 4\varphi - \frac{D}{6}\sin 6\varphi + ...) \quad (2)$$

$$\Delta \lambda = \lambda - \lambda_0$$

$$\alpha = 6378245m$$

$$e^2 = 0.006693421623$$

$A = 1.0050517739$

$B = 0.00506237764$

$C = 0.00001062451$

$D = 0.00000002081$

2 Linear differential equation

$$\begin{cases} x = \dfrac{(s-s_1)\cdot(x_2-x_1)}{(s_2-s_1)} + x_1 \\ y = \dfrac{(s-s_1)\cdot(y_2-y_1)}{(s_2-s_1)} + y_1 \end{cases} \tag{3}$$

3 Plane coordinates are transformed into latitude and longitude coordinates[6]

Calculated by gauss projection: according to plane rectangular coordinates of gauss projection (x, y), geodetic coordinates of the point of the ellipsoid (λ, φ) are calculated.

The projection process is described as follows,

According to x, ordinate projections of ellipsoid bottom point latitude is calculated.

$$\begin{cases} \varphi = \varphi_f - \dfrac{t_f}{2M_fN_f}y^2 + \dfrac{t_f}{24M_fN_f^3}(5+3t_f^3+\eta_f^2-9\eta_f^2t_f^2)y^4 \\ \qquad - \dfrac{t_f}{720M_fN_f^5}(61+90t_f^2+45t_f^2)y^6 \\ l = \dfrac{1}{N_f\cos\varphi_f}y - \dfrac{1}{6N_f^3\cos\varphi_f}(1+2t_f^2+\eta_f^2)y^3 \\ \qquad + \dfrac{1}{120N_f^5\cos\varphi_f}(5+28t_f^2+24t_f^4+6\eta_f^2+8\eta_f^2t_f^2)y^5 \end{cases} \tag{4}$$

When converting longitude is 0.01", it can be simplified to the following formula:

$$\begin{cases} \varphi = \dfrac{t_f}{24M_fN_f^3}(5+3t_f^3+\eta_f^2-9\eta_f^2t_f^2)y^4 \\ \qquad + \varphi_f - \dfrac{t_f}{2M_fN_f}y^2 \\ l = \dfrac{1}{120N_f^5\cos\varphi_f}(5+28t_f^2+24t_f^4)y^5 \\ \qquad + \dfrac{1}{N_f\cos\varphi_f}y - \dfrac{1}{6N_f^3\cos\varphi_f}(1+2t_f^2+\eta_f^2)y^3 \end{cases} \tag{5}$$

Where,

$$N_f = \alpha(1 - e^2 \sin\varphi_f)^{-1/2}$$

$t = \tan\varphi$

$\eta^2 = (e')^2 \cos(\varphi)$

4 IMPLEMENTATION OF INTERACTIVE INTERFACE IN GIS SYSTEM

4.1 Interactive in the GIS system

In the GIS system, interaction determines the effect of simulation software. The software has good interactive features, which can be achieved through the interaction of the map to zoom, roaming operations. And the state has a view of the vehicle features, easy to understand running state of the vehicle. It has the ability to view the vehicle status, which is easy to understand the status of the vehicle.

4.2 System interface function definition and implementation

Map is to be achieved operational by adding a map operation toolbar such as zoom in, zoom out. The vehicle status information is to be achieved by adding a mouse message response function to achieve click the pop-up information.

4.2.1 Software implementation process

A flowchart of the vehicle's GIS display is shown in Figure3:

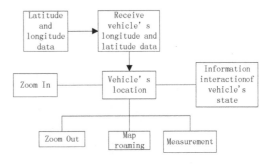

Figure 3. Flowchart of the vehicle's GIS display.

4.2.2 Data interface

In vehicles GIS map display, the latitude and longitude of the vehicle are needed to get. The method of getting the latitude and longitude of the vehicle is described in the second section. The program interface of receiving information for the latitude and longitude data is reserved. The location of multiple vehicles is displayed, and status information of the vehicles is also displayed.

5 GIS SIMULATION SYSTEM BASED ON MAPX OCX

GIS functions of the system:

1 Map Zoom: the current map can zoom in and out in order to understand the details of the location of a moving target, larger regional or global or understand the overall situation.
2 Map Roaming: map roaming can be achieved by moving the mouse
3 Information interaction: longitude, latitude, speed and other information can be got.

5.1 Zooming and roaming

The basic function of electronic map is realized by using MapX in the system, including map zoom, roaming, etc. These features are achieved by calling the appropriate method in MapX:

```
m_map.SetCurrentTool(miZoomInTool);//Zoon In
m_map.SetCurrentTool(miZoomOutTool);//Zoom
Out
m_map.SetCurrentTool(miPanTool); // roaming
```

5.2 Real-time displaying of vehicle location

The latitude and longitude of the vehicle are obtained through calculating. The location of the vehicle is displayed on the map. Clearing the historical point is needed in order to show the current momentary position of the vehicle.

5.3 Interaction operation of the system

By clicking the vehicle's location on the map, a dialog box is pop-up, which displays vehicle's status information, including latitude, longitude, speed. The implementation method:

1 The screen coordinates are transformed into latitude and longitude coordinates
2 The latitude and longitude coordinates are converted to plane coordinates
3 Mouse message is responded when less than a certain distance by calculating the position between two points.

6 THE GIS SIMULATION INTERFACE

By setting simulation parameter, the simulation of vehicle in different configurations is achieved and it generates GPS NMEA format statement through the serial port output, while the statement is displayed. When the vehicle is moving on a six-ring position, the vehicle status information is displayed by mouse clicking. The simulation result is shown in Figure 4.

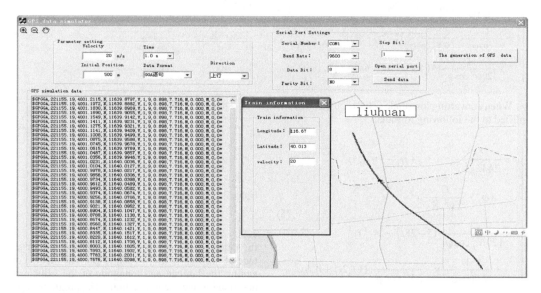

Figure 4.   The GUI interface of the GIS simulation system.

7 CONCLUSION

In this paper, based on OLE Customer Control (OCX) provided by MapX, an offline GPS data generation method and a GIS simulation system are implemented. Based on the database obtained before, and the initial configurations, such as the initial position, velocity, direction of the vehicle, and the frequency of data output, the offline GPS data generation can output the standard GNSS data in real-time, just like the vehicle is driving on site. The implemented GIS simulation system can be used for displaying the position of the vehicle in the map, some sample interactions of GUI interface based are provided to make the operation more flexible and effective.

The simulation results show that the proposed method is practical and cost effective, which can be used in many applications.

REFERENCES

Sang W.G.,etc. The Application of GPS Technology in the Road Electronic Pricing System[J]. GNSS World of China,2004.4.

Xu A.G.. Research on a New GPS/GIS Based on ERP System. Intelligent Transportation Systems, 2005.

Lishengle,luyuan zhong,cheshi.Mapinfo Secondary development examples of geographic information systems[M]. Beijing: Electronics Industry Press,2004.

Liuyao lin, Miaozuo hua, Design and Implementation of Collective Land Registration and Certification Issue System Base on Map Objects[J].bulletin of Surveying and Mapping,2005(7):48–51.

Zhang C D, Hsu H T, Wu X P, et al. An alternative algebraic algorithm to transform Cartesian to geodetic coordinates[J]. Journal of Geodesy, 2005, 79(8): 413–420.

Civicioglu P. Transforming geocentric Cartesian coordinates to geodetic coordinates by using differential search algorithm[J]. Computers & Geosciences, 2012, 46: 229–247.

Computational Intelligence in Industrial Application – Ling (ed.)
© *2015 Taylor & Francis Group, London, ISBN: 978-1-138-02818-0*

Temperature error compensation for fiber optic gyroscope based on CPSO-BP neural network

Ming. Jiang & Zhi Jun. Liu
Beijing Institute of Technology, Beijing, China

ABSTRACT: Fiber optic gyroscope is a temperature sensitive device, in order to solve the influence of temperature on the accuracy of fiber optic gyroscope, this paper presents a model based on Chaotic Particle Swarm Optimization (CPSO) BP neural network, which uses internal temperature measured value of fiber optic gyroscope as the input, the calculated values of bias and scale factor as the output. Using this model can compensate the gyroscope temperature error. The CPSO-BP model solved the problems of general BP neural network model slow iteration speed and easily leads to local optimal, improving the ability of particle swarm optimization algorithm to get rid of local extremum, raising the calculation precision and the convergence speed of BP network. Experimental results show that the temperature error compensation scheme of fiber optic gyroscope can well fit the zero bias and scale factor changing with temperature, greatly reducing the influence of gyroscope bias and scale factor temperature error on the standard error of the angular velocity, to ensure the accuracy of measurement of fiber optic gyro total in different temperature environment.

KEYWORDS: Fog; Particle Swarm Optimization (PSO); BP network; Temperature compensation.

1 INTRODUCTION

Fiber optic gyroscope has the advantages of simple structure, no moving parts, quick start, low power consumption, small volume, light weight, impact resistance, wide coverage of accuracy, large dynamic range. It has been widely used in the field of aviation, aerospace, weapons, oil exploration and geological prospecting.[1-2] Due to the core components of fiber, optic gyroscopes are more sensitive to temperature, so the change in temperature is one of the important factors affecting the performance of fiber-optic gyro. At present, many scholars at home and abroad put forward some modeling compensatory method through research on reasons of the temperature drift error of fiber optic gyro, such as polynomial model, neural network models, wavelet variance model [3], fuzzy model [4], controlled Markov chain model, etc. [5]. In recent years, the application of neural network for the temperature compensation for fiber optic gyro has become a research hot spot.

The neural network is commonly used in BP neural network to realize the identification function. Because of BP neural network in the training error function is along the direction of the error function gradient descent, it can easy to fall into a local minimum, unable to perform global search, and low search speed, which makes the output of trained neural network inconsistency and unpredictability, reduces the accuracy and reliability of the trained neural network [6].

PSO algorithm also exists the contradiction between improving the convergence speed and adding the possibility of falling into local extremum. This paper uses chaotic particle swarm optimization algorithm (CPSO) to train the BP neural network, through the chaos optimization on the optimal particle, accelerating the speed of particle swarm evolution, improving the capability of particle swarm optimization algorithm escaping from local extremum, enhancing the algorithm global search ability, improving the convergence speed and accuracy, which makes the trained neural network has higher accuracy and reliability, and through the experiment verify the effectiveness of the algorithm.

2 CHAOTIC PARTICLE SWARM OPTIMIZATION ALGORITHM

2.1 *Particle Swarm Optimization (PSO) algorithm*

Particle swarm optimization algorithm is a kind of collective optimization algorithm, which originated as a simplified model of the social simulation, proposed by Kennedy and Eberhart in 1995 [7]. It is a typical cluster intelligent method which is inspired by the bird flock motion mode. In the particle swarm algorithm, particle swarm initialized to a set of random values, each particle represents a candidate solution of optimization problems, and are covered by two items information of speed and position, the space of the i particles in N dimensional position solution can be expressed as: $X_i = (x_{i1}, x_{i2}, \dots x_{iN})$; velocity can be expressed as: $V_i = (v_{i1}, v_{i2}, \dots v_{iN})$; V_i decided the direction and distance of particle movement. All particles have a fitness value (f_i) decided by fitness function, using the fitness to measure particle current position is good or not. The ith particle so far to searched the optimal position is: $P_i = (p_{i1}, p_{i2}, \dots p_{iN})$; The whole particle swarm so far to searched the optimal position is: $P_g = (p_{g1}, p_{g2}, \dots p_{gN})$, particle by tracking individual and group optimal position to adjust and update their own position and velocity, the evolution of the particle update process as shown in Figure 1.

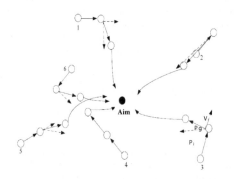

Figure 1.　Schematic diagram of particle movement principle and evolution.

Particle swarm optimization according to the formula (1) and the formula (2) to update the particle's velocity and position.

$$v_{id}^{k+1} = w v_{id}^k + c_1 r_1 (p_{id}^k - x_{id}^k) + c_2 r_2 (p_{gd}^k - x_{gd}^k) \qquad (1)$$

$$x_{id}^{k+1} = x_{id}^k + v_{id}^k \qquad (2)$$

In the formula, k is the number of iterations, w is the inertia weight, c_1, c_2 are learning factors, r_1, r_2 are random numbers interval on [0,1].

v_{id}^k is the Kth iteration of particle i article D component search velocity vector, the particle velocity in each dimension can not exceed the seted search algorithm maximum speed V_{max}; x_{id}^k is the Kth iteration of particle i article D component search position vector; p_{id}^k is the ith particle best individual position; p_{gd}^k is the position groups best position.

As shown in Figure 1, each particle through tracking the optimal position of individuals and groups to adjust and update their own position and velocity, so as to continuously to the advantages of evolutionary. But when we have the least number of particles or the speed of adjustment is an improper selection, it can easily make the particles falling into local extreme point and can not jump out, thus the particle swarm evolutionary comes stagnation.

2.2　Chaotic particles swarm optimization algorithm

To prevent the particles iteration appear stagnation, improve the ergodicity of the particle search and increase the diversity of particle groups, chaos theory was introduced into the PSO algorithm. Because of chaos motion have randomness, ergodicity, sensitivity to initial conditions and other characteristics [8], we use chaotic motion to improve the ability of the PSO algorithm escape from local extremum, and improve the convergence speed and calculation accuracy of the BP neural network algorithm. We call this embedding chaotic series into the particle swarm optimization algorithm as CPSO [9].

Using chaos optimization to deal with the kth times search update particle's optimal position p_g^k, p_g^k by formula (3) mapping to the Logistic equation domain definition [0, 1]:

$$y_1^k = \frac{p_g^k - p_{g\,min}^k}{p_{g\,max}^k - p_{g\,min}^k} \qquad (3)$$

Let y_1^k M times iteration through the Logistic equation $y_{n+1}^k = \mu y_n^k (1 - y_n^k)$, we obtain the chaotic sequences of $y^k = (y_1^k, y_2^k, \dots, y_M^k)$. Let chaotic sequence we obtained by the formula (3) mapping back to the original solution space (m=1, 2,…M):

$$\overline{p}_{gm}^k = p_{g\,min}^k + (p_{g\,max}^k - p_{g\,min}^k) y_m^k \qquad (4)$$

Then generate a chaotic feasible solution sequence:

$$\overline{p}_g^k = (\overline{p}_{g1}^k, \overline{p}_{g2}^k, \dots, \overline{p}_{gM}^k) \qquad (5)$$

Calculating the fitness of each feasible solution in chaotic feasible solution sequences, use the optimal fitness feasible solution to randomly replace one of the particle positions in the particle swarm, continue iteration according to formula (1) and (2) until getting a satisfactory optimal solution or reaching the maximum number of iterations.

In order to test the performance of chaotic particle swarm optimization algorithm, we tested with two typical Benchmark function, these two functions are:

2.2.1 *Sphere function*

$$f_1(x) = \sum_{i=1}^{D} x_i^2, |x_i| \le 100, D = 30 \tag{6}$$

The optimal solution and the optimal value is:

$min\ f_1(x) = 0.$

2.2.2 *Griewank function*

$$f_2(x) = \frac{1}{4000} \sum_{i=1}^{D} x_i^2 - \prod_{i=1}^{D} \cos(\frac{x_i}{\sqrt{i}}) + 1, D = 10 \tag{7}$$

The optimal solution and the optimal value is:

$min\ f_4(x) = 0.$

The simulation results as shown in figure 2 and figure 3, figure 2 obtained by the standard PSO algorithm and CPSO algorithm act on Sphere function simulation comparison curve, figure 3 obtained by the standard PSO algorithm and CPSO algorithm act on the Griewank function simulation comparison curve.

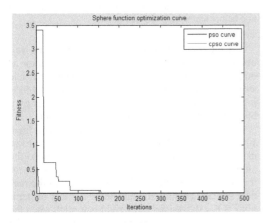

Figure 2.   Sphere function optimization curve.

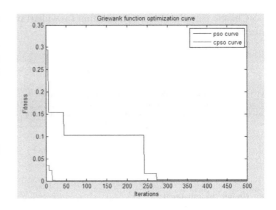

Figure 3.   Griewank function optimization curve.

Simulation results show that the chaotic particle swarm optimization algorithm (CPSO) has fast convergence speed and can take into account both global optimization and local optimization, effectively avoid the premature convergence, significantly improved particle optimization ability, which has better adaptability.

2.3 *Based on CPSO BP neural network training method*

BP neural network training is along the error function gradient descent direction, so inevitably there are easy to fall into a local minimum, with the problem of slow convergence speed and long training time. Using CPSO algorithm instead of the BP algorithm gradient decreased method in training neural network parameters, can improve the performance of the BP neural network, to speed up the search speed, prevent the premature convergence of the whole algorithm, so that it is not easy to fall into a local minimum, to enhance the generalization performance of the network. In CPSO BP network training, the connection weights and the neuron threshold in a BP neural network are respectively corresponding to one of element in individual particle vector position. Let each of the BP network outputs mean square error function as the fitness function, as shown in the formula (8):

$$f_i = \frac{1}{M+N} \sum_{j=1}^{M} \sum_{k=1}^{N} (\overline{y}_{jk} - y_{jk})^2 \tag{8}$$

In the formula, M is the number of samples, N is the output of BP network layer nodes $\overline{y}_{mn}$ is the ideal output for the Nth node under the mth sample, y_{mn} is the actual output for the nth node under the mth sample.

117

Chaotic particle swarm optimization flow chart of the BP neural network as shown in Figure 4.

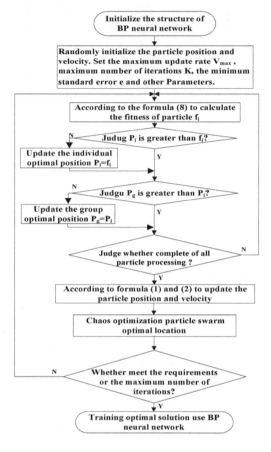

Figure 4. Implementation process of CPSO optimization BP neural network.

$$K = \frac{\sum_{j=1}^{M} \omega_j F_j - \frac{1}{M} \sum_{j=1}^{M} \omega_j \sum_{j=1}^{M} F_j}{\sum_{j=1}^{M} \omega_j^2 - \frac{1}{M} (\sum_{j=1}^{M} \omega_j)^2} \qquad (9)$$

$$F_0 = \frac{1}{M} \sum_{j=1}^{M} F_j - \frac{K}{M} \sum_{j=1}^{M} \omega_j \qquad (10)$$

In the formula K is the scale factor, F0 is virtual zero position, M is the number of selected input speed; Fj is the input angular velocity measurement values.

Let the scale factor and zero bias under the room temperature(25°C)as the standard values, calculated out without temperature compensation gyro output angular velocity ω0, use the values of ω minus ω0 as the gyro angular velocity error before compensation output. Thus the angular velocity error output without temperature compensation curve as shown in Figure 5.

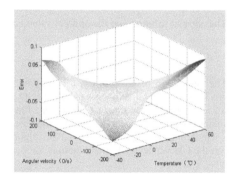

Figure 5. Fiber optic gyro angular velocity output deviation before temperature compensation.

3 BASED ON CPSO-BP NEURAL NETWORK FIBER OPTIC GYRO TEMPERATURE ERROR COMPENSATION EXPERIMENT

In order to synthetically research the fiber optic gyroscope temperature error characteristics , The experiment used a type of fiber optic gyro biaxial turntable with temperature control box, at the temperature range of −40~60°C, every 10 °C as a testing temperature point conduct the constant temperature environment test. Input angular rate respectively is 0°/s, ±1 °/s, ±2 °/s, ±5 °/s, ±10 °/s, ±15 °/s, ±20 °/s, ±25 °/s, ±30 °/s, ±40 °/s, ±50 °/s, ±100 °/s, ±180 °/s, collect the gyro angular velocity and internal temperature output after the gyro internal temperature and rotating speed stabilized, according to the formula (9) and (10) calculate gyro bias and scale factor:

Let the gyro internal temperature measuring values as input, computed zero bias and scale factor as output, according to the process of chaotic particle swarm optimization algorithm BP network, establishing the neural network model of the optical fiber gyro bias and scale factor changing with temperature.

The network model using single hidden layer structure, and the hidden layer contains 10 neurons, using the hyperbolic tangent transfer function, the output layer contains 2 neurons, using the linear transfer function, the number of particles in the particle swarm is N=60, the maximum number of iterations is K=300, the inertia weight is w=0.9, when the velocity at the maximum and learning factor in the beginning stages let V_{max}=100, c_1=0.5 and c_2=0.6, when the particle optimal position fitness less than 0.1, let V_{max}=0.1, c_1=0.8 and c_2=0.2, that is after whole situation sufficiently close to the optimal position, Strengthening the local search.

The fitting curves between output from chaotic particle swarm optimization the neural network and the sampling data as shown in Figures 6 and 7, Asterisks in the figure are the zero bias and scale factor calculating from experimental data, solid lines are the neural network output fitting curve .

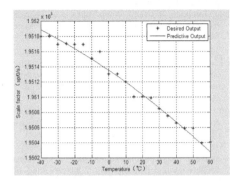

Figure 6.   Fitting curve of gyro scale factor changing with temperature.

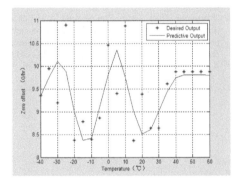

Figure 7.   Fitting curve of gyro bias changing with temperature.

Then the scaling factor error average value before the temperature compensation is 455.183533sp/0/s. Using the absolute value of the difference between actual scaling factor K_i and the scale factor under the 25°C Km as the before compensation scale factor error , then the scaling factor error average value before the temperature compensation is 455.183533sp/0/s. Using the absolute value of the difference between actual scaling factor K_i and the neural network output scale factor KBP as the after compensation scale factor error, then the scaling factor error average value after the temperature compensation is 39.5876.39sp/0/s.

In the same way, using the absolute value of the difference between actual zero bias B_{oi} and the zero bias under the 25°C B_{om} as the before compensation

zero bias error , then the zero bias error average value before the temperature compensation is 0.764785°/hr. Using the absolute value of the difference between actual zero bias B_{oi} and the neural network output zero bias BBP as the after compensation zero bias error, then the zero bias error average value after the temperature compensation is 0.004612°/hr.

Using the scaling factor after neural network model compensation and zero bias to calculate gyro output angular velocity w_1, the values between turntable input rotational angular velocity w minus w_1 as the gyro output angular velocity error after temperature compensation, gyro output angular velocity error after temperature compensation as shown in figure 8.

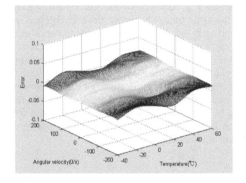

Figure 5.   Fiber optic gyro angular velocity output deviation after temperature compensation.

Under different temperatures before and after temperature compensation fiber optic gyroscope output angular velocity deviation mean as shown in Table 1

Table 1.   Angular velocity output deviation before and after temperature compensation.

Temperature (°C)	Error before compensation (°/s)	Error after compensation (°/s)
−40	1.038217e-06	−2.074432e-08
−30	1.035741e-06	−9.192794e-08
−20	1.258943e-06	−1.438029e-08
−10	1.041540e-06	5.272021e-08
0	2.326803e-07	−5.21369e-08
10	9.625369e-08	2.67602e-09
20	1.122657e-07	3.188254e-08
30	−1.701590e-07	9.492119e-08
40	−3.063065e-07	5.069231e-08
50	−5.124551e-07	−2.194397e-08
60	−5.517569e-07	5.108430e-08

119

From Figures 6 and 7, we can see that the chaotic particle swarm optimization BP neural network can better fitting fiber optic gyro bias and scale factor varies with the temperature, the Table 1 and comparison results between figure 5 and figure 8 shows that the algorithm reduces angular velocity error caused by the fiber optic gyroscope zero bias and scale factor temperature error.

4 CONCLUSIONS

This paper adopts the fog temperature compensation algorithm based on Chaos Particle Swarm Optimization BP neural network, in assurance the convergence accuracy of particle swarm algorithm, at the same time, to improve the convergence speed, improving the defect of BP neural network easy to filling into a local minimum. Experimental results show that this algorithm can faster and accurately fit the role of optical fiber gyro bias and scale factor changing with temperature, let the output angular velocity error caused by the temperature error, reducing by one order of magnitude, ensuring the optical fiber gyro measurement accuracy under total temperature environment.

REFERENCES

Jin Jing, Li Min, Zhang Zhonggang, et al. Analysis of temperature errors in digital closed-loop fiber optic gyroscope [J]. Infrared and Laser engineering, 2008, 37 (3):521–524. (in Chinese).

Wang Wei, Yang Qingsheng, Wang Xuefeng. Application of fiber-optic gyro in space and key technology [J]. Infrared and Laser engineering, 2006, 35(5):509–512(in Chinese).

Song Ningfang, Chen Jing, Jin Jing. Wavelet variance analysis of random error properties for fiber optic gyroscope [J]. Infrared and Laser engineering, 2010, 39(5):924–928. (in Chinese).

Zhang Hongxian, Wu Yanji, Wang Yuhui, et al. Temperature compensation for FOG based on fuzzy logic [J]. Journal of Chinese Inertial Technology, 2007, 15(3):343–346(in Chinese).

Shen Jun, Miao Lingjuan, Wu Junwei, et al. Application and compensation for startup phase of FOG based on RBF neural network [J]. Infrared and Laser engineering, 2013, 42(1):119–124. (in Chinese).

Zhou Chi, Gao Haibing, Gao Liang, et al. Particle swarm optimization (pso) algorithm [J]. Application Research of Computers, 2003, 20(12):7–11. (in Chinese).

Xu Xiangfan, Ming Jun. Method of singularity target signal multiscale orientation and recognition [J]. Computer Technology and Development, 2006, 16(3):63–65. (in Chinese).

Liu Wei, Wang Kejun, Shao Keyong. Predicting chaotic time series using hybrid particle swarm optimization algorithm [J]. Control and Decision, 2007, 22(5):562–565. (in Chinese).

Bai Chuanfang, Zhao Panlin, Sun Shujing. A switch algorithm and its OPNET simulation based on network performance analysis of DWCS [J]. System Simulation Technology and Application, 2012(2):28–32.Larch, A.A. 1996a. Development.

Computational Intelligence in Industrial Application – Ling (ed.)
© 2015 Taylor & Francis Group, London, ISBN: 978-1-138-02818-0

A hybrid method using aerial photograph to extract forest land area

C. Ran
Landscape Architecture Corporation of China, Beijing, China

ABSTRACT: In this paper, the feasibility of using spectral information and textural information derived from aerial photographs for extracting forest area was investigated in Enshi of Hubei Province, in the center of China. Torrential rains or dense mist throughout the year characterizes these forests, often shrouded in heavy clouds. For this reason, a great deal of investigation is focused on aerial photograph. The analysis was performed with the following two approaches: (i) the use of spectral information for constructing new Normalized Difference Vegetation Index (NDVI), (ii) the use of morphological texture features for forest discrimination. A combination of spectral responses and image textures improves the performance to extract forest area. Forest information was extracted from aerial photographs. The investigation showed that the extracting accuracy of forest area fulfilled the requirements compatible with detailed forest resources maps. Thus, the use of aerial photographs can be considered as primary source for mapping detailed forest resources in similar environments of China, where the climatic condition is limited.

KEYWORDS: Color aerial image, Forest land area extraction, NDVI, Morphological texture.

1 INTRODUCTION

Forest ecosystems are the main terrestrial which is distributed and has many functions. The natural forest succession or disturbance activities have the characteristics of spatial distribution which makes the distribution of trees with a high degree of spatial and temporal heterogeneity.

Forest-area estimation has been an important application of remote-sensing techniques since the 1970s. The forest area(% of land area) in China was last reported at 22.18 in 2010, according to a World Bank report released in 2011(Cliffe, 2011). The remote sensing has been used generally to reduce the number of samples of ground surveys, making them less expensive. Several approaches have been proposed for this objective, where the role of remote sensing varies from main data source to auxiliary support for ground surveys (Gallego, 2004). When the two data sources are used together, the results are more accurate than either the ground survey or the image classification separately(Gallego and Delincé, 1991). In any case, the convenience of using remote-sensing data depends on how much the costs of the survey can be reduced with respect to a pure ground survey achieving a certain grade of accuracy (Taylor et al. 1997).

The use of high-resolution images for the updating of land-cover maps at the regional scale is usually very expensive (Annoni and Perdigao, 1997). Thus, low-resolution images have been used for area measurements and land-cover-change analysis (Stern et al. 2001). In these cases, conventional classification techniques are not directly applicable to area estimation due to the mixed nature of almost all pixels (Rembold and Maselli, 2006). Fuzzy classifiers, which are capable of providing only approximate estimates of sub-pixel cover types and are unsuitable for quantitative area assessment, have therefore been proposed (Foody, 1996). Similar estimates can be obtained through unmixing techniques that, however, are very sensitive to variations in multispectral/multi-temporal signatures caused by external factors(Settle and Drake,1993). For example, when using multitemporal Normalized Difference Vegetation Index(NDVI) profiles, amplitude greenness variations can arise from different weather patterns during the growing seasons (Gaston et al.1994). Conventional univariate or multivariate linear-regression techniques, which are based on similar statistical principles, generally suffer from the same drawbacks (Maselli, 2002).

In spite of these limitations, relatively inaccurate area estimates derived from remote-sensing data can also be used as an auxiliary tool to improve statistics computed from reference samples over relatively large land units. In these cases, remotely statistics computed from the ground measurements. Among the existing methods, those based on the direct expansion and regression estimator are among the most popular and used (Gallego 2004).

2 STUDY SITE

2.1 Study region

The methodology was tested using data for the Enshi region (center of China). Enshi is located at 109° 29′E longitude and 30°17′ E latitude, and covers about 2.4 × 108ha. The region is abundant in forest resource and is one of the producing bases for Artificial Commercial Forest. According to the forest inventory statistics in 2011, woodland covers 1.5 × 108ha, among which 0.99 × 108ha. The forest coverage is 66%.

The land use is predominantly agricultural in the plains and mixed agricultural and forestry on the hills and mountains. The upper hill and mountain zones are mostly covered by forests, which occupy about half of the regional surface. Among broadleaves, the most widespread forest formations are conifers. Among conifers, the most abundant forest formations are dominated by Chinese red pine、China fir、dawn redwood.

Figure 1. Sketch map showing the location of Enshi city in China.

2.2 Dataset

Aerial photographs at a scale of 1:50,000 were acquired in June 2003 using Airborne Digital Sensor 40 (ADS40) camera.The ADS40, the digital aerial pushbroom scanner from Leica Geosystems, Heerbrugg, Switzerland, was first launched in 2000.

A detailed description may be found in (Sandau et al., 2000). The ADS40 is a single optic system with a focal length of 62.7mm and a field of view of 64o across track. The sensor head has individual CCD lines for pan, red, green, blue, and nir bands with 12'000 pixels and 6.5μm pixel size (Sandau et al., 2000; Beisl, 2006). The spectral bands are narrow, non-overlapping and have a response characteristic with almost a rectangular shape and a radiometric resolution of 12bit (Beisl, 2006).

In this study, flight campaign was performed with SH40 sensor heads. The SH40 head consists of four bands in nadir direction (red, green, blue and panchromatic), three bands in forward direction (nir, red and green) and two additional oblique panchromatic bands. The oblique acquisition angle of the nir band, compared to the nadir RGB bands affects the SH40 data interpretation adversely. The images were saved as multiband TIFF files having red(R)、green(G) and blue(B) channels.

3 METHODOLOGICAL APPROACH

Figure 2 shows the main steps in the forest area mapping. The methodology can be divided into analysis of aerial photograph, spectral features extraction, textural features extraction, speckle filter, mask matching.

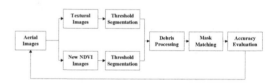

Figure 2. Flowchart of hybrid method.

3.1 Analysis of aerial photograph

Traditionally, vegetation monitoring by remotely sensed data has been carried out using vegetation indices, which are mathematical transformations designed to assess the spectral contribution of green plants to multispectral observations. The potentials and limits of different vegetation indices have been extensively discussed in the literature (see for instance Bannari, Morin, Bonn, & Huete, 1995; Baret & Guyot, 1991). Vegetation indices are mainly derived from reflectance data from discrete red (R) and near-infrared (NIR) bands. They operate by contrasting intense chlorophyll pigment absorption against the high reflectance of leaf mesophyll in the near infrared. Such is the case of the well-known normalized difference vegetation index NDVI=[NIR−R]/[NIR+R] (Bannari et al., 1995), which is the most widely used index especially when analyzing data taken from satellite platforms. In practice, NDVI is indicative of plant photosynthetic activity and has been found to be highly related to the green leaf area index (LAI) and the fraction of photosynthetically active radiation absorbed by vegetation (FAPAR) (Bannari et al., 1995; Baret & Guyot, 1991; Veroustraete, Sabbe, & Eerens, 2002).

3.2 New NDVI

In this study, the image lacks near infrared (nir) band. We use green (G) channel instead of near infrared (nir) band，thus the new NDVI is given:

$$NNDVI = \frac{G - R}{G + R} \tag{1}$$

Using New NDVI profiles can obtain approximately forest NDVI value.

3.3 Morphological texture extraction

Mathematical morphological is an effective tool for extracting image components that are useful in the representation and description of region features. Morphological features are based on some basic operators, such as opening and closing, used to remove small bright (opening) or dark (closing) details while leaving the overall features relatively undisturbed. The morphological reconstruction is another commonly used technique that has been proven to have better shape preservation than classical morphological operator (R. Bellens, S. Gautama，2008. P.Soille and M. Pesaresi. 2002). Morphological features have been successfully applied to urban VHR images (R. Bellens, S. Gautama, 2008. M. Pesaresi and J.A. Benediktsson, 2001). In particular, some operations, such as gradients, top hats, and opening/closing by reconstruction, have been proved to be efficient for texture extraction and segmentation of natural landscapes (P. Soille and M. Pesaresi, 2007). However, few studies have been reported where morphological features have been used for forest species discrimination. In this letter, several operators are presented for texture analysis.

1 Erosion and dilation of f by SE:

$$\varepsilon^{\mathrm{SE}}(f) \text{ and } \delta^{\mathrm{SE}}(f). \tag{2}$$

2 Opening and Closing of f by SE:

$$\gamma^{\mathrm{SE}}(f) = \delta^{\mathrm{SE}}\left(\varepsilon^{\mathrm{SE}}(f)\right) \text{ and } \varphi^{\mathrm{SE}}(f) = \varepsilon^{\mathrm{SE}}\left(\delta^{\mathrm{SE}}(f)\right) \tag{3}$$

Some commonly used morphological textures are also utilized for comparison, such as morphological gradient (MG).

$$MG^{\mathrm{SE}}(f) = \delta^{\mathrm{SE}}(f) - \varepsilon^{\mathrm{SE}}(f). \tag{4}$$

3.4 Spectral_textural information fusion

Due to the similar spectral response between forest land and non-forestry land, it is necessary to integrate spectral and textural information for accurately mapping.

3.5 Threshold segmentation

Forest land areas of the processed images are extracted using threshold segmentation and the mask of forest land is obtained. If each pixel of image fills the followed function, then they are viewed as forest land areas.

$$\mathrm{Mask}(i,j) = \begin{cases} B_{Forest} & A(i,j) \geq T_{Forest} \\ 0 & A(i,j) < T_{Forest} \end{cases} \tag{5}$$

T_{Forest} is threshold of forest land area, A(i, j) is a mask point location of (i, j), B_{Forest} is forest land area.

Two obtained images through band combination have merged into a mask in the setting of threshold, and then fill in the blanks of mask.

3.6 Debris processing

Many thin blanks exist often in mask of forest land area, after some processing patches are always in the non-forest area. Morphological reconstruction can be used to fill in the blanks of mask. Reconstruction of images g and f by SE: Re^{SE}(g, f) where g, the marker, is the starting point for the transformation and f, the mask, constrains the transformation. The reconstruction aims to preserve the shapes of features since some morphological operators deform the objects(Huang, 2009). In this framework, the Opening by reconstruction(OBR) and the Closing by reconstruction(CBR) can be expressed as

$$\gamma^{\mathrm{SE}}(f) = \mathrm{Rec}^{\mathrm{SE}}\left(\varepsilon^{SE}(f), f\right) \tag{6}$$

$$\varphi^{\mathrm{SE}}(f) = \mathrm{Rec}^{\mathrm{SE}}\left(\delta^{SE}(f), f\right) \tag{7}$$

Thin patches in non-forest area can be left out in setting of threshold. If object areas of continuous pixels are bigger than threshold Ta, and the object width is bigger than threshold Tw, and decides that the object is forest land.

if area > Ta and width > Tw then class_label = Forest

A forest land image I_{Forest} can obtained after matching of processed mask f_m and original image I.

3.7 Performance evaluation

This section presents a numerical assessment of the results and comparison of the developed method.

To enable numerical evaluations of the developed method, test regions with different land cover types were cropped from the scenes. Manual interpretation of forest land regions was made in order to forest land regions (true forest land masks). The true forest land masks were used in the evaluation of the precision of the forest land area extracting methods.

The performance of forest land area extracting methods is usually assessed by several metrics, among errors of commission and errors of omission. These metrics are appropriate for an assessment of still forest land area extracting methods and depend on the true positives (TPs; the number of forest land areas that are classified correctly), false negatives (FN; the non-forest land areas classified as forest land areas), false positives (FP; the forest land areas are detected as non-forest land areas).

$$Ce = \frac{FN}{TP + FN} \tag{8}$$

$$Oe = \frac{FP}{TP + FP} \tag{9}$$

The Individual Classification Success Index (ICSI) applies to the classification effectiveness for one particular class of interest (Sotirios Koukoulas et al. 2001). The formula for this index is

$$ICSI = 1-(Error\ of\ Omission\%\ +\ Error\ of\ Commission\ \%) \tag{10}$$

In the study, error of omission and error of commission are considered as equal importance, it used of normal classification. If ICSI was used in forest resource classification system, errors of omission are more important than errors of commission. Because the forest land areas are detected as non-forest land areas, followed post-processing can correct the error difficultly. Therefore, the Weighted Individual Classification Success Index (WICSI) introduces a new concept of accuracy estimation that refers to measure the final extracting error of forest land area.

$$WICSI=1-[\ wxCe+(1-w)xOe] \tag{11}$$

In the study, weight w is 0.2.

4 EXPERIMENTS AND DISCUSSION

4.1 *Study area and data*

4.2 *Comparative algorithms in experiments*

The new NDVI approach and the morphological textures approach are adopted for comparison purposes. The New NDVI approach is used to simulate NDVI

Figure 3.   Input color aerial image of the study area.

for extraction of forest land area. The morphological textures approach is also used to adaptively model the spatial context of each pixel.

4.3 *Results and analysis*

The new NDVI image is shown in Fig.4(a). After threshold segmentation, debris processing and mask matching, the forest fraction image is shown in Fig.4(b). Red boxes that are really the non-forest land areas are mistakenly classified as forest land areas.

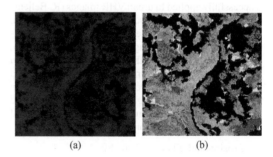

(a)   (b)

Figure 4.   Results using new NDVI method (a) new NDVI image (b) Forest-cover fraction image.

Fig.5. show the results using morphological textures method. The morphological gradient image is shown in Fig.5(a). The threshold value in threshold segmentation is 0.07. After debris processing and mask matching the forest fraction image is shown in Fig.5(b). Red boxes of non-forest land are mistakenly classified as forest land areas.

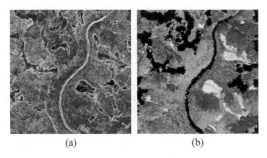

(a)   (b)

Figure 5. Results using morphological textures method (a) morphological gradient image (b) Forest-cover fraction image.

Fig.6 show results using hybrid method. The threshold segmentation image is shown in Fig.6(a), the threshold value is 0.07. After debris processing and mask matching, the forest fraction image is shown in Fig.6(b).

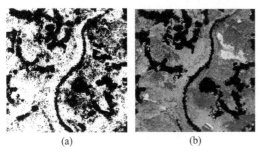

(a)   (b)

Figure 6. Results using hybrid method (a) threshold segmentation image (b) Forest-cover fraction image.

4.4 Comparison of different methods

Table 1. Accuraccies for different methods.

	Ce(%)	Oe(%)	WICSI(%)
New NDVI	3.34	24.01	80.12
MT	4.5	4.92	95.16
Hybrid	3.95	2.31	97.36

The accuracies of three different method are reported in Tab.1 compared with the other two methods, the hybrid method substantially improved the mapping accuacry, the improvement of WICSI accuracy are 17.2% and 2.2% for extraction of forest land area. The hybrid method is a version of combination betwee new NDVI and morphological texture, and can reduce the spectral variation and noise in homogeneous regions and preserve discriminative information simultaneously.

5 CONCLUSION

This letter evaluated the feasibility of hybrid method with spectral features and morphological textures for extraction of forest land area image using color aerial image. It found that the hybrid method can give promising results for forest-cover fraction (97.36%), which significantly outperformed the new NDVI or the morphological textures method.

ACKNOWLEDGMENTS

The acquisition of remote sensing imagery was supported by a grant to X. Chen form the Wuhan Scientific Bureau Project (200970634267). The author would thank the associate editor and reviewers for helpful suggestions which significantly improved this paper.

REFERENCES

Bannari, A., Morin, D., Bonn, F., & Huete, A. R. 1995. A review of vegetation indices. Remote Sensing Reviews, 13, 95–120.

Baret, F. 1995. Use of spectral relectance variation to retrieve canopy biophysical character. In F. M. Danson,& S.E. Plumer, Adavances in environmental remote sensing. Chichester:Wiley (chap.3).

Baret, F.,& Guyot, G. 1991. Potentials and limits of vegetation indices for LAI and PAR assessment. Remote Sensing of Enviroment, 35, 161–173.

Beisl, U., 2006. Absolute spectroradiometric calibration of the ADS40 sensor. International Archives of Photogrammetry, Remote Sensing and Spatial Information Sciences. XXXVI (part 1), on CD-ROM, 5 pages.

Bellens, R. & Gautama,S., L.Martinez-Fonte, W.Philips, J.C.Chan, and F. Canters, 2008."Improved classification of VHR images of urban areas using directional morphological profiles," IEEE Trans. Geosci. Remote Sens., vol. 46, no. 10, pp. 2803–2813, Oct.

Cliffe S. et. al. 2011. World development report 2011: conflict, security, and development.

Epifanio I. & Soille, P. 2007. "Morphological texture features for unsupervised and supervised segmentations of natural landscapes," IEEE Trans. Geosci. Remote Sens., vol. 45, no. 4, pp. 1074–1083, Apr.

Foody, G. M.,& Arora, M. K. 1996. Incorporating mixed pixel in the training, allocation and testing stages of supervised classification. Pattern Recognition Letters,17,1389–1398.

Gallego, F.J., 2004, Remote sensing and land cover area estimation. International Journal of Remote Sensing, 25,pp. 3019–3047.

Huang X., L. Zhang, 2009. "Evaluation of morphological texture features for mangrove forest mapping and species discrimination using multispectral ikonos imagery, vol.6, no.3,pp.393–397. Jul.

Leica Geosystems, 2007. Leica ADS40 2nd Generation Airborne Digital Sensor, www.leica-geosystems.com, (accessed 20 Oct. 2007).

Koukoulas, S. Geogreg Alan Blackburn. 2001. Introducing new indices for accuracy evaluation of classified images representing semi-natural woodland environments. Photogrammetric Engineering & Remote Sensing. 67(4): 499–510.

Perdigo V., Annoni A., 1997. Techinal and Methodological Guide for Updating CORINE Land Cover Data Base. JRC-EC.

P.Soille and M. Pesaresi, 2002. "Advances in mathematical morphology applied to geoscience and remote sensing," IEEE Trans. Geosci. Remote Sens., vol. 40, no. 9, pp. 2042–2055, Sep.

Pesaresi M. and Benediktsson, J. A. 2001. "A new approach for the morphological segmentation of high resolution satellite imagery," IEEE Trans. Geosci. Remote Sens., vol. 39, no.2, pp. 309–320, Feb.

Rembold, F.; Maselli, F. 2006. Estimation of inter-annual crop area variation by the application of spectral angle mapping to low resolution multitemporal NDVI images. Photogrammetric Engineering and Remote Sensing, v.72, p.55–62.

Sandau, R., Braunecker, B., Driescher, H., Eckhardt, A., Hilbert, S., Hutton, J., Kirchhofer, W., Lithopoulos, E., Reulke, R.,Wicki, S.,2000. Design principles of the LH Systems ADS40 airborne digital sensor. International Archives of Photogrammetry, Remote Sensing and Spatial Information Sciences, 33(Part B1), 258–265.

Stern, A.J.; Doraiswamy, P.C.; Cook, P.W. 2001. Spring wheat classification in an AVHRR image by signature extension from Landsat TM classified images. Photogrammetric Engineering and Remote Sensing, Reston, v.67, n. 2, p.207–211.

Taylor, G.J., Bagby, R.M., & Parker, J.D.A.1997. Disorders of affect regulation. Cambridge, England: Cambridge University Press.

Settle,J.,& N.A.Drake,1993:Linear mixing and the estimation of ground cover proportions. Int. J. Remote Sens.,14,1159–1177.

Veroustraete, F., Sabbe, H., Eerens, H. 2002. Estimation of carbon mass fluxes over Europe using the C-Fix model and Euroflux data, Remote Sensing of Enviroment 83(3). 376–399.

Computational Intelligence in Industrial Application – Ling (ed.)
© 2015 Taylor & Francis Group, London, ISBN: 978-1-138-02818-0

Topology relations based on the integrative data model of vector and raster

Y.T. Jin, X.F. Yang & Y.C. Wang
North China Institute of Astronautic Engineering, Langfang, China
Hebei Collaborative Innovation Center for Aerospace Remote Sensing Information Processing and Application, Langfang, China

Z.W. Zhang
Institute of Remote Sensing and Digital Earth Chinese Academy of Sciences, Beijing, China

ABSTRACT: In Geographic Information System (GIS), the exploration of topology spatial relations, the research on the metric description, which has been paid more attention to, are directly affected by the spatial data model. Vector and raster data model are the two types of basic spatial data model, and these two data models have distinguished advantages in terms of describing spatial relations between objects respectively. The integrative data model of vector and raster is derived from the integration of the advantages of vector and raster data model. Firstly, this paper defines qualitative topology relations by using the 9-intersection model. Secondly, the ratio of the grid number of intersections from two objects to the two objects is used to determine the intersect component. Thirdly, according to the statistical characteristic of the grid, the maximum and minimum distances are used to determine the closeness component. Finally, a triple group, including qualitative topology relations, intersect component and closeness component, is proposed to describe the topology spatial relation. Because of two advantages of integrated data models of vector and raster, the metric description of topology between different types of objects can be realized more effectively in this paper.

KEYWORDS: Topology relation; The integrative data model of vector and raster; Intersect component; Metrization.

1 INTRODUCTION

The spatial relationship is based on spatial data model, spatial analysis process modeling, spatial query and space operation. From the point of view of the psychological perception, Freeman (1975) and other scholars researched the theory of the spatial relationship in the mid of 1970s. In the 1980s, the spatial relationship got the full attention of the experts in the field of GIS, and Boyle et al (1983) proposed and developed the basic framework of the theory.

In the field of topological relationship, Egenhofer et al (1991) proposed 4-intersection model and 9-intersection model, then Randell (1992) proposed the RCC model (regional connectivity calculus), after that, J Chen et al (2000, 2001) proposed 9-intersection model based on Voronoi diagram, as well as 4-intersection model proposed by M Deng (2005). Recently, because the traditional model cannot describe the feature object of the topological relationship between quantitative information, scholars also conducted a series of quantitative research based

on the vector model (Egenhofer, 1998; Shariff, 1996; Nedas, 2007; Isl, 1999; Moratz, 2003), thus we can use quantitative information to describe the topological relationship.

However, many scholars, based on the vector model study. Take 9-intersection model, for example, Shariff (1996), Egenhofer (1998) and Nedas (2007), respectively, studied topology relationship between face and face, line and face, line and line. But in the space of R2, there exist differences in the concept of border and internal between point, line and face, for example, line objectives take inconsecutive points in 0-D as the border while continuous line in 1-D as the internal, and face objectives take continuous line in 1-D as the border while continuous in 2-D face as the internal. The three above scholars failed to achieve formal unity of expression about topological relations degree quantitative parameter in their line and line, line and face, face and face. As a result, there exist shortages in the degree of vector spatial data model.

JY Gong (1992) suggested a way of two-Morton code to build vector and raster integrative model in the early 1990s. Based on UML language, Kjenstad (2006) proposed parameterized geographic object model (PGOModel) for expressing vector and raster model. Cova et al (2002) applied a series of mappings to integrative vector and raster together. Goodchild (2007) integrated vector and raster data model from an atomic form of basic data structures (geo-atom). Thus, this paper study the quantitative expression of topological relationship by means of vector and raster integration model. We will use the vector expression of the model and 9-intersection model to determine the qualitative topological relationship, and the raster expression to create a topological relationship of overlapping components and neighboring components, as well as, describe the topology information. Because the points, lines and faces can be described by raster, it can formally picture the topological relationship between different types of geographic targets (point and point, point and line, point and face, line and line, line and area, face and face).

2 INTEGRATIVE DATA MODEL OF VECTOR AND RASTER

In this paper, we use an integrated model of vector and raster (JY Gong, 1992) as the basic object, that is to say, use two Morton code to record the intersection of points, lines and faces.

In order to take full use of the advantages of vector data model, we introduce precise positioning information of the vector model:

1 For points, we will record the position of raster and precise positioning information of points in the space of R2;
2 For lines, we will record the starting point, ending point, and intermediate nodes of various arcs in R2 space.
3 For faces, we will record the intermediate nodes of arcs on the face domain boundary in R2 space.

Hence, Spatial data model describes the geospatial (Figure 1) from both R2 and Z2 spatial characteristics. In the data model of vector and raster, all geographic targets combines both vector and raster.

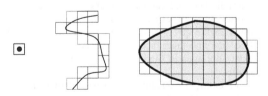

Figure 1. The description of point, line and area in the integrative data model of vector and raster.

3 DETERMINATION OF TOPOLOGICAL RELATIONSHIP

A qualitative topological relationship can be reformed by the 9-intersection model. In Figure 2, $A1$, $B1$ are the vector objects A and B, respectively. According to the point set topology knowledge, internal($A°$), external(Ae) and boundary(αA), we establish the topology relationship matrix, namely:

$$tr_9 = \begin{bmatrix} \partial A_1 \cap \partial B_1 & \partial A_1 \cap B_1^0 & \partial A_1 \cap B_1^e \\ A_1^0 \cap \partial B_1 & A_1^0 \cap B_1^0 & A_1^0 \cap B_1^e \\ A_1^e \cap \partial B_1 & A_1^e \cap B_1^0 & A_1^e \cap B_1^e \end{bmatrix} \quad (1)$$

According to the formula, we can express two kinds of relationship between point and point, three kinds of relationship between point and line, three kinds of relationship between between the relationship between point and face, twenty-three kinds of relationship between between line and line, nineteen kinds of relationship between line and face, eight kinds of relationship between face and face.

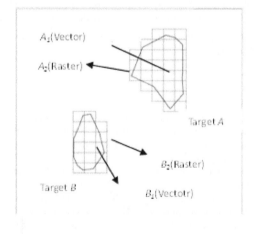

Figure 2. The vector and raster of the integrative data model.

We may meet the following questions in the study of integrative vector and raster data model about qualitative topological relationship: in the vector model, the two geographic targets can disjoint or touch, but the two raster goals show overlap phenomenon; in the raster model, the two geographic targets meet, but two goals intersect in the vector model of topological relationship, in Figure 3. For this phenomenon, this paper presents the following settings:

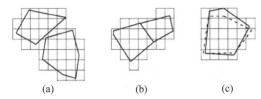

(a)   (b)   (c)

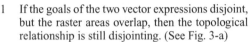

Figure 3. The difference of topological relationship of the vector and raster of the integrative data model.

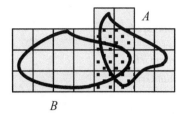

Figure 4. Raster of intersection between object *A* and *B*.

1 If the goals of the two vector expressions disjoint, but the raster areas overlap, then the topological relationship is still disjointing. (See Fig. 3-a)

2 If the goals of the two vector expressions touch, but the raster areas overlap, then the topological relationship is still touching. (See Fig. 3-b)

3 If the goals of the two vector expressions intersect, but the raster areas equal, then the topological relationship is still intersecting. (See Fig. 3-c)

The qualitative topological relationship is determined by the vector model in the integrative model, while the raster model participates the determination of quantitative parameter.

Topological relationship is divided into two categories in this paper: The first category includes touch, tangent, intersection, and containment; the second category is disjoint. For the first type, the inevitable overlapping raster model, this paper introduces the overlap component to complement the description of the metric information; For the second category, this paper mainly uses neighboring component metric information supplementary description.

4 OVERLAPPING OF COMPONENT

When the two goals intersect, the intersecting portion size and how to intersect characteristics can not be obtained from the qualitative topological relationship. As a result, the measure of the topological relationship is necessary. With the help of the integrative vector and raster model, we introduce the overlap component pairs (CI_A, CIB) to supplement the description of topological relationship.

As in the integrative data model of vector and raster, points, lines, faces, and other objects are described as one or more of the raster. Therefore, this paper proposes overlap component, which can be used to describe all types of topological relationship between point and point, point and line, point and face, line and line, line and face, face and face.

Where denotes the number of overlapping part Given source target B and the reference target A (see in Fig. 4), overlapping raster number of the two objects is 6. We define the overlap as the raster number of intersecting parts of the two target components, respectively, and it shares the same ratio of the total number of raster. Overlapping component is represented by the pairs, where and are the ratio of the target number and total number. And are in the range of [0, 1], when t or disjoint, the values are not necessarily to 0 (Fig. 3-a, 3-b); however, he topological relationship is intersecting, the values are in the range of (0,1); when A contains B, is in the range of (0, 1], and is 1, and the opposite is also right. And can be described as follows: and denotes the raster number of A and B.

In addition to the topological relationship shown in Fig. 4, another special case is as follows, because the effect of intersecting parts, it makes intersecting portions separate, see Fig. 5: the role of the wound due to two goals,: (a) multiple intersections formed by line and line; (b) two non-contact formed by line and face; (c) two separate face formed by the face and face.

In line and line, the multiple intersections are formed with two goals; in line and face, two separate arcs are formed with two goals; to face and face, two separate faces are formed with two goals.

To better distinguish whether the intersection area is consistent, this paper uses more pairs to express the overlaps between the two target components.

First, we define the intersecting area as a separate point, or internal communication arc, or region. When the two targets tangent or touch, the overlapping parts of the raster may form more than one area. Overlapping components can be expressed as $\{(CI_{A_1}, CI_{B_1}), ..., (CI_{A_n}, CI_{B_n})\}$, where n represents the number of the intersecting areas. Formula (2) (3) becomes:

$$CI_A = \frac{numGrid_{AB}}{numGrid_A} \qquad (2)$$

129

$$CI_B = \frac{numGrid_{AB}}{numGrid_B} \tag{3}$$

$$CI_{A_i} = \frac{numGrid_i}{numGrid_A} \tag{4}$$

$$CI_{B_i} = \frac{numGrid_i}{numGrid_B} \tag{5}$$

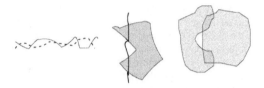

Figure 5.   The topological relationship of winding targets.

Where $numGrid_i$ denotes the raster number of i-the intersecting part.Overlapping component is mainly used for topological relationship of intersect, tangent, consistency, or containing. When the topological relationship between the two targets is disjoint, overlapping component is 0. At this time, we can not give appropriate metric information. Hence, this paper introduces the neighboring component to better complement the description of the degree of disjointing.

5   ADJACENT COMPONENT

We can use pairs to describe the adjacent component, where the two variables denote the maximum distance and minimum distance.

Raster data satisfies the two most important characteristics of the statistics, independence and representation. In Fig. 1, a point in the raster is expressed by independent raster, while line and face is a set of raster to express. Each raster can be provided by a set of quantized coordinates to express uniquely (sign the rows and columns, or calculate the coordinates of the center of the raster), and there was no overlapping phenomenon between each raster, so each raster is independent. Each raster has its attributes, which represent the characteristics of the occupied space, while all rasters perform overall (geographic objects) structure and edge nature, so it is representative. Thus, this paper takes the raster data as a statistical sample, and determines the topological relationship.

In this paper, we will use the Euclidean distance to measure the distance between each raster, which can be casted as:

$$dis_{ij} = \sqrt{(x_i - x_j)^2 + (y_i - y_j)^2} \tag{6}$$

Where i、 j denote the rasters, disij denotes the distance beween i and j.

Let the raster number of target A and target B be m and n. Select a distance approach to the feature the distance dis-ij between object A-i and B-j -th, then we get a m × n distance matrix D.

$$D = \begin{bmatrix} dis_{11} & dis_{12} & \cdots & dis_{1n} \\ dis_{21} & dis_{22} & \cdots & dis_{2n} \\ \cdots & \cdots & \cdots & \cdots \\ dis_{m1} & dis_{m2} & \cdots & dis_{nn} \end{bmatrix} \tag{7}$$

After that, we apply the system clustering method in multivariate statistics to denote the long distance and the shortest distance, then the distance between A and B can be expressed as in Equation (8) and (9): The longest distance:

$$D_{min} = \min_{i \le m, j \le n} \{dis_{ij}\} \tag{8}$$

The shortest distance:

$$D_{max} = \max_{i \le m, j \le n} \{dis_{ij}\} \tag{9}$$

Then, the adjacent component is (D_{min}, D_{max}).

6   CASE STUDY

Based on overlapping component and adjacent component, we will use triple to describe the topological relationship between the two goals of qualitative and metric information. The formula (10) is shown below:

Where TR (A, B) denotes topological relationship between target A and B; Tr9 (A, B) denotes 9 intersection model under qualitative topology; the pairs (CIA, CIB) and (D_{min}, D_{max}) denote overlapping component and adjacent component, respectively. The following gives 16 representative topological relationships between points, lines and faces (Fig. 6).

$$TR(A,B) = <tr_9(A,B), \{(CI_{A_1}, CI_{B_1}), (CI_{A_2}, CI_{B_2}), \ldots, \\ (CI_{A_3}, CI_{B_3})\}, (D_{min}, D_{max})> \tag{10}$$

The topological relationship between A and B in Fig.6 is the same as in Table 1. The above 31 kinds of instances covers different types of ground goals, and taking into account of the different qualitative topological relationship. In Fig. 6-1,6-4,6-8,6-12,6-19 and 6-25, topological relationship of the 9-intersection model is disjointing, overlapping component is (0,0), while the adjacent component, through the maximum and minimum distance, describes the disjointing situation between the two goals. When the two goals are points, the maximum and minimum are the same (Fig. 6-1); In Fig. 6-3,6-5,6-9,6-13,6-20 6-26, though the target topological relationship is disjointing, raster shows overlapping, and the overlapping component is not (0,0);

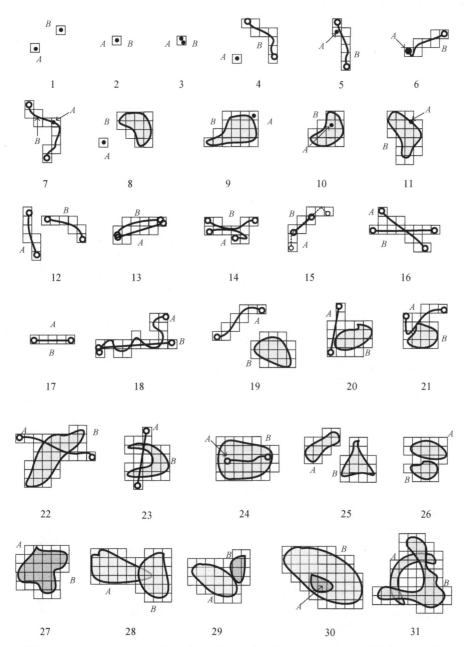

Figure 6. Thirty-one representative topological relationship based on the integrative data model of vector and raster.

131

When the topological relationship between the two goals is containing (Fig. 6-7,6-l0, 6-15,6-24,6-30), in overlapping component pairs must have CIA=1 or CIB=1, and the adjacent component can be expressed as $(0, Dmax)$, and when Dmax is also 0, topological relationship is shown in Fig. 6-2,6-17,6-27, namely equal. In Fig. 6-16,6-22 and 6-28, topological relationship is intersecting, and overlapping components are not zero, Dmin=0, Dmax$\neq$0; If the topological relationship is tangent (Fig. 6-6,6-11,6-14,5-21,6-29), although the intersection is a point, the number of overlapping raster may not be 0, and may be greater than 1, so the overlapping component CIA$\neq$0 and CIB$\neq$0, while Dmin=0, Dmax$\neq$0.The above description of the topological relationship is basic and simple. In tangent or intersecting relationship, the intersection is a connected area. But in many cases, due to the twinning of the two targets, its intersection may be composed by independent sub-regions, as shown in Fig. 6-18, 6-23 and 6-31. As described in section 4, we use a set of pairs to represent the overlapping components, and the number of pairs can be learned from overlapping components, so we can fully acknowledge topological relationship like intersecting or tangent.

7 CONCLUSION

Based on the topological relationship of 9-intersection model, this paper introduces the overlapping component and the adjacent component to complement the description of the topological information, thus makes a topological relationship describe more effectively. In view of the different dimensions of point, line, and face (zero-dimension, one-dimension, two-dimension), We, based on the R2 space of vector spatial data models, cannot effectively describe the relationship information. Therefore, this paper utilizes the integration of vector and raster spatial data model to the research topological relationship. We use overlapping ratio between the target components to determine the overlap, and calculate the shortest distance and the maximum distance to determine the adjacent component. In this paper, We, through analysis of the results, present triples descriptive model of topological relationship to describe more detailed and effective. We will study the effects of vector and raster integrative data model in the next work.

REFERENCES

Boyle A R, Dangrmond J, Marble D F, et al. 1983. Final Report of a Conference on the Review and Synthesis of Problems and Directions for Large Scale Geographic Information System Development. National Acronautics and Space Administration, Contract NAS2–11246.

Cova T. J, Goodchild M.F. 2002. Extending Geographical Representation to Include Fields of Spatial Objects. International Journal of Geographical Information Science,, 16:509–532.

Chen J., Li, C. Li Z. et al. 2001. A Voronoi-based 9-intersectionModel for Spatial Relations. International Journal of Geographical Information Science, 15 (3): 201–220.

Chen J. , Li C., Li Z., et al. 2000. Improving 9-intersection model by replacing the complement with voronoi region. Geo-Spatial Information Science, 3(1):1–10.

Deng M., Liu W., Feng X.. 2005. Generic Model of Topological Relations between Spatial Regions in GIS. Acta Geodaeticaet Cartographica Sinica, 34 (1):85–90.

Egenhofer M J, Shariff R. 1998. Metric Details for Natural-Language Spatial Relations. ACM Transactions on Information Systems, 16(4): 295–321.

Egenhofer M J, Matthew P D. 2009. Topological relations from metric refinements. Proc. of the 17th ACM SIGSPATIAL Int. Conference on Advances in GIS, 2009.

Egenhofer M J, Franzosa R. Point-set 1991. Topological Spatial Relationships. International Journal of Geographical Information Systems, 5(2):161–174.

Egenhofer M J, Herring J. 1991. Categoring Binary Topological Relationships between Regions, Lines, and Points in Geographic Databases. Oronoi Department of Surveying Engineering, University of Maine.

Freksa C, Brauer W, Habel C, et al. 2003. Spatial Cognition III. Berlin: Springer - Verlag, 385–400.

Freeman J. 1975. The modeling of spatial relations. Computer Graphics and Image Processing, 4:156–171.

Gong Jian-ya. 1992. An Unified Data Structure Based on Linear Quadtrees. Acta Geodaeticaet Cartographica Sinica, 21(4):259–266.

Goodchild M F, Yuan M, Cova T J. 2007. Towards a general theory of geographic representation in GIS, International Journal of Geographical Information Science, 21(3):239–260.

ISL I A, MORATZ R. 1999. Qualitative Spatial Representation Reasoning: Algebraic Models for Relative Position. University at Hamburg, FB Informatik, Hamburg.

Kjenstad K. 2006. On the integration of object-based models and field-based models in GIS, International Journal of Geographical Information Science, 20(5):491–509.

Moratz R, Nebel B, Freksa C. Qualitative spatial reasoning about relative position: The tradeoff between strong formal properties and successful reasoning about route graphs.

Nedas K, Egenhofer M. 2007. Metric Details of Topological Line-Line Relations. International Journal of Geographic Information Science,21(1):21–48.

Randell D, Cui Z, Cohn A. 1992. A spatial logical based on regions and connection[A]. In: Kaufmann M, San Mateo (eds.). Proceedings of 3rd International Conference on Knowledge Representation and Reasoning, New York: Springer-Verlag: 165–176.

Shariff R. 1996. Natural-Language Spatial Relations: Metric Refinements of Topological Properties. Ph.D. Thesis. Department of Spatial Information Science and Engineering, University of Maine, Orono, ME, USA.

Virtual machine migration strategy based on multi-objective optimization

ZhenGuo Chen & XiaoJu Wang

Department of Computer Science and Technology, North China Institute of Science and Technology, East Yanjiao, Beijing, China

ABSTRACT: The virtual machine has important applications in the cloud computing environment. The seamless migration of virtual machines needs to be resolved in its application process. In this paper, multi-objective optimization algorithm based on particle swarm optimization is introduced for virtual machine migration strategy in the selection process. Meanwhile, combined with the performance and state of the virtual machine, we will divide the virtual machine on a different list. By this division, the question of the virtual machine with the excellent performance was wrong; migrated can be avoided. By comparison with the traditional method of simulation, we can see that this algorithm has faster convergence and selection speed, and improves the performance of virtual machines work.

KEYWORDS: Virtual machine; Migration strategy; Particle swarm optimization.

1 INTRODUCTION

In recent years, with the fast development of cloud computing, virtualization technology has got considerable progress, which has also provided a solid foundation for the rapid development of cloud computing. After the large-scale use of virtualization technology by IDC (Internet Data Center), the number and load of virtual machine will change frequently along with the demand of users, so it has provided challenges for the resource scheduling of the virtual machine cluster. Virtual machine migration technology [1] can migrate the whole operation environment (including the operating systems) from one physical server to another physical server, so as to effectively manage the data center resources and get convenient for the maintenance and energy saving of the physical server.

During the migration process of virtual machine, it should select the suitable virtual machine to reduce system spending caused by virtual machine migration. How to find the best program to migrate the virtual machine to the target node, in order to make the summation of all the resources used by the virtual machine in each node does not exceed the resource limit of the node server, besides, the problem of using the least servers number can be understood as the MOP (Multi-objective Optimization Problem). In fact, most of the science and engineering problems are multi-objective optimization problems, they usually do not have the unique optimal solution, but a group of optimal solution set, so the current focal point of research is how to get this optimal solution set. At present, there is no optimal solution algorithm for the polynomial, in general, people use some heuristic algorithms and make simple rules, for example, suboptimal matching, priority matching and optimal matching.

Because of the advantages of simple calculation, strong robustness, fast convergence and less experience dependent parameters, Particle swarm optimization [2] shows the stronger superiority in solving complex problems, which makes the application on MOP gradually becomes the hot spot problem of research. Literature [3] puts forward a multi objective particle swarm optimization based on the adaptive selection on global optimal location, it also introduces the crowding distance mechanism, so as it has better convergence and distribution. Literature [4] puts forward a kind of improved multi objective particle swarm optimization by increasing the hop count change mechanism to change the method of particle searching and make the non-inferior excellent solution be more converge to the front of Pareto with better astringency.

Literature [5] [6] describes the dynamic migration of virtual machine into optimization problem, the optimization objective is to minimize the energy consumption. Literature [7] decomposes the virtual machine placement problem into two parts of combinatorial optimization problems and multi-objective optimization problems in a box-packing. Firstly, it uses a genetic algorithm to solve the combinatorial optimization problem of placing the virtual machine on the nodes, and then combine the fuzzy logic to optimize multiple objectives, including the overall resource waste, energy consumption and heat dissipation cost minimization. But the literatures above only consider the energy consumption problem, do not consider the influence of load balance between all the performances and the server.

Aiming at the problem of virtual machine migration selection strategy, this paper puts forward a kind of modified particle swarm optimization, through the definition to match the distance, make the quantification on a variety of performances and add the servers whose residual performance cannot meet the migration of virtual machine into the avoid list, which reduces the subsequent particle swarm search range and increase the convergence speed.

2   PARTICLE SWARM OPTIMIZATION

Particle swarm optimization is a new kind of EA (Evolutionary Algorithm) developed in recent years. PSO is similar to the genetic algorithm, it is an optimization algorithm based on iteration. It is initialized by the system into a set of random solutions and search for the optimal value by iteration. But is has no crossover and mutation used by the genetic algorithm, it is just searching for the optimal particle in the solution space. Compared with the genetic algorithm, the advantage of PSO is the simple and easy realizing method, and it does not have so many parameters to be adjusted. At present, it has been widely used in function optimization, neural network training, fuzzy system control and other application fields of genetic algorithm.

The particle can update its speed and location according to the following formula (1) and (2).

$$v_i^{k+1} = \omega \times v_i^k + c_1 \times rand_1 \times (pbest - x_i^k)$$

$$+ c_2 \times rand_2 \times (gbest - x_i^k) \tag{1}$$

$$x_i^{k+1} = x_i^k + v_i^{k+1} \tag{2}$$

In the formula: k is the number of iterations, c1 and c2 are the learning factors, they are also called as the acceleration constant, r1 and r2 are the uniform random number in the range of [0,1].

3   VIRTUAL MACHINE MIGRATION SELECTION STRATEGY

3.1   *Virtual machine migration selection strategy*

The performance characteristics of virtual machine and server mainly include CPU, memory, network bandwidth and so on, when using k performance characteristics to describe the performance characteristics of virtual machine, it can establish the performance demand vector for each virtual machine $(Vm_{n1},...,Vm_{nk})$, and establish performance vector for each server $(Ser_{m1},...,Ser_{mk})$. If the value of is $(Ser_{mk} - Vm_{nk})$ negative, it shows that the server cannot satisfy the performance demand of the virtual machine; if the value is positive, it needs to measure the fitness degree between all the virtual machines and the servers.

Because the dimensions of different performances are different, so the magnitude grade of server performance rest and virtual machine performance need also have differences, therefore, it makes the normalization processing on the server performance rest and virtual machine performance need [8].

Assume $\{Ser_{1p},...,Ser_{mp}\}$ is the set of same the kind of remaining properties p in server Ser_m, $\{Vm_{1p},...,Vm_{np}\}$ is the set of same kind need properties p corresponding to Vm_n. The normalization algorithm of server performances is:

$$S_{mp} = \frac{Ser_{mp} - Ser_p^{\min}}{Ser_p^{\max} - Ser_p^{\min}} \tag{3}$$

Among them, $Ser_p^{\max}$ is the maximum value in set of $\{Ser_{1p},...,Ser_{mp}\}$ in the same kind of remaining properties p in server, $Ser_p^{\min}$ is the minimum value. The normalization algorithm of virtual machine needs performances is:

$$V_{np} = \frac{Vm_{np} - Vm_p^{\min}}{Vm_p^{\max} - Vm_p^{\min}} \tag{4}$$

In order to define the fitness of virtual machine and the server, this paper puts forward the concept of matching distance $Mat_{Vm_nSer_m}$ (the followings is abbreviated as). The matching distance is mainly obtained by the Euclidean distance of the migration of virtual machine performance needs and the remaining properties of server:

$$Fit_{nm} = Mat_{nm} = \sqrt{\sum_{p=1}^{k} (Snp - Vmp)^2} \tag{5}$$

The smaller of the matching distance, the remaining properties of server is more fit to the migration of virtual machine performance needs, in this paper, the smaller the value of fitness is, it shows that the server is more fit to the migration of virtual machine.

3.2 Server avoid list

This paper lists the definition of server avoid list:

Definition 1: After the virtual machine migrating to the server, if the remaining performance of the server cannot meet the performance requirements of some virtual machine, it should add the server into the particle swarm avoid list of the migration of virtual machine server.

In order to make the summation of the use resources of virtual machine in each node does not exceed the resource limit of node server, and minimize the used number of servers, when the servers are migrated to another virtual machine, if the remaining performance cannot meet the remaining performance of virtual machine, it should add it into the particle swarm avoid list of the migration of virtual machine server.

3.3 Algorithm of modified particle swarm optimization

The algorithm flow of MPSO is as follows:

1 Initialize the particle swarm, include the group size N, the position Xi and speed Vi of each particle;
2 Initialize the avoid list and clear the avoid list;
3 Calculate the fitness value f(i) of each particle according to the objective function and the formula (3);
4 Calculate the extreme value gBest of each individual for each particle according to the target function;
5 Calculate the whole extreme value gBest for f1(x)...fn(x) under the target function;
6 Update the avoid list according to the whole extreme value, if the remaining performances of the server cannot meet the performance demand of virtual machine, it shall be added into the particle swarm avoid list of the migration of virtual machine server.
7 Update the speed Vi and position Xi of the particle according to formula and (2);
8 If the end condition is satisfied (all the virtual machines have migrated into the most appropriate server or reached to the maximum cycle number) to quit, or return to (2).

4 SIMULATIONS AND RESULTS ANALYSIS

4.1 Parameter setting

This paper uses the internet laboratory in Melbourne University in Australia and the cloud simulation platform – CloudSim put forward by Gridbus project [9] as the test simulation tool. Through the modification

on Vm and Datacenter class, re-compile the CloudSim to obtain the simulation environment. And then compile the simulation program on this base.

Firstly, establish a Datacenter, including 15 servers (Host), select 3 performances of CPU, and memory and network bandwidth as the matching distance measure parameter, as listed in table 1. And then submit the migration task request of 50 virtual machines (Vm), as listed in table 2.

As for the set of algorithm parameters in this paper: the number of particles usually uses 20-40. In fact, for most of the problems, 10 particles are enough to get the excellent result. But as for the more difficult problems or some certain classes of problems, the number of particles can take 100 or 200. The learning factor $c1=c2=1.49$, $w=0.7$, conduct 100 times of iteration for each function, and conduct 200 times of iteration to get the average for the optimal value.

Table 1. List of servers.

Num	CPU/num	CPU/MIPS	MEM/Gb	BPS/Mb/s
0	2	2000	4	1000
1	2	2500	4	1500
2	2	2000	4	1000
3	4	2500	6	1500
4	4	3000	6	2500
...	...	...	...	...
15	6	7000	8	4000

Table 2. List of submitted virtual machines.

Num	-CPU/num	CPU/MIPS	MEM/Gb	BPS/Mb/s
0	1	200	1	100
1	1	250	0.5	150
2	1	250	1	250
3	1	300	0.5	250
4	1	250	1	350
5	2	350	1.5	400
6	2	350	1	400
7	2	250	1.5	450
8	2	300	0.5	350
...	...	...	...	...
50	6	800	3	600

4.2 Comparative analysis of algorithms

In order to verify the convergence of the algorithm, compare the based PSO with the MPSO(modified particle swarm) put forward in this paper, the convergence speed test results of the algorithm in this paper and the basic particle swarm optimization algorithm in each 50 times of iterations are shown in Figure 1. It can be seen that the algorithm in this paper, it is better than the basic particle swarm optimization algorithm in the aspect of convergence speed and can get the search results faster.

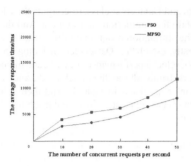

Figure 1. Compare of the convergence speed with MPSO and PSO.

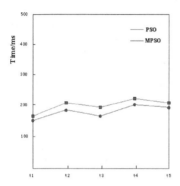

Figure 2. Selection speed comparison condition in each time.

It can be seen from the test result that, with the increase of virtual machine migration request number in each second, both the convergence time of the two algorithms increase gradually, but the convergence time of MPSO algorithms put forward by this paper is always lower than the basic particle swarm optimization algorithm, it is lower for about 13% to 19%%.

In order to compare the selection speed of selection strategy in this paper, it has designed the following scenes: it select 5 times (t1...t5), the time interval is 30 minutes, the requested virtual machine number at t1 time is 10, the number is 17 in t2 time, it is 15 in t3 time, 20 in t4 time and 16 in t5 time, compare the time from the requested virtual machine migration task to the server of the algorithm in this paper and the basic particle swarm optimization algorithm, the comparison result is shown in Figure 2.

From the results above, it shows that d could be measure of weigh up the accuracy and the recall rate to meet various practical needs.

It can be seen from the test results that, when processing the same virtual machine migration task requested number, the selection speed of selection strategy based on MPSO algorithm is lower for 15% to 30% than the selection strategy of optimization algorithm based on particle swarm.

5 CONCLUSIONS

This paper puts forward a kind of virtual machine migration selection strategy based on the particle swarm optimization. By defining the fitness of matching distance quantitative server or virtual machine, adding the avoid list to avoid the resources occupied by the virtual machine exceeding the server resources top-limit and reach the target of combinatorial optimization, the test proves the validity of that algorithm and improve the convergence speed of the algorithm.

ACKNOWLEDGMENTS

The work was supported by the Fundamental Research Funds for the Central Universities (3142013098, 3142013070) and Natural Science Foundation of Hebei Province of China (F2014508028).

REFERENCES

Michael Nelson, Beng-Hong Lim, Greg Hutchins. Fast Transparent Migration for Virtual Machines[C]. In Proceedings of USENIX ATC,2005.

Kennedy, J. And Eberhart, R. Particle swarm optimization. In Proceedings of IEEE International Conference on Neural Networks, volume 4,1995.

Huang Min, Jiang Yu, Mao An and so on. The Multi-Objective Particle Swarm Optimization Algorithm Based on the Overall Optimal Position and Adaptive Selection and Local Search. Computer Application. 2014.34(4):1074–1079.

Feng Jinzhi, Chen Xing, Zheng Songlin. A kind of Algorithm and Its Application of Improved Multi-Objective Particle Swarm Optimization. The Research on Computer Application. 2014.31(3):1001–3695.

Wang Xiao-Rui,Wang Ye-Fu.Coordinating power control and performance management for virtualized server clusters.IEEE Transactions on Parallel and Distributed Systems,2011,22(2):245–259.

Jung G, Hiltunen M, Joshi K et al. Mistral: Dynamically managing power, performance, and adaptation cost in cloud in Irastructures//Proceedings of the 30th IEEE International Conference on Distributed Computing Systems (ICDCS'2010).Genoa, 2010: 62–73.

Xu J, Fortes J. Multi-objective virtual machine placement in virtualized data center environments//Proceedings of 2010 IEEE/ACM International Conference on Green Computing and Communications (GreenCom'2010). Hangzhou,2010;179–188.

Cheng Liming, Wu Jiang, Zhang Yulin, Operational research model and method of the tutoria[M].Bei Jing:Tsinghua university press, 2002.

Calheiros R N,Ranjan R,De Rose C A F,et al.cloudsim:A novel Framework for Modeling and Simulation of Cloud Computing Infrastructures and Services[R]. GRIDS-TR-2009-1,Grid Computing and Distributed Systems Laboratory.The University of Melbourne. Australia.March 13.2009.

Computational Intelligence in Industrial Application – Ling (ed.)
© 2015 Taylor & Francis Group, London, ISBN: 978-1-138-02818-0

Band selection for hyperspectral image based on differential evolution algorithm

J. Wu & X. Li
Computer College, China University of Geosciences, Wuhan, China

ABSTRACT: Most of the traditional band selection algorithm uses the exhaustive search, and search efficiency is low, not easy for practical application. Differential Evolution (DE) algorithm, as a new kind of random optimization algorithm, is based on group differences, with global convergence speed, parallelism, robustness and other characteristics, and will be suitable for band selection for hyperspectral image. But because the dimension of the decision space is too high, the population convergence speed is slow, the standard of the DE algorithm of the optimal solutions to search would happen. To solve above problems, this paper proposes a kind of differential evolution algorithm based on the population initialization of opposition-based learning to apply to band selection for hyperspectral image, used to speed up the convergence rate populations and improve the overall efficiency of the search for the optimal solution. Experiments show that the algorithm is feasible and effective in the band selection for hyperspectral image.

KEYWORDS: Hyperspectral image; Band selection; Differential evolution algorithm; Population initialization of opposition-based learning.

1 INTRODUCTION

Hyperspectral image has a number of multi-band, high spectral resolutions, narrow bandwidth the huge amount of data and other characteristics, and the strong correlation between their bands, higher data redundancy, this for hyperspectral image classification and recognition has brought great difficulties. Band selection for hyperspectral image can be reduced dimensionality of hyperspectral image data on the basis of possible reserves object recognition rate, and reduces the computational complexity of the image classification, improve the precision of classification, which is currently an important direction of processing hyperspectral dimensionality reduction study [1]. At present more mature traditional band selection algorithm is divided into two categories:

1 Optimal band selection based on the amount of information, such as entropy and joint entropy, covariance matrix eigenvalues, optimal index factor, adaptive band selection and other methods.
2 Optimal band selection based on the separability between classes, such as standard distance between the mean, discrete degree, Bhattacharyya distance, Jeffries-Matusita distance, mixing distance, Euclidean distance, Markov distance, spectral angle mapping, spectral correlation coefficient and other methods.

Most of the traditional band selection algorithm uses an exhaustive search, but it will consume large amounts of search operation time, not easy to practical application. Differential evolution (DE) algorithm as a new kind of random optimization algorithm based on group differences, when dealing with high dimensional data, has a faster convergence rate, the parallelism of information processing, the robustness of the application, simplicity of operation etc. [2], which is suitable for hyperspectral image processing and analysis. However, the standard DE algorithm, because the decisions spatial dimension is too high, will appear difficult to search for the optimal solution, so that the population evolution is not only easy to fall into local optimum, and also can lead to a slow convergence rate of population and long computing time. Based on the above issues, this paper proposes a kind of DE algorithm based on the population initialization of opposition-based learning [3] to apply to band selection for hyperspectral image, used to speed up the convergence rate populations and improve the overall efficiency of the search for the optimal solution.

2 DIFFERENTIAL EVOLUTION ALGORITHM

DE algorithm is an evolutionary algorithm based on the real number encoding, have similar a framework with other evolutionary algorithms, such as population initialization, individual fitness evaluation, and from one generation of population to the next generation of population after mutation, crossover, and selection operation etc. DE algorithm mainly involves the following four parameters: (1) the population size N; (2) the individual dimension D (i.e., the length of the chromosome); (3) the variability factor F; (4) crossover probability CR.

2.1 Generate the initial population

Randomly generated to meet the constraints of the N chromosome in D dimensional space, the specific processes such as formula (1):

$$x_{ij} = rand(0,1)*(x_j^U - x_j^L) + x_j^L (1 \leq i \leq N, 1 \leq j \leq D) \quad (1)$$

The x_j^U and x_j^L in formula (1) are respectively the lower and upper bounds on the first j variables, $rand(0,1)$ returns a random number between [0,1].

2.2 Mutation operation

For individuals $x_{r1}(1 \leq r1 \leq N)$ in the population, thereby generating new individual $x_{r1}'(1 \leq r1 \leq N)$ satisfies the formula (2).

$$x_{r1}' = x_{r1} + F \cdot (x_{r2} - x_{r3}) \quad (2)$$

In formula (2), $r2, r3 \in [1, N], r1 \neq r2 \neq r3, F > 0$ is a scaling factor.

2.3 Crossover operation

The crossover operation is performed between each of the population of an individual whereby the $x_{r1}(1 \leq r1 \leq N)$ and $x_{r1}'(1 \leq r1 \leq N)$ after a new individual variability resulting, x_{r1} and x_{r1}' after crossover operation produce offspring of candidate individual v, through the back of the selection operation, determine the individual x_{r1} or individual v is reserved for the next generation, such as calculation of formula (3).

$$v_j = \begin{cases} x_{r1,j}', rand(0,1) \leq CR \\ x_{r1,j}, rand(0,1) > CR \end{cases} \quad (3)$$

In formula (3), $rand(0,1)$ returns a random number between [0,1], $CR(0 \leq CR \leq 1)$ is the crossover probability.

2.4 Selecting operation

After crossover and mutation populations produced a new individual v, according to the objective function value, select a genetic from the x_{r1} and v to the next generation. Using the minimum function as an example, the calculation such as formula (4).

$$x_{r1} = \begin{cases} x_{r1}, f(x_{r1}) \leq f(v) \\ v, else \end{cases} \quad (4)$$

3 DIFFERENTIAL EVOLUTION ALGORITHM BASED ON POPULATION INITIALIZATION OF OPPOSITION-BASED LEARNING

3.1 Basic idea of population initialization of opposition-based learning

In the process of algorithm calculation, when get to a solution, at the same time through calculation to get the symmetric solution of the solution in the decision space, in this article is called mirror individual, which is the core idea of opposition-based learning initialization.

Mirror individual is defined as follows: set a gene in individual P to x, $x \in [a,b]$, then the individual P in the decision space exists and only a mirror individual P, its corresponding gene x can be obtained by $\breve{x} = a + b - x$.

3.2 Algorithmic process

In the DE algorithm, the initial size of the population will be randomly generated N, the ultimate goal is to make the population as close as possible global optimal solution, but the idea of opposition-based learning initialization is create the mirror individual in accordance with the decision space center as a symmetric point for each individual in the population, and then choose a better between the individual and its mirror individual, making it closer to the global optimal solution. Through a screening for each individual, this will make the entire initial population closer to the global optimal solution. In fact, each mirror of individual has a 50% probability of more outstanding than the original individual, and therefore opposition-based learning initialization will accelerate the convergence of the population.

For any one individual $P(x_1, x_2, ..., x_n)$ in the population Pop(t), assuming that each of its components to meet $x_i \in [a_i, b_i](i = 1, 2, ..., n)$, the individual P in the decision space has one and only one mirror individual $\breve{P}(\breve{x}_1, \breve{x}_2, ..., \breve{x}_n)$, where $\breve{x}_i = a_i + b_i - x_i (i = 1, 2, ..., n)$.

Opposition-based learning initialization steps are as follows:

Step1: Randomly generate a population $Pop(t) = \{x^1, x^2, ..., x^N\}$ of size N, where $x^i = (x_1^i, x_2^i, ..., x_n^i)(i = 1, 2, ..., N)$, n is the dimension of the decision space.

Step2: Calculate and get the mirror individual $\breve{x}^i = (\breve{x}_1^i, \breve{x}_2^i, ..., \breve{x}_n^i)$ of each individual

$x^i = (x_1^i, x_2^i, \ldots, x_n^i)$ in the population, forming a mirror population $\text{Pop}(t) = \{\breve{x}^1, \breve{x}^2, \ldots, \breve{x}^N\}$.

Step3: Merge two populations, namely $\text{Pop}(t) \cup \text{Pop}(t)$, then sorted according to the fitness function values, the first N individuals inherited to the next generation.

4 EXPERIMENT AND ANALYSIS

4.1 Experimental data

The experimental data are 242 band 256 × 400 hyperspectral image data which obtained by the sensor of the hyperspectral imager Hyperion, the spectral resolution near 10nm, and the spectral range of 360nm-2570nm, as shown in Figure 1. The experimental object categories and the number of samples are as shown in table 1.

Table 1. Sample space category distribution table.

Category	Number of samples
Mountains	1550
Vegetation	1380
Water	620
Ground	650

Figure 1. Hyperspectral image data sample selection area.

4.2 Results of three band combination

Table 2 and table 3 are permutation tables of three band combination that obtained by various traditional algorithms, which only selecting related index value in the top 10 results.

Table 4 is the optimal experimental results of three band combination that obtained by the improved DE

algorithm. Comparison of Table 2 and table 3 using enumeration methods to calculate the evaluation criteria results can be found, the improved DE algorithm can calculate the optimal band combination enumeration method results in a relatively quick period of time. It is further proved that the improved DE algorithm is feasible and effective in the hyperspectral image band selection.

Table 2. Selection results of three band combination of various traditional algorithms.

No.	Joint entropy	Covariance matrix eigenvalues	Optimal index factor	Discrete degree	Bhattacharyya distance
	Band combination/ Value	Band combination/ Value	Band combination/ Value	Band combination/ Value	Band combination/ Value
1	120,166,167/0.569106	32,77,87/1.45249e+17	32,33,77/854.407	9,37,129/19.0628	35,39,47/14.9904
2	100,101,112/0.557731	33,77,87/1.43956e+17	30,32,77/853.991	21,97,101/18.3314	35,39,48/14.7158
3	100,101,115/0.541415	32,77,88/1.43263e+17	30,33,77/852.032	9,21,97/18.3194	29,39,47/14.6753
4	100,101,104/0.532693	32,77,85/1.42544e+17	26,32,77/850.515	16,21,97/18.1911	35,39,41/14.5973
5	100,101,113/0.528093	33,77,88/1.42018e+17	26,33,77/848.587	21,97,102/17.7562	29,39,48/14.5755
6	100,101,108/0.518998	33,77,85/1.41307e+17	26,30,77/848.038	18,21,97/17.7434	30,39,48/14.3068
7	100,101,109/0.518615	30,77,87/1.41176e+17	29,32,77/847.922	20,21,97/17.6767	30,39,47/14.2193
8	100,101,103/0.514258	30,77,88/1.3936e+17	27,32,77/846.268	9,37,166/17.6228	32,39,48/14.113
9	100,101,102/0.506548	30,77,85/1.38603e+17	29,33,77/845.988	21,97,100/17.6157	32,39,47/14.1057
10	100,101,111/0.495005	26,77,87/1.35809e+17	29,30,77/845.482	8,21,97/17.6111	33,39,48/14.0381

Table 3.   Selection results of three band combination by various traditional algorithms (Continued table2).

No.	Mixing distance	Euclidean distance	Markov distance	Spectral angle mapping	Spectral correlation coefficient
	Band combination/ Value	Band combination/ Value	Band combination/ Value	Band combination/Value	Band combination/Value
1	30,32,33/9400.06	30,32,33/5427.16	78,93,217/541994	120,182,183/0.436492	26,40,93/1.1228e-05
2	26,32,33/9359.96	26,32,33/5404.17	84,107,161/535842	120,182,184/0.426074	30,55,96/4.2353e-05
3	26,30,32/9357.54	26,30,32/5402.77	39,46,158/467368	120,183,184/0.422381	29,107,112/4.91636e-05
4	29,32,33/9337.55	29,32,33/5391.41	39,46,158/399601	33,41,183/0.412904	24,110,115/5.74507e-05
5	26,30,33/9337.13	26,30,33/5390.9	167,184,190/391776	33,41,131/0.412478	19,44,47/8.29085e-05
6	29,30,32/9335.13	29,30,32/5390.0	167,184,190/390094	33,41,188/0.41239	48,53,103/0.000111075
7	27,32,33/9315.99	27,32,33/5379.19	84,107,161/331921	33,41,134/0.412141	19,27,31/0.00011664
8	29,30,33/9314.71	29,30,33/5378.1	77,117,130/288255	33,41,189/0.412011	10,32,44/0.000151303
9	27,30,32/9313.57	27,30,32/5377.78	80,103,159/282084	33,41,184/0.411736	119,139,144/0.000157459
10	26,29,32/9295.04	26,29,32/5366.86	78,93,217/227168	33,41,182/0.411687	18,24,95/0.000172225

Table 4.   Optimal three band combination table by the improved DE algorithm.

Evaluation criteria	Band combination	Fitness function value
Joint entropy	120,166,167	0.5691056499
Covariance matrix		
Eigenvalues	32,77,87	1.45249e+17
Optimal index factor	32,33,77	854.4072641284
Discrete degree	9,37,129	19.0628072483
Bhattacharyya distance	35,39,47	14.9903921191
Mixing distance	30,32,33	9400.0585607940
Euclidean distance	30,32,33	5427.1553350273
Markov distance	78,93,217	541994.4180718781
Spectral angle mapping	120,182,183	0.4364923556
Spectral correlation Coefficient	26,40,93	1.1228e-05

5   CONCLUSIONS

For hyperspectral image band selection to remove redundant information is the key step in the hyperspectral image classification and recognition. Because the traditional band selection algorithm uses the exhaustive search, search efficiency is low, and then this paper proposes a kind of DE algorithm based on the population initialization of opposition-based learning to apply to band selection for hyperspectral image. The improved algorithm takes full advantage of global convergence speed faster and solution efficiency, higher of the DE algorithm, and overcomes the shortcomings of population convergence rate slower of the standard DE algorithm. By contrast with traditional band selection algorithm, the proposed algorithm is effective. Therefore, the proposed algorithm provides a new method for band selection of hyperspectral image.

ACKNOWLEDGEMENT

In this paper, the research was supported by the Fundamental Research Funds for the Central Universities of China University of Geosciences (Project No. CUGL120291).

REFERENCES

Mojaradi, B. & Abrishami, M. 2009. Dimensionality Reduction of Hyperspectral Data via Spectral Feature Extraction, *IEEE Transactions on Geoscience and Remote Sensing, 2009.* 47(2): 2091–2105.

Storn, R. & Price, K. 1995. Differential evolution-A simple and efficient adaptive scheme for global optimization over continuous spaces, Technical Report: TR-95-012, 1995.

Tizhoosh, H. R. 2005. Opposition-based learning: A new scheme for machine intelligence. *Int. Conf. on Computational Intelligence for Modeling Control and Automation-CIMCA'2005, Vienna, Austria, 2005.* pp: 695–701.

Computational Intelligence in Industrial Application – Ling (ed.)
© 2015 Taylor & Francis Group, London, ISBN: 978-1-138-02818-0

Demeter satellite electric field signal mutation point detecting based on wavelet transform

ZhiAn Pan
Department of Disaster Information Engineering, Institute of Disaster Prevention, Sanhe Hebei, China

LingLing Zhao
Department of Basic Courses, Institute of Disaster Prevention, Sanhe Hebei, China

ABSTRACT: While understanding and analyzing Demeter satellite electric field data, we first experiment on data preprocessing and information extraction. Then we do research on the signal mutation point detecting method based on wavelet transform, and implement it in VC++6. 0 and Matlab platform, making use of the corresponding relationship between the extreme maximum value of the wavelet transform modulus and mutational point of the signal. Experimental results show that using a low pass nature of Gaussian function as the generating function of wavelet transform can have a good effect on satellite data mutation point detection. It has a good application prospect on Demeter satellite electric field data processing.

KEYWORDS: Wavelet transform; Demeter satellite; Mutation detection Process neuron.

1 INTRODUCTION

Extracting earthquake ionosphere precursor information by satellite observation data has been drawn greater attention in earthquake prediction research in recent years. Precursory phenomena in the ionosphere[1] by Satellite are mainly: electromagnetic radiation, plasma parameters and energetic particle precipitation. This is called Lithosphere-Atmosphere-Ionosphere coupling phenomenon. The coupling possible ways include: chemical way, electromagnetic way, and acoustic way. Theoretical research and practical observations show that extracting earthquake ionosphere precursor information[2-3] by satellite observation data has a good application prospect.

France Demeter satellite[4] is a seismic monitoring satellite in the world. It was launched on June 29, 2004. It is in a synchronous solar quasi-circular orbit at a low-altitude of 710 km (changed into a 660 km since December 2005), with an inclination of 98. 23° and weighs of 130 kg. Satellite's payload consists of IMSC(search-coil magnetometer instrument, it can lead to several bands of electric and magnetic field data), ICE (electric field instrument), IAP (The Plasma Analyzer Instrument), ISL[5] (Langmuir probes) and IDP(high-energy particle detection instruments).

Mutation signal is also called singular signal. The mutation point of mutation signal often carries important information, and is one of the important characteristics of the signal. It has an important role and position in digital signal processing and digital image processing. It also can be used in Demeter satellite data processing. Traditional signal mutation detecting method is based on Fourier transform. We can infer whether a function has the mutability by the certain speed of Fourier transform of the function to near zero. It can only reflect the overall mutation of the signal, while the local mutation is beyond description. Therefore, we introduce wavelet transform algorithm.

2 DEMETER SATELLITE DATA PREPROCESSING

Demeter satellite data are divided into 4 levels, namely, Level-0, Level-0', Level-1, Level-2. Level-1 experimental data are standardized format data, mainly for scientific research. Level-1 data are divided into 6 categories, namely, electric field data, the magnetic field data, electron density data, ion density data, ion-temperature-density data, et al. electric field data are divided into 6 frequency bands. We use the electric field data [6] of the Level-1 ULF frequency band whose file identifier is 1129 for analysis.

The electric field data file format of ULF frequency is shown in Figure 1.

Time T_1				Time T_2				
block 1	block 2	block 3	block 4	block 1	block 2	block 3	block 4	...

Figure 1. Electric field data format of ULF frequency.

Each Level-1 data file is divided into several chunks at a time interval (Time Tn). Each chunk is divided into 4 smaller blocks of data (block1,block2,block3, block4), and the extension name of the file is .dat.

Because the encoding format for the source data file is Big Endian, the original data must be changed by encoding conversions and information extraction. The data files after transformation and extraction are shown in Figure 2.

2008.1.1 1:45:37 876ms
track number:18693
subtrack type:0
geocentric latitude:70.214806
geocentric longitude:148.896088
ground level:673.282166
local time:11.686927

Data coordinate system:-1.580993 2.141368 1.016956 -1.080875 0.111914
0.001986 -0.093847 0.009717 1.003935 sampling frequency:39.062500
Ex:92.096558 92.872658 91.894020 93.146500 92.054672 92.642044
91.585777 92.328156 91.748283 92.365341 91.860321 92.504349
91.590286 92.159615 91.377548 92.144371 91.489388 92.262047
91.711578 91.853302 90.845139 91.189606 90.982193 91.497147
91.085312 91.252983 90.905823 90.950302 90.856674 90.954010

Figure 2.   The preprocessed data file format.

3   WAVELET MUTATION POINT DETECTING METHOD

Step 1: Set h(t) is convolution of function f(t) and g(t), that is:

$$h(t) = f(t) \otimes g(t) \tag{1}$$

Step 2: According to the nature of the Fourier transform:

$$h'(t) = f'(t) \otimes g(t) = f(t) \otimes g'(t) \tag{2}$$

$$F[h'(t)] = j\omega F[f(t) \otimes g(t)] = j\omega \hat{f}(\omega)\hat{g}(\omega)$$
$$=[j\omega \hat{f}(\omega)]\hat{g}(\omega)=\hat{f}(\omega)[j\omega \hat{g}(\omega)] \tag{3}$$
$$= F[f'(t)] \otimes F[g(t)] = F[f(t)] \otimes F[g'(t)]$$

If we regard the function f(t) as a signal, g(t) as a filter, then the result of the derivative of the signal and the filter convolution is regarded as the result of the derivative of the filter and the signal convolution. For example, if we choose g(t) as the Gaussian function, then Morlet wavelet and Maar wavelet can be constructed using their derivative, so extreme value point and mutational point using Wavelet transform has a corresponding relationship at the extreme value point and mutational point of signal f(t), and mutational signal can be detected using Wavelet signal.

Step3: Setting θ (t) is a smooth, low-pass function, and satisfies the condition:

$$\int_{-\infty}^{\infty} \theta(t)dt = 1 \tag{4}$$

$$\lim_{|t|\to\infty} \theta(t) = 0$$

Choosing θ(t) as a Gaussian function.

$$\theta(t) = \frac{1}{\sqrt{2\pi}} e^{-t^2/2} \tag{5}$$

Step 4: Assuming θ(t) is twice differentiable, and define it as follows:

$$\psi^{(1)}(t) = \frac{d\theta(t)}{dt} = -\frac{1}{\sqrt{2\pi}} t e^{-t^2/2} \tag{6}$$

$$\psi^{(2)}(t) = \frac{d^2\theta(t)}{dt^2} = \frac{1}{\sqrt{2\pi}} (1-t^2) e^{-t^2/2} \tag{7}$$

The function $\psi^{(1)}(t)$ and $\psi^{(2)}(t)$ meet the conditions of admission of wavelet:

$$\int_{-\infty}^{\infty} \psi^{(1)}(t)dt = 0, \quad \int_{-\infty}^{\infty} \psi^{(2)}(t)dt = 0 \tag{8}$$

Therefore it can be used for wavelet generating function.

Step 5: $\psi^{(1)}(t)$, $\psi^{(2)}(t)$ is convolution type of wavelet transform for wave function.

$$w_s^{(1)} f(t) = f * \psi_s^{(1)}(t) = f * \left(s\frac{d\theta_s}{dt}\right)(t) = s\frac{d}{dt}(f * \theta_s)(t) \tag{9}$$

$$w_s^{(2)} f(t) = f * \psi_s^{(2)}(t) = f * \left(s^2\frac{d^2\theta_s}{dt^2}\right)(t) = s^2\frac{d^2}{dt^2}(f * \theta_s)(t) \tag{10}$$

Wavelet transforms $w^{(1)}f(t)$ and $w^{(2)}f(t)$ are first derivative and second derivative of the function f(t) by smoothing of θ(t) with scaling s. When s is smaller, the $\theta_s(t)$ smoothing results to f(t) have little effect on mutation position. When s is larger, the smoothing process will strip some minor mutations of f(t), leaving only the large size of mutations. We know, as the wavelet function can be seen as first derivative of a smoothing function, the signal local maxima of the wavelet transform modulus is corresponding to signal mutation point(or edges).When the wavelet function may be thought of as second derivative of a smoothing function, signal Wavelet transform zero-crossing

142

point is corresponding to signal mutation point(or edges).This is principle of detecting signal mutation point(or edges) using the zero-crossing of wavelet transform modulus and local maxima point.

4 EXPERIMENT

In the paper, we develop a software called DemeterProject in VC++6.0 and matlab platform, analyzing wavelet mutation point of the satellite electric field data of ULF frequency in the first half of 2008.

The electric field curve diagram without anomalous change is shown in Figure 3 and Figure 4. Figure 3 is source data waveform diagram. A time unit of the horizontal axis is second, and electric field strength unit of the vertical axis is mV/m. Figure 4 is waveform diagram of wavelet transform to Figure 3. The basic parameter information of Figure 3 and Figure 4 is as follows.

UT time:2008.1.1 1:45:37 876ms
track number:18693
subtrack type:0
geocentric latitude:70.2148
geocentric longitude:148.896
ground level:673.282
local time:11.686927
total number of file blocks:309
block number:0

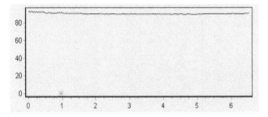

Figure 3. Source data waveform diagram without anomalous change.

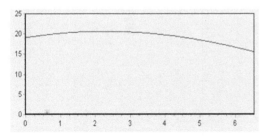

Figure 4. Waveform diagram without anomalous change.

Figures 3 and 4 show the electric field signal, which is steady and no abnormal phenomenon happens.

The electric field curve diagram having anomalous change is shown in Figure 5 and Figure 6. Figure 5 is source data waveform diagram. A time unit of the horizontal axis is second, and electric field strength unit of the vertical axis is MV/m. Figure 6 is a waveform diagram of wavelet transform to Figure 5. The basic parameter information of Figures 5 and 6 is as follows.

UT time:2008.5.9 11:16:46 373ms
track number:20593
subtrack type:1
geocentric latitude:-67.021
geocentric longitude:181.333
ground level:688.486
total number of file blocks:363
block number:6

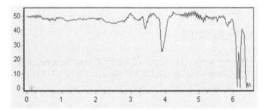

Figure 5. Source data waveform diagram having anomalous change.

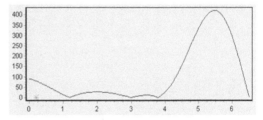

Figure 6. Waveform diagram having anomalous change.

From Figures 5 and 6, we can see an obvious anomalous change in the electric field curve diagram. There is a distinct mutation of the electric field between 5.2 and 5.3 seconds, corresponding to modulus maxima of wavelet analysis diagram, which is a mutation point. It will have an important role on extracting and analyzing earthquake ionosphere precursor anomalous information by satellite observation data.

5 CONCLUSION

In the paper, the method using wavelet transform to detect signal mutation has a high sensitivity and characterization of mutation point abilities, but also noting that due to the diversity of mutant signal, how to choose appropriate wavelet function can make a calculation minimal, and not reducing the detection capability. That is an important part of wavelet transforms. By the wavelet method we find many suspicious anomalies of the electric field signal, and achieve more efficient results. It will lay a good foundation for feature extraction of abnormal changes and seismo-electromagnetic precursor forecast.

ACKNOWLEDGMENT

This research is supported by funding program of the Youth Science Fund for Disaster Prevention and Reduction (No. 201203). The Author (Zhian Pan) thanks the financial support for this funding program.

REFERENCES

Pulinets, Sergey, Davidenko, Dmitry. Ionospheric precursors of earthquakes and Global Electric Circuit.Advances in Space Research, 53: 709–723, 2014.

Sharma, A.K, Patil, A.V, Haridas, R.N, Parrot, M. Detection of ionospheric perturbations associated with earthquake using data of IAP and ISL instruments of DEMETER satellite. General Assembly and Scientific Symposium: 1–4,2011.

Parrot, M, et al. Examples of unusual ionospheric observations made by the DEMETER satellite over seismic regions. Physics and Chemistry of the Earth. 31:486–495,2006.

Cussac,Thibery, Clair, Marie-Anne, et al.The Demeter microsatellite and ground segment.Planetary and Space Science, 54:413–427, 2006.

Lebreton, J.P, Stverak, S, Travnicek, P, Maksimovic, M, Klinge, D, et al. The ISL Langmuir Probe experiment processing onboard DEMETER: scientific objectives, description and first results. Planetary Space Science,54:472–486,2006.

Berthelier, J.J, Godefroy, M, et al. ICE, the electric field experiment on Demeter, Planetary and Space Science,54:456–471,2006.

Computational Intelligence in Industrial Application – Ling (ed.)
© *2015 Taylor & Francis Group, London, ISBN: 978-1-138-02818-0*

Essential protein identification by a bootstrap k-nearest neighbor method based on improved edge clustering coefficient

YaQun Jiang, Yan Wang*, GuiShen Wang, Ge Ou, Chao Su & Lan Huang*
College of Computer Science and Technology, Jilin University, China

ABSTRACT: Essential proteins are vital for living life. Meanwhile, essential protein identification based on protein-protein interaction network is a hot topic for analysis of the vital nodes in complex networks currently. In this paper, we proposed an Improved Edge Clustering Coefficient (IECC) and a new Node Edge Clustering (NEC) method based on IECC for essential protein prediction, which has integrated both node and edge topological properties of protein-protein interaction network. Furthermore, we employed a Bootstrap K-Nearest Neighbor model (Bootstrap-KNN) by combining NEC and several features from other essential protein identification methods to further improve the prediction performance. The experimental results in yeast protein-protein network show that NEC is more efficient than many traditional methods and the Bootstrap- KNN process gets even better results, which may provide some guidance information on essential protein detection for biologists.

KEYWORDS: Essential protein, Complex network, Edge clustering coefficient, Bootstrap, K-nearest neighbor, Protein-protein interaction.

1 BACKGROUND

Essential proteins play an important role in living life. They are closely related to the organism's survival and reproduction, which are kinds of vital nodes in protein-protein interaction network. In general, there are many traditional experimental methods that can get essential protein information, such as single gene knockouts[1], RNA interference[2] and conditional knockouts[3]. However, these experimental techniques are always laborious and time-consuming.

Meanwhile, the detection of vital nodes has become one of the hot topics in the research field of complex network analysis. As we all know that, the whole world, with people and things, are on different network unconsciously, such as scientific co-network, security network, information-sharing network, social network, computer network, transportation network, disease transmission network, *etc.*. And the vital nodes in these networks are very important in real life. For examples: (1) Detect the authorities in the scientific co-network. (2) Identify precautionary measures point in secure network. (3) Find out which leave will lose more information in the information-sharing network. (4) Point out the leader of terrorism, drug trafficking, fraud and gang crime in criminal network and terror network, and then guide the police to monitor and disintegrate these criminal

organizations rapidly. (5) Target isolated pathogenic in infectious disease and vital network, and then prevent the spread of the virus diffusion effectively. (6) Discover the initiator inthe rumor spread network, and then curb the spread of rumors, and so on.

At present, there are many classical methods for vital node detection in complex networks. These traditional methods mainly focus on calculating the score of a node by the impact of the relationship between the node itself and the associated nodes, such as Degree Centrality(DC)[4], Closeness Centrality(CC)[5], Eigenvector Centrality(EC)[6], Sub graph Centrality(SC)[7], *etc.* Though getting some efficient results in vital node detection, these methods are only focused on the importance of the node itself and ignore the edge property which is the correlation between nodes. With the rapid development of network analysis theory, edge property has been paid more attention, especially in community detection research field.

In 2011, Wang *et al.* proposed an edge clustering coefficient (ECC) which introduced the edge property in the area of vital nodes detection, and a new centrality (NC) which is based on ECC to calculate node centrality [8]. The performance of NC on yeast protein-protein interaction network is better than many traditional methods. And then the concept of edge clustering coefficient is employed

---

*Corresponding authors: *wy6868@hotmail.com; huanglan@jlu.edu.cn*

in the many community detection field of complex network analysis[9-11]. However, NC ignores the importance of the node itself and the correlation between node and edge. Furthermore, the prediction accuracy of NC is unstable for different network structure.

Nevertheless, many existing methods have different effects on different networks, but it is difficult to determine which method is the best choice. Actually, the accuracy rate of the top part is more concerned and importance to direct and guide experiments. With the accumulating of prediction nodes, the predicted vital nodes will increase, but the accuracy of the top nodes is not satisfied enough.

Based on all the above, we introduced the node clustering coefficient to ECC, and proposed an improved edge clustering coefficient (IECC), and then presented a new node edge clustering method (NEC) based on ECC, which integrated both node and edge topological properties. After that, we employed a bootstrap k-nearest neighbor (Bootstrap-KNN) strategy, and integrated NEC and several other methods as the feature set to further improve and optimize the prediction performance. Experimental results in yeast protein–protein interaction network show that NEC achieves better prediction performance than many traditional methods and the Bootstrap-KNN model gets even better.

2 METHOD

2.1 Edge clustering coefficient

Edge Clustering Coefficient (ECC)was proposed by Wang et al. in 2011 which is derived from node clustering coefficient and a new centrality (NC) which is based on ECC was applied to identify essential proteins in protein–protein interaction (PPI) networks of DIP database[12].

The PPI network may be considered as an undirected graph $G = (V, E)$, where V is node set which represent all proteins and E is edge set which represent interactions between proteins. The node clustering coefficient $C(u)$ of node u is defined as follows:

$$C(\mathrm{u}) = \frac{2E(\mathrm{u})}{k(\mathrm{u})(k(\mathrm{u}) - 1)} \qquad (1)$$

where $k(\mathrm{u})$ is the degree of node u, $E(\mathrm{u})$ is the number of edge samong node u and its $k(\mathrm{u})$ neighbors. Node clustering coefficient characterizes the tightness between the node and its neighbors. The node clustering coefficient C of the whole network is the average of all individual node clustering coefficients. It reflects the trend of clustering capability for interaction nodes.

Derived from node clustering coefficient, edge clustering coefficient (ECC) of edge (u,v) is defined as follows:

$$ECC(u, v) = \frac{t_{u,v}}{\min(d_u - 1, d_v - 1)} \qquad (2)$$

where $t_{u,v}$ denotes the number of triangles that include the edge (u,v) within the whole network, d_u is the degree of node u, $\min(d_u - 1, d_v - 1)$ is the number of triangles in which the edge (u,v) may possibly participate at most. Edge clustering coefficient characterizes the tightness between the edge of two nodes and the nodes around them.

2.2 The whole process

The whole improved process in this paper mainly consists of two parts. (1) We combined the node clustering coefficient and original ECC to get an improved edge clustering coefficient (IECC), and then proposed Node Edge Clustering(NEC) method, which is based on IECC. (2) We introduced the k-nearest neighbor model by combining six different methods as features and then used bootstrap re-sampling strategy as Bootstrap-KNN to build the N models. Finally, we summarized the results of N times'Bootstrap-KNN model as the final prediction.

2.2.1 Node edge coefficient
The original ECC only considers the number of triangles that actually contains the edge to measure importance. It does not consider the information of two endpoints on the edge. So we introduced the node clustering coefficient to the original ECC to consider the impact of the node and edge clustering coefficient in the same time, and formed a new edge clustering coefficient IECC:

$$IECC(u, v) = \frac{t(u, v) * C(u)}{d(u) - 1} * \frac{t(u, v) * C(v)}{d(v) - 1} \qquad (3)$$

where $t_{u,v}$ denotes the number of triangles that includes the edge of the network, $C(u)$ is the node clustering coefficient of node u, d_u is the degree of node u.

And then we presented a new method Node Edge Clustering (NEC) based on IECC to detect vital nodes as follows:

$$NEC(\mathrm{u}) = \sum_{v \in N_u} IECC(\mathrm{u}, v) \qquad (4)$$

where $t_{u,v}$ denotes the number of triangles that includes the edge of the network, $C(u)$ is the node clustering coefficient of node u, d_u is the degree of node u, N_u denotes the set of all edges actually in the network.

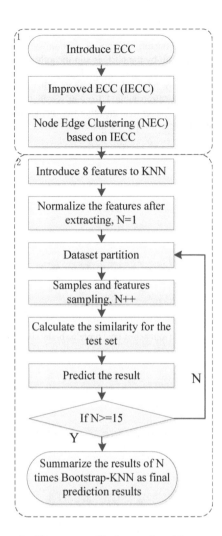

Figure 1.   The process of the improved model.

2.2.2 *Bootstrap-KNN model*

K-nearest neighbor (KNN) is a popular model in machine learning research field [13]. The main idea of KNN is that as ample belongs to the category which contains the major part of the nearest k samples. In KNN method, the classification results of selected neighbors should be known.

As K-nearest neighbor model is relatively simple, it is prone to produce over-fitting results. The selecting of training dataset may affect the effeteness directly[14]. In order to overcome the potential over-fitting problem, we employed a bootstrap strategy to further improve the KNN model as Bootstrap-KNN.

1 We used Manhattan distance[15] to calculate the k-nearest neighbor similarity, and the weight of each node in testing set is the accuracy rate of six methods respectively.

2 We introduced the bootstrap to resample the samples and features, using N sub-train samples after sampling to build N models. Finally, we summarized the results of N times Bootstrap-KNN results and then sort them to get the final order scores.

The improved k-nearest neighbor combined eight evaluations as features[16], including five methods of evaluations and three evaluations of NEC method, including DC score, EC score, SC score, node clustering coefficient score, NC score, NEC score, the average score of all the nodes that connected with the node, and the max score in all the nodes that connected with the node.

The experimental steps of Bootstrap-KNN are as follows:

1 Feature normalization: We normalized the features after extracting from different methods.

2 Dataset partition: We randomly selected 2500 samples as training set, and the remained 2537 samples as testing set.

3 Data sampling: In order to reduce the impact from over-fitting or under-fitting of KNN, we randomly selected negative samples similar to a positive samples number for sub-training sets, due to the number of positive samples are few.

4 Features sampling: Similar to data sampling as (3), in order to obtain the importance of nodes from various of aspects, we sampled the features in sub-training sets that is obtained from (3), which means there are six selected in eight features totally. Besides the two features that NEC and NC in these features must be contained to ensure the prediction results, and then we experiment the four which are selected randomly in the other six features.

5 Model established: We iterated step (3) and (4) for N times to establish N prediction models.

6 Results combination: After putting the testing dataset to N prediction models of step (5), we got N prediction results. Finally, we summarized all these predicted results to rank the final order.

Using Bootstrap-KNN model we can further reduce the over-fitting and under-fitting risk, and improve the generalization ability of the method. Combining sub-models can predict the importance of nodes by various analytical perspectives, which may make the predicted results more objective and get a higher accuracy rate.

3 EXPERIMENT

3.1 *Data source*

In the following experiment, we used the yeast protein-protein interaction network in DIP (dataset of interacting proteins) database [12], which has 5037proteins and 22061 protein-protein interactions in total. From DEG database [17], we obtained the essential protein information that there are 978 essential proteins, 3322 non-essential proteins and 737 unknown proteins.

3.2 *Order evaluation*

To evaluate the results, we used several common statistical standards, such as sensitivity, specificity, positive predictive value, negative predictive value, F-measure and accuracy rate[8].

1 Sensitivity(SN):

$$SN = \frac{TP}{TP + FN} \qquad (5)$$

2 Specificity(SP):

$$SP = \frac{TN}{TN + FP} \qquad (6)$$

3 Positive Predictive Value(PPV):

$$PPV = \frac{TP}{TP + FP} \qquad (7)$$

4 Negative Predictive Value(NPV):

$$NPV = \frac{TN}{TN + FN} \qquad (8)$$

5 F-measure(F):

$$F = \frac{2 \times SN \times PPV}{SN + PPV} \qquad (9)$$

6 Accuracy(ACC):

$$ACC = \frac{TP + TN}{P + N} \qquad (10)$$

where TP(truepositives) is essential proteins predicted correctly as essential, FP(falsepositives) means non-essential proteins predicted in correctly as essential, TN(truenegatives) refers to non-essential proteins predicted correctly as non-essential and FN(falsenegatives) is essential proteins predicted incorrectly as non-essential.

3.3 *Results and analysis*

First, we tested the six methods mentioned above respectively. From Figure 2, we can see that in the top 978 nodes, the number of vital nodes hit by NEC is

larger than the other five methods. From Figure 3, we can see that in the top 3322 nodes, the number of non-essential nodes hit by the NEC is significantly larger than the other five methods. And from Figure 4, the predicted accuracy rate of NEC is higher than the other five methods. Similarly, from table 1 we can see that the NEC method is also better than the other five methods by each evaluation order for the whole testing dataset.

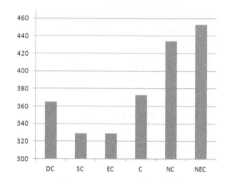

Figure 2.   The number of vital proteins hit by six methods.

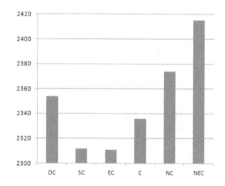

Figure 3.   The number of non-essential proteins hit by six algorithms.

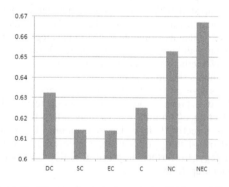

Figure 4.   The total accuracy rate of essential proteins and non-essential proteins.

Table 1. Six methods comparison with evaluation order.

Methods	SN	SP	PPV	NPV	F	ACC
DC	0.4790	0.8214	0.4162	0.8557	0.4454	0.6323
SC	0.4273	0.8078	0.3743	0.8398	0.3990	0.6142
EC	0.4267	0.8078	0.3743	0.8394	0.3988	0.6140
C	0.4681	0.8091	0.3898	0.8538	0.4254	0.6251
NC	0.5351	0.8374	0.4849	0.8630	0.5088	0.6530
NEC	0.5458	0.8426	0.5011	0.8650	0.5225	0.6670

Finally, we compared the six methods, including NEC, with Bootstrap-KNN model as shown in figure 5. The Bootstrap-KNN method and the other six methods sorted the predicted values of 2537 testing samples and selected different proportions of top predicted number. Among them, the accuracy rate of NEC algorithm is obviously higher than other traditional algorithms. However, Bootstrap-KNN integrates the advantages of a variety of predicted methods and predicts the node importance in several different analysis aspects, and thus gets the best performance than any other six methods.

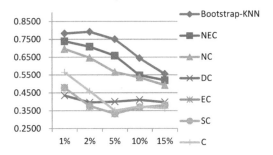

Figure 5. The comparison of individual accuracy rate among Bootstrap-KNN and six other methods.

4 CONCLUSION

To sum up, the accuracy rate of NEC algorithm is higher than several traditional vital node detection methods. And the Bootstrap-KNN model integrates the characteristics and advantages of the six methods in all, which has further improved the performance. Experimental results show that our strategy can get better prediction results effectively and may guide the biological experiment.

For future work, we will try to integrate several other machine learning algorithms, such as logistic regression and random forests, to improve the accuracy rate further. Meanwhile, we will try to integrate more biological information, such as the protein's genetic information in the Bootstrap-KNN strategy.

ACKNOWLEDGEMENTS

This work was supported by the Natural Science Foundation of China (61472159), Jilin Innovation Team Project (20122805), Jilin Province Science, Technology Development Project (20140101180JC) and Jilin University Graduate Innovation Fund (2014092).

REFERENCES

Giaever.G. 2002. Functional profiling of the Saccharomyces cerevisiae genome. Nature. 418 (6896):387–391.
Cullen.L.M & Arndt. G.M. 2005. Genome-wide screening for gene function using RNAi in mammalian cells. Immunology and Cell Biology.83(3):217–223.
Roemer.T.2003. Large-scale essential gene identification in Candida albicans and applications to antifungal drug discovery. Molecular Microbiology.50(1):167–181.
Jeong.H & Mason.S.P.2001.Lethality and centrality in protein networks.Nature.411(6833):41–42.
Wuchty.S. & Stadler.P.F.2003.Centersof complex networks. Journal of Theoretical Biology. 223(1):45–53.
Bonacich.P.1987. Powerand centrality:A family of measures.American Journal of Sociology.92(5):1170–1182.
Ernesto.E & Juan.A.S2005.Subgraph centrality in complex networks. Physical Review E,71(5):1–29.
Wang.J.X & Li.M.2011.Identification of Essential Proteins Based on Edge Clustering Coefficient.IEEE.147:1–12.
Radicchi.F & Castellano.C. 2004. Defining and identifying communities in networks. PNAS.101(9):2658–2663.
Hu,J & Dong,Y,H.2008. Community detection algorithm in large complex network. Computer Project. 34(19):92–93.
Zhang.P & Wang.J.2008.Clustering coefficient and community structure of bipartite networks.Physic A:Statistical Mechanics and its Applications. 387(27):6869–6875.
http://dip.doe-mbi.ucla.edu/dip/Main.cgi.
Harringto. P. 2013.Machine Learning in Action. POSTS& TELECOM PRESS.
Bi.X.M & Bi.R.H.2009.The overview of KNN. Science and Technology Innovation Herald.14–31.
Krause. Eugene F.Taxicab Geometry. Dover. 1987.
Jin,D & Liu,J.2010.k-nearest neighbor network based data clustering algorithm. Journal Of Pattern Recognition and Artificial Intelligence.23(4):546–551.
http://www.essentialgene.org/.

Computational Intelligence in Industrial Application – Ling (ed.)
© *2015 Taylor & Francis Group, London, ISBN: 978-1-138-02818-0*

Research of customer churn warning model for a mobile internet company

Chun Gui

Mathematics and Computer Science College, Northwest University for Nationalities, Lanzhou, China

ABSTRACT: We use a dataset of 450.000 data of a mobile internet company in Gansu province. In the process of modeling, according to a 7:3 proportion, the dataset is divided into a training set and test set. We choose the classical classification algorithms: decision tree C4.5 algorithm, Radical Basis FunctionNetwork and Logistic algorithm to model, and get three different models. In the evaluation phase, we choose three industry general evaluation criteria to evaluate the quality of the model: precision, recall and run-time. Finally, we get the conclusion that the decision tree model is the relative optimal one in these three algorithms for mobile internet company customer churn prediction.

KEYWORDS: Decision tree; Radical Basis function network; Logistic algorithm; Model.

1 INTRODUCTION

Mobile Internet company each month will produce the loss of customers, especially now market competition become fierce , enterprises to obtain new customers costs rise, mobile internet company operators from their own point of view, customer retention is required for the survival and development of enterprises. A set of data to illustrate the problem very well: the cost of developing a new customer is 4 times to retain an old customer; marketing products to new customers is the success rate for 15%, selling products to existing customers success rate is 50%, to new customers to sell 6 times more expensive than to existing customers to promote sales spend [1]. The goal of this dissertation is to use machine learning classification algorithm, carries on the analysis to the customer over a period of time call behavior and payment behavior and other information, for mobile Internet company find a good customer churn prediction model.

2 EXPERIMENTAL PREPARATION

2.1 *Dataset*

We used the dataset which was randomly collected from an operator call-center's database of December month. The dataset contains 450000 customer data such as USER_ACT_TYPE, INNET_MONTH, TOTAL_FEE, TOTAL_FLUX, IS_LOST and so on. Obviously the class variable is IS_LOST. Looking into the data, we saw that there were 100000 records with the class label IS_LOST and the rest 350000 records, were non-IS_LOST. For the sake of keeping our experiments with proper randomization environment, for each single algorithm we randomly select 70% of dataset for the training set and the rest of test process. In another word, our every single experiment was done with a training set to 245000 records labeled non-IS_LOST and 70000 records labeled IS_LOST ones (totally 315000 records). Also, all of this paper test process has been performed by 105000 non-IS_LOST and 30000 IS_LOST classes (totally 135000). By the aforementioned process, we prepared 3 single algorithms in order to run and compare experiments. The size of the dataset is 64.5M. The name and meaning of 21 explanatory variables and1 class variable are shown in Table 1. The data preprocess stage mainly analyzes the variable importance (VI henceforth) of each explanatory variable with class variable IS_LOST. It is obviously that MONTH_ID, PROV_ID, USER_ID have no association with class variable. So we remove these three explanatory variables and then followed by the analysis of the VI between remaining 18 explanatory variables and class variable IS_LOST.

2.2 *The experimental operation environment and the attribute selection*

By deploying the PC machine with R software, hardware properties for Intel® Core(TM) i5-3210M CPU @ 2.50GHZ, Memory is 2.38G, the operating system is WINDOWS 732 bit.

First, we analyze the relationship of each explanatory variable with class variable. For the large variable number, here we just list the graphs of four variables, the horizontal axis as the explanatory variable, the vertical axis as class variable IS_LOST. The graphs a, b and cin Fig.1 show the distribution of part of the explanatory variables which churn rate is obvious. While in Figure d, loss rate is stable, the trend is not obvious , a weak impact on the class variable. The smooth trend means the explanatory variable has little effect on the class variable. Figure 1 is the relationship graph of four explanatory variables with the customer churn rate, part of the codes are:

```
>mydata<-read. table('D:\\wen.txt',sep=',')
>is_lost<-as. vector(unlist(mydata["V4"]))
>in_m<-as. vector(unlist(mydata["V22"]))
>is_lost=as. integer(is_lost)
>in_m=as. integer(in_m)
>is_lost[1]
>in_m[1]
>a<-vector();b<-vector()
>for(i in 1:57){a[i]<-0;b[i]<-0}
>for(i   in   1:450000){a[in_m[i]]=a[in_m[i]]+1;
if(is_lost[i]==1){b[in_m[i]]=b[in_m[i]]+1}}
>c<-vector()
>for(i        in       1:55){     if(a[i]==0){c[i]<-0}
else{c[i]=b[i]/a[i]}}
>z=1:55
>plot(z,c)
>plot(z,c,type="o")
>for(i       in      1:55){      if(a[i]==1){d[i]<-0}
else{d[i]=(a[i]-b[i])/a[i]}}
>plot(z,d,type='o',col='red',xlab='FLUX_
BHD',ylab='ISNOT_LOST_RATE')
```

The code is written and run in R.Through the graph, if the trend is basically the same , these explanatory variables are combined into one explanatory variable. We select ten explanatory variables to build the final model, they are: TOTAL_FEE_RATE, OWE_FEE, TOLL_TIMES, OLL_TIMES_RATE, INNET_MONTHS, USER_ACT_TYPE, ROAM_TIMES, FLUX_BHD, TOTAL_FLUX_RATE and CALL_DURA_BHD. In order to demonstrate the accuracy of the data analysis result, we use the method of logistic regression to check the classification ability of explanatory variables, and on the basis of the initial model, using the stepwise regression to remove the attribute which is not significant in order to get a more optimal model. Data analysis results has no difference with before analysis.

3   EXPERIMENT ANALYSIS

3.1   *Selection of dataset and modeling algorithm*

The experiment selected three kinds of classic data mining algorithm [2][3], which are decision tree 4.5 algorithm[4][5][6], Radical Basis Function Network(RBFNetwork)[7] and Logistic regression algorithm[8]. The original dataset according to the proportion of random 7:3 is divided into two parts, 70% of them as training set, 30% as test set.

3.2   *Model evaluation criteria*

As we all know, if a true non-churner is classified as a churner by the model it is not a big mistake. On the other hand, it will cause a steep loss is a true non-churner is classified as a churner by depending on the outcome of the model. In mobile internet company, there are some known evaluation criteria to check the performance of a classification model. As precision and recall, which can show more consequential information about the model. In additional, run-time

Table 1.   Meaning of explanatory variables.

NAME	MEANING	NAME	MEANING
MONTH_ID	ID of month	ZHUJIAO_TIMES	total calling time
PROV_ID	ID of province	TOTAL_FEE_RATE	total cost growth rate
USER_ID	ID of user	TOTAL_FLUX_RATE	total flow growth rate
IS_LOST	whether loss next month	LOCAL_TIMES_RATE	local call time growth rate
USER_ACT_TYPE	type of user	TOLL_TIMES_RATE	total call time growth rate
INNET_MONTH	in net time	ROAM_TIMES_RATE	roaming call time growth rate
TOTAL_FEE	total fee	ZHUJIAO_TIMES_RATE	calling time growth rate
TOTAL_FLUX	total flow	OWE_MONTH	owe month
LOCAL_TIMES	local call time	OWE_FEE	owe fee
NAME	MEANING	NAME	MEANING
TOLL_TIMES	total call time	CALL_DURA_BHD	an index of calling time
ROAM_TIMES	roaming call time	FLUX_BH	an index of flow

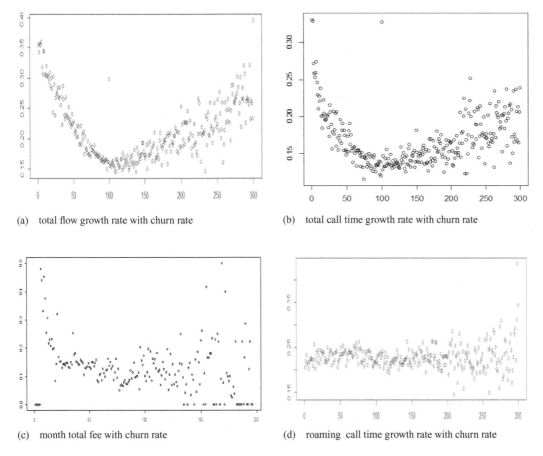

(a)   total flow growth rate with churn rate

(b)   total call time growth rate with churn rate

(c)   month total fee with churn rate

(d)   roaming call time growth rate with churn rate

Figure 1.   Relationship diagram of explanatory variable with churn rate.

is also an important merit to evaluate the model. Now I'll make a short description of the criteria of precision and recall.

Precision and recall are computed by four terms of measure which can be seen from table 2. They are TP, TN, FP and FN. TP means true positive, which refers to the positive case that is correctly classified by the model. TN means true negative, which refers to the negative case that is correctly classified by the model. FP means false positive, which refers to the negative case that is incorrectly classified as positive by the model. FN means false negative, which refers to the positive case that is incorrectly classified as negative by the model. Table 2 is the confusion matrix of these four terms of the measure. Eqs. (1)-(2) respectively express the formulas of precision and recall measurements.

$$precision = \frac{TP}{TP + FP} \tag{1}$$

$$recall = \frac{TP}{TP + FN} \tag{2}$$

Table 2.   Confusion matrix of four terms of measure.

Prediction outcome	Actual value		Total
	True Positive	False Positive	P'
	False Negative	True Negative	N'
Total	P	N	

4   EXPERIMENT RESULT ANALYSIS

The experiment separately calculated precision, recall and run-time(unit: second) of three algorithms, the contrast diagram is shown in Figure 2. The test conclusion is: contrast the run time, recall and precision for three models of decision tree, RBFNetwork and logistic, we found that the decision tree model is optimal in the coverage and precision. In the aspect of runtime, the decision tree is worse than the RBFNetwork model, but better than logistic model. A comprehensive comparison of various aspects of performance, the decision tree model of customer churn warning for a mobile Internet company is the best algorithm.

Comparison Chart of Three Algorithms for Precision

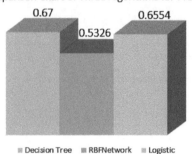

Comparison Chart of Three Algorithms for Recall

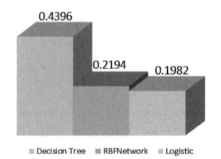

Comparison Chart of Three Algorithms for Run-time

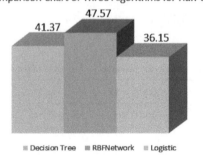

Figure 2. Comparison of three algorithms for precision, recall and run-time.

5 FUTURE WORK

In the later study, I plan to compare and analyze a variety of machine learning algorithms to find the most suitable customer churn prediction model for a mobile Internet company in a big data situation. On the basis of these models, dig out a model with higher accuracy, lower time complexity.

ACKNOWLEDGEMENT

This work is supported by the Fundamental Research Funds for the Central Universities 2014: 31920140089.

REFERENCES

Liu Jinjing. The outlook for 2013 mobile internet companymunicationservice trends[N]. China's information industry network - people's post and telecommunications. 2013-1-6.
Zhou Hongbo. The application of off-network analysis use data mining in a China Mobile branch[D]. Shanghai: Shanghai Jiao Tong university. 2009:27–30.
Xiong Ping. Data mining algorithm and Clementine practice[M]. Beijing: Tsinghua University Press, 2011:15–45.
Tan Junlu, WuJianhua. Classification algorithm research based on decision tree rules[J]. Beijing: Computer engineering and design, 2010, 31(5): 1017–1019.
Deng Quan. Decision tree and customer churn analysis[D]. Xi'an:Journal of Xi'an University of post and Telecommunications, 2013,18(3):49–51.
Hao Mer. Model construction and control of telecom customer loss based on CART two binary decision tree[J]. Hangzhou: bulletin of science and technology, 2012,28(6):103–105.
Dong Jin-zhao,Xie Yong-cheng,HuangJie-yin,Zhai Yingwen.2013,Application of Radical Basis Neural Network in Fault Diagnosis of Rectifier[C],International Conference on Advanced Computer Science and Electronics Information(ICACSEI),Beijing,PEOPLES R CHINA,Advances in Intelligent Systems Research,41:700–703.
Yu Qingsheng. Design and implementation of logistic regression model algorithm based on cloud platform[J]. Hangzhou: bulletin of science and technology, 2013, 29(6):137–139.

Computational Intelligence in Industrial Application – Ling (ed.)
© 2015 Taylor & Francis Group, London, ISBN: 978-1-138-02818-0

An energy consumption model for the cooperative communication in wireless ad hoc networks

J. Tang, X. Zou, L. Dong & W. Wang
National Computer Network Emergency Response Technical Team/Coordination Center, Beijing, China

ABSTRACT: Cooperative Communication has emerged as a promising technique to enhance system reliability and performance in wireless ad hoc networks. In this paper, we propose a novel analytical framework to evaluate the energy consumption of any station in saturated wireless ad hoc networks with the cooperative media access control protocol. With an accurate performance analysis, we provide a more comprehensive analysis of different operations in cooperative transmissions, by taking into account the different operational modes. Thus, an energy consumption model for each station is built. With the proposed model, we develop a case study and provide interesting indications to make a tradeoff between throughput and lifetime of wireless ad hoc networks.

KEYWORDS: Energy consumption model; Cooperative communication; AD hoc networks.

1 INTRODUCTION

Recently, multimedia becomes increasingly popular for the users of ad hoc networks. In wireless channels, however signal attenuation is a major factor of performance degradation if not well treated. In order to mitigate the deteriorated effects of signal fading, the cooperative communication is presented in wireless communications. Recently, cooperative MAC (shorted as CoopMAC), as a novel MAC protocol using cooperative diversity, was proposed to improve the performance of wireless local area networks[1]. The main idea of CoopMAC is to select an intermediate station, called helper, to assist a low rate source station by incorporating a new two-hop link from the source to its designated destination in the transmission. And, either the link between the source and the helper or the link between the helper and the destination suffers much weaker signal fading with both the source and the destination as compared to the direct link between the source and the destination. Thus, CoopMAC can achieve much better throughput relative to the conventional MAC[2,6].

In the short-range wireless ad hoc network, terminal stations are energy limited, which is treated by the MAC protocol. The energy consumption is determined by the MAC protocol; therefore to improve the utilization of energy is one of key problems. Understanding the energy consumption at all network layers, as well as energy consumed resulting from inter-layer interactions, is a fundamental step in the design of power efficient protocols for cooperative MAC protocols. For these reasons, the study of energy consumption in the cooperative ad hoc networks represents an important step towards the design of future energy-efficient protocols.

The rest of this paper is organized as follows: section 2 introduces definitions and assumptions. Section 3 gives a specific analysis of energy assumption of any station in the duration ΔT of CoopMAC protocol. Section 4 gives simulation and result. Section 5 gives the final conclusion of this paper and the future work.

2 DEFINITION AND ASSUMPTION

In order to convenient for analyzing the model of energy consumption, the universal assumption is given:

1 The network is saturated, and the condition of channel is ideal
2 The energy consumption is ignored of controlling frame above the layer of MAC transmitting
3 All of the competitive stations have limited energy consumption.

The working state of stations in the multi-rate network is defined to four types:

1 Transmission power P_{tr}
   The station turns to transmitting state when the control frame (RTS, CTS, HTS, ACK) or packets is transmitting. P_{tr} is nonlinear and determined by RF Power Amplifier.
2 Receiving power P_{re}
   A receiving station can be one of the relay stations of the source station, or one of the control frames (RTS, CTS, HTS, ACK) in the process of

overhearing handshaking. Pre is depends on design of transmitter and transceiver and influenced of types of modulation.

3 Sensing power P_{se}

The station judges the channel state by the sensing channel to decide the reaction to the backoff counter. The output power of station is same as P_{se} and closer to Pre when channel is idle.

4 Sleeping power P_{sl}

The station which is not involved in the process of transmitting and receiving turns to the sleeping state when the packets are transmitting according to the analysis of transmitting duration. P_{sl} is the least energy consumption in the working state.

3 ENERGY CONSUMPTION ANALYSIS IN THE DURATION ΔT

3.1 State probabilities

First we give the probabilities of idle state and successful state, denoted by P_i and P_s. Since the state probability is independent of choosing station, and it can be obtained by the result of CD-MA model [3]:

$$P_s = b_{2,0,0} + \sum_{k=1}^{W_0-1} \tilde{b}_{2,0,k} = W_0 b_{2,0,0} \tag{1}$$

$$P_i = \sum_{j=0}^{m}\left(b_{0,j,0} + \sum_{k=1}^{W_j-1} \tilde{b}_{0,j,k} \right) = \sum_{j=0}^{m}\left(\frac{W_j}{2} b_{0,j,0} \right) \tag{2}$$

$b_{2,0,0}$, $b_{0,j,0}$ and W_j are satisfied with the condition:

$$W_j = \begin{cases} 2^j (CW_{min}+1), & j \in \left[0, \log_2^{(CW_{max}+1)/(CW_{min}+1)} \right] \\ CW_{max}+1, & j \in \left(\log_2^{(CW_{max}+1)/(CW_{min}+1)}, m \right] \end{cases} \tag{3}$$

3.2 Energy consumption model

According the deferent backoff phase during interval of transmitting twice ΔT, the energy consumption of station k is the sum of transmitting and counting process. In the phase of transmitting, the energy consumption comes from the assumption of transmitting packets and receiving control frame. In the phase of counting process, the energy consumption comes from the idle and dormant state of channel. Meanwhile, the station k demand receiving the handshaking control frame to confirm the accessing channel fully or the during time set NAV[3].

Assuming the duration of transmitting one frame (including the head and payload) by direct rate R_i is T_i, and the duration of transmitting one packet by two hops routing path is $T_{i,j}$, and the straight rate from source station to relay station is R_i, the straight rate of relay station is R_j. It can be obtained:

$$T_i = \frac{L_{PHY\_h}}{R_K} + \frac{L_{MAC\_h}+L}{R_i} \tag{4}$$

$$T_{i,j} = \frac{2L_{PHY\_h}}{R_K} + \frac{L_{MAC\_h}+L}{R_i} + \frac{L_{MAC\_h}+L}{R_j} \tag{5}$$

L_{PHY_h} and L_{MAC_h} is described the length of frame head of physical layer and MAC layer. And the same time, the energy consumption of two stage of station k during ΔT is e_k^{TX} and e_k^{BF} respectively. That is $e_k = e_k^{TX} + e_k^{BF}$. If the cooperative matrix is $Q = (q_1, \cdots, q_k, \cdots, q_n)$, $q_k = (q_{k1}, \cdots, q_{kh}, \cdots, q_{kn})$, we can get the two stage of energy consumption by analysis the during time of working state when given Q.

1) transmitting stage energy e_k^{TX}

The station k can be divided to successful energy $e_k^{(TX,s)}$ and the failing energy $e_k^{(TX,c)}$.

1.1) energy consumption of station k with a successful transmission $e_k^{(TX,s)}$

The duration of successful transmission of station k includes control frame, frame interval, frame head and payload duration. The fixed energy consumption of transmitting duration is $e_k^{(TX,s,1)}$ independent of transmitting mode. Review the process of CoopMAC protocol, it can be got:

$$e_k^{(TX,s,1)} = T_{RTS}P_{tr} + (T_{CTS}+T_{ACK})P_{re} + (3T_{SIFS}+T_{DIFS})P_{id} \tag{6}$$

The energy consumption of variable duration is $e_k^{(TX,s,2)}$. The difference of station transmitting once in cooperative mode compared to basic mode is that the HTS frame is involved and the interval of two SIFS is added and the duration of transmitting is decreased. If the potential relay station of station k doesn't existed, that is transmitting straightly, then transmitting duration of transmitting is $T_{l(k)}$ every station; If station k have best relay stations, one of those is station h ($h \in H^{(k)}$). Then when station k and station h transmit cooperatively, the duration of transmitting from station k to station h is $T_{l(k,h)}$, duration of station h to a destination station is $T_{l(h)}$. Noticed the cooperative probability of between station k and station h is q_{kh}, and then the consumption of variable part is described as weighted average of two modes, that is:

$$e_k^{(TX,s,2)} = \left[\left(1 - \sum_{h \in H^{(k)}} q_{kh} \right)T_{l(k)} + \sum_{h \in H^{(k)}} q_{kh}T_{l(k,h)} \right]P_{tr}$$
$$+ \left[\sum_{h \in H^{(k)}} q_{kh}T_{HTS} \right]P_{re} + \left[\sum_{h \in H^{(k)}} q_{kh}T_{SIFS} \right]P_{se}$$
$$+ \left[\sum_{h \in H^{(k)}} q_{kh}\left(T_{l(h)}+T_{SIFS} \right) \right]P_{sl} \tag{7}$$

The consumption of energy of station k is:

$$e_k^{(TX,s)} = e_k^{(TX,s,1)} + e_k^{(TX,s,2)}$$

$$= \left[T_{RTS} + \left(1 - \sum_{h \in \mathbf{H}^{(k)}} q_{kh} \right) T_{l(k)} + \sum_{h \in \mathbf{H}^{(k)}} q_{kh} T_{l(k,h)} \right] P_{tr}$$

$$+ \left[T_{CTS} + T_{ACK} + \sum_{h \in \mathbf{H}^{(k)}} q_{kh} T_{HTS} \right] P_{re} \qquad (8)$$

$$+ \left[\left(3 + \sum_{h \in \mathbf{H}^{(k)}} q_{kh} \right) T_{SIFS} + T_{DIFS} \right] P_{se}$$

$$+ \left[\sum_{h \in \mathbf{H}^{(k)}} q_{kh} \left(T_{l(h)} + T_{SIFS} \right) \right] P_{sl}$$

1.2) energy consumption for station k with a failing transmission $e_k^{(TX,c)}$

The energy consumption of station k includes transmitting energy of RTS and follow-up intercepting energy. Then the consumption is $T_{RTS} P_{tr} + T_{EIFS} P_{se}$. On the other hand, the failing probability is p for any station, the success probability is $(1-p)$. So the average times of failure is $p(1-p)$ for station k before succeeding. Then the consumption of failing energy in ΔT duration is:

$$e_k^{(TX,c)} = \frac{p}{1-p} \left(T_{RTS} P_{tr} + T_{EIFS} P_{se} \right) \qquad (9)$$

During the ΔT for transmitting, the consumption of station k is:

$$e_k^{(TX)} = e_k^{(TX,s)} + e_k^{(TX,c)}$$

$$= \left[\frac{1}{1-p} T_{RTS} + \left(1 - \sum_{h \in \mathbf{H}^{(k)}} q_{kh} \right) T_{l(k)} + \sum_{h \in \mathbf{H}^{(k)}} q_{kh} T_{l(k,h)} \right] P_{tr}$$

$$+ \left[T_{CTS} + T_{ACK} + \sum_{h \in \mathbf{H}^{(k)}} q_{kh} T_{HTS} \right] P_{re} \qquad (10)$$

$$+ \left[\left(3 + \sum_{h \in \mathbf{H}^{(k)}} q_{kh} \right) T_{SIFS} + T_{DIFS} + \frac{p}{1-p} T_{EIFS} \right] P_{se}$$

$$+ \left[\sum_{h \in \mathbf{H}^{(k)}} q_{kh} \left(T_{l(h)} + T_{SIFS} \right) \right] P_{sl}$$

2) energy consumption for station k in backoff stage e_k^{BF}

According the variable of working state, the consumption can be divided to three parts.

2.1) The energy consumption of station k for a successful transmission from any other station s $e_k^{(BF,s)}$: the station s we choose is deferent from station k($s \neq k$), we can get by fairness of system in a long time, the station s can transmit once averagely during station k transmitting twice consecutively, ΔT.

First, the consumption of fixed duration of station k is $e_{k,s}^{(BF,s,1)}$ which is independent from successful transmission mode of station s once. It can be derived:

$$e_{k,s}^{(BF,s,1)} = \left(T_{RTS} + T_{CTS} \right) P_{re} + \left(T_{SIFS} + T_{DIFS} \right) P_{se} \qquad (10)$$

Secondly, the consumption of variable duration of station k is considered which is dependent on successful transmission mode of station s once. If station s has direct transmission mode, the consumption of variable duration part is $e_{k,s}^{(BF,s,2)}$, that is :

$$e_{k,s}^{(BF,s,2)} = \left[\left(1 - \sum_{h \in \mathbf{H}^{(s)}} q_{sh} \right) \left(T_{l(s)} + 2T_{SIFS} + T_{ACK} \right) \right] P_{sl} \qquad (11)$$

If station s has cooperative transmission mode which is independent from station k, the consumption of variable duration part is $e_{k,s}^{(BF,s,3)}$, that is :

$$e_{k,s}^{(BF,s,3)} = \sum_{h \in \mathbf{H}^{(s)}, h \neq k} \left\{ q_{sh} \left[\begin{array}{c} T_{HTS} P_{re} + T_{SIFS} P_{se} \\ + \left(T_{l(s,h),l(h)} + 3T_{SIFS} + T_{ACK} \right) P_{sl} \end{array} \right] \right\} \qquad (12)$$

If station s has cooperative transmission mode which is dependent on station k, the station k need response HTS after receiving RTS; And the station k receives the packets from station s in rate $R_{l(s,k)}$ after three times handshaking control frame, then forward that frames to the destination station in rate $R_{l(k)}$ after a SIFS. The variable consumption of station k is $e_{k,s}^{(BF,s,4)}$, that is :

$$e_{k,s}^{(BF,s,4)} = q_{sk} \left[\begin{array}{c} \left(T_{HTS} + T_{l(k)} \right) P_{tr} \\ + \left(T_{l(s,k)} + T_{ACK} \right) P_{re} + 4T_{SIFS} P_{se} \end{array} \right] \qquad (13)$$

Therefore, the energy consumption of station k in backoff phase for a successful transmission from station s, that is:

$$e_k^{(BF,s)} = \sum_{s=1,s \neq k}^{n} \sum_{j=1}^{4} e_k^{(BF,s,j)}$$

$$= \sum_{s=1,s \neq k}^{n} \left\{ \begin{array}{c} \left(T_{RTS} + T_{CTS} \right) P_{re} + \left(T_{SIFS} + T_{DIFS} \right) P_{se} \\ + \left(T_{l(s)} + 2T_{SIFS} + T_{ACK} \right) P_{sl} \end{array} \right\} \qquad (14)$$

$$+ \sum_{s=1,s \neq k}^{n} \left\{ q_{sk} \left[\begin{array}{c} \left(T_{HTS} + T_{l(k)} \right) P_{tr} + \left(T_{l(s,k)} + T_{ACK} - T_{HTS} \right) P_{re} \\ + 3T_{SIFS} P_{se} - \left(T_{l(s,k),l(k)} + 3T_{SIFS} + T_{ACK} \right) P_{sl} \end{array} \right] \right\}$$

$$+ \sum_{s=1,s \neq k}^{n} \sum_{h \in \mathbf{H}^{(s)}} \left\{ q_{sh} \left[\begin{array}{c} T_{HTS} P_{re} + T_{SIFS} P_{se} + \\ \left(T_{SIFS} + T_{l(s,h),l(h)} - T_{s,l(s)} \right) P_{sl} \end{array} \right] \right\}$$

2.2) the energy consumption of station k with a fail transmission from others $e_k^{(BF,c)}$

The station k in counting phase has duration time of single collision including receiving RTS and

157

waiting idle channel. So the station k "see" the consumption of single collision $T_{RTS}P_{re} + T_{EIFS}P_{se}$.

P_c is the probability of system in collision state, P_s is the probability of system in successful transmission state. Since the number of competition stations is n, the average failing times is nP_c / P_s during consecutive successful transmission of station k. And the collision probability is p in RTS transmission of station k, so the average collision times of station k involved is $p/(1-p)$. The average failing transmission times of others are $nP_c / P_s - p/(1-p)$ in backoff phase of station k.

The consumption of station k from failing transmitting of others can be derived:

$$e_k^{(BF,c)} = \left(\frac{nP_c}{P_s} - \frac{p}{1-p} \right) \left(T_{RTS}P_{re} + T_{EIFS}P_{se} \right) \quad (15)$$

2.3) the energy consumption of station k in idle state $e_k^{(BF,i)}$

P_i is the probability of system in idle, P_s is the probability of system in successful transmission state. The average backoff time slots numbers are nP_i / P_s during two consecutive successful transmission of station k. The length of backoff time slot is σ, then :

$$e_k^{(BF,i)} = \frac{nP_i}{P_s} \sigma P_{se} \quad (16)$$

The consumption of station k during ΔT in counting phase can be describe as:

$$e_k^{BF} = e_k^{(BF,s)} + e_k^{(BF,c)} + e_k^{(BF,i)}$$

$$= \left(\frac{nP_c}{P_s} - \frac{p}{1-p} \right) \left(T_{RTS}P_{re} + T_{EIFS}P_{id} \right) + \frac{nP_i}{P_s} \sigma P_{id}$$

$$+ \sum_{s=1,s\neq k}^{n} \left\{ \begin{array}{l} (T_{RTS} + T_{CTS})P_{re} + (T_{SIFS} + T_{DIFS})P_{id} \\ +(T_{l(s)} + 2T_{SIFS} + T_{ACK})P_{sl} \end{array} \right\} \quad (17)$$

$$+ \sum_{s=1,s\neq k}^{n} \left\{ q_{sk} \left[\begin{array}{l} (T_{HTS} + T_{l(k)})P_{tr} + (T_{l(s,k)} + T_{ACK} - T_{HTS})P_{re} \\ +3T_{SIFS}P_{id} - (T_{l(s,k),l(k)} + 3T_{SIFS} + T_{ACK})P_{sl} \end{array} \right] \right\}$$

$$+ \sum_{s=1,s\neq k}^{n} \sum_{h\in H^{(s)}} \left\{ q_{sh} \left[\begin{array}{l} T_{HTS}P_{re} + T_{SIFS}P_{id} \\ +(T_{SIFS} + T_{l(s,h),l(h)} - T_{l(s)})P_{sl} \end{array} \right] \right\}$$

$q_{kk}=0$, $k\in[1,n]$. After simplification, the sum consumption of station k e_k can be described:

$$e_k = e_k^{TX} + e_k^{BF}$$

$$= \frac{1}{1-p} T_{RTS}P_{tr}$$

$$+ \left[\left(n + \frac{nP_c}{P_s} - \frac{1}{1-p} \right) T_{RTS} + nT_{CTS} + T_{ACK} \right] P_{re}$$

$$+ \left[\frac{nP_c}{P_s} T_{EIFS} + \frac{nP_i}{P_s} \sigma + (n+2)T_{SIFS} + nT_{DIFS} \right] P_{se}$$

$$+ \left[(n-1)(2T_{SIFS} + T_{ACK}) + \sum_{s=1}^{n} T_{l(s)} \right] P_{sl} \quad (18)$$

$$+ T_{l(k)}(P_{tr} - P_{sl}) + \sum_{h=1}^{n} q_{kh} \left[(T_{l(k,h)} - T_{l(k)})(P_{tr} - P_{sl}) \right]$$

$$+ \sum_{s=1}^{n} \sum_{h=1}^{n} q_{sh} \left[\begin{array}{l} (T_{l(s,h),l(h)} - T_{s,l(s)})P_{sl} + T_{HTS}P_{re} \\ +T_{SIFS}(P_{se} + P_{sl}) \end{array} \right]$$

$$+ \sum_{s=1}^{n} q_{sk} \left[\begin{array}{l} T_{l(k)}(P_{tr} - P_{sl}) + (T_{l(s,k)} + T_{ACK})(P_{re} - P_{sl}) \\ +T_{HTS}(P_{tr} - P_{re}) + 3T_{SIFS}(P_{se} - P_{sl}) \end{array} \right]$$

4 SIMULATION AND RESULTS

We adopt the proposed energy consumption model to compare the life time with the 802.11b protocol[4], and the simulation is run with a discrete event-diven kernal[5]. The relay station (also is source station) with rate 11Mbps is included in network, and other source stations with same rate 1Mbps are set. The number of network stations is varied from 2 to 9. All of stations have the same beginning energy, and the work-state power setting is same with [2].

The figure 1 shows the result about the throughput and lifetime compared to 802.11b. The throughput of CoopMAC keeps beyond 85% compared to 802.11b. However, as we can see, the lifetime is decreased with the increasing numbers of source stations. The lifetime is decreased close 75% when the number of source stations is 8. As we know, the throughout is same for stations at the low rate and stations at the high rate in single hop network, that is the accessing successfully and collision times in any duration for all stations. The transmission delay of low rate station is bigger, and the consumption of unit payload is bigger, so the system lifetime can be determined by these low rate stations. However, the stations at the high rate help stations at the low rate to transfer packets lead to increase consumption rapidly. Thus, we shall know the crucial stations are these stations at the high rate in CoopMAC systems.

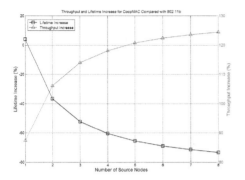

Figure 1. Throughout and lifetime comparison.

5 CONCLUSIONS

The theory and the simulation result show that mechanism of CoopMAC protocol has increased the throughput of system obviously by cooperative mode, the energy of high rate stations as relay station consumes rapidly in limited energy state. That character influences the fairness of the consumption and the lifetime of system.

The energy utilization can be improved by balancing the independent mode and fully cooperative mode, adopting the partial cooperative transmission mode to increase the throughout and give up the lifetime problem.

ACKNOWLEDGMENT

We shall express our thanks to National Science and Technology Major Project of the Ministry of Science and Technology of China (2012BAH45B00).

REFERENCES

Liu P., Tao Z., Narayanan S., Korakis T., et al. 2007. CoopMAC: A Cooperative MAC for Wireless LANs. IEEE Journal on Selected Areas in Communications (JSAC), Feb. 2007, V25(2): 340–354.
Feeney L. M., Nilsson M. 2007. Investigating the Energy Consumption of a Wireless Network Interface in an Ad Hoc Networking Environment. INFOCOM 2001. Anchorage, AK, April 2001: 1548–1557.
Tang Jiqiang. 2012. Performance Analysis of the Cooperation at MAC Layer for Multi-rate Wireless Networks. Doctoral dissertation, Jun. 2012:18–36.
S. Narayanan and S. S. Panwar. 2013. To forward or not to forward – that is the question, to appear in Special Issue on Cooperation in Wireless Networks, Springer - Wireless Personal Communications.
MacDougall M. 1987. Simulating computer systems: Techniques and tools. The MIT Press.
F. Alizadeh-Shabdiz and S. Subramaniam, 2004. MAC layer performance analysis of multi-hop ad hoc networks, in *Proc. IEEE GLOBECOM*, Nov. 2004:2781–2785.

Section 2: Application of computer science and communication

Computational Intelligence in Industrial Application – Ling (ed.)
© 2015 Taylor & Francis Group, London, ISBN: 978-1-138-02818-0

Research on a hierarchical solution to security of cloud computing

Ling Zhang

Beijing Information Technology College, ChaoYang District, Beijing, China

ABSTRACT: This paper analyzed the security issues of four aspects, the cloud, the clients, the transmission, and the data storage. Then, one solution strategy has been proposed based on hierarchical thinking, from five-level procedures, the independent data layer, the database access layer, the server access layer, the network access layer and the access layer to solve security problems in cloud computing.

KEYWORDS: cloud computing, security, database, network, hierarchical solution.

1 INTRODUCTION

Cloud computing is the further development of distributed computing. The purpose of Cloud computing is to realize the large-scale computing, through better use of distributed resources and integration of resources to improve the throughput. [1] The application of cloud computing is shown in figure 1.

Figure 1. The application of cloud computing.

At present, the industry has a lot of companies gathered in cloud computing, such a new computing concept, use different techniques to implement cloud computing. Many scholars have put forward the security problems in the cloud, and given its own solution to the safety. This paper analyzed the security of cloud computing in each link of the whole process from a new angle, and put hierarchical solution into cloud computing security problems. It provided an idea and method for cloud computing security.

2 CONCEPTION OF CLOUD COMPUTING

Cloud computing is an emerging model of Business Computing, is development of the parallel computing, distributed computing and grid computing, is results of mixing, evolution and progress of Virtualization, Utility computing,, IaaS (Infrastructure as a service), PaaS (Platform as a service), SaaS (Software as a service) .[2] It is a new computing method of sharing infrastructure which can make a large number of computer resources work together through network, and different from the locale computing and remote server computing. The nature of cloud computing is a service platform that can get computing power, storage space and a variety of software service according need of application systems. There are many similarities in the goal, structure and technology of Cloud computing and grid computing, but in the security, programming model, business model, data model, application and abstraction and so on are different. [3]

3 SECURITY OF CLOUD COMPUTING

The nature of cloud computing is a service platform base on network communication, so some network safety problems still exist in cloud computing. In addition, the structural layers of cloud computing include: client, cloud and transmission terminal, and the cloud include the server, database and so on. Therefore, from the analysis of the cloud computing structure itself, the security issues of cloud computing includes the following aspects:

3.1 *Security of cloud*

3.1.1 *Security of privilege user logon*
In the cloud computing platform, to assign different permissions for different users, we must ensure that no one user can be accessed.

3.1.2 *Security of data recovery*
In the cloud computing platform, user data are stored in the same physical environment, when a user data

is damaged, cloud computing providers must have a very fast data recovery measures that can't affect the normal use of other users.

3.1.3 Security of check

In the cloud computing, security of check occupies a very important position. Both the user's cross-border operations and the server's abnormal behavior must have the necessary check measures, so that we can quickly find problems and make a response.

3.2 Security of client

3.2.1 Security of users

The user is not only cloud computing experience, but also the final service object, so the system security and software security of user are very important for the whole application system. [4]

3.2.2 Security of send data

In the cloud computing platform, users and cloud are an interactive process, so the interaction process must be safe.

3.3 Security of transmission

3.3.1 Security of service continuity

Cloud computing services depend on the extent of the network is very strong, if there are faults in the network, the user cannot use the cloud service, that has a greater impact on its business continuity.

3.3.2 Security of persistence service

In the process of cloud computing, cloud must be persistent, otherwise the user's data will be loss in a short period of time, and it is the adverse consequences.

3.3.3 Security of transmission route

In cloud computing, the transmission route is a very important link. We must prevent an attacker from the middle of the interception.

3.4 Security of data storage

3.4.1 Security of data location

In the cloud computing platform, to the user, the specific location doesn't concern about their data; just download the data from the cloud. Service providers make sure that the data memory address security, which can provide stable services.

3.4.2 Security of data isolated

In cloud computing, a large number of users' data are in the shared environment, even if the data encryption methods used to ensure data security, it cannot

be guaranteed, and perhaps lose part of efficiency. [5] Therefore, there is a safety problem needing to be considered that How to ensure that a large amount of user data are independent and isolated from each other in a shared environment.

4 SOLVING STRATEGIES FOR CLOUD COMPUTING

For above mentioned cloud computing security issues, with considering of the technical characteristics, the strategy of establishing the layered security system could be adopted during the solving process. As shown in Figure 2, the five security layer system will be established: the data layers, the database access security layer, services access security layer, network access security layer and program access security layer, to realize cloud security.

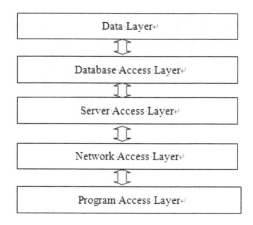

Figure 2. Five layer security architecture of cloud computing.

4.1 Program access layer

Program accessing requires strict measures to ensure standardize the access process. Access program focus on interface's control and usage, for the interface of the development program, there must be strict monitoring measures. And for a accessed program in system, it must have a completed procedure for behavioral tracking and reviewing, once the access program have exceptional behavior, the operation will be terminated [6].

4.2 Network access security layer

The network is the base of cloud computing, any data interaction is achieved and transmitted through the network. For security control of network access is a kind of method from the source, which only secure the network access, in order to secure the next transmission. Adopting VPN technology, or building a dedicated fiber-optic line, that all can improve the security of network access. But building the dedicated fiber-optic line is clearly incompatible with the characteristics of the cloud computing. Therefore, a VPN can be used to secure the network access channel, so that the data is transmitted under security.

4.3 Server access security layer

For secure server access layer, dual authentication can be set with, one is the server authentication and the other is a trusted third-party authentication, which can prevent malicious privileged users to modify data or copy [6]. The dual authentication approach is maturely applied in the field of online banking, so it can try to use in the cloud computing.

4.4 Database access security layer

The database is a core of the cloud, which stores all user information, including data and some personal information. For access to the database to have good control, the specific methods are included of identity authentication and behavior tracking and review, and through a series of protocol to regulate database access.

4.5 Data layer

The data security is very important, because data is the foundation of user and server interaction. Cryptographic algorithms are critical to ensure data security. For all sending data from client and server will be encrypted, in particular at server side, will adopt a relatively complicated algorithm, and transmits the content data to build a message digest to ensure the integrity of the information. In addition, the data stored in the database will also be encrypted. Nowadays, a popular technology is integrated in the computer hardware encryption module, once the data is out of this set of hardware systems, data will not be used.

5 CONCLUSION

Cloud computing technology proposed to promote the development of data storage technology, and to improve the efficiency of information exchange, promoting the changes of information technology to "big data" era, but also brings cloud computing information security challenges. In addition to the technical aspects of solving strategies, cloud computing security control measures should also include the establishment of relevant standards, laws and regulations to protect and service quality monitoring and other means [7], making cloud computing as a platform for big data processing, both in technology and service aspects of security.

REFERENCES

Takahiro Miyamoto, Michiaki Hayashi, Hideaki Tanaka, 2009.Customizing Network Functions for High Performance Cloud Computing,". Eighth IEEE International Symposium on Network Computing and Applications: 130–133.
JianZhang.2009.Analysis of cloud computing concept and influence. Telecommunications network technology. (01):15–18.
Peichang Shi, Huaimin Wang, Jie Jiang, Kai Lu,2009. Research and implementation of network platform based on Cloud Computing. Computer engineering and science No. A1 thirty-first:136–137.
Lixia Luo.2009. The service platform base on Cloud Computing...Information and Computer. NO.8:98–99.
Detian Tong, Xudong Liu, Taofeng Guo, Jiewen Liang, 2013. Cloud computing and information security analysis and practice. Telecommunication science (2): 135–136.
Xiaoming Lu, Weihua Cao, Yongchang Yu,2010. To explore the network of telecom operator in development of cloud storage service.Telecom Science (6) :107–109.
Wenjuan Li, Lingdi Ping.2009.Trust Model to Enhance Security and Interoperability of Cloud Environment.M.G. Jaatun, G. Zhao, and C. Rong (Eds.): CloudCom, LNCS 5931:69–79.

Computational Intelligence in Industrial Application – Ling (ed.)
© 2015 Taylor & Francis Group, London, ISBN: 978-1-138-02818-0

Exploration of improving graduates' English scientific and technological writing

Yan Ling Yang
Foreign Languages College, Northeast Dianli University, Jilin, P.R. China

Han Li
College of Science, Northeast Dianli University, Jilin, P.R. China

ABSTRACT: The graduates, as national high-level talents, should have a strong capability of English academic writing in order to let the world know their research achievements timely and achieve international academic exchanges. Therefore, in scientific and technological writing teaching, we should respect the cognitive characteristics of graduate students. Find a way to let students involve them in the study. To enhance the writing ability of the postgraduates, this paper gives the key points of scientific and technological writing, analyzes the problems which graduates may often encounter and puts forward the promotion strategies.

KEYWORDS: English scientific and technological writing; Graduates; Promotion strategies.

1 INTRODUCTION

As early as 1999, linguists Flowerdew pointed out: "In the age of globalization, articles published in any language other than English means to insulate oneself from the international society, on the other hand, bring an adverse effect to the further development of the carrier." Some colleges and universities have taken the construction of the international renowned university as their development goal. "The internationalization of education is not only becoming the impetus for international communication and cooperation, but also providing a historic opportunity for the development of higher education in China." The number of papers published in international journals, especially indexed by SCI or EI, has become an important parameter to measure the academic research level of universities. The graduates, as national high-level talents, should have a strong capability for English academic writing in order to let the world know your research achievements timely and achieve international academic exchanges of professional information.

In recent years, great focus has been on the educational reform of graduate English. Just as the National Research Council of Graduate Student Foreign Language Teaching in 2005 explicitly said, "Graduate English teaching must be need-oriented". And different kinds of ways and channels should be trying to improve the students' English application ability, especially the communication ability in spoken English and written words in the major

and related professional fields. In order to meet the new requirements, many colleges and universities in our country have offered graduate English optional courses for non-English majors, among which English academic writing is regarded as an indispensable and an important part in improving students' English comprehensive application ability. However, it is often a relatively few elective class with a lot of students. And it is a big class for students from different majors. What's worse, most teachers use the traditional all hall explanation mode and the old teaching material. Therefore, students can't fully understand the relevance between teaching contents and their major. So the participation and enthusiasm are not high, with the lack of learning motivation and interest. And it directly affects the study efficiency.

As the reform of college English teaching goes into its depth and the rapid recruitment of graduate students. The English level of today's graduate students cannot be compared with it before 20 years. A great number of graduate students had already passed the CET-4 and CET-6, with solid English foundation. On the other hand, as a high-level academic elite, they have known a mass of professional knowledge. They have strong abilities in cognition, comprehension and self-learning. If we just put emphasis on the interpretation of language, the cramming education is hard to mobilize their enthusiasm. We should respect the cognitive characteristics of graduate students. Find a way to let students involved them in the study. Learn to observe, practice, communicate, learn and summarize with the teacher's guidance. Update the knowledge structure of teaching mode in practice. To enhance the writing ability of the postgraduate, it gives the points of science and technology thesis writing of English, especially for the problems which graduates may often encounter. We'll have further discussion and put forward the promotion strategy.

2 THE BASIC PRINCIPLES OF ENGLISH SCIENTIFIC AND TECHNOLOGICAL WRITING

2.1 *Simplicity*

The first principle is concise in English scientific and technological writing. Most readers when reading scientific articles will first browse the full text, they will read the details of the article if they find it is helpful for their scientific research. When reading scientific articles most readers first browse the full text, and read the details of the article if they find it is helpful for their scientific research. Even deciding to take the time to read the full text, they will only spend fragmented time in reading it in most cases because of the fast pace and habits of modern society.

Graduate students, therefore, should use concise sentences as many as possible in scientific and technological writing in order to attract readers and better pass information to readers. Lengthy and winding expressions don't only cover the core idea of your article, but also waste readers' time and then make them lose patience. Take the following sentences as an example:

Graduate students during the period of school most of the energy will be used for academic research and the graduate student's academic research process is of academic information collection and processing process, in the accumulation of academic information and conducted on the basis of digesting and absorbing the process of scientific research and innovation.

It can be rewritten as: The academic research graduate students focus their attention on during graduate study is to innovate on the basis of the collection, procession, digestion and absorption of academic information.

2.2 *Logic*

The second basic principle is clear and logical ideas. Scientific papers are different from prose, poetry, news, documents and whatever in writing style. Scientific papers are aimed at convincing readers of experimental results, understanding research ideas and then identifying with the research conclusions by describing the experimental contents and analyzing experimental processes. To achieve this goal, scientific writing should be clear, logical and understandable. Only in this way can the readers understand the author's point of view accurately. For example:

Understanding information law and information ethics, getting rid of the bad habits of information, avoiding violating information law and information moral event are the things that students should be known. It can be rewritten as: It should be known to understand information law and information ethics, get rid of the bad habits of information, avoid violating information law and morality.

2.3 *Integrity*

Finally, an outstanding scientific paper should have accurate data and clear charts, which makes it readable and reliable. If contributing hastily without well preparing data and charts, a graduate student does not receive an objective evaluation of his research from the reviewers. They may well reject his paper when they believe it is confusing.

Based on their experience, editors think that the author' scientific research is not likely to be high-level and his experimental operation and data are even doubtful if his manuscript is careless and perfunctorily. The manuscript may well not be submitted for review and rejected directly if there are missing materials, rough charts and scratchy typeset in it. So graduates must carefully read instructions for authors, prepare your manuscript as perfectly as possible, in which graphics and text are elaborated in order to leave a good first impression on editors and reviewers before submitting the manuscript.

3 ESSENTIALS OF ENGLISH SCIENTIFIC AND TECHNICAL WRITING

3.1 *Title*

Readers first read the title of an article no matter where they find it, either from the search engines or from a directory of periodicals. The title of an article can contribute to whether it is worth reading. If it is not proper, readers who need such an article do not read it, which becomes useless. Besides, index and abstract depend on the accuracy of the title. If the title is untrue, the readers may not search it. Using keywords for the retrieval may search for a large number of articles in the internet or other searching systems, readers decide whether the article is used or not through its title. It is clear that the title plays an important role in the scientific papers.

The title of a scientific paper should be the most clear, concise and easily-searched phrases. It can summarize the main idea of the article with simple, suitable, clear and accurate means, and catch readers' attention. In foreign countries, titles of scientific papers should have the characteristics: brevity, clarity, indexibility, specificity.

The academician of Zhou Cheng Lu claims that if a sentence can generalize the conclusions or main findings, it can be taken as the title for it is eye-catching and vivid. For this reason, a title should manifest your opinions and conclusions rather than the description of the research areas. A large amount of articles put the opinions and experimental designs into the title, such as "the comparative experiment of using several natural solid materials to cultivate tomatoes.

3.2 Abstract

The abstract is the basis for the reader to judge the value of the paper and determine whether it is worth reading. As for the conference, the abstract is used to judge whether the paper is accepted or not. Writing Abstract is an essential basic skill in scientific paper writing. Abstract generally consists of three parts: research purpose, research methods, research results or conclusions. Many graduate students lose the chance of being accepted when they miss something in their abstract, even neither research purposes nor research conclusions.

The trick of writing abstracts is to highlight the new contribution, which covers new technologies, new theories, new methods, new viewpoints and new rules, correcting others' mistakes, resolving the dispute, supplementing and developmental achievements of the predecessors, etc. Some graduate students often do not reflect the important research results in the abstract, but at the end of the paper after analyzing and reasoning, which is not correct. New contributions should be introduced incisively, not ambiguously. We should clearly point it out rather than let readers analyze "newness" of the contributions. We also should avoid self-assessment.

3.3 Introduction

The introduction aims at explaining the reasons for this research and the purposes we want to achieve. Therefore, background information should be introduced in order to make readers who aren't very familiar with this field understand the significance of this research, what has been achieved, the problems which have been solved and unsolved, the problems that will be solved and technical solutions that will be used in this research. When writing the introduction, we should use the descriptive statements and get straight to the points instead of wordiness. What's more, if some research results have been published, we should explain it at the end of the introduction.

3.4 Text

3.4.1 Experiment methods
Graduate students should write experimental methods according to the real sequence of experimental operations, which makes the processes of your research understood easily. If this method has been published, graduates describe the procedures of the experiment briefly, as well as list the corresponding references. While in other cases the author must clearly describe the main steps of the experimental method, so that other researchers can repeat the operations according to the author's description. For graduate students who first write a scientific paper, the common problem is not to provide less information, but to write lengthily and describe something useless to the readers. Usually, graduates have the difficulty keeping balance in describing experimental methods when they are a green hand in scientific writing. It is recommended to consult experienced teachers to help modify and delete redundant experimental details.

3.4.2 Experiment results
When describing experimental results, we should state the important results rather than all the results. Graduates may well make such a mistake – repetitive description. In general, tables and charts can demonstrate experiment results in detail, so the information shown in tables and charts should not be described in words. The key research results should be written in words and ranked in importance, from most to least significant. If the information in tables and charts should be written in words, only the key points are stated.

3.4.3 Discussion
Researchers should read a large number of scientific papers for being informative in modern scientific research, which results in browsing fast the title and abstract, then judging quickly whether the paper is related and decide whether to read it further. If he decides to read the full text, a reader will read the discussion section in order to know the whole research conclusion. Therefore, discussion is an important part of the whole article.

In the discussion section, the results obtained is compared and discussed horizontally or vertically, including the comparison between your own results and comparison between your own results with others. If there are differences in comparison, we should analyze and discuss the causes of differences. This difference analysis further reinforces the importance of this study. In writing discussion section, there are three situations: first, get the same results with others (meaningless); second, get different results, but does not discuss the causes of differences (need improving); obtain different results, and analyze the causes of the difference (high-level).

4 IMPROVEMENT STRATEGIES OF ENGLISH SCIENTIFIC AND TECHNICAL WRITING

4.1 Reading for writing

Scientific and technical writing starts with scientific and technical reading. The problem of writing source or contents can be solved in material reading and critical reading.

Although academic writing is quite different from general writing in vocabulary, grammar, structural arrangement and so on, it is bound to have substantial academic contents. Therefore, extensive reading, listening to the lectures, then focusing on writing should be done before English scientific and technical writing.

4.2 Develop thinking of writing in group activities

Group and pair work are common in academic English writing class, its theory is based on constructivism, according to Vygotskypoint of view, the development of human beings is rooted in the social behavior, the learner's cognitive development needs in the social interaction "scaffolding" type of assistance.

According to Donato, Storch, "learning support" also plays an important role in group activities or in cooperative learning during the study. Storch studies have shown that cooperative learning plays an important role in writing。 It is better and more effective to complete the writing task, enhance the accuracy and appropriateness of the grammar and expand the writing thoughts and feedback in time

4.3 Focus on the writing process and promote writing results

The process and the result are in a dialectical and unified relationship. The first writing scientific article is the existence of the process, it can say that, what kind of process determines the results. In general the process of science and technology thesis writing is in four stages: prewriting, drafting, revising and editing.

Prewriting is the preparation of the writing, Students through reading, group communication and consultation, cooperation collect information, form thoughts and purpose, select a genre, determine the readers and draw up the outline of the thesis.

The drafting stage, where graduate students can actively participate in it, is that thoughts and ideas take shape in student's theses. And also the student must know that even a professional author isn't perfect in every respect. While correcting wrong concepts should be kept in mind all the time.

Gradually consummate and effectively revise the thesis in Revising period. Meanwhile, smooth out all problems of thinking, richness of content, abundance of thought through the peer review and so on.

It is the last stage to edit phases, that proceed literal proofreading, wording polishing and further perfect expression.

5 CONCLUSIONS

At present, our foreign language education is faced with the transition from quality education to applied education which conforms to the international development in all industries. To successfully participate in the international academic communication depends on our graduate students' academic English writing ability.

There are still many problems for academic English writing teaching in our country, but if we know the actual needs of graduate students, take the student as the center, get rid of traditional teaching modes, pay attention to the training of students' academic communication, our outstanding graduate students will have more opportunities to show their talents in the international academic circles.

REFERENCES

Aragon, S.R. & Johnson, E.S. 2008. Factors influencing completion and noncompletion of community college online courses. *American Journal of Distance Education* 22: 146 – 458.

Bergendahl, C. & Tibell, L. 2005. Boosting complex learning by strategic assessment and course design. *Journal of Chemical Education* 82: 645–651.

Bienkowski, I. & Feng, M. & Means, B. 2012. Enhancing teaching and learning through educational data mining and learning analytics: an issue brief. *Washington, D.C. Office of Educational Technology,* 9-10. http://en.wikipedia.org/wiki/Big-data.

Hung, J. & Hsu, H. C. & Ricc, K. 2012. Integrating data mining in program evaluation of k-12 online education. *Educational Technology & Society* 15(3): 27–41.

Jiyuan, Li. 2013. The idea of MOOC. *China Education Network* 4:3 9–41.

Mullen, G.E. & Tallent – Runnels, M.K. 2006. Student outcomes and perceptions of instructors' demands and support in online and traditional classrooms. *The internet and Higher Education* 9:257.

Saettler. P.L. 1990. *The evolution of american educational technology*. Englewood, CO: Libraries Unlimited.

S. Papert. 1980. *Mindstorms*. New York: Basic Books.

Viktor, Mayer-Schönberger. & Kenneth, Cukier. 2013. *Big data*. Hangzhou: Zhejiang People's Publishing House.

Wei Shi.&Jing-peng Hou.2011. Specialized-English education mode for graduate students. *Journal of Northeast Dianli University* 31: 169–171.

Yufan, Fu. 2013. "Pathfinder" of China online education. *China Education Network* 4: 26–28.

Computational Intelligence in Industrial Application – Ling (ed.)
© *2015 Taylor & Francis Group, London, ISBN: 978-1-138-02818-0*

MOOC and advanced education reform in China

YanLing Yang
Foreign Languages College, Northeast Dianli University, Jilin, P.R. China

Han Li
College of Science, Northeast Dianli University, Jilin, P.R. China

ABSTRACT: MOOC has brought new vitality and a new connotation to education. It does not only breed the new technological system, but also new teaching paradigms. This paper, first, analyzes the characteristics- large scale, openness, non-structured course contents and the application of big data, and the influence of MOOC on Chinese higher education. Then, it puts forward suggestions for the development of MOOC in China.

KEYWORDS: MOOC; Higher education; Teaching paradigm; MOOC platform.

1 INTRODUCTION

In 2011, "MOOC", which comes from the U.S.A., is developing rapidly and has great influence on advanced education. New York Times said that 2012 was "the Year of MOOC". Many experts believe that MOOC has been the greatest educational innovation since the invention of printing.

The spread of the internet has had an effect on all walks of life to some degree. By the end of 2011, IT elites and educational experts have been devoting themselves to the combination between internet and education. Although there is an e-learning platform that includes online learning, online examination, learning assessment, course and training management previously, it is only the supplement to traditional education and does not overturn traditional "blackboard + chalk" mode. Now, MOOC that mixes modern information technology with education may well give a new opportunity to educational reform. Online education is not a new thing, while the "newness" of MOOC lies in its large scale and its revolutionary potentiality of development. The popularity of MOOC derives from people's desire for high-quality education resources.

2 CHARACTERISTICS OF MOOC

MOOC is the abbreviation of Massive Open Online Course. Theoretically, "massive" means the number of registrations is limitless and order of magnitude surpasses ten thousand; "open" refers to anyone can take part freely; "online" means the learning activities are mainly online; "course" is contents with a series of learning goals in a specific field. MOOC has brought new vitality and a new connotation to education. The technologies applied to MOOC include high-quality cataloging videos, data collection and analysis, and the platform with social functions, which facilitates the teaching and learning based on the internet; from a cultural perspective, MOOC adheres to the learning culture on the basis of the communication, collaboration and knowledge discovery online. MOOC does not only breed the new technological system, but also new teaching paradigms.

2.1 Large scale

The learning environment of online learning before class includes self-learning platform and social media, for instance, QQ, We chat and Microblog. First of all, the teacher prepares teaching materials and designs tasks for students before class according to teaching goals, objects and contents. Based on the teaching goals of each lesson, the teacher uploads the teaching materials and tasks to network disk so as to make students download them conveniently. Students receive the teaching tasks and finish them independently. In the independent learning, students can choose learning materials and arrange learning time and contents freely, which reflects individualized learning. After that, students should sum up what they have learned and what they don't work out, and then tell their difficult points to the teacher online. The teacher subsequently collects students' problems and then analyzes and summarizes in order to explain them in class.

2.2 Openness

Openness, first, reflects that MOOC is open to the learners from all around the world. Region, age, language, culture, race, capital, incomes and whatever which are barriers in traditional education are no longer to be the obstacles in the age of MOOC. Taking Coursera as an example, among the 3 million learners who registered in Coursera, Americans account for only 27.7%, Indians for 8.8%, Brazilians for 5.1%, Englishman for 4.4%, Spanish for 4%, Canadians for 3.6%, Australians for 2.3% and Russians for 2.2% which makes up over 70% and shows the wide regional origins according to some statistics (in March, 2013).

The teaching and learning mode is open. The co-founder of Coursera, pointed out that the courses offered in Coursera were composed of ten minutes or so micro lesson, high-frequency retrospective test, interaction with learning resources, homework that has deadline, homework correction, a platform in which questions can be asked and answered and so on based on the ideas of active learning and deep learning.

2.3 Non-structured course contents

The content organization in traditional online courses is systematic and takes chapter set as the basic framework. It consists of teaching goals, teaching contents, teaching designing, teaching activities, teaching practice and teaching evaluation. There, however, are not standardized teaching contents but changeable, uncertain, non-structured contents in MOOC. The main part in MOOC is that the course-executer arranges and classifies answers that learners have to the questions on topic discussion and learners' blogs, learning notes. Non-structured course contents can assure the informal learning pattern taken in MOOC. A study shows that individual structured activities account for only 10%-30%, while non-structured activities for 70%-90% of the whole learning activities.

2.4 Application of big data

The super-large scale learning visits, global collaboration and communication and generative resources are bound to generate complicated big data. The massive and real-time data are stored in MOOC platform. In the course of learning in MOOC, learners do not only watch videos and answer questions, but communicate with their teachers and others through social network, forum, blog and QQ, which can leave long and rich "footprints", that is, data. They are analyzed with technologies and become meaningful. The results of data analysis can reveal tendency and models which cannot be attained in traditional teaching practice. It can help us know how learners learn, what they understand, what accounts for learner's success in MOOC learning.

Smart analytics aim at mining "connotation" of data and providing services for guidance, academic assessment, predicting the future and finding potential problems.

3 INFLUENCE OF MOOC ON CHINESE HIGHER EDUCATION

The rise of MOOC has had a great impact on traditional universities. It is impossible that a person could learn online with a computer or other web devices and even learn with people everywhere in the past. MOOC brings both opportunity and challenge to our higher education. If we way the influence of MOOC on the survival of universities and the integration between educational cost and quality is on the macroscopical aspect, the effect of MOOC on the teaching patterns in classroom teaching. The MOOC model is a little bit different from excellent courses and traditional network educational pattern. MOOC asks for higher requirements for teachers and students, while it also brings a new life to classroom education.

3.1 Promote the internationalization and popularization of higher education

The majority of MOOC in U.S.A. is free of charge, which gives more students, especially students in poverty, access to advanced education. Almost all the MOOC are run by America's or world famous universities are run jointly by universities and enterprises, which guarantee the quality of education in MOOC. More people can receive high-quality advanced education. In traditional higher education, students should pay tuition that ranges from several thousand dollars to tens of thousand dollars, while they should not pay tuition or pay only a little for MOOC. Daphne Koller, a computer scientist at Stanford University, calls it "true democracy".

Higher education internationalization is the inevitable tendency of economic globalization. The integration of the global economy is an irreversible trend. Take America, for instance. American universities pay much attention to the enrollment of students from all over the world and for sending their students abroad to have short-term learning or long-term learning. The international students at Yale University, Duke University and University of California at Berkeley account for 17.7%, 13.6% and 9.3% respectively. Therefore, MOOC can promote higher education internationalization.

3.2 Revolutionize the university teaching paradigms

The development of economy and scientific technology promotes the improvement of teaching paradigms. At present, it is the tendency for educational research to improve learning efficiency and optimize teaching process with new technologies, among which MOOC is. Different from traditional courses, MOOC allows students to learn according to their own time and place, which provides opportunities for students who cannot receive higher education for the reasons of time, money, region and whatever. Course developers in Coursera and Udacity optimize MOOC according to people's online learning habit. A lesson is divided into several videos, each of which includes lectures and quiz and last about 15 minutes. Teachers can communicate and interact with their students through social media, such as e-mail, QQ, blog, micro-blog and We chat. Compared with previous course videos which only have lectures provided by online education companies, MOOC has had great improvement.

3.3 Bring challenges to the survival and development of universities in China

First, MOOC will promote the university's disintegration and reorganization of educational resources. The first-class universities have taken over the lead in providing high-quality MOOC and attracting students to register their MOOC owing to their excellent teachers and educational resources. The second and third-rate universities will be in a disadvantageous position and face existence crisis on accounts of their limited educational resources. It is probable that the second and third-rate universities will be reduced to be teaching laboratories of the first-class universities. Of course, it can promote the integration of educational resources. The second challenge is about whether education costs can accord with education quality. The registered numbers in MOOC have reached hundreds of thousands, which leads to a significant reduction in costs. Meanwhile, except for traditional, fundamental courses, the innovative courses that meet market demand are offered in MOOC. MOOC makes the integration between education costs and education quality possible. With the development of MOOC, the leaders in universities should catch chances as well as meet the challenges by providing high-quality and novel courses with advanced network informational technologies.

4 SUGGESTIONS TO THE DEVELOPMENT OF MOOC IN CHINA

4.1 Aspect of ideology

Different from the previous video public class, MOOC reflects unprecedented openness, impartiality and accessibility of high-quality educational resources. Universities should fully recognize practical significance and necessity of resource sharing, and realize the importance of open sources in the internet age to push forward the quality in education and improve the university's reputation and competitiveness. Universities can have pilot exploration on resource sharing and course opening from encouraging teachers and students to take part in the application of open resources to opening its part of resources gradually.

4.2 Aspect of teaching models

Joycean believed that teaching model was equal to learning model. The difference of the two models is the perspective. The learning model focuses on what the learners do, while teaching model centers on what teachers do to help learners promote their learning. Models of learning and teaching try to find all the elements which contribute to learning and then organize them according to the simplicity of operation in specific learning environment. Now, "flipped classroom" becomes quite popular, which means what's been done traditionally in a classroom is now done at home, and what's been done traditionally for homework is now done in the classroom. This model transfers the learning right from teachers to students. After class, students can arrange their time schedule in accordance with their pace and choose learning resources at their understanding level, which reflects the individualization of learning. In class, they share their acquisition and learning experience with their teachers and classmates, and internalize knowledge through various learning activities.

In the teaching design, learning activities and main contents should be arranged in accordance with learners' experience. The design can be in details, that is, learners are led to follow the way the teacher designs for them; or it can be broad, namely, learners can choose what and where they learn.

In the assessment, the diversified evaluation methods, such as e-file pack and peer assessment, can be used to replace traditional methods including exams and writing papers.

In the roles, the teacher does change from lecturer to guider and designer, which leads to the change of students' role. Students should be beyond the scope of universities, institutions and courses and communicate in a wide community. They also need to develop lifelong learning skills. That is more important than learning for exams.

In so many open courses to a large number of registered learners, if we want the online learning achieves success, we should innovate in the teaching methods. How to teach should accord with learners' needs.

4.3 Aspect of construction of systematic learning analysis

Learning analysis means that learning models, learning preferences and behavior performance are got and potential problems are recognized by dealing with the data related to learners and various algorithms in order to provide learners with high-quality, individualized learning support and services. When a learner has difficulties in learning, the system will offer constructive measures and suggestions. In this way, the learner almost does not have time in depression and frustration. MOOC platform records massive data on learning behaviors. If we mine the big data, with new technology, models and algorithms, we can easily attain the learning patterns from big data, which cannot be done traditionally.

4.4 Aspect of construction of MOOC platform

MOOC has had great influence on Chinese higher education. Collaboration with MOOC in western countries is inevitable. We should construct our MOOC which is suitable for Chinese learners. The first-class universities in China have begun to explore the new model of cooperation with three MOOC educational institutions. Peking University and Tsinghua University have joined edX and become the first Asian member. Fudan University Shanghai and Jiaotong University have joined Coursera and started to offer excellent courses in Chinese or English.

Collaboration is just the beginning. To learn western experiences and improve ourselves is our purpose. Universities have begun to explore to build up our MOOC. For example, specialized agencies have been found and 30 member universities have been promoted to develop and share excellent instructional resources in Shanghai. Chongqing University has helped establish eastern and western university course sharing alliance, among which 19 universities are at present. As a matter of fact, we have had experience in open courses and developed instructional resources. Taking national quality courses as an example, video open courses of university have reached 286 courses up to 2013. Our goal of "12th Five-Year Plan" is to open 1000 video public courses and 5000 resource-sharing courses, which helps build up MOOC.

4.5 Aspect of the support of effective policies to MOOC

In the construction of MOOC, the institutions MOOC relies on being mainly university consortiums and affiliates which are almost non-profit. However, they begin to commercial exploration and cooperate with investment corporations in western countries. In China, most universities are public institutions and depend on the investment of the government. MOOC is undertaken by universities, but also need the support of the government's policies and financial aid. In the course of the development of MOOC, there might well be behaviors and things in a lack of standardization, which needs the government to figure out the regulations and rules to secure the healthy development of MOOC.

5 CONCLUSIONS

Although there is debate about MOOC, the value of MOOC is to make universities rethink online learning and open education from the highest point. MOOC presents a new perspective on the business pattern of higher education and teaching models and will boost the quality of online learning and school education. With the change of the external environment of advanced education, MOOC with innovation will have a deep effect on the present online education and higher education, which brings learners more choices that cover the new learning styles, time arrangement of learning and selection of learning content. MOOC, however, does not replace the traditional education completely. With the development of MOOC, universities should keep balance between online learning and face-to-face learning so as to cope with the challenges and opportunities in the future.

REFERENCES

Christensen C. M. & Anthony S. D. & Roth E. A. 2004. *Seeing what's next: Using the theories of innovation to predict industry change.* Harvard Business school: 227–250.

Christensen C. M. & Raynor M.E. 2003. The innovator's solution: creating and sustaining successful growth. Harvard Business School Press.

Downes, S. 2013. MOOC – The resurgence of community in online learning [EB/OL]. http://halfanhour. Blogspot. co.uk/2013/05/mooc-resurgence-of-community-in-online.html.

Geoff Maslen. 2012. MOOCs challenge higher education's business models [EB/O]. http://www.universityworld news.com/article.php? story=20120831103842302 &query=MOOC

Li Yuan & Stephen Powel & Hongliang Ma. 2013. International analysis of MOOC. *Open education research* (3): 56–62.

Li Yuan & Stephen Powel & Hongliang Ma.etc. 2013. Influence of MOOC on higher education: destructive and innovative perspective of theory *Modern Distance Education Research* (2): 3–9.

Nicholas Carr. 2012. Crisis for higher education [EB/OL] http://www.techreviewchina.com/story/3948.htm

Powell S. & Tindal I. & Millwood R. 2008. Personalized learning and the ultraversity experience. *Interactive learning environments* 16(1): 63–81.

Siemens G. & Downes S. 2009. Connectivism and connective knowledge[EB/OL].http://ltc.umanitoba.ca/connectivism/

Shuti Yu. & Xuemei Tang. & Rongwei Fu.2012.Thinking and enlightens of group cooperation learning for college English teaching. *Journal of Northeast Dianli University* 32(6):103–106.

Tamar Lewin. 2012. *Education site expands slate o universities and courses.* The New York Times.

Computational Intelligence in Industrial Application – Ling (ed.)
© 2015 Taylor & Francis Group, London, ISBN: 978-1-138-02818-0

Survey of information literacy of foreign language teachers in universities

Yan Ling Yang
Foreign Languages College, Northeast Dianli University, Jilin, P.R. China

Han Li
College of Science, Northeast Dianli University, Jilin, P.R. China

ABSTRACT: In the information age, foreign language teachers in university should have a good information literacy, which is essential for them to adapt themselves to the modern education. In this paper, the investigation on foreign language teachers' information literacy is conducted and the result shows that they have relatively low levels of information awareness and knowledge and know little about information morality and law. Therefore, the information literacy of language teachers needs to be improved.

KEYWORDS: Information literacy; Foreign language teachers; Survey.

1 INTRODUCTION

"University English teaching requirements" (2007) clearly pointed out, "colleges and universities should make full use of modern information technology, using English teaching model based on computer and classroom to change the single teaching mode." With the development of modern information and communication technology (ICT), the education informatization upsurges and includes various English learning and teaching disciplines of teaching revolution in the whole world. Therefore, foreign language teachers must have a good command of information literacy in order to adapt to the development of educational informationization.

2 RESEARCH FOUNDATIONS

The term "information literacy" was proposed by PaulZurkowski who is the American Information Industry Association (ILA) chairman in 1974. He thinks that the information literacy means a lot of information and main information source used to utilize techniques and skills. In 1979 American Information Industry Association gave the explanation of the information literacy: people know how to solve problems with information technology and skills. USA information expert Paterieia Breivik said: information literacy is an understanding of information systems and to identify the value of information, the best choice of channels to obtain information, the basic skills of acquiring and storing information, such as a database, spreadsheet, and word processing skills. The USA

Library Association to information literacy explains: "One needs to have the information literacy and be able to recognize it when information is needed, and have the ability to find, evaluate, and use effectively the needed information. People who have the information literacy know how to learn. They know how to learn, because they know how to organize knowledge, how to find information, and how to use the information, so that others can learn from them. They have been ready for lifelong learning." The definition of information literacy at home and abroad related to this concept is various. Nature scientific interpretation is: all kinds of information in the intersection, technology is highly developed in the society and the practical skills of information processing are to make people screen, identify and use information.

To sum up, the author thinks foreign language teachers' information literacy covers Information awareness, information knowledge, information competence, the integration competence and the information ethics, information security and the information moral etc.

2.1 Information consciousness

Information consciousness is the awareness of obtaining information and the basis of the information literacy education. In modern society where the electronic information is developed rapidly, the cultivation of information awareness is particularly important.

Without information consciousness, it is impossible to learn the knowledge, do research, much less innovate. First is to concern interdisciplinary and search the innovation point. Many of today's innovations are to find a breakthrough in the intersection between disciplines, so innovators must understand multi-disciplinary knowledge in order to avoid the bottleneck of the whole subject of the study caused by the lack of subject knowledge.

Just as information literacy is required in innovation, to know how to obtain the knowledge beyond your field seems to be very important. Students should be educated to develop a good habit of getting information, and no matter what the research project they do, firstly, check what others have done, have not done, and what should be done continuously. Secondly, acquire information randomly and stick to it. That is, without mental preparation, we can grasp the information needed and make full use of it from what we don't care about or what happens and quickly disappears. This behavior is a typical process: potential demand—involuntary attention—conscious attention—to obtain information and to do it. Thirdly, determine the target information and deliberately retrieve it. Students should be cultivated to make every effort to get the information after they have identified a target, and have the information awareness of never giving up until they reach their goal. The fourth point, consult and work out when you meet with the difficulties. If someone doesn't have the ability or the time to solve the problems encountered in the information retrieval, it is necessary to ask others for help. Only in this way can they cultivate good habits to get the information unceasingly and consciously, and as time passes the improvement of information consciousness will naturally follow.

2.2 *Information knowledge*

Information knowledge is the comprehensive knowledge which covers information definition, contents, characteristics, types, regulations, source of information, information channel and whatever, as well as the knowledge of the information technology and information retrieval, method, path, tools of access to information etc. It is the key of the information literacy education. Information is everywhere and graduate students must cultivate information awareness itself constantly so that they can identify information better, have access to and make use of the information better. Cultivating complete information knowledge, which is an important aspect of graduate education, is crucial in acquiring knowledge, doing research and innovating.

2.3 *Information ability*

Information ability education is the core of the information literacy education. Information ability means the ability to acquire, analyze, process, create, transfer, use and evaluate the information, including the information acquisition ability, information processing ability, information using ability and information evaluating abilities and whatever. Information acquisition ability refers to obtain, process and send information from various sources, such as the Internet. Information ability education should break through the traditional classroom teaching of literature retrieval course and help students know the new ways of obtaining information and acquiring skills. Let students know the library collections and online databases available, master how to use them and how to obtain useful knowledge, and then, on this basis, know how to screen, manage, manage, process and publish information. Students can come to apply the refined information to learning exchange, scientific research and social practice activities. Information capabilities include information cognitive ability, ability to obtain information, the information processing capability and information utilization capacity. It means to find something of value out of a large amount of complex information and deep ideological connotations from commonplace matters; discover and express effectively information needs, retrieval, procession, analysis and judgment and show the value of information. In this way, obtain information quickly and effectively, improve the rationality and necessity of the scientific research project, effectively avoid repetition and low level of scientific research in order to improve the level of innovation.

2.4 *Information morality and law*

Information morality and law education is an important aspect in information literacy education, including the ability of differentiating the useful or useless, healthy or unhealthy information, capable of resisting the temptation and contamination of unhealthy Internet information. It should be known to understand information law and information ethics, get rid of the bad habits of information, avoid violating information law and morality. As the creator of the information, we should guarantee what we have done scientifically. We should realize that they must be responsible for the authenticity and scientificity of their published results and what they have done are bound to be valid. When quoting, we should analyze others results them so as not to spread false information. What is more, as a disseminator of information, we should ensure what we spread is in line with the human ethics and the development of human culture instead of being harmful to the cultural development of human beings, such as not to spread unscientific and incorrect fantastic talk, make and disseminate computer virus and so on.

3 ANALYSIS OF THE CAUSES OF LOW INFORMATION LITERACY OF FOREIGN LANGUAGE TEACHERS IN COLLEGES AND UNIVERSITIES

In order to be familiar with the current foreign language teachers' information literacy, a questionnaire was designed in five aspects about information literacy, which investigated foreign language teachers' information literacy, surveyed a college, six undergraduate colleges and four of among which are third-level. We send 220 questionnaires and collect 208 copies, 205 copies effective. After finishing statistics, we found that the current information literacy of foreign language teachers in universities have the following problems:

1 Aspect of information consciousness: through the survey, we find that there are still 58% (see Table 1) teachers are lack of understanding the importance of teachers in the application of computer network technology in the university English teaching, the teaching mode. The general effect on the result is bad and it should be popularized in large areas. The author also chat network to further understanding with peers and friends, these teachers have to worry, doubt and even have the attitude of rejection, the university English teaching network spreading The machine will replace teachers and in this way they will be faced with the unemployment, so they are worried that the teacher's role will be weakened or they worry about the implementation of the informatization teaching model. Some teachers do not have confidence or even have a "computer phobia" owing to their low ability of computer and information technology. They are afraid that operational errors may have a negative influence on their teaching, or the soft fault cannot process equipment, which makes them embarrassed in front of students, so as for information technology they usually have a reverse psychology. In addition, a lot of time and energy will be used for the integration of information technology and curriculum teaching required that many teachers are not interested in.

2 Technical aspects: the survey shows that the teacher who consider their current computer network technology can fully meet the requirement of network teaching accounts for only 26%, that is to say, only about a quarter of the teachers can be fully qualified for network teaching, while three fourth of computer network technology teachers need to improve. As for the computer virus prevention and information security, only 38 percent of the teachers know a lot and can satisfy the need of network teaching. No matter how the operation of computer network technology is, most foreign language teachers still can not meet the need of network teaching. Therefore, lack of information literacy restricts popularization network English teaching and multi-level, multiform, multi specification.

3 The theoretical knowledge of the network English teaching methods: the survey shows that only 21% of the teachers has access to network information reasonably and design network teaching plan and tasks when teaching in the network according to course requirements, and only 24 percent of the teachers can manage the students' learning process, carry on the reasonable appraisal and analysis on learning behavior students' online effectively. That is about three fourth of the teachers know how to effectively use the technical ability of the overall design of the classroom teaching.

Table 1. Information literacy of foreign language teachers in universities.

problems	Results		
Effect of application of computer network technology in English Teaching in University and its importance	Good effect Need strengthen	General effect Not need	Bad effect Need limiting
	People 56	People 37	People 29
	Proportion 58%	Proportion 27%	Proportion 15%
Can your information ability meet the need of the network teaching?	Fully meet the needs	To meet the basic needs	Can't meet the needs
	People 71	People 49	People 87
	Proportion 37%	Proportion 21%	Proportion 42%
In network teaching, you can integrate the class teaching materials, have access to network information reasonably and design network teaching plan and tasks?	No problem	Occasionally can	Can't
	People 37	People 57	People 102
	Proportion 11%	Proportion 32%	Proportion 57%
In network teaching, can you have control of the whole process of students' learning, behavior and give e reasonable appraisal and analysis?	No problem	Occasionally can	Can't
	People 29	People 51	People 67
	Proportion 27%	Proportion 31%	Proportion 42%
In network teaching, do you know the computer virus prevention and online information security knowledge?	Much	General	No
	People 56	People 54	People 29
	Proportion 38%	Proportion 37%	Proportion 25%

This shows that most of the teachers lack of the network knowledge, theoretical knowledge needed to learn the effective integration of technology and curriculum. Through the network chat we can further understand, although many teachers participated in computer technology training organized by the school, but they are limited to the basic operation ability of the computer. The computer used for teaching knowledge is rarely involved. In addition, many teachers reflect, in the new teaching mode, the original curriculum teaching principles, teaching plan design theory will need to do some adjustments and changes, but how to adjust in order to make the technology courses services is placed in front of many network English teachers.

4 CONCLUSIONS

With the arrival of the information age, network technology, multimedia technology has brought a new revolution in the field of higher education, getting access to information, processing information; dissemination of information ability becomes the essential ability of foreign language teachers in the twenty-first Century. Teachers in Colleges and universities are facing a profound reform: update the concept of education, improve the educational technology, explore new modes of teaching, and improve teaching efficiency and effectiveness. This requires the teacher to come out as soon as possible from the traditional mode of teaching in the university English teaching reform, which

is actually the reform of teachers' teaching quality and consciousness of the. Only with a team of highly qualified teachers, we can build up the teaching mode, to experiment, to exchange, to promote, to get to the development of teaching reform, to enable students to become the biggest beneficiary. The change of teachers' subjective consciousness and the objective conditions of the training are crucial. As the teachers keep pace with the times in the teaching process, a clear understanding of their responsibilities can improve and develop their information literacy in the development of education in the educational concept, and have the high-level information literacy.

REFERENCES

Kai Liu.&Jia Zhong.&Haitai Gao.&Jun Yue. 2008. An investigation on the influence of internet on college students'ideological and political quality. *Journal of Northeast Dianli University* 28(3): 75–79.

Naipeng Cao. 2010. Research on the university students' using behavior of search engineer. *Library and information service* 54(24): 50–55.

Xiaoli He. 2003. Analysis of teacher's information behavior in IT environment. *Academic Journal of Ningxia University* 25(1): 125–128.

Xinmin Sang.& Aianwei Zhang. 2000. *Learning theory and practice in information age*. Beijing: Central radio & university press: 25–27.

Hong Zhou. 2003. *Cultivation and development of information literacy*. Beijing: People's education press.

Heming Huang. 2004.*Moden educational technology*. Beijing: High education press.

Computational Intelligence in Industrial Application – Ling (ed.)
© 2015 Taylor & Francis Group, London, ISBN: 978-1-138-02818-0

Fast single image dehazing method based on physical model

Haibo Liu
Key Laboratory of Fiber Optic Sensing Technology and Information Processing, Ministry of Education, Wuhan University of Technology, Wuhan, China
School of Electrical and Information Engineering, Hunan Institute of Technology, Hengyang, China

Jie Yang & Zhengping Wu
Key Laboratory of Fiber Optic Sensing Technology and Information Processing, Ministry of Education, Wuhan University of Technology, Wuhan, China

Qingnian Zhang
School of Transportation, Wuhan University of Technology, Wuhan, China

ABSTRACT: In this paper, we propose a simple but effective single image dehazing method based on physical models. The physical properties of the optical reflectance imaging and morphology operation are adopted to obtain the coarse medium transmission. Then, a fast joint bilateral filter is introduced to refine the transmission medium. The global atmospheric light can be estimated by using a two-step approach. By using the physical model of hazy image, a high-quality haze-free image can be recovered, which is optimized by tone mapping. Compared with existing algorithms, the complexity of the proposed algorithm is a linear function of the number of input image pixels and obtains a faster processing speed. Experimental results demonstrate that the proposed algorithm outperforms some state-of-the-art methods.

1 INTRODUCTION

The processing of fog-degraded images has been an important topic in the areas of computer vision and computer graphics, which has wide potential applications. For real-time situation, the study of fast image dehazing is extremely important.

Recently, the research of single image dehazing obtains a big breakthrough based on stronger priors or assumptions. Tan recovered the fog image by maximizing the local contrast of the image. However, it may not be physically valid. Fattal assumed that the surface shading and the transmission function were locally statistically uncorrelated and proposed a fast dehazing algorithm based on color image. The performance of the algorithm largely depends on the statistical features of the input data. He *et al.* proposed a valid image dehazing method based on dark channel prior, which was effective in most cases. However, the algorithm has high time and space complexity because of the soft matting. Tarel *et al.* proposed a fast image dehazing algorithm based on the median filtering by analyzing the scope of atmospheric scattering light. However, the atmospheric scattering light is not always consistent with the depth information of a scene. When the depth information changes suddenly, the method can not obtain the desirable effect. In this paper, we propose a new method in a single image dehazing based on physical models. To this end, grayscale opening operation is introduced to obtain the coarse medium transmission. Take-in account of the impact of grayscale opening operation, a two-step approach is adopted to estimate the global atmospheric light, and a fast joint bilateral filter is used to refine the transmission medium. Then, a haze-free image based on the physical model can be recovered, which is optimized by tone mapping. The experimental results indicate that the proposed algorithm is simple and effective.

The rest of the paper is organized as follows. Section 2 explains the problem of single image dehazing. In Section 3, the proposed image dehazing algorithm is described. Experimental results and analysis are presented in Section 4, followed by the conclusions in Section 5.

2 BACKGROUND

The following linear interpolation model is widely used to explain the formation of a haze image:

$$I(x) = J(x)t(x) + A(1 - t(x)) \qquad (1)$$

where $I(x)$ is the observed image, $J(x)$ is the scene radiance, A is the global atmospheric light, and $t(x)$ is the medium transmission. When the atmosphere is homogeneous, the transmission $t(x)$ can be expressed as

$$t(x) = e^{-\beta d(x)} \qquad (2)$$

where β is the scattering coefficient of the atmosphere and $d(x)$ is the scene depth. The goal of haze removal is to recover $J(x)$ by estimating two unknown factors A and $t(x)$ from $I(x)$, which is an ill-posed problem. So, many researches have been done to deal with the problem in estimating the factors A and $t(x)$.

3 PROPOSED ALGORITHM

In order to estimate the global atmospheric light and the medium transmission, the minimum operators are put on both sides of (1):

$$
\begin{aligned}
I^{dark} &= \min_{C \in \{R,G,B\}} I^C(x) \\
&= \min_{C \in \{R,G,B\}} J^C(x) t(x) + A(1 - t(x)) \\
&= J^{dark} t(x) + A(1 - t(x))
\end{aligned}
\qquad (3)
$$

where $I^C(x)$ is a color channel of $I(x)$, $\min_{C \in \{R,G,B\}}$ is performed on each pixel to get the minimum of its (R, G, B) values, I^{dark} is defined as the minimum channel figure.

3.1 Atmospheric light estimation

Taken into consideration the influence of white objects of hazy image, gray-scale opening operations are adopted to put on both sides of (3) to estimate the global atmospheric light. Then, (3) can be changed into

$$\hat{I}^{dark} = \hat{J}^{dark} t'(x) + A(1 - t'(x)) \qquad (4)$$

where $\hat{I}^{dark}$, $\hat{J}^{dark}$ and $t'(x)$ are the results, which are dealt with I^{dark}, J^{dark} and $t(x)$ by using gray-scale opening operation. The value of $t'(x)$ is no more than the value of $t(x)$ and the value of $\hat{J}^{dark}$ is close to zero.

Then, (4) can be rewritten as

$$\hat{I}^{dark} = A(1 - t'(x)) \qquad (5)$$

In (5), it is easy to get

$$t'(x) = 1 - \frac{\hat{I}^{dark}}{A} \qquad (6)$$

Here, (6) is defined as the coarse medium transmission.

Since the value scope of $t'(x)$ is the closed interval $[0,1]$, from (6), the first estimated value of A can be expressed as

$$A \geq \max(\hat{I}^{dark}) \qquad (7)$$

In general, the value of A is no more than the maximum value of hazy image $I(x)$. So, the second estimated value of A can be described as

$$A \leq \max \left(\max_{C \in \{R,G,B\}} I^C(x) \right) \qquad (8)$$

Therefore, A can be expressed as

$$A = \alpha \cdot \max(\hat{I}^{dark}) + (1 - \alpha) \cdot \max \left(\max_{C \in \{R,G,B\}} I^C(x) \right) \qquad (9)$$

where α is a parameter to adjust the value of A, which can be defined as

$$\alpha = mean\,2(\hat{I}^{dark}) \qquad (10)$$

where $mean\,2(\cdot)$ means the average of matrix elements.

3.2 Medium transmission estimation

Firstly, the mean filtering and median filtering are adopted to improve the edge features of $t'(x)$, which can be described as

$$t''(x) = median_{s_v}(mean_{s_v}(t'(x))) \qquad (11)$$

where s_v is the size of the square or disc window, and the value of s_v is determined by the minimum size m of $\hat{I}^{dark}$, which can be defined as

$$s_v = floor\left(\frac{m}{50}\right) \qquad (12)$$

where $floor(\cdot)$ means a round number toward negative infinity. If the value of s_v is not an odd number, (12) can be rewritten as

$$s_v = floor\left(\frac{m}{50}\right) + 1 \qquad (13)$$

Secondly, from (3), (4), (6) and (11), $t''(x)$ and I^{dark} are structural similar, and I^{dark} has lots of detail information of texture and structure the same as $I(x)$. So, a fast joint bilateral filter, which is proposed by Paris et al. is adopted to refine $t''(x)$ by using I^{dark}. For images $t''(x)$ and I^{dark}, the joint bilateral filter is defined as

180

$$t^{jb}(x) = \frac{1}{W_x^{jb}} \sum_{x \ y \in S} G_{\sigma_s}\left(\|x-y\|\right) G_{\sigma_r}\left(\|t''(x) - I^{dark}(y)\|\right) I^{dark}(y) \quad (14)$$

where σ_s defines the size of the spatial neighborhood used to filter a pixel, and σ_r controls the weights based on the intensity difference. W_x^{jb} is used to the standardization of weights, which is expressed as

$$W_x^{jb} = \sum_{y \in S} G_{\sigma_s}\left(\|x-y\|\right) G_{\sigma_r}\left(\|t''(x) - I^{dark}(y)\|\right) \quad (15)$$

Thirdly, the bright areas of scene can be estimated by defining the absolute value of the difference of atmospheric light A and dark channel I^{dark}, which are used to refine the medium transmission $t^{jb}(x)$. Then, $t^{jb}(x)$ can be modified into

$$t_{rf}(x) = \begin{cases} t^{jb}(x) & |A - I^{dark}| > K \\ \dfrac{K}{|A - I^{dark}|} \cdot t^{jb}(x) & |A - I^{dark}| < K \end{cases} \quad (16)$$

where K is the threshold to estimate the bright areas. When the absolute value is less than K, the area is regarded as the bright area and $t^{jb}(x)$ would be revised. On the contrary, the area is not bright area and the value of $t^{jb}(x)$ remains the same.

Finally, in (16), some values of $t_{rf}(x)$ may be bigger than 1. So, the value of $t_{rf}(x)$ should be truncated to [0,1]. And the value of $t_{rf}(x)$ should be no less than 0.1 [3]. Therefore, $t_{rf}(x)$ can be changed into

$$t(x) = \min(\max(t_{rf}(x), 0.1), 1) \quad (17)$$

3.3 Image restoration

From (9) and (17), the atmospheric veil can be given by

$$\begin{aligned} V(x) &= A(1 - t(x)) \\ &= [\alpha \cdot \max(\hat{I}^{dark}) + (1-\alpha) \cdot \max(\max_{C \in \{R,G,B\}} I^C(x))] \quad (18) \\ &\quad \cdot (1 - \min(\max(t_{rf}(x), 0.1), 1)) \end{aligned}$$

In, in order to retain part of the mist in the distant scenery, a constant γ is introduced and the atmospheric veil is redefined as $\gamma \cdot V(x)$. The empirical value of γ is 0.95. However, by the calculation of (16), the value of $t(x)$ in the proposed algorithm is frequently bigger. To reduce the effect of the fog, the value of γ is modified to 1 in this paper. According to (1), the scene radiance $J(x)$ can be recovered by

$$J(x) = \frac{I(x) - V(x)}{\min(\max(t_{rf}(x), 0.1), 1)} \quad (19)$$

3.4 Tone mapping

Generally, the color of restored image $J(x)$ is dark. Tone mapping, which can display high contrast scenes and result in better preservation of details and contrast, is adopted to improve the effectiveness of the restored image.

Firstly, in (19), in order to decrease the impact on the tone mapping, some values of $J(x)$, which are less than 0, should be set to 0.

Then, the tone mapping function, which is proposed by Drago *et al.*, is adopted in this paper. For any channel of the restored image $J(x)$, the tone mapping function can be expressed as

$$J_d(x) = \frac{L_{d\max} \cdot 0.01}{\log_{10}(J_{\max}(x)+1)} \cdot \frac{\log(J^c(x)+1)}{\log\left(2 + \left(\left(\dfrac{J^c(x)}{J_{\max}(x)}\right)^{\log(b)/\log(0.5)}\right) \cdot 8\right)} \quad (20)$$

where $J_d(x)$ is an output channel of $J(x)$, $L_{d\max}$ is the maximum luminance capability of the displaying medium, whose typical value is 100. b is the bias parameter to adjust compression of high values and visibility of details in dark areas. The proposed value of b is 0.85. $J^c(x)$ is a color channel of $J(x)$. $J_{\max}(x)$ is the maximum value of $J^c(x)$, which can be expressed as

$$J_{\max}(x) = \max(J^c(x)) \quad (21)$$

3.5 Time Complexity

For an image of size $s_x \times s_y$, the time complexity of the proposed algorithm is $O(s_x s_y)$, which mainly comes from the calculated amount of (11), (14) and (16). So, it is a linear function of the image pixels and greatly improves the speed.

4 EXPERIMENTAL RESULTS AND ANALYSIS

To demonstrate the effectiveness of the proposed algorithm, two indicators e and $\bar{r}$ are adopted as the quantitative assessment, which are proposed by Hautière *et al.* e is the rate of new visible edges, and $\bar{r}$ is the ratio of the gradient of the visible edges after and before restoration. In fact, if visibility restoration algorithms must increase the contrast, artificial edges must not become visible. It means that the value of e and $\bar{r}$ must be balanced.

4.1 Algorithm assessment and analysis

Four groups of typical images are selected as representatives of hazing pictures. Fig. 1 illustrates the experimental results before and after restoration.

(a)   (b)

(c)   (d)

Figure 1. The first column are hazing images, the second column are recovered images. (a) building; (b) mountain; (c) tiananmen; (d) tree.

For the sake of assessment, Table 1 shows the quantitative assessment of four typical images with e and $\bar{r}$.

Clearly, the proposed algorithm improves the visibility from the value of $\bar{r}$, increases the artificial edges to varying degrees from the value of e, and has a good color fidelity.

Table 1. Performance with e and $\bar{r}$.

Name	building	mountain	tiananmen	tree
e	0.111	2.044	0.470	1.420
$\bar{r}$	1.973	2.628	1.862	2.395

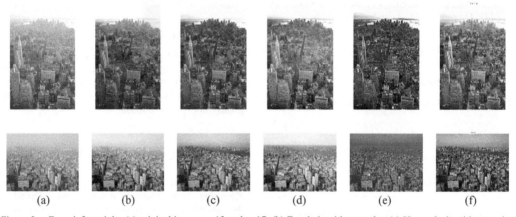

(a)   (b)   (c)   (d)   (e)   (f)

Figure 2. From left to right. (a) original image, ny12 and ny17; (b) Fattal algorithm results; (c) He *et al.* algorithm results; (d) Kopf *et al.* algorithm results; (e) Tan algorithm results; (f) the proposed algorithm. results.

Table 2. Performance with different algorithms.

Name	Fattal		He *et al.*		Kopf *et al.*		Tan		our	
	ny12	ny17	ny12	ny17	ny12	ny17	ny12	ny17	ny12	ny17
e	-0.06	-0.12	0.06	0.01	0.05	0.01	-0.14	-0.06	0.055	0.041
$\bar{r}$	1.32	1.56	1.42	1.65	1.42	1.62	2.34	2.22	1.694	1.827

4.2 Algorithm comparison and analysis

Two typical images are quantitatively assessed and rated by Fattal, He *et al.*, Kopf *et al.*, Tan [1] and our algorithms with e and $\bar{r}$ from Table 2.

It is clear from Table 2 that our algorithm increases the visible edges compared with Fattal, He *et al.*, Kopf *et al.* and Tan algorithms. And the proposed algorithm is only inferior to Tan algorithm from the value of $\bar{r}$. However, the high value of $\bar{r}$ in Tan algorithm maybe come from strongly improving the contrast in the recovered image and decreasing the value of e from Table 2.

5 CONCLUSIONS

In this paper, a fast image dehazing algorithm based on a physical model is proposed, which effectively improves the performance of haze removal. In order to obtain the best performance of the proposed algorithm, the size of the structural element in the gray-scale opening operation should be bigger than the white object in the image, and the threshold K used to estimate the bright areas of the scene should be balanced. Experimental results demonstrate the effectiveness of the proposed algorithm. In the future, we will extend the algorithm to the video image processing system.

ACKNOWLEDGMENTS

The authors would like to thank the anonymous reviewers for their insightful comments and valuable suggestions. This work was supported by National Science Foundation of China (NSFC) under Grand 51479159 and Soft Science Project of China's Ministry of Transport under Grand 2013-322-811-470.

REFERENCES

R. T. Tan(2008). "Visibility in Bad Weather from a Single Image". *Proceedings of IEEE Conference on Computer Vision and Pattern Recognition*, 1–8.

R. Fattal(2008). "Single Image Dehazing". *ACM Transactions on Graphics*, 27(3):1–9.

Kaiming He, Jian Sun, Xiaoou Tang(2009). "Single Image Haze Removal Using Dark Channel Prior". *Proceedings of IEEE Computer Society Conference on Computer Vision and Pattern Recognition*, 1956–1963.

J.P. Tarel, N. Hautière(2009). "Fast Visibility Restoration from a Single Color or Gray Level Image". *Proceedings of IEEE International Conference on Computer Vision*, 2201–2208.

Wei Sun(2013). "A New Single Image Fog Removal Algorithm Based on Physical Model". *International Journal for Light and Electron Optics*, 4770–4775.

M. Paris, D. Fredo(2006). "A Fast Approximation of the Bilateral Filter Using a Signal Processing Approach". *Springer–Verlag, Graz, Austria, Heidelberg*, 568–580.

Xiaoming Sun, Junxi Sun, Lirong Zhao, Yonggang Cao(2014). "Improved Algorithm for Single Image Haze Removing Using Dark Channel Prior". *Journal of Image and Graphics*, 0215–0220.

F. Drago, K. Myszkowski, T. Annen, N. Chiba(2003). "Adaptive Logarithmic Mapping for Displaying High Contrast Scenes". *Computer Graphics Forum*, 419–426.

N. Hautière, J.P. Tarel, D. Aubert, Eric Dumont(2008). "Blind Contrast Restoration Assessment by Gradient Ratioing at Visible Edges". *Image Analysis and Stereology*, 27(2):87–95.

J. Kopf, B. Neubert, B. Chen, et al.(2008). "Deep Photo: Model-Based Photograph Enhancement and Viewing". *ACM Transactions on Graphics*, 27(5):116:1–10.

Computational Intelligence in Industrial Application – Ling (ed.)
© 2015 Taylor & Francis Group, London, ISBN: 978-1-138-02818-0

The study of visual elements design based on user experience in web design

Chao Gao
Student, Kunming University of Science and Technology, China

KunQian Wang
Teacher, Kunming University of Science and Technology, China

ABSTRACT: With the rapid development of Internet technology, the user experience becomes the criterium of Internet products. From the perspective of the user's experience, this paper analyses the users use habits on a website, and ultimately explores the principles of usage of visual elements in web design based on the user experience.

1 INTRODUCTION

The term user experience is the first to be widely recognized in the mid-1990s, by a user experience designer Donald? Norman (Donald Norman). However, in recent years, computer technology has made great advances in graphics technology and other aspects of operating experience, Human-Computer Interaction (HCI) technology has permeated into almost all areas of human activity. These technological advances and expansion led to a huge shift, the evaluation index of system has extended to a broader range of user experience from simple usability. In human-computer interaction technology development, user experiences are becoming more and more important.

In ISO9241-210 standard, the user experience is defined as "People's cognitive impressions and response for the product, system or service which they use or expect to use." In other words, it means "Is this thing easy to use or convenient." Therefore, the user experience is the subjective.

ISO has defined the following explanation: the user experience, that user before using a product or system, all feel during use and after use, including emotions, beliefs, preferences, cognitive impression, physiological and psychological reactions, behavior and achievement and other aspects. Based on the actual use situation of the website, this paper analyze the visual elements in web design and summarize the principles of visual element usage based on user experience in web design.

2 THE ANALYSIS OF USER EXPERIENCE IN USING WEBSITE

Some designers might think, the user's actual usage situation can be very complex and it may depend on the type of the page, or what the user intends to do. But the experimental results are very simple and close to reality. One study showed that there is a considerable gap between the way users actually use the website and the way designers think.

When a designer creates a website, they tend to think that users will stare at each page, carefully read the text, even understand the way designers organized the page, and then determine which link they should click on.

In fact, most of the time, users actually do is glance at each page, swept some text, and then click on a link which is interesting or roughly accord with targets they are looking for. So they ignore most of the information on the page. This means that the rational and focused user is an idealized person which website designer envisaged when they doing design.

A column called "How Users Read on the Web" which Jakob Nielsen published in 1997 discussed a phenomenon by experiment that the user does not read word-by-word Web page, they just visit the website and pick out what they need. They just scan the page to find links which attract their attention. The news webpage is exceptional. However, despite the news page will attract user a lot of time, when they encounter a longer length of the article, they still will use the scan mode to read.

1 The user is always in busy. In most cases, the user is using the web to save time, therefore, web users do not have time to read those unwanted contents. The internet was born to information exchange and sharing of resources, to obtain valuable information in the shortest time, which has sped up the pace of work, study and life. So, from this perspective, people hoping more efficient and direct when they are using the web.

2 Users know that they do not have to read all the content. In the vast majority of the web page, the user

is actually only interested in a small part of the content, users will find the content they are interested in or the content which relevant to the task, they do not care about the rest of the content. Scan the page is the way that users get the content they need.

3 Users are good at scanning. In daily life, the environment the user lived in has a huge amount of information. They have been scanning newspapers, magazines, books and so on to find the part they are interested in, so they follow this habit to read the web pages. What they see on the web page depending on what they want to see.

3 WEB DESIGN ELEMENTS

The main web Design visual elements, including text, color, graphics, image, icon, constitute of points and lines, three-dimensional shape and other elements. However, in the design of various elements of the web page, form is the one most easily understood and striking for people, so that all designers are exhausted in the user interface design. The form in interface design usually closely linked with the composition, structure, material, color, space and function. The form in the web page design, is a non-materialistic digital form which interact with people. All the form, no matter how complex it is, can be summarized as point, line, surface, body. It can express different personalities and rich connotation.

4 USING PRINCIPLES OF VISUAL ELEMENTS IN WEB DESIGN

Based on the analysis of user experience and the extraction of visual elements extraction, we can find out how to enhance the site's user experience by rational use of visual elements.

4.1 Contrast principle

Contrast refers to the principle of comparing two things. By contrast, the characteristic has become more obvious. In website design, contrast can make a visual center of the web page, making the topic more distinctive and stand out from the background.

Contrast principle has different types, including: size contrast, the brightness contrast, the comparison of horizontal and vertical lines, the texture contrast, and other multiple comparisons.

Size contrast is the most important one in the interface. Using different size in different area of a web page will come out a different visual experience. The relationship between the size of the area determines the user's basic impression on the web page: small differences in the size will make a moderate impression;

big difference will make a clear and shocking impression. This will help designer to highlight the thematic content by using contrast.

1 Chiaroscuro is a basic element of color sense, by using chiaroscuro can achieve outstanding thematic content. With the dark background, the bright important part of the page can be enhanced.
2 Font contrast is one of the easiest methods. The thicker font is, the more prominent the text is.
3 Curve makes people feel soft, linear makes people feel sharp.
4 In web design, the texture is also a very important element. We usually are able to have access to texture: smooth feeling, concavity, rough and so on. Texture can communicate with the user to generate emotion.

4.2 Coordination principles

Contrast principle can make the page vivid and distinctive; coordination can make pages gentle and kind. Without coordination, a web page will have stiff feeling. Without contrast, a web page will have flat and dull feeling. With repeated use of the same shape things can make the interface to generate a sense of coordination.

The so-called coordination is a unified treatment and a reasonable match to make it in harmony unified whole. It includes the same interface, coordination of various elements, including coordination between the various elements of different interfaces, there are several aspects: Lord and from the dynamic and static, out of and into. It includes coordination of various elements, including coordination between the various elements of different interfaces, there are several aspects: main and subordinate, dynamic and static, out of and into.

1 Web design must be clear "hero" and "supporting role", only a clear relationship, can make a user get the important information. If the relationship fuzzy, it will make the user do not know what to do.
2 Interface design needs dynamic and static. Dynamic part includes dynamic picture and process of things, static part usually refers to the button on the interface or text. Between Dynamic and static, it needs an appropriate blank to emphasize their independence.

4.3 Principles of fun

The use of the image, intuitive, vivid graphics optimized interface can improve computer interface software interesting, so as to enhance the beauty of the user interface. It needs to make a good use of: the proportion, stressed, cohesion and diffusion, the intention of mind, a sense of change in the rate law, orientation, blank area, the text on the screen.

1 In web design, we must determine the header size according to the content. The ratio between the title and the text is the rate of change. In general, the greater the rate of change is, the livelier interface.
2 When the shape of a common impression is repeated or occurs, it will generate a sense of the law. Not have to use the same interface in the shape of things, as long as allowing users to generate strong impression on it.

5 CONCLUSION

The establishment of user experience is to analysis user experience in the use of a system well. Its research focuses on the performance and value system brought a sense of pleasure degree, rather than the system. Research departure from the actual use situation is an effective way to improve the user's experience. In web design, the designer only reduces the user's thinking, clarity of expression page content, to be able to provide users with a better user experience.

REFERENCES

Alan Cooper, Robert Reiman, David Cronin. About Face3. Beijing: Electronic Industry Press, 2012.: 2–255.

Steve Krug. Don't Make Me Think. America: New Riders Press, 2005: 55.

Donald Norman. Design Psychology. Beijing: China CITIC Press 2007: 42–98.

Jakob Nielsen. How Users Read on the Web. America: Academic Press. 2004: 10–38

Christina Wodtke, Austin Govella. Information Architecture: Blueprints for the Web. Beijing: Posts and Telecom Press, 2009-11: 40.

ROBERT HOEKMAN. Designing the Obvious: A Common Sense Approach to Web and Mobile Application Design, Second Edition. Beijing: China Machine Press, 2008-1.37–64.

Peter Merholz / Brandon Schauer / David Verba / Todd Wilkens. The Power of Design. Beijing: China Machine Press, 2009: 3–13.

Lukas Mathis. Designed for Use: Create Usable Interfaces for Applications and the Web. Beijing: Posts & Telecom press, 2012: 32–36.

Computational Intelligence in Industrial Application – Ling (ed.)
© 2015 Taylor & Francis Group, London, ISBN: 978-1-138-02818-0

Weight calculation of the railway signal infrastructure fault attributes based on rough set

Y.Q. Ai & H.B. Zhao
Beijing Jiaotong University, Haidian District, Beijing, P. R. China

ABSTRACT: Railway signal infrastructure is the guarantee of the safety and efficiency of the railway transportation. When the signal infrastructure gets into trouble, it's very important to find the fault quickly and solve it. CBR could help diagnose the fault by retrieving the case base and find the most similar case, then using the same solution to solve the fault. The case attribute weight determines the accuracy of case similarity. This paper makes a study on calculating the attribute weight by rough set and compares the calculation result with the other methods.

KEYWORDS: Weight calculation, Railway, Rough set.

1 INTRODUCTION

Due to its transportation capacity, transportation speed, high safety and low transportation costs, railway transportation has become the main transportation in our country. At the same time, signal infrastructure has played an important role in ensuring the safety and improving the efficiency of the railway. The development of signal infrastructure maintenance system which can ensure the operation without failures is a major issue currently [1]. At present, domestic and foreign researchers have made lots of research on the fault diagnosis of railway signal infrastructure [1-4]. However, the railway signal infrastructure fault has various characters, such as diversity, fuzziness, randomness, hierarchy, radioactive [5] and the complexity of the composition, as well as the complicated factors at the scene [6]. For the complexity and variety of the railway fault cases, putting forward a more effective and appropriate diagnostic method is still worth studying. While diagnosing the faults at the scene, people usually make analysis and judgment by their abundant experience. Also, the operation department has accumulated a lot of experience of emergency rescue and restored many successful cases.

Case-Based Reasoning is a problem solving paradigm which solves new problems by adapting previously successful solutions to similar problems. It is suitable for the field which has rich experience and records, while lacking of complete and precise mathematical model. It has significant advantages in solving the complex unstructured decision-making problems [7]. We consider use CBR to diagnose the railway signal infrastructure fault.

Conceptually CBR is commonly described by the CBR-cycle which comprises four activities: retrieve, reuse, revise and retain [8]. Among that, retrieve plays a key role because it directly affects the efficiency and quality of CBR [9]. Nearest Neighbor techniques are perhaps the most widely used technology in the procedure of retrieve. The similarity of the target case to a case in the case-library for each case attribute is determined. This measure may be multiplied by a weighting factor [10]. That factor led to the research of this paper. The target of this paper is determining the weighting factor of the case attribute.

Currently, most studies give the attribute weight by using the expert manually setting method, which is excessively relied on subjective judgment and experience. That is difficult to obtain reasonable case solution sometimes [9].

In this paper, rough set is applied to make reduction of the complicated and various railway signal infrastructure fault case attributes, as well as quantitatively determine the weight of the attributes after reduction.

2 ROUGH SET

Rough set, first described by Polish computer scientist Zdzislaw I. Pawlak, is a theory about data analysis and data reasoning. It is a new mathematical tool for dealing with uncertain and imprecise problem, including knowledge reduction, data meaning assessment, approximate classification and so on.

Let $I = (U, A)$ be an information system, where U is a non-empty set. While R is an equivalent relation about U.

$\forall A \subseteq U$, the upper and lower approximation set of A is:

$$\overline{R}(A) = \{x \mid [x]_R \cap A \neq \varphi$$
$$\underline{R}(A) = \{x \mid [x]_R \subseteq A$$

(1)

If $\overline{R}(A) \neq \underline{R}(A)$, then we define that A is the rough set about R based on U [11].

2.1 The rough set representation of case

In rough set, the knowledge representation system is defined as KRS= (U, A, V, f). U is a non-empty set of finite objects. A= C∪D is also a nonempty finite set of attributes, including condition attribute set C and decision attribute set D, while the value of A could be quantitative or qualitative. V is the range of the whole attributes. f represents a mapping function.

Generally, the knowledge representation system mainly consists of information tables and decision tables. In this paper, the fault case base of railway signal infrastructure uses the type of decision table, which is T= (U, A, V, f). In the table, each row is a fault case and each column represents the character-istic attributes C and D. In the fault diagnosis system, fault symptom belongs to C, such as the environment, the type of railway track and the voltage, while fault mode belongs to D, such as the unbalanced traction current, breakdown capacitors and damaged insulated joints. V is the range of those attributes. Above is dis-played in Table 1.

Table 1. The decision table of the fault cases.

Fault Case (U)	Condition attriute(C)				Decision Attribute (D)
	C_1	C_2	...	C_m	
X_1	V_{11}	V_{12}	...	V_{1m}	d_1
X_2	V_{21}	V_{22}	...	V_{2m}	d_2
...	...	...	...	...	...
X_n	V_{n1}	V_{n2}	...	V_{nm}	d_n

2.2 Reduction of case attributes

Knowledge reduction is a key function of rough set. In general, there are attributes in the information sys-tem, which are more important to the knowledge rep-resented in the equivalence class structure than other attributes. So knowledge reduction is to delete those redundant attributes while keeping the original rela-tionship between C and D.

In railway, the attributes of the fault case are com-plex and various. Without reduction, the speed of case retrieve and the efficiency of fault diagnosis would directly be affected.

Definition1: Let the knowledge base K= (U, S). There are two equivalent relationship sets called P and Q and both belong to S.∀ R∈P, if

$$pos_{IND(P)}(IND(Q)) = pos_{IND(P)-\{R\}}(IND(Q))$$

Then we call that R is unnecessary for Q in P.

The condition attribute set is C= $\{c_1, c_2,..., c_m\}$, meaning that the number of fault symptoms are m. Here come the detailed procedures of the attribute reduction.

Input: Decision table T=(U,C∪D ,V ,f).
Output: B, a relative reduction of C.

Step 1. Let B=C.

Step 2. Calculate the equivalent relation U/B and U/D.

Step 3. Calculate the negative region POS$_B$(D),

using the formula $pos_B(D) = \bigcup_{X \in U/D} \underline{B}(X)$.

Step 4. k=1:m, and delete the number k attribute. If POS$_{B-\{bk\}}$(D)=POS$_B$(D), then the attribute b_k is redundant, retaining the deletion. However, if POS$_{B-\{bk\}}$(D)≠POS$_B$(D), then it means that b_k is necessary, we cannot delete it.

Step 5. After the whole analysis of each attribute, the algorithm comes to end and then we output the reduction set B.

2.3 Determination of the attribute weight

The value of the weight means the relative importance of the attribute when calculating the case similarity.

Definition 2

Let the knowledge base K=(U, S). There are two equivalent relationship sets called P and Q and both belong to S. Then the dependency of Q on P, $\gamma_P(Q)$ is given by

$$\gamma_P(Q) = k = \frac{|pos_P(Q)|}{|U|} = \frac{\left| \bigcup_{X \in U/Q} P(X) \right|}{|U|}$$

(2)

Definition 3

A decision table DT=(U,C∪D ,V ,f).∀ $\alpha \in$ C, the importance of α to C relative to D, sig(α,C;D) is given by

$$sig(\alpha, C; D)$$
$$= \gamma_C(D) - \gamma_{C-\{\alpha\}}(D)$$
$$= \frac{|pos_C(D)| - |pos_{C-\{\alpha\}}(D)|}{|U|}$$

(3)

According to the definition above, we know that if sig(α_i,C;D) > 0, then the attribute α_i is necessary in C.

If there is any $sig(\alpha_i, C; D) = 0$, then the attribute α_i is redundant and it should be deleted. After that, we calculate the value of attribute weight in the decision table after reduction again. That wouldn't stop until all the $sig(\alpha_i, C; D) > 0$. And we can get the detailed procedures of calculating the attribute weights.

Input: Decision table T=(U,C∪D,V,f), C= {c_1, c_2,…, c_m}.

Output: The weight of each attribute in set C, ω_i, i=1:m. The C here has been got reduction.

Step 1. Calculating the importance β_i of each attribute c_i according to formula (3).

Step 2. Judging all the β_i. If there is $\beta_i=0$, then the corresponding attribute c_i should be deleted and a new set C' is formed. Then back to step1. But if all $\beta_i > 0$, get into the next step.

Step 3. When we get all β_i, they should be standardized according to the formula (4),

$$\omega_i = \frac{\beta_i}{\sum \beta_i} \qquad (4)$$

3 CASES AND ANALYSIS

Railway signal infrastructure mainly consists of relay, signal, track circuit, turnout, console, power supply panel and so on. According to the analysis of the historical signal fault data, we can find that the track circuit fault accounts for about 40% to 50% of the signal system fault. The frequent track circuit fault has decreased the reliability of the whole signal interlocking system. Among the track circuit fault, red zone and bad shunting are the most common fault types. So in this paper, red zone was chosen as the research object.

The reasons that lead to red zone are various. After analyzing the case base, we come to the conclusion that the reasons include the type of track, the frequency of trains, the weather, the voltage of sending and receiving end flowing on the track and so on.

In this paper, we select 10 fault cases from the red zone case base to verify the rough set method. The 10 cases are numbered from X_1 to X_{10}, and the set U={X_1,…,X_{10}}. The case condition set C={C_1,…,C_9} and they are the type of track circuit, the performance of red zone, the frequency of train, weather, air humidity, voltage of the sending end, voltage on the sending track, voltage on the receiving track and voltage of the receiving end.

The decision attribute D is the fault mode of red zone. Those are displayed in table2. Since the detailed description of the cases make the table a little big, so we just display some of it.

We can see that values of some attributes are qualitative and some are quantitative, so we should do some discretization to them so that it could be easy to deal with them by using rough set. The results of discretization are showed in table3.

Now all the values are discrete. By using formula (1), (2), (3) and (4) introduced before, the calculation process are showed below.

$$U/IND(D) = \{ \{X_1,X_2,X_5,X_8\}, \{X_3\}, \{X_4\}, \{X_6\}, \quad (5)$$
$$\{X_7\}, \{X_9, X_{10}\} \}$$

$$U/IND(C\text{-}\{C_1\}) = \{ \{X_1\},\{X_2\}, \{X_3\}, \{X_4\}, \{X_5\}, \quad (6)$$
$$\{X_6, X_9\}, \{X_7\}, \{X_8\}, \{X_{10}\} \}$$

$$POS_C(D) = \{X_1, X_2, \cdots, X_{10}\} \qquad (7)$$

$$POS_{C\text{-}\{C_1\}}(D) = \{X_1, X_2 \cdots X_5, X_7, X_8, X_{10}\} \qquad (8)$$

So the importance value of C_1 is

$$\beta(C_1) = \frac{|POS_C(D)\text{-}POS_{C\text{-}\{C_1\}}(D)|}{|U|} = 0.2.$$

We can also get the other two attributes values of importance: $\beta(C_3) = \beta(C_5) = 0.2$. While the other six attributes values of importance are zero.

After the process of standardization, the β_i turns into the weight of each attribute, and the values are: $\omega(C_1) = \omega(C_3) = \omega(C_5) = 0.333$, while the others are zero.

The results indicate that attribute C_1, C_3, C_5 are more important than others in the classification of red zone's fault modes because the negative region of U/D is changed more by C_1, C_3, C_5 than the other six attributes. However, the attribute C_2, C_4, $C_6 \sim C_9$ do little in the change of the negative region, so they can be eliminated when calculating the case similarity. Besides, the function of the case attribute table after reduction is the same as before, but it gets fewer condition attributes and becomes simpler to deal with.

4 CONCLUSIONS

This paper elaborates how to extract the key attributes of track circuit red zone fault case by rough set, as well as how to make reduction to the attributes and calculate their weights. And there are cases to show the dealing process. What's more, rough set provides a new way to calculate the similarity of track circuit fault cases. Compared with the subjective experience methods like experts hand-set method, rough set is totally based on data itself, so it effectively avoids the over-reliance on subjective experience and improves the objectivity and credibility of case retrieve.

Table 2. The fault case attribute table of red zone.

Case	C_1	C_2	C_3	C_4	C_5 (%)	C_6	C_7	C_8	C_9	Fault mode D
1	25Hz	red zone	frequent	sunny	55	normal	normal	0	0	The cable nail is virtual connected.
2	UM71	red zone	average	sunny	40	normal	0	0	0	The cable is broken.
3	25Hz	red zone	frequent	rainy	91	0	0	0	0	The porcelain fuse is struck by lightning.
4	ZPW2000	Adjacent red zone	frequent	sunny	51	normal	normal	normal	0	The lead wire is loose or disconnected.

Table 3. The fault case attribute table of red zone after discretization.

Case	C_1	C_2	C_3	C_4	C_5	C_6	C_7	C_8	C_9	D
X_1	1	1	2	0	1	2	2	0	0	1
X_2	2	1	1	0	0	2	0	0	0	1
X_3	1	1	2	1	2	0	0	0	0	2
X_4	3	2	2	0	1	2	2	2	0	1
X_5	3	2	1	0	0	2	2	1	1	3
X_6	3	1	0	1	2	2	2	1	1	4
X_7	1	1	2	0	0	2	2	0	0	5
X_8	1	1	1	1	2	0	0	0	0	1
X_9	2	1	0	1	2	2	2	1	1	6
X_{10}	3	1	2	0	0	2	1	1	1	6

REFERENCES

Zhang Xi, Du Xusheng, Liu Chaoying. Development of Railway Station Signaling Control Equipment Fault Diagnosis Expert System [J].*Journal Of The China Railway Society*, 2009, 31(3): 43–49.

J. Chen, C. Roberts, P.Weston. Fault Detection and Diagnosis for Railway Track Circuits Using Neuro-fuzzy Systems [J]. *Control Engineering Practice*, 2008, 16(5): 585–596.

Zhou Wenjuan, Wang Qiang. Research on fault diagnosis for railway signal equipment based on Agent[J]. *Railway Computer Application*, 2009, 18(11): 41–44.

Sun Shangpeng, Zhao Huibing. The Method of Fault Detection of Compensation Capacitor in Jointless Track Circuit Based on Phase Space Reconstruction[J]. *Journal Of The China Railway Society*, 2012, 34(10): 79–84.

Cui Lina, Zhang Xi. Design and Implementation of Knowledge Database for Diagnose Expert System of Railway Station Signal Equipment [J] *Railway Computer Application*. 2005, 14(12): 13–16.

Zhang Zhenhai, Wang Xiaoming, Dang Jianwu, Min Yongzhi. Research on CBR Decision Method of Railway Emergency Rescue Based on Integral Similarity Degree [J] *Journal Of The China Railway Society*. 2012, 34(11): 49–53.

Li Qing, Shi Yaqin, Zhou Yang. CBR Methodology Application in Fault Diagnosis of Aircraft[J]. *Journal of Beijing University of Aeronautics and Astronautics*, 2007, 33(5): 622–626.

Sun Ling, Zhang Jinlong, Chi Jiayu. Study on Weighing Coefficient of Case Feature Attributes in CBR System Based on Rough Set Theory [J]. *Computer Engineering and Application*, 2003, 39(30): 44–46.

Zhou Ping, Ding Jinliang, Yue Heng, Chai Tianyou. A New Approach Based on Similarity Rough Set for Determing Case Feature Weights [J]. *Information and Control*, 2006, 35(3): 329–334.

I. Watson. Case-based reasoning is a methodology not a technology [J]. *Knowledge Based Systems*, 1999, 12(5): 303–308.

Miao Duoqian. *Rough Set Theory, Algorithm and Application*[M]. Beijing: Tsinghua University Press, 2008.

Computational Intelligence in Industrial Application – Ling (ed.)
© 2015 Taylor & Francis Group, London, ISBN: 978-1-138-02818-0

Data fusion of cyber physical systems

L. Chen, L.X. Shi & L.L. Kong
School of Computer and Information Engineering, Shanghai Second Polytechnic University, Shanghai, China

ABSTRACT: The tight integration of computing, control and communication is one of the key features of Cyber Physical Systems (CPS). In this paper, we investigate it for Vehicular Cyber Physical Systems (VCPS), and show how to comprehensively fuse data from different areas. In VCPS, vehicles nodes periodically broadcast packets to notify its own existence and allow other nodes to track them and consists of a networked system. As a parameter of communication components of VCPS, the periodical message transmission rate has great influence on the performance of VCPS, however, it is tightly related with physical and computing processes, elements and parameters. Speed, distance and other physical characters, track errors and other computing characters have all had an important impact on it. Thus, a data fusion method based on Rough Sets Theory (RST) for VCPS is presented to obtain optimum or near-optimum periodical packet rates. Furthermore, a detailed case is also presented to verify its effectiveness and efficiency.

KEYWORDS: Cyber Physical Systems; Data Fusion; Vehicular Cyber Physical Systems; Rough Sets Theory.

1 INTRODUCTION

In the traditional concepts of computing systems and physical systems, the cyber space separates from the physical world (Rajkumar 2010). With the development of embedded computing, sensor monitoring, wireless communication, data processing and other technologies, the physical processes, computing processes and communication processes can be highly integrated which leads to the Cyber-physical system (CPS). The CPS represents a system which tightly integrates computation, communication and physical processes (Wen 2012). In CPS, through dynamic sensing of physical environments and resources, real-time reliable information transmission, comprehensive computing processing of data, with feedback cycle, the physical process is effectively controlled. Moreover, the whole process is completed automatically with people in the auxiliary position

CPS is involved in computer science, network communication and control theory and multiple disciplines which are widely applied to real systems, including transportation, aerospace, telemedicine, power control and many large infrastructure systems (Li 2011, Li 2012 & Sha 2008), and other local or fine systems such as precision agriculture, electronic endoscope. The impact of CPS will far bypass the IT revolution at 20th, and it will change the interactive mode between human and physical world just as the Internet has changed it among humans.

At present, Mobile Cyber Physical System (MCPS) has attracted wide attention of scholars and research institutions (Edward 2008). It refers to a system composed of computing and physical entities that are mobile and can communicate with each other over a network. The integration and fusion of different domains of information are one of the most crucial challenges of CPS. It usually includes the integration of computing processes and physical processes, that of computing, control and communication, that of discrete and continuous data. Especially for vehicular cyber physical system (VCPS), it is a kind of mobile cyber physical system where vehicles move the with time change, meanwhile their state changes constantly, and they communicate with each other, exchange different cyber or physical information. Thus, it needs to tightly integrate the information from different domains which can be further different areas.

In the paper, the information integration and fusion method of the vehicular cyber physical system (VCPS) are presented. It comprehensively considers the impact of computing, communication and control. Moreover, the rough set theory is utilized to fuse data from physical, computing and communication processes which is further applied to optimize the rate of packet transmission. The remainder is organized as follows: The motivation is detailed in section 2; RST is introduced in section 3; section 4 details the fusion methodology; the conclusion is presented in the last section.

2 MOTIVATION

As a specific MCPS, in VCPS, vehicles move on the road, communicate each other, deliver warning messages or routine road status data to drivers and may even directly take control of the vehicle to carry out evasive maneuvers which is shown in Figure 1.

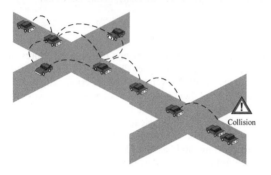

Figure 1. Vehicular cyber physical system (VCPS).

When an accident occurs, such as a sudden collision, then the emergency message will be transmitted to the vehicles behind which can take a brake on time and avoid more collisions. To do so, for VCPS, vehicles should track accurately the positions of neighbor vehicular nodes, and then transfer the messages one by one at the emergent time. In VCPS, vehicles periodically broadcast packets including their own information such as position coordinates and ID. When a vehicle receives packets, it then adds the senders as its neighbor nodes. Thus, the periodical packets are very important for VCPS. However, the message rate is tightly connected with other factors, such as physical states. For example, how fast vehicle position and states change has great influence on the rate. It is obvious that faster moving speed requires higher message rates. On the other hand, too more periodical packets will result in network traffic congestion. Then, the emergent information cannot be transmitted to the affected vehicles on time. So, the optimum message rate is one of the most important parameters of communication component of VCPS.

Since the transmission message rate is also tightly coupled with physical and computing components of VCPS, it needs a comprehensive consideration. Rough set theory (RST) is applied to fuse all physical, computing and communication information to obtain the optimum or near-optimum message rate.

3 ROUGH SETS THEORY

As a way to handle uncertainty, fuzziness tool, RST has obtained more and more attention. For the inaccuracy problem, its advantages are to remove the redundant input information, simplify the express space without additional information such as the prior probability and membership degree (Xie 2005).

DEFINITION 1 (Pawlak 1997). Information System is an four-tuple $S = (U, A, V, F)$, where $U = (x_1, x_2, \cdots, x_n)$ is a non-empty finite set of objects, A is a non-empty finite set of attributes, if $A = C \bigcup D$ and $C \bigcap D = \Phi$, then S is also called decision table, where C is the set of condition attributes, D is the set of decision attributes, and usually D may be simplified to a single decision attribute, that is $A = C \bigcup \{d\}$; V is the range of attribute A; F is mapping from $U \times A \rightarrow V$, and is called information function, which specifies the unique value for each object attribute in U, and for $\forall_x \in U, a \in A, f(x, a) \in V_a$.

The objects that are identified with the same information are indiscernibility relation, which constitutes the mathematical basis of RST.

4 MODEL AND INFUSION METHOD

The infusion method to obtain the optimum or near-optimum message rate based RST for VCPS is composed of the following steps:

(1) *Fusion information table construction.* Based on experience and expert knowledge, different data from different views of VCPS are selected to construct the fusion information table.
(2) *Attribute Reduction.* From the point of view of the entire information system, remove the redundant attributes on the basis to ensure that the amount of information dose not reduces.
(3) *Eliminate duplicate rows.* Duplicate rows can be considered as the same state, the same decision.
(4) *Reduce Value.* For each decision, in the case does not affect the classification and decision ability, the values of certain attributes can be omitted.
(5) *Organize rules.* Select the minimum set of rules.

4.1 Fusion decision table

Through experience and expert knowledge, Density, speed, distance, error, U, the five data are obtained which is considered to have influence on the message rate. Then, after the RST knowledge sample set, the above data feature values are extracted and discretization for continuous parameters. Rules of sample sets are divided into five condition attributes $a_1 \sim a_5$, which correspond to the described parameters, where 0 means high, 1 for moderate and 2 low. For The decision attribute is used to determine the periodical message rate, and its value 0 represents fast, 1 for normal, and 2 for slow. The quantitative condition and decision attributes forms a two-dimensional table, and each row describes a specific determination object for message rate, each column describes a kind of attribute of objects, which is shown in Table 1.

Table 1. Fusion information table.

Object Set	a1	a2	a3	a4	a5	Decision attribute
1	1	0	2	1	1	0
2	0	1	2	0	2	0
3	2	2	1	2	0	2
4	1	2	0	0	0	2
5	0	1	0	1	1	1
6	2	0	2	0	0	1
7	0	1	0	2	0	2
8	2	0	1	0	1	0
9	2	1	1	0	1	1
10	1	0	2	0	1	0

The condition attributes are a1, a2, a3, a4, a5.

4.2 Attribute reduction and eliminate duplicate rows

The objective of reduction computation is to remove the redundant attributes, and delete the according rows. The discernibility matrix method is adopted to carry out condition attributes reduction. The reduction steps are as follows:

1 Compute the discernibility matrix $M(S)$ of system S. $M(S)$ represents matrix $[c_{ij}]_{n \times n}$ with order $n \times n$, where $[c_{ij}]_{n \times n} = \{a \in A; a(u_i) \neq a(u_j) \wedge u_i, u_j \in U, i, j = 1, 2, \cdots, n\}$ (symbol $\wedge$ means the conjunctive operation), and then we get the discernibility function matrix of the decision matrix table. Since the matrix is symmetric, only the lower triangular part should be listed.

2 Calculate the discernibility function $f_{M(S)}$ related to the discernibility matrix. The expression of $f_{M(S)}$ is the conjunctive of all $\vee c_{ij}$, where symbol $\vee$ means the disjunctive operation, and by computation, we obtain $f_{M(S)}(a_1, a_2, a_3, a_4, a_5) = (a_2 \wedge a_4 \wedge a_5)$

3 Calculate the minimum disjunctive normal form of $f_{M(S)}$. Take $\{a_2, a_4, a_5\}$ as the condition attributes, the reduced decision table is shown in Table 2 which remove the redundant condition attribute a_1 and a_3.

4.3 Attribute value reduction

According to the condition attribute reduction table, the attribute value reduction table of decision rules can be further calculated out. The attribute value reduction table of decision rules are obtained which is shown in Table 3.

Table 2. Decision table with removing redundant condition attributes.

Object Set U	a2	a4	a5	Decision attribute
1	0	1	1	0
2	1	0	2	0
3	2	2	0	2
4	2	0	0	2
5	1	1	1	1
6	0	0	0	1
7	1	2	0	2
8	0	0	1	0
9	1	0	1	1
10	0	0	1	0

Table 3. Value reduction.

	a2	a4	a5	Rate
1	0	1	*	0
1'	0	*	1	0
2	*	*	2	0
3	2	*	0	2
3'	*	2	*	2
4	2	*	*	2
5	1	1	*	1
5'	1	*	1	1
6	0	*	0	1
7	*	2	*	2
8	0	0	*	0
8'	0	*	1	0
9	1	*	1	1
10	0	0	*	0

4.4 Formation of minimum decision rule table

Merge from Table 3, the minimum decision rule condition reduction table can be obtained, which is a minimum solution shown in Table 4.

Table 4. One attribute simplification merge of decision rule.

Object set U	a2	a4	a5	Decision attribute D
1	0	1	-	0
2	-	-	2	0
4	2	-	-	2
5	1	1	-	1
6	0	-	0	1
8	0	0	-	0

195

Data sensing and fusion rules are adopted to represent knowledge based RST as"mode-action", such as $if\left(p_1, p_2, ..., p_m\right)$ then $if\left(q_1, q_2, ..., q_n\right)$, where $p_1, p_2, ..., p_m$ are the condition attributes of rapid occupant classification system, $q_1, q_2, ..., q_n$ are the decision attributes. According to the minimum solution from table 5, the fusion rules are as follows:

If the speed is fast *and* the error is moderate, *then* the message rate is high.

If the channel utilization is low, then the rate is high.

If the speed is low, then the rate is low.

If the speed is moderate and the error is moderate, then the rate is moderate.

If the speed is high, *and* the channel utilization is high, *then* the rate is moderate.

If the speed is high *and* the error is high, *then* the rate is high.

5 CONCLUSION

CPS is a system of collaborating computing, control and communication elements and entities. These entities have influence on each other. As a specific application, for VCPS, three different components should be tightly considered. In the paper, data of different computing, control and communication components are collaboratively fused so as to obtain the optimum or near optimum periodically broadcasting message rate. As a tool to handle uncertainty, RST is adopted to carry out data fusion, and a case is also presented to illustrate the process.

In fact, when vehicles nodes move on the road, geography has also a great impact on the message rate or other parameters of VCPS. In the future, the road topology factor will be introduced to select the suitable rate.

ACKNOWLEDGMENTS

This research was supported by the National Natural Science Foundation of China (Grant No. 61202368), Shanghai Natural Science Foundation of China (Grant No. 12ZR1411500) and Innovation Program of Shanghai Municipal Education Commission (Grant No. 13ZZ141).

REFERENCES

Edward A.L. 2008. Cyber Physical Systems: Design Challenges. *Proceedings of ISORC*. USA:

Li R.F., Xie Y., Li R. & Li L. 2012. Survey of Cyber-Physical Systems, *Journal of Computer Research and Development* 49(6):1149–1161.

Li Z.P., Zhang T.C. & Zhang J.2011. Survey on the Research of Cyber-Physical Systems (CPS), *Computer Science* 38((9): 25–31.

Pawlak, Z. 1997. Rough Set Approach to Knowledge based Decision Support. *European Journal of Operational Research*, 99:48–57.

Rajkumar R, Lee I & Sha L, Stankovic J 2010. Cyber-physical systems: the next computing revolution. *Proceedings of the 47th ACM/IEEE .Design Automation Conference.* USA: Anaheim.

Sha L., Gopalakrishnan S., Liu X., Wang Q. X. 2008. Cyber-physical Systems: a new frontier. *Proceedings of the IEEE International Conference on Sensor Networks, Ubiquitous, and Trustworthy Computing.* China: Taichuang.

Wen J. R., Wu M.Q & Su J.F. 2012. Cyber-physical System, *ACTA Automatica Sinica* 38(4): 507–517.

Xie, N., Li A.P., Xu L.Y. 2005. Research on Diagnosability Technique for Reconfigurable Manufacturing System (RMS). *Chinese Mechanical Engineering* 16(17):1545–1549.

Computational Intelligence in Industrial Application – Ling (ed.)
© 2015 Taylor & Francis Group, London, ISBN: 978-1-138-02818-0

Convergence analyses for the FOM and GMRES algorithm

L. Li & G.L. Wu

Faculty of Science, Guilin University of Aerospace Technology, Guilin, China

ABSTRACT: Iterative methods are generally adopted for solving large sparse linear system, FOM and GMRES algorithm are two very important Krylov subspace methods. Based on the residual analysis of FOM and GMRES algorithm, this paper deduces a formula of the residual of GMRES on the adjacent two Krylov subspace, gives their relationship of residual vectors of FOM and GMRES algorithms and further proves a formula of residual norms between the two algorithms.

KEYWORDS: Linear equation; Krylov subspace; FOM; GMRES; Convergence.

1 INTRODUCTION

As is well known, many a scientific computation attributes to solving large sparse linear Systems

$$AX = b, \qquad (1)$$

in which $A \in R^{n \times n}$ is nonsingular, and $X, b \in R^n$. It is a large scale of this kind of equations. When adopting the direct method to solve them, the demand for the computer hardware is too high to cause problems, such as insufficient memory, and so iterative method is the most common choice to solve this kind of equations. In the known iteration methods, Krylov subspace methods on the Krylov subspace iteration, with less computing and storage, has abstracted more and more attention. In recent years, people carry on serious research on Krylov subspace methods and put forward all kinds of algorithms such as CG [1], FOM [2], GMRES [3], GMRESR [4], FGMRES [5], GCR [6] and so on. In Krylov subspace Methods, full orthogonalization method (FOM) and generalized minimum residual methods (GMRES) are very important, because many current developed methods learn from their thoughts. Therefore, in order to analyze the convergence property and to improve the convergence rate of other algorithms of Krylov subspace, it is of great importance to make an analysis and research on the convergence property and the relationship of the two methods. Over the past 20 years, many scholars have done a lot of work about the convergence of the GMRES algorithm. Here are examples: Zhong Baojiang [7] has analyzed the relationship between the convergence rate of the GMRES algorithm and the approximate degree of Ritz value for eigenvalues during the projection process. Cao Zhihao [8] and w. Joubert [9] have explored that the GMRES convergence rate is affected by the roots of GMRES polynomial and the spectrum of the coefficient matrix A. Wu guolin [10] has analyzed that the GMRES convergence rate is affected by the projection of residual vectors in Krylov subspace. The less famous paper also focuses on relationship of convergence of FOM and GMRES algorithms. Based on the error analysis of FOM and GMRES algorithms, this paper deduces a formula of the residual of GMRES on the adjacent two Krylov subspace, gives their relationship of residual vectors of FOM and GMRES algorithms, and further proves a relationship of residual norms between the two algorithms. The results show that on the same Krylov subspace, GMRES algorithm solution is better than FOM algorithm.

2 FOM AND GMRES ALGORITHMS

Iteration of FOM and GMRES algorithms take place on the Krylov subspace, so we first introduce a property of Arnoldi algorithm finding an orthogonal basis on the Krylov subspace. Details about Arnoldi algorithm and its property proof can be seen in reference [11].

Define X_0 as an initial guess solution to equations (1.1), $r_0 = b - AX_0$, and $K_m(A, r_0) = span$ $\{r_0, Ar_0, Ar_0^2, \cdots, Ar_0^{m-1}\}$ as a Krylov subspace expanded by the vector r_0. Set $\beta = \|r_0\|_2$, $v_1 = r_0 / \beta$, the vectors $v_2, v_3, \cdots, v_{m+1}$ formed by the m steps Arnoldi algorithm, and then:

Property 2.1 Denote by V_m the $n \times m$ matrix with column vectors $v_1, v_2, \cdots, v_m$, by $\overline{H}_m$ the $(m+1) \times m$ Hessenberg matrix whose nonzero entries $h_{i,j}$ are defined by Arnoldi algorithm, and by H_m the matrix obtained from $\overline{H}_m$ by deleting its last row. Then the following relations hold:

$$AV_m = V_{m+1}\overline{H}_m, \qquad (2)$$

$$V_m^T A V_m = H_m. \tag{3}$$

FOM algorithm is an orthogonal projection method. This method seeks an approximate solution X_m from the affine subspace $X_0 + K_m(A, r_0)$ by imposing the Galerkin condition

$$X_m = X_0 + V_m Y_m, \tag{4}$$

$$Y_m = H_m^{-1}(\beta e_1), \tag{5}$$

in which $\beta = \|r_0\|_2$, e_1 is the first column of the $m \times m$ identity matrix, V_m, H_m are matrices created by Arnoldi algorithm. As for Approximate solution X_m, the following relations hold:

Lemma 2.1 [2]. The residual vector of the approximate solution X_m computed by the FOM algorithm is such that

$$b - AX_m = -h_{m+1,m} e_m^T Y_m v_{m+1}, \tag{6}$$

or

$$\|b - Ax_m\|_2 = h_{m+1,m} \left| e_m^T Y_m \right|. \tag{7}$$

GMRES algorithm is also an orthogonal projection method. Its approximate solution is given by the formula (2.3). Unlike the FOM algorithm, Y_m is not given by formula (2.4), but is obtained by minimizing norm $\|\beta e_1 - \overline{H}_m Y\|_2$. That is

$$X_m = X_0 + V_m Y_m, \tag{8}$$

$$Y_m = \min \|\beta e_1 - \overline{H}_m Y\|_2, \tag{9}$$

where Y_m is the least-squares solution of over-determined equation $\|\beta e_1 - \overline{H}_m Y\|_2$. A common technique to solve the least-squares problem min $\|\beta e_1 - \overline{H}_m Y\|_2$ is to transform the Hessenberg matrix into upper triangular form by using Givens plane rotations. Let $\overline{H}_m^{(m)}$ be an upper triangular matrix obtained by which $\overline{H}_m$ make m times Givens plane rotations. Define $P_{m+1}^{(i)}$ as the ith Givens matrix,

$$P_{m+1}^{(i)} = \begin{bmatrix} 1 & & & & & & \\ & 1 & & & & & \\ & & \ddots & & & & \\ & & & c_i & s_i & & \\ & & & -s_i & c_i & & \\ & & & & & \ddots & \\ & & & & & & 1 \end{bmatrix} \begin{matrix} \\ \\ \\ i \\ i+1 \\ \\ \\ \end{matrix}, \tag{10}$$

$$s_i = \frac{h_{i+1,i}}{\sqrt{(h_{ii}^{(i-1)})^2 + h_{i+1,i}^2}}, c_i = \frac{h_{ii}^{(i-1)}}{\sqrt{(h_{ii}^{(i-1)})^2 + h_{i+1,i}^2}}. \tag{11}$$

Define

$$Q_{m+1}^{(m)} = P_{m+1}^{(m)} P_{m+1}^{(m-1)} \cdots P_{m+1}^{(1)}, \tag{12}$$

$$\overline{H}_m^{(m)} = Q_{m+1}^{(m)} \overline{H}_m, \tag{13}$$

And

$$\overline{\Delta}_m^{(m)} = Q_{m+1}^{(m)}(\beta e_1) = (\delta_1, \delta_2, \cdots, \delta_{m+1})^T. \tag{14}$$

Since $Q_{m+1}^{(m)}$ is unitary, we have

$$\min \|\beta e_1 - \overline{H}_m Y\|_2 = \min \|\overline{\Delta}_m^{(m)} - \overline{H}_m^{(m)} Y\|_2. \tag{15}$$

Similar to notation H_m, define $H_m^{(m)}$ and $\Delta_m^{(m)}$ as $m \times m$ an upper triangular matrix and m vector which are obtained from $\overline{H}_m^{(m)}$ and $\overline{\Delta}_m^{(m)}$ by deleting their last row respectively. Then the least-squares solution of over-determined equation $\|\beta e_1 - \overline{H}_m Y\|_2$ hold:

$$Y_m = (H_m^{(m)})^{-1} \Delta_m^{(m)} \tag{16}$$

As for the error of GMRES, we have the following lemma.

Lemma 2.2 [3]. Let X_m be an approximate solution of equations (1.1) computed by GMRES algorithm, then

$$\begin{aligned} b - AX_m &= V_{m+1}(\beta e_1 - \overline{H}_m Y_m) \\ &= V_{m+1}(Q_{m+1}^{(m)})^T \delta_{m+1} e_{m+1}, \end{aligned} \tag{17}$$

And, as a result

$$\|b - AX_m\|_2 = |\delta_{m+1}|. \tag{18}$$

3 CONVERGENCE RERLATIONSHIP OF FOM AND GMRES ALGORITHM

In order to expediently explore the convergence relationship between FOM and GMRES algorithms, FOM and GMRES iterates are denoted by the superscripts F and G, respectively. For relations of the residual vectors obtained by the two methods, we have the following theorem.

Theorem 3.1. Let X_{m-1}^G, X_m^G be Approximate Solutions GMRES algorithm iterate on the two subspaces $X_0 + K_{m-1}(A, r_0)$ and $X_0 + K_m(A, r_0)$, respectively, and X_m^F be Approximate Solution FOM algorithm iterate on the subspace $X_0 + K_m(A, r_0)$, then

$$X_m^G = s_m^2 X_{m-1}^G + c_m^2 X_m^F, \tag{19}$$

or

$$r_m^G = s_m^2 r_{m-1}^G + c_m^2 r_m^F, \tag{20}$$

in which s_m, c_m are elements coming from the mth Givens matrix.

Proof. From what has been discussed above, we know that the GMRES approximation is obtained by solving least-squares solution of over determined equations. If we search GMRES approximation of equation (1.1) on the subspace $X_0 + K_{m-1}(A, r_0)$, least-squares solution of over determined equations $\left\| \beta e_1 - \overline{H}_{m-1} Y \right\|_2$ will be obtained with $m-1$ Givens transformations. By (2.15), we infer that

$$Y_{m-1}^G = (H_{m-1}^{(m-1)})^{-1} \Delta_{m-1}^{(m-1)}, \tag{21}$$

or

$$H_{m-1}^{(m-1)} Y_{m-1}^G = \Delta_{m-1}^{(m-1)},$$

$$\Leftrightarrow H_m^{(m-1)} \begin{pmatrix} Y_{m-1}^G \\ 0 \end{pmatrix} = \begin{pmatrix} \Delta_{m-1}^{(m-1)} \\ 0 \end{pmatrix}, \tag{22}$$

in which

$$H_{m-1}^{(m-1)} = \begin{bmatrix} r_{11} & r_{12} & \cdots & r_{1,m-1} \\ & r_{22} & \cdots & r_{2,m-1} \\ & & \ddots & \\ & & & r_{m-1,m-1} \end{bmatrix}, \Delta_{m-1}^{(m-1)} = \begin{pmatrix} \delta_1 \\ \delta_2 \\ \vdots \\ \delta_{m-1} \end{pmatrix}.$$

$$H_m^{(m-1)} = \begin{bmatrix} r_{11} & r_{12} & \cdots & r_{1,m-1} & r_{1m} \\ & r_{22} & \cdots & r_{2,m-1} & r_{2m} \\ & & \ddots & & \\ & & & r_{m-1,m-1} & r_{m-1,m} \\ & & & & h_{mm}^{(m-1)} \end{bmatrix}.$$

So the GMRES approximation X_{m-1}^G on the subspace $X_0 + K_{m-1}(A, r_0)$ is shown below:

$$\begin{aligned} X_{m-1}^G &= X_0 + V_{m-1} Y_{m-1}^G \\ &= X_0 + (V_{m-1}, v_m) \begin{pmatrix} Y_{m-1}^G \\ 0 \end{pmatrix} \\ &= X_0 + V_m \begin{pmatrix} Y_{m-1}^G \\ 0 \end{pmatrix} \end{aligned} \tag{23}$$

Secondly, we search GMRES approximation of equation (1.1) on the subspace $X_0 + K_m(A, r_0)$. $\overline{H}_m, (\beta e_1)$ of over determined equations $\left\| \beta e_1 - \overline{H}_{m-1} Y \right\|_2$ are turned into the following matrix and vector by $m-1$ Givens transformations,

$$\overline{H}_m^{(m-1)} = \begin{bmatrix} r_{11} & r_{12} & \cdots & r_{1,m-1} & r_{1m} \\ & r_{22} & \cdots & r_{2,m-1} & r_{2m} \\ & & \ddots & & \\ & & & r_{m-1,m-1} & r_{m-1,m} \\ & & & & h_{mm}^{(m-1)} \\ & & & & h_{m+1,m} \end{bmatrix}, \overline{\Delta}_m^{(m-1)} = \begin{pmatrix} \delta_1 \\ \delta_2 \\ \vdots \\ \delta_{m-1} \\ \delta_m \\ 0 \end{pmatrix}.$$

Define

$$P_{m+1}^{(m)} = \begin{bmatrix} 1 & & & & & \\ & 1 & & & & \\ & & \ddots & & & \\ & & & \ddots & & \\ & & & & c_m & s_m \\ & & & & -s_m & c_m \end{bmatrix} \begin{matrix} \\ \\ \\ \\ m \\ m+1 \end{matrix} \tag{24}$$

as the mth gives transformation matrix, in which c_m, s_m are given by (2.10). After the mth gives transformation, $\overline{H}_m^{(m-1)}, \overline{\Delta}_m^{(m-1)}$ are turned into the following matrix and vector:

$$\overline{H}_m^{(m)} = \begin{bmatrix} r_{11} & r_{12} & \cdots & r_{1,m-1} & r_{1m} \\ & r_{22} & \cdots & r_{2,m-1} & r_{2m} \\ & & \ddots & & \\ & & & r_{m-1,m-1} & r_{m-1,m} \\ & & & & r_{mm} \\ & & & & 0 \end{bmatrix}, \overline{\Delta}_m^{(m)} = \begin{bmatrix} \delta_1 \\ \delta_2 \\ \vdots \\ \delta_{m-1} \\ c_m \delta_m \\ -s_m \delta_m \end{bmatrix}, \tag{25}$$

in which $r_{mm} = \sqrt{(h_{mm}^{(m-1)})^2 + (h_{m+1,m})^2}$. Then the over-determined equations $\left\| \beta e_1 - \overline{H}_{m-1} Y \right\|_2$'s least-squares solution $Y_m^G = (H_m^{(m)})^{-1} \Delta_m^{(m)}$ is obtained by the following formulas,

$$\begin{bmatrix} r_{11} & r_{12} & \cdots & r_{1,m-1} & r_{1m} \\ & r_{22} & \cdots & r_{2,m-1} & r_{2m} \\ & & \ddots & & \\ & & & r_{m-1,m-1} & r_{m-1,m} \\ & & & & r_{mm} \end{bmatrix} [Y_m^G] = \begin{bmatrix} \delta_1 \\ \delta_2 \\ \vdots \\ \delta_{m-1} \\ c_m \delta_m \end{bmatrix} \tag{26}$$

$$\Leftrightarrow \begin{bmatrix} r_{11} & r_{12} & \cdots & r_{1,m-1} & r_{1m} \\ & r_{22} & \cdots & r_{2,m-1} & r_{2m} \\ & & \ddots & & \\ & & & r_{m-1,m-1} & r_{m-1,m} \\ & & & & c_m r_{mm} \end{bmatrix} [Y_m^G] = \begin{bmatrix} \delta_1 \\ \delta_2 \\ \vdots \\ \delta_{m-1} \\ c_m^2 \delta_m \end{bmatrix}$$

$$\Leftrightarrow \begin{bmatrix} r_{11} & r_{12} & \cdots & r_{1,m-1} & r_{1m} \\ & r_{22} & \cdots & r_{2,m-1} & r_{2m} \\ & & \ddots & & \\ & & & r_{m-1,m-1} & r_{m-1,m} \\ & & & & h_{m,m}^{(m-1)} \end{bmatrix} [Y_m^G] = \begin{bmatrix} \delta_1 \\ \delta_2 \\ \vdots \\ \delta_{m-1} \\ c_m^2 \delta_m \end{bmatrix},$$

equivalent to

$$H_m^{(m-1)}Y_m^G = \begin{pmatrix} \Delta_{m-1}^{(m-1)} \\ c_m^2\delta_m \end{pmatrix}. \tag{27}$$

Finally, we search FOM approximation of equations (1.1) on the subspace $X_0 + K_m(A, r_0)$. By (2.4), we can deduce that

$$H_mY_m^F = \beta e_1 \Rightarrow Q_m^{(m-1)}H_mY_m^F = Q_m^{(m-1)}\beta e_1,$$

$$\Leftrightarrow H_m^{(m-1)}Y_m^F = \overline{\Delta}_{m-1}^{(m-1)},$$

$$\Leftrightarrow H_m^{(m-1)}Y_m^F = \begin{pmatrix} \Delta_{m-1}^{(m-1)} \\ \delta_m \end{pmatrix}. \tag{28}$$

By multiplying both sides of (3.4) and (3.10) by s_m^2 and c_m^2 respectively, we can obtain that

$$H_m^{(m-1)}\begin{pmatrix} s_m^2 Y_{m-1}^G \\ 0 \end{pmatrix} = \begin{pmatrix} s_m^2\Delta_{m-1}^{(m-1)} \\ 0 \end{pmatrix},$$

$$H_m^{(m-1)}(c_m^2 Y_m^F) = \begin{pmatrix} c_m^2\Delta_{m-1}^{(m-1)} \\ c_m^2\delta_m \end{pmatrix},$$

thus,

$$H_m^{(m-1)}\begin{pmatrix} \begin{pmatrix} s_m^2 Y_{m-1}^G \\ 0 \end{pmatrix} + c_m^2 Y_m^F \end{pmatrix} = \begin{pmatrix} \begin{pmatrix} s_m^2\Delta_{m-1}^{(m-1)} \\ 0 \end{pmatrix} + \begin{pmatrix} c_m^2\Delta_{m-1}^{(m-1)} \\ c_m^2\delta_m \end{pmatrix} \end{pmatrix},$$

$$H_m^{(m-1)}\begin{pmatrix} \begin{pmatrix} s_m^2 Y_{m-1}^G \\ 0 \end{pmatrix} + c_m^2 Y_m^F \end{pmatrix} = \begin{pmatrix} \Delta_{m-1}^{(m-1)} \\ c_m^2\delta_m \end{pmatrix},$$

in which $s_m^2 + c_m^2 = 1$. By (3.9), the following equation is obtained,

$$Y_m^G = \begin{pmatrix} s_m^2 Y_{m-1}^G \\ 0 \end{pmatrix} + c_m^2 Y_m^F. \tag{29}$$

Now we prove relation (3.1) satisfied by (3.5), (2.1) and (2.7). Relation (3.2) follows immediately. QED.

According the proof of theorem 3.1, GMRES algorithm residual is δ_m on the subspace $X_0 + K_{m-1}(A, r_0)$. If subspace $X_0 + K_{m-1}(A, r_0)$ is expanded to $X_0 + K_m(A, r_0)$, the algorithm residual become $-s_m\delta_m$. So we obtain the following useful relation for δ_{m+1}:

$$\delta_{m+1} = -s_m\delta_m, \quad (m = 1, 2, \ldots, n-1). \tag{30}$$

Because of $|s_m| \le 1$, GMRES algorithm residual gradually decreases as Krylov subspace dimensions

increase. In particular, if $s_m = 0$, then the residual norm must be equal to zero, which means that the solution is exactly at step m.

Let γ_m^F, γ_m^G be FOM with GMRES algorithm residual norms respectively. As for the two residual norms relation, we have the following theorem.

Theorem 3.2. Assume that m steps of the Arnoldi process have been taken and that H_m is nonsingular. Then the residual norms produced by the FOM and the GMRES algorithms are related by the equality

$$\gamma_m^G = |c_m|\gamma_m^F. \tag{31}$$

Proof. By making inner product both sides of (32) by r_m^G, we can obtain the following equation,

$$(r_m^G, r_m^G) = s_m^2 (r_m^G, r_{m-1}^G) + c_m^2 (r_m^G, r_m^F). \tag{32}$$

By (2.16) and

$$\begin{aligned} (r_m^G, r_{m-1}^G) &= e_m^T\delta_m Q_m^{(m-1)}V_m^T V_{m+1}(Q_{m+1}^{(m)})^T\delta_{m+1}e_{m+1} \\ &= e_m^T\delta_m Q_m^{(m-1)}\begin{pmatrix} v_1^T \\ v_2^T \\ \vdots \\ v_m^T \end{pmatrix}\begin{pmatrix} v_1 & v_2 & \cdots & v_m & v_{m+1} \end{pmatrix}(Q_{m+1}^{(m)})^T\delta_{m+1}e_{m+1} \\ &= e_m^T\delta_m Q_m^{(m-1)}(I_m, 0)(Q_{m+1}^{(m)})^T\delta_{m+1}e_{m+1}. \end{aligned} \tag{33}$$

By equation (2.9) and (2.11)

$$Q_m^{(m-1)} = P_m^{(m-1)}P_m^{(m-2)}\cdots P_m^{(1)},$$

$$\begin{aligned} (Q_{m+1}^{(m)})^T &= (P_{m+1}^{(1)})^T (P_{m+1}^{(2)})^T \cdots (P_{m+1}^{(m)})^T \\ &= \begin{pmatrix} (P_m^{(1)})^T & 0 \\ 0 & 1 \end{pmatrix}\begin{pmatrix} (P_m^{(2)})^T & 0 \\ 0 & 1 \end{pmatrix}\cdots\begin{pmatrix} (P_m^{(m-1)})^T & 0 \\ 0 & 1 \end{pmatrix}(P_{m+1}^{(m)})^T. \end{aligned}$$

So

$$\begin{aligned} (r_m^G, r_m^G) &= e_m^T\delta_m Q_m^{(m-1)}(I_m, 0)(Q_{m+1}^{(m)})^T\delta_{m+1}e_{m+1} \\ &= e_m^T\delta_m P_m^{(m-1)}\cdots P_m^{(1)}(I_m, 0)\begin{pmatrix} (P_m^{(1)})^T & 0 \\ 0 & 1 \end{pmatrix}\cdots\begin{pmatrix} (P_m^{(m-1)})^T & 0 \\ 0 & 1 \end{pmatrix}(P_{m+1}^{(m)})^T\delta_{m+1}e_{m+1} \\ &= e_m^T\delta_m P_m^{(m-1)}\cdots P_m^{(2)}(I_m, 0)\begin{pmatrix} (P_m^{(2)})^T & 0 \\ 0 & 1 \end{pmatrix}\cdots\begin{pmatrix} (P_m^{(m-1)})^T & 0 \\ 0 & 1 \end{pmatrix}(P_{m+1}^{(m)})^T\delta_{m+1}e_{m+1} \\ &= \cdots = \\ &= e_m^T\delta_m(I_m, 0)(P_{m+1}^{(m)})^T\delta_{m+1}e_{m+1} \\ &= e_m^T\delta_m\begin{pmatrix} 1 & & & \\ & 1 & & \\ & & \ddots & \\ & & & c_m & -s_m \end{pmatrix}\delta_{m+1}e_{m+1} \\ &= \delta_m\begin{pmatrix} 0 & \cdots & c_m & -s_m \end{pmatrix}^T e_{m+1}\delta_{m+1} \\ &= -s_m\delta_m\delta_{m+1}. \end{aligned}$$

Pluging (3.12) into the above equation, we obtain $(r_m^G, r_{m-1}^G) = \delta_{m+1}^2$, and namely

$$(r_m^G, r_{m-1}^G) = \delta_{m+1}^2$$

$$= \left\| r_m^G \right\|_2^2 \tag{34}$$

$$= (r_m^G, r_m^G).$$

As for (r_m^G, r_m^F), by lemma 2.1 and 2.2,

$$
\begin{aligned}
(r_m^G, r_m^F) &= -h_{m+1,m} e_m^T Y_m^F v_{m+1}^T V_{m+1} (\beta e_1 - \overline{H}_m Y_m^G) \\
&= -h_{m+1,m} e_m^T Y_m^F e_{m+1}^T (\beta e_1 - \overline{H}_m Y_m^G) \\
&= -h_{m+1,m} e_m^T Y_m^F (0 - h_{m+1,m} e_m^T Y_m^G) \\
&= h_{m+1,m}^2 e_m^T Y_m^F e_m^T Y_m^G \\
&= h_{m+1,m}^2 e_m^T Y_m^F e_m^T \left\{ \begin{pmatrix} s_m^2 Y_{m-1}^G \\ 0 \end{pmatrix} + c_m^2 Y_m^F \right\} \\
&= c_m^2 h_{m+1,m}^2 e_m^T Y_m^F e_m^T Y_m^F,
\end{aligned}
$$

that is

$$(r_m^G, r_m^F) = c_m^2 \left\| r_m^F \right\|_2^2 = c_m^2 (r_m^F, r_m^F). \tag{35}$$

Formulas (3.16) and (3.17) are taken to formula (3.14), then

$$(r_m^G, r_m^G) = c_m^2 (r_m^F, r_m^F)$$

$$\Leftrightarrow \left\| r_m^G \right\|_2 = |c_m| \left\| r_m^F \right\|_2,$$

and namely $\gamma_m^G = |c_m| \gamma_m^F$. QED.

Easily to discovery $|c_m| \leq 1$, so $\gamma_m^G \leq \gamma_m^F$. Therefore, the GMRES algorithm residual is lower than FOM algorithm in the same subspace $X_0 + K_m(A, r_0)$, namely the GMRES algorithm solution is better than FOM algorithm.

ACKNOWLEDGMENT

This work was partially supported by Research Project of Guangxi Zhuang Autonomous Region Department of Education(201106LX717).

REFERENCES

M.R. Hestenes, E.L. Stiefel. Methods of Conjugate Gradients for Solving Linear Systems. Journal of Research of the National Bureau of Standards, 1952, 49(6): 409–436.

Y. Saad. Krylov subspace methods for solving large unsymmetric linear systems. Mathematics of Computation, 1981, 37(155): 105–126.

Y. Saad and M.H. Schultz. GMRES: A generalized minimal residual algorithm for solving nonsymmetric linear systems. SIAM J. Sci. Stat. Comput., 1986, 7(3): 856–869.

H.A. Van der Vorst, K. Vuik. a family of nested GMRES methods. Numerical Linear Algebra with Applications, 1994,1(4): 369–386.

Y. Saad. A flexible inner-outer preconditioned GMRES algorithm. SIAM Journal on Scientific Computing, 1993,14(2): 461–491.

E. de Sturler. Nested Krylov methods based on GCR. Journal of Computational and Applied Mathematics, 1996, 67(1): 15–41.

Zhong Baojiang. The rate of convergence of gmres. Numerical Mathematicas a Journal of Chinese Universities, 2003, 25(3): 253–260. (in Chinese).

W. Joubert. On the convergence behavior of the restarted GMRES algorithm for solving nonsymmetric linear systems. Numerical Linear Algebra with Applications, 1994, 1(5): 427–447.

zhi-hao Cao. A note on the convergence behavior of GMRES. Applied Numerical Mathematics, 1997, 25(1): 13–20.

Wu Guolin, Wang Sheng. Influence of residual vector and Krylov subspace on convergence Velocity of GMRES algorithm. Guangxi Sciences, 2011, 18(3): 214–219. (in Chinese).

W.E. Arnoldi. The principle of minimized iteration in the solution of the matrix eignvalue problem. Quart Appl. Math, 1951, 9: 17–29.

Computational Intelligence in Industrial Application – Ling (ed.)
© 2015 Taylor & Francis Group, London, ISBN: 978-1-138-02818-0

Centralized log server construction and application in network management

Jianning Yang
Business School, Yunnan Normal University, Kunming, China
Business School Network Center, Yunnan Normal University, Wuhan District, Kunming, China

Jiankun Yang
Kunming branch of China Telecom Co. Ltd, Kunming, China

Kun Lin
Dehong Vocational College, Yunnan Province, China

ABSTRACT: Centralized log server for collecting and storing logs generated by a variety of network equipments and application servers, network faults and performance problems can be a timely and effective discovery by an on-line analysis of a log. This article first introduces the function for the centralized log server in the network management, and then gives a description of how to establish a centralized log server in the network, introducing with emphasis the system log protocol to establish as a centralized log server. Finally, a specific application of the centralized log server is introduced in network management.

KEYWORDS: Centralized log server, SYSLOG protocol, Network management.

1 THE ROLE OF A CENTRALIZED SERVER, LOG IN NETWORK MANAGEMENT

The log is a network device to various events that occur during the running process of the recording. The log is network troubleshooting, network security audit and network performance optimization analysis of the important reference basis, especially in the log alarm information to discover and eliminate network faults, plays a very important role in network security. In large networks, there exist a variety of equipment, including a variety of switches, routers and a variety of application servers, these devices will produce a large amount of log every day. These devices will be generated log collection and centralized storage has the following role in the unified central log server:

1.1 *The unified log query interface, shielding equipment difference, improving the efficiency of the log management*

There are a large variety of devices and servers in the network of these devices, landing need to consume a large amount of time and energy. Landing equipment, improper operation error will cause the new security threat to the equipment. At the same time, different brands, different types of equipment log view mode and command also each are not identical,

in log analysis, network management personnel need for different equipment or access to the corresponding command to remember to log view. The establishment of centralized log server, only from a single log server visualization interface will be able to see all the equipment log view mode, unified, intuitive, simple, greatly improving the efficiency of network management, but also reduces the requirements for network maintenance and management personnel of technology, to avoid landing equipment to equipment may produce false operational risk.

1.2 *Can set flexibly log saved time, prevent the loss of important log information*

Because the log buffer storage space of network equipment is usually small, logging using FIFO queue processing mode, when the storage space is full, the old log will be automatically deleted, resulting in some important logging possible in were analyzed before treatment is covered. Using the log server, the log is saved in a log server on your hard disk, and the log server hard disk space is big and can be flexibly expanded, when necessary, can also log dump to the mediums of preservation. So you can log on the server in any flexible setup log save time, avoid important log information due to inadequate storage space and loss.

1.3 *It is easy to manage and query logs*

Fast switching network equipment is mainly responsible for the information and transfer, to other information processing ability is weak, simple text sequentially stored only on the log, only support simple log chronological view, it is difficult to carry out screening and filtering of log complex. The establishment of the log server, the log is saved in a log server database, according to time, various combinations of keywords for precise or fuzzy query, can be extracted easily log information need, provides great convenience for the management, log analysis. Log the information stored in the database can also through the WEB interface for remote way sharing technical personnel-related consulting, and analysis can carry out real-time on-line analysis of the log, and can in time discover equipment failure or hidden trouble and real time alarm

Log server with information analysis and processing ability strong, the log collection process, can carry on the online analysis of log information received, can immediately send out sound, light, text messages or other means of alarm information to meet the conditions of the first time alarm log, so that the administrators can in the event of failure of timely fault condition and immediately know for processing, processing time is shortened fault centralized log server, application of the most important is that the real time monitoring and management network fault.

2 THE ESTABLISHMENT OF A CENTRALIZED LOG SERVER

2.1 *The remote log acquisition mode*

To achieve the device log centralized in network storage, the remote data acquisition to achieve log. There are many ways to be on a remote device log collection to the log server, commonly used acquisition mode, text mode with SNMP Trap mode and Syslog mode. The text is the FTP from a remote device download log files to the log server, need to download the active devices, and every time is to download the entire log files, low efficiency, poor real-time performance. SNMP Trap is based on event driven, when the equipment fault occurs immediately alarm log information is sent to the log server, real-time. But the fault information SNMP Trap send only a limited, non fault information is not sent to the management system, and SNMP Trap to generate the message format for device dependent, different manufacturers, different types of devices, the format of the message is larger, parsing the SNMP message is more complex, higher requirements for the supervision server performance, the corresponding implementation cost is higher. Syslog overcomes the shortcomings of the two acquisition mode, remote log collection is the ideal mode, the following highlights a system based on log (Syslog) method for realizing protocol to establish a centralized log server.

2.2 *The system log (SYSLOG) protocol*

The system log (SYSLOG) protocol is composed of RFC3164, RFC3195 and other documents defined industry standard protocol. It is better than an SNMP protocol application is more extensive, the commonly used network equipment and all of the LINUX, the UNIX server will follow the agreement. The default WINDOWS server does not support the Syslog protocol, but through certain settings, can also be carried out in the event log according to the protocol of Syslog store and forward.

Syslog protocol provides equipment generates logs can be stored in the Syslog protocol format in the local, but also can be transmitted to a remote server on the specified. In long-distance transmission, Syslog uses UDP as the transport protocol, the destination port 514 will log information is transmitted to the remote server specified. The operation mechanism of Syslog protocol and the process is simple and reliable, in between the message sender and receiver does not require the coordination of strict, equipment only need to send in accordance with the requirements of the Syslog protocol log messages, receive log server is normal or not will not affect the normal operation of equipment.

A complete Syslog message consists of 3 parts, respectively is PRI, HEADER and MSG, composed by all the printable characters. The PRI part is composed of digital contains angle brackets, this figure includes programming module (Facility) and incident severity (Severity) two parts; part HEADER includes two fields, the time and the host name (or IP); the MSG part is divided into two parts of TAG and Content.

In practical application, because the SYSLOG protocol related document definitions appear later, packaging and storage is not all devices are in strict accordance with the unified format proposed SYSLOG protocol log messages, especially in the HEADER part of the time format and the MSG part of the TAG description is very different. Therefore in the message analysis to consider when full fault tolerance.

2.3 *Based on the SYSLOG protocol to establish a centralized log server*

According to the above description of SYSLOG protocol, can create very convenient for centralized log server in the network. The specific implementation methods are as follows:

2.3.1 Log messages

Log Syslog agreement the network equipment can be generated using UDP as the transport protocol is sent to a remote server. But what is sent to which server? The need for related settings on the device. Different equipment type setting method is different, the following is the related setup command on CISCO7206 router:

Log host A.B.C.D //A.B.C.D for a log server IP address

Log trap severity / / severity 1-7 digital, on behalf of the event severity (small values more serious).

Through the two command set, as long as there are events, severity and event to reach the required level, the event log will be sent to the set with "log host A.B.C.D" command on the server.

2.3.2 Receiving and parsing the log message

Equipment log is 514 port using UDP protocol to send the log server from a remote server, so in the preparation n of a responsible for the log collection procedures to eavesdrop on the 514 port of the server net card, and capture is sent to the UDP data of the ports on the package, you can realize remote receiving log messages.

UDP data packet received by the server contains the log message content packaging, remove the bag body and the analysis in accordance with the Syslog protocol format, it can be decomposed into various parts of the log, but the proposed Syslog protocol message format is not completely unified all the equipment, in order to simplify the procedures for handling and fault tolerance, the program only from the packet parsing out PRI and MSG part of the message body, the two part is now known in network equipment log will be included, and for differences in HEADER large format contained in time and the host name not analytic, for these two important log parameters can be achieved using another way. First of all, the host name can be obtained through the analysis of the UDP header, UDP header in the remote device source address is the IP address of the sending log. Secondly, generating, sending the Syslog log is real-time, therefore, the time of receipt of UDP messages can be regarded as the log generation time.

The PRI part contains a program module (Facility) and severity (Severity) of two parts, constitute a contains angle brackets number, this number is composed of Facility multiplied by 8, then add Severity to the digital, with 2 hexadecimal, 3 bit low expressed Severity, the rest of the high part right 3, is Facility. In the log management practice, usually only need to know Severity, but does not care about the Facility, therefore, the program only need from the log message part PRI of Severity can be calculated, the specific algorithm is the PRI numbers into binary and binary number 00000111 bit by bit to do with operation, the results is Severity.

Through the process above, it can realize the receiving remote event log send messages in the log on the server, and from the received event log message parsing out the event occurs for equipment IP address, time of occurrence of events, event, event severity level of details and other information.

3 SPECIFIC APPLICATIONS, CENTRALIZED LOG SERVER

Kunming Telecom has established the centralized log server in 2010, mainly to the core equipment of IP metropolitan area network generated log collection, centralized storage, and realize the real-time monitoring of network fault online through the log analysis, in order to achieve to the timely detection and timely treatment of equipment fault, fault link, as far as possible to shorten the fault processing time.

The establishment of a centralized log server does not require special network construction, only the log server access to monitoring network, the allocation of a fixed IP address. Log server is running on the log collection and the author's own development log online analysis processing program. When receiving a meet the preservation conditions of stored in log log database, receive meet alarm conditions and log according to the preset conditions immediately sends out the sound, light and short message alarm, tell the staff on duty is the occurrence of the fault what kind of which device, which link, the timely discovery so as to realize the network fault and accurate positioning.

Because of the network element equipment log log on the server are sent to the centralized storage, analysis and processing of the log is easy to achieve, through software design appropriate can be convenient, reliable to monitor the network running status. Kunming Telecom IP man network fault monitoring system provides for acousto-optic alarm room staff on duty, for the WEB interface processing window and SMS remote alarm functions. With the aid of the WEB interface, network administrators can in any network accessible place, open the WEB monitoring interface for real-time monitoring of network fault, log remote viewing and analysis etc.. Through the message function, when there are significant failure and satisfy the message sending condition, automatic fault information will be sent to the relevant mobile phone set in advance of the. In actual operation, there are many important link broken network, power down device (fault through the adjacent normal equipment reporting) and other major faults have been response timely and accurately, showed good reliability system.

4 CONCLUDING REMARKS

The establishment of a centralized log server is not complicated, the server hardware performance requirements are not high, is applied to the network running status and fault monitoring has the characteristics of simple implementation, low cost, stable and reliable. But in the building still need to fully consider the processing capability of log server. When the regulatory network equipment is more, log sudden and distribution will have higher. Make some restrictions on log transmission conditions on the device, set the appropriate severity value, under the premise of ensuring the log regulatory requirements, and to minimize the number of log transmission, processing can reduce the pressure of log server.

The log collection procedures using multi thread technology can improve the concurrent processing capability and response ability of log collection. In addition, the log log online collection, analysis and alarm processing, log publishing functions separated by different server for processing can greatly improve the processing and response ability of the centralized log server system.

REFERENCES

RFC: 3164 The BSD syslog Protocol,Category: Informational. C.Lonvick Cisco Systems 2001.8.

Gambetta D, Can we trust trust In: Gambetta D, ed. Trust: Making and Breaking Cooperative Relations, Basil Blackwell: Oxford Press, pp.213–237, 1990.

Chen Ruiqing. A fast algorithm to determine the relationship between point and polygon. Journal of Xi'an Jiao Tong University,.2007.1 Vol.41 No1.

Korsakov A., Popov B, Terziev I., Manov D., and Ognyanoff, D, "Semantic Annotation, Indexing and Retrieval", Journal of Web Semantics, vol. 2, no.1, pp.49–79, 2005.

Eschenauer L,Gligor V D., "A key-management scheme for distributed sensor networks", Proceedings of the 9th ACM Conference on Computer and Communication Security, pp.41–47, 2002.

Bouhafs F,Merabti M,Mokhtar H.A, "Semantic Clustering Routing Protocol for Wireless Sensor Networks", IEEE Consumer Communications and Networking Conference, pp. 351 – 355, 2006.

Computational Intelligence in Industrial Application – Ling (ed.)
© *2015 Taylor & Francis Group, London, ISBN: 978-1-138-02818-0*

Intrusion detection and response based on data mining and intelligent agents

Yong Hao Wu & Feng Li Zhou

City College, Wuhan University of Science & Technology, Wuhan, China

ABSTRACT: Intrusion detection is an issue that computer scientists try to tackle without complete success. This paper introduces data mining and intelligent agent technology, which presents an architecture consisting of sensors, detectors, data warehouse, model generators and analysis engines. This architecture facilitates the sharing and storage of audit data, automates the generation and distribution of new models, intelligentizes the handling of large, heterogeneous data, and improves the accuracy, efficiency and usability.

KEYWORDS: Data warehouse; Data mining; Intelligent agent; Instrusion detection.

1 INTRODUCTION

With the increase of sensitive information that needs to be stored and processed by network, security issues gradually become the most important issue in the network and system. As the complexity of new system increased, the existing defense technology is not enough, intrusion detection and response is a necessary "firewall" that can ensure the security of network and system [1]. The traditional intrusion detection and response is based on past experience and uses manual signature and encoding. In this way there are some flaws:

1 Need to hand-code for large amounts of data collection, analysis, model generation.
2 Easy to lose data model because of comprehensive experience-based approach and inaccurate considering.
3 Cannott automatically detect the statute of new intrusion when the system has no matching mode.
4 When intrusion is found, the reaction was not quick enough, so that the intruder can achieve at least part of invasion purpose easily.

From a data-centric point of view, the core of detection is data analysis and communication. Data mining technology is an excellent tool for data analysis that combines statistics, pattern recognition, machine learning and database technology. It not only can classify and query the past data, but also can identify the potential links between them, thereby promoting the transfer of information [2]. Then the system can use that information to detect new intrusion and generate new detection mode. For data modules' management and communication, system and network's monitoring, intelligent agent is the appropriate choice. It can

apply the same invasion and response model to heterogeneous multi-subsystem without administrator's intervention, which greatly improves the efficiency and speed of detection and of response.

Assume: (1) We can distinguish the aggression invasion and the legitimate user's normal behavior by their data set characteristics. (2) Collect the feature data of events after the end of related activities. For this assumption, we present intrusion detection and response system based on data mining and intelligent agent technology (DIIDR System).

2 ARCHITECTURE OF DIIDR

Misuse detection and anomaly detection are divided into two separate issues in traditional intrusion detection. Anomaly detection is also known as behavior-based intrusion detection, it can establish the behavior model under user's normal circumstances in learning phase, and then compare it with the existing behavior model. If the deviation is greater than the credibility threshold, it is considered an invasive. The basic principle is that any behavior that does not meet with the known model is intrusion. Misuse detection is also known as knowledge-based intrusion detection, the known intrusion will be significant for building intrusion feature, then the current user behavior and system status will be pattern matched with the existing intrusion behavior. The basic principle is [3] that any behavior that can meet with the known behavior model is intrusion. We can integrate these two modes and store them in a system, thus the new intrusion detection's basic principle is that the normal behavior can meet with the normal models (behavior models), and the intrusion can meet with the abnormal behavior models (intrusion features). If the behavior does not match the principle, the analyzer can generate a new detection model by automatically determining mechanism built according to the threshold criteria, and add it to the historical detection model warehouse. Whether the new model meets the requirements or not, we can determine it by comparing the support rate and trust of calculation results with the given minimum support and minimum trust.

On the basis of these principles and integrating the two modes of distributed and centralized, this paper presents intrusion detection system model as shown in Fig. 1. This system is distributed the plurality of sensors and detectors in different position, detectors can exchange information reciprocally and calculate cooperatively through the intelligent agents, and then we can get the results or latest intrusion detection model in the shortest time. Analyzer integrates global information and analysis results. This system uses the current mature data mining technology as well as variety of other algorithms, the core is data warehouse system, and it can extract and analyze data features, generate and match data model. The local detector agent that reside in the host can complete original data collection, data feature extraction, information exchange and analysis of model data feature at the same level independently. Then each detector agents can submit characteristic data and results information to top layer analyzer through the communication proxy mechanism. Finally, we can get the conclusion by analyzer based on the system global data information.

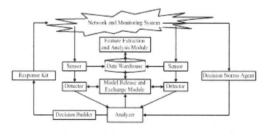

Figure 1. Architecture.

2.1 Data warehouse

Data warehouse can centrally store the data collected from sensors, the data set returned by feature extraction and analysis, the normal and abnormal models generated from model generator [4]. The normal and abnormal models are distributed to each detector located in different positions through the model release module. In order to interact with the data warehouse, we use XML language to represent data. The XML technology is used to view and access

relational data that is data interaction. XML-based data is self-describing, so we can exchange and process the date without any internal description, its scalability and flexibility is much better. The same time, we use a SQL Server 2005 database, which supports the W3C standards, that can collect the relational and online analytical processing data, and can map XML elements and attributes to a relational schema. Thus, each detector in the same host or different hosts can exchange information by XML format, analyzer can converge data information of each agent, and analyze severity of the event behavior from a global perspective of the system, and select an appropriate program.

2.2 Sensing and detection agents

The sensor is composed by two sub-modules: basic audit module and communication module. The audit module can be regarded as an interface of monitoring system, it send the data information derived from data stream to the communication module, the communication module is responsible for encoding information into the XML format and sending it to the data warehouse [5]. The detector gets characteristic data of current event by sensor firstly, then match it with the existing data model to determine whether its intrusion or not.

In most cases, traditional intrusion detection often bind the sensors and detectors, now we separate these two components, integrate them with other analyzers and subassemblies. So when specific detector needs to evaluate too much audit data stream, bottlenecks may cause. Detection agents can distribute the calculation to other machines and break the computational bottleneck.

2.3 Analyzer and decision-making norms agent

Decision-making norms are simple and effective rule set mainly for the new model, they are hand-coded by system administrator and based on experience and expert knowledge. The purpose is that implement human intervention if necessary, improve the efficiency and coverage of the automated system. The Analyzer collects the results of each detection agents and its credibility, then adds them to the standard rules of decision-making, analyze and evaluate the integrated information from a global perspective of system. If the pattern matching is successful, response to the request will be sent to the decision builder; otherwise, a new detection model will be generated.

2.4 Decision builder

According to the given circumstances and based on the plan submitted by the analyzer, the decision builder can select and execute a series of specific implementation steps from the plans. When a subsystem is under attack, analyzer can decide what kind of reaction programs and strategies will be taken, decision builder decides which steps of specific programs and strategies will be executed, then corresponding response strategies in the system response toolkit will be activated, some appropriate response measures will also be taken for intrusion.

3 KEY TECHNOLOGIES

3.1 Data mining

Data mining, also known as knowledge discovery in the database, can mine previously unknown, valid and practical information from data warehouse.

3.2 Structural model of data mining

Data mining in the intrusion detection is mainly used to formal data collection, classification and integration of the system, it also can generate and distribute the detection model, its pattern is shown in Fig. 2. Its core is data warehouse, so it can store data obtained from all sensor, as well as the data features and models through the processing of different parts [6]. Feature extractor input the raw data from the data warehouse and output feature data with additional information through extracting and processing to the data warehouse, a single data records set may not make sense, but it can be integrated into a sequence with other records set, then it presents an intrusion event. Analysis mechanism selects interest feature datasets by the algorithm from the data warehouse, chooses appropriate mining algorithm to find suspicious activity in the dataset, then uses SQL statement to add a column for the data table, the data needs to be indicated whether is normal or not. Model generator gets feature data from the data warehouse, after processing it can generate a model, we store it back to data warehouse. Model release distributes each mode of normal model and abnormal model to the detector, detector will carry out the patter matching between it and the data from sensor, then it can determine the new behavior is normal or intrusion.

3.3 Main algorithm

Four categories of data mining algorithms are often used: association analysis, series model analysis, classification analysis and cluster analysis, and this system uses the first two. Sensor uses series model analysis in the data collection and audit; it can get high-frequency sequence information. Analysis mechanism applies association analysis and can mine the relationship hidden in the data[7]. Association rules are implications like $X \Rightarrow Y$, among it, $X \subset I$, $Y \subset I$, ($I = \{i_1, i_2, \ldots i_m\}$, I is a collection of binary text). $X \cap Y = \Phi$, two main attributes of association rules $X \Rightarrow Y$ in a transaction database D are as follows: support and confidence. The number of transactions X, Y and all their ratios contained in the transaction set is support, it is recorded as support(XY), that is

support($X \Rightarrow Y$)=|: {T: $X \cup Y \subseteq T, T \in D$ }|/|D|

The number of transactions X, Y and X's ratios contained in the transaction set is confidence, it is recorded as confidence (X Y), that is

confidence (X Y)= |: {T: $X \cup Y$ T,T $\in$ D }| / |: {TX T, $T \in D$}|

If transaction set is given, the question of mining association rules is how to generate support and confidence, in addition to this, the support and confidence should be greater than user-specified minsupp and minconf.

3.4 Intelligent agents

Cohen's research shows that the intrusion success depends on the interval from detection to intrusion response, if the interval is 10 hours, then a skilled intruder will have 80% chance of success, if it is 30 hours, the intrusion success is almost certain. If the system administrator monitors system and network at the same time alone, determines the proper evaluation algorithm and appropriate response measures, it is difficult to detect the intrusion and response.

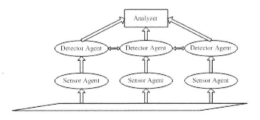

Figure 3.   Structure of agent.

We use distributed intelligent agents to improve the detection and reaction speed, as it shown in Fig. 3. At the user terminal, sensor will act as agent; it can

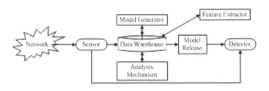

Figure 2.   Data mining application mode.

209

obtain the original data from registry file, network protocols and activity monitor. Then these data will be submitted to the upper layer of detector agent after processing and extracting, the detector will compare feature data to known pattern independently. If the data stream is too large and the calculation is too complex, in order to avoid bottlenecks, detector can distribute the calculation to other hosts' detectors. All results acquired by detector agents will be submitted to the high level analyzer agent resided on the server-side, then the analyzer gives the conclusion.

4 CONCLUSION

This paper builds a system framework based on data mining as well as intrusion detection and response of intelligent agents. The application of data mining technology makes intrusion detection and response system more automatic in data collection, analysis, model generation and distribution. Intelligent agents can monitor various subsystems at the same time collect full data and response timely. The introduction of reaction norms makes up the automatic system response strategy incompletely. However, the determination of data feature extraction and normal data feature or not is still insufficient.

REFERENCES

B. H. Yu, L. M. Gong. "A method for intrusion detection based on intelligent prediction to network traffic", *Microcomputer & Its Applications*, 2011, (1):d50–52, 56(in chinese).

Lee W, Stolfo S. J. Data Mining Approach for Intrusion Detection. In: Proceedings of the 7th USENIX Security Symposium, San Antonio, TX. 1998, pp.78–82.

Lee W, Stolfo S.J, Philip K, et al. Rea-time data mining-based intrusion detection. In: Proceedings of DISCEX II. 2001, pp. 271–350.

Honig A, Howard A, Eskin E, et al. An Architecture for Deployment of Data Mining-based Intrusion Detection Systems. To Appear in Data Mining for Security Applications, Kluwer. 2002, pp. 234–238.

Helmer G G, Wong J S K, Honavar V, et al. Intelligent agents for intrusion detection. In: Proceedings of IEEE Information Technology Conference, Syracuse, NY. 1998, pp. 121–124.

P. Chen, W. F. Lv, Z. Dan. "Network-based intrusion detection approach: a survey", *Computer Engineering and Applications*, 2001, (19): 44–48, 60 (in chinese).

H.Y. Liu, J. Chen, G.Q. Chen. "Review of classification algorithms for data mining", *Journal of Tsinghua University*, 2002, (6):30–35 (in chinese).

Computational Intelligence in Industrial Application – Ling (ed.)
© 2015 Taylor & Francis Group, London, ISBN: 978-1-138-02818-0

A strong-security protocol based on AES algorithm for passive RFID tags

LiQuan Han, Fang Yuan & ZhengChao Xu

College of Computer Science and Engineering, Changchun University of Technology, Changchun, P.R. China

ABSTRACT: The Radio Frequency IDentification (RFID) system may suffer from serious threats that may result in security problems, which becomes a critical issue for RFID systems and applications. As a consequence, a security protocol for RFID tags is needed to ensure privacy and authentication between each tag and their reader. In order to achieve this purpose, this paper briefly analyses the current security protocols, such as Hash-Lock protocol, randomized Hash-Lock protocol, Hash chain, LCAP RFID protocols, and distributed inquiry reply authentication protocol, etc. We analyze their features and performances, what is more important, this paper proposes a new encryption security authentication protocol, which takes full advantages of advance encryption standard (AES) algorithms and combines it with Hash function. To a certain extent this protocol greatly improved the security strength. The research of the subject includes the improvement of AES arithmetic and the details of the authentication protocol.

KEYWORDS: Security protocol; RFID; AES; Hash function.

1 INTRODUCTION

Radio frequency identification (RFID) systems are expected to replace optical barcodes due to many important advantages, such as their low cost efficiency, tiny size, fast identification, and invisible implementation within objects. An RFID system consists of three parts: RFID tags, an RFID reader, and database.

In the passive radio frequency interfaces, frequency range at 13.56 MHz the maximum means current consumption without reducing the operation range of the tags is 15 µA. Due to the limited available chip performance, the limited power and the limited time, an algorithm is allowed to execute, the selection of appropriate security algorithms and protocols are very crucial. However, the use of public key cryptography is out of range with present semiconductor process technologies.

The paper [10] essentially described several common protocols between tag and reader. We have analyzed their advantage and disadvantage.

1.1 Hash-Lock protocol

Hash-Lock protocol in paper [1] is proposed by scholar Sarma in order to prevent information in case of traced up, which mainly use Meta ID instead of the real ID. There is no ID refresh mechanism, so metaID also has no change. ID in the form of plaintext transmits in an insecure channel. As a consequence, the scheme is vulnerable to replay attack, and location tracing attack forward security, as the ID does not change every session.

1.2 Randomized Hash-Lock protocol

Weis [2] proposed a simple, formal definition of strong privacy hash-lock protocol. In the randomized hash-lock scheme, a reader sends a random number Rr then a tag transmits the value H (Rr||Rt||ID), where Rt is a random number generated by the tag. This protocol provides strong privacy and can protect against a replay attack. Their scheme is also strong against a replay attack and location tracing attack. Nonetheless, the scheme is still vulnerable to forward security, as the ID does not change every session. Plus, their protocol is inefficient in terms of the computational load, as the database is required to perform on N/2 hash operations for an ID search, and N is the number of IDs. Our strong-security protocol is mainly based on this protocol.

1.3 Hash-Chain protocol

In [3], Okubo et al. proposed hash-chain based authentication protocol which protects users' location privacy and anonymity. They claim that their scheme provides strong forward security. However, hash-chain calculation must be a burden on low-cost RFID tags and gives back-end servers heavy calculation loads. Similarly, hash chain is a one-way agreement protocol function to label the identity authentication, not the reader authentication.

1.4 Based on the Hash ID change protocol

In paper [4], the dynamic refresh mechanism is used to refresh ID. Every time the reply ID is different, so it is very effective to protect against a replay attack. However, the time when the mechanism refresh ID has lagged behind the tag will receive the message. But before that, the database has completed the data change. During this period if an attacker counterfeit a false message or commit malicious interference, destroy the tag, the tag cannot receive the message. It means that when the backend database and data updates are not synchronized between tags, results in the next legal tag cannot be received by the back-end database. What is more this protocol cannot be applied in the distributed database system.

1.5 LCAP RFID protocol

In paper [5], the proposed scheme needs only two one-way hash function evaluations and hence is quite efficient. Leakage of information is prevented in the scheme since a tag emits its identifier only after authentication. By refreshing an identifier of a tag in each session, the scheme also provides a location privacy and can recover lost massages from many attacks such as spoofing attacks.

1.6 David digital library RFID protocol

According to paper [6], there are no obvious security vulnerabilities in the proposed scheme. However, in order to achieve this performance, the tag circuit must contain random number generation, and security pseudo-random function two major function modules, so it need the support from the hardware circuit of tags. The cost of tags is relatively higher than the others.

2 AES ALGORITHM USED IN THE PROTOCOL

The AES module requires only a little chip area and power. The AES algorithm is a symmetric block cipher with a variable block length and a variable key length. Therefore AES is a flexible algorithm for hardware implementations. AES hardware implementations can be tailored for low die-size demands in embedded systems or can be optimized for high throughput in server applications. This flexibility of the AES algorithm was intended by its creators. They paid attention that the algorithm can be implemented on systems with different bus sizes. Efficient implementations are possible on 32-bit, 64-bit, 128-bit and 256-bit, platforms [8]. AES module for RFID tags regard low size and low power consumption requirements.

We use 8-bit architecture for our low power AES circuit. This approach for a low power AES cryptographic circuit implementation is motivated by two reasons. First, RFID system offers very strict circuit design environment such as circuit area and limited power. Second, an 8-bit architecture enables to decrease the number of S-box to save silicon area. To compute AES encryption, 128-bit data block is divided into sixteen 8-bit data state.

Each common data block of AES is modified by several predefined rounds of processing, where each round involves four functional steps. As Fig. 1 indicates, the four steps in each round of data encryption are called SubBytes, ShiftRows, MixColumns, and AddRoundKey [9].

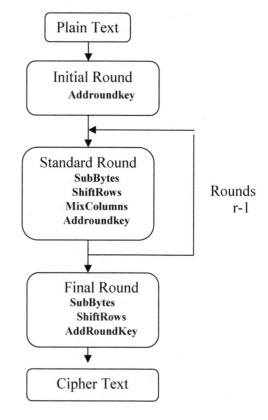

Figure 1.   Operating steps of AES algorithm.

1 SubBytes substitutes each byte of the State. This operation is non-linear. It is often implemented as a table look-up. Sometimes the SubBytes transformation is called S-Box operation.
2 ShiftRows rotates each row of the state by an offset. The actual value of the offset equals the row index, e.g. the first row is not rotated at all; the last row is rotated three bytes to the left.

212

3 MixColumns transforms columns of the State. It is a multiplication by a constant polynomial in an extension field of 128-bit.
4 AddRoundKey combines the 128-bit State with a 128-bit round key by adding corresponding bits mod 2. This transformation corresponds to a XOR-operation of the State and the round key.

M. Feldhofer [13] designed a data structure to reduce these four steps to three ones. The first step of each round operation is SubByte function. During the execution of SubByte, the controller addresses the data memory to operate ShiftRow function at the same time. Then MixColumn and AddRoundKey functions are executed.

Our innovative method to optimize our AES algorithm is the reordering and modifying of the AES round operation steps. The primitive functions SubByte and ShiftRow are based on byte arithmetic, and AddRoundKey is a simple 128-bitwise XOR operation, which can be substituted by any available length. Their operating order is not important because SubByte operates on one single byte, and ShiftRow reorders byte data without changing them. We use these arithmetic features of AES algorithm to reduce AES round operation to two functional steps by reordering and merging AddRondKey, SubByte, and ShiftRow into a single step, step 1, which are more efficient to combine with Hash function during the process of authentication.

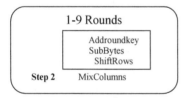

Figure 2.   The efficient method of AES.

The MixColumn multiplies the input polynomial by a constant polynomial c(x), given by

$$c(x) = \{03\}x^3 + \{01\}x^2 + \{01\}x + \{02\} \qquad (1)$$

As shown in equation 2, the MixColumn operation for one column is written as output bytes q_i of MixColumns are calculated by the same function just the order of the input column bytes a_i differs.

$$q(x) = a(x) \cdot c(x) = \bmod n(x) \qquad (2)$$

$$q0 = (a0 \otimes \{02\}) \oplus (a3 \otimes \{01\}) \oplus (a2 \otimes \{01\}) \oplus (a1 \otimes \{03\})$$

$$q1 = (a1 \otimes \{02\}) \oplus (a0 \otimes \{01\}) \oplus (a3 \otimes \{01\}) \oplus (a2 \otimes \{03\})$$

$$q2 = (a2 \otimes \{02\}) \oplus (a1 \otimes \{01\}) \oplus (a0 \otimes \{01\}) \oplus (a3 \otimes \{03\})$$

$$q3 = (a3 \otimes \{02\}) \oplus (a2 \otimes \{01\}) \oplus (a1 \otimes \{01\}) \oplus (a0 \otimes \{03\}).$$

The equation shows that all output byte data of MixColumn are calculated using the same function except the order of the input column bytes. For our low power AES architecture, we designed a sub-block which calculates one fourth of the full MixColumn operations

An encryption of a plaintext block works as follows. Before encryption is started the plaintext block has to be loaded into the RAM of the AES module. In the RFID tag application, the plaintext block is the 128-bit challenge which was received from the reader. The communication between the reader and tag is byte-oriented which fits nicely into the 8-bit architecture of the AES module: every received byte can be stored in the AES module.

3   AUTHENTICATION PROCESS DESCRIPTION

The main purpose of authentication technology is to guarantee the accuracy of the tag reader to identify the ID of the reader, and prevent the storage data information without permission is illegal to read and malicious tampering, at the same time also must ensure that the Reader correct identification Tag the Tag's identity, to prevent the Reader read the fake data information mutual authentication between Reader and Tag. Accordingly, after the research and analysis on the existing RFID security protocols and encryption algorithm, we design a new hybrid ncryption security authentication protocol, combining the AES and Hash arithmetic. This article applies the international standard ISO9798-2 certification process, this is the method of user authentication protocol.

3.1   *Design is based on the following prerequisite and conditions*

1 Each tag has a unique starter, which can generate and protect the safety of the tag parameters.
2 The reader has a pseudo random number generator, able to generate a random number;
3 Each tag has a pseudo random number generator, able to generate random number
4 Each tag is able to Hash arithmetic and AES matrix operations.
5 The communication channel between the database and the reader is safe and reliable, the database and the reader may be regarded as a whole;

3.2 The detail of protocol

Table 1. Parameters.

Parameter	Description
ID	the identification of tags
PID	ID saved in the database
N	the number of tags in RFID system
$\oplus$	XOR computation
H(x)	hash function in reader
F(x)	hash function in tag
||	concatenation of two inputs
AES()	AES encryption algorithm
AES()-1	AES decipherment algorithm
K	secret key that is safely shared among readers, tags and database
R1	random number generated by reader
R2	random number generated by tag
Q	modify signal from the trigger

1 There is a secret key K between Tag and Reader, which is regarded as an encryption key of the AES algorithm. K is saved in both in the Reader and the Tag. At the same time, the reader defines a data Q as trigger sending the signal to modify the ID.

2 By means of the pseudo-random number generator, the Reader generates random number R1. R1 and key K do AES encryption algorithm together, and generate the encrypted data block AES(R1\|\|K).

3 The reader query to the tag, and the generated encrypted data blocks AES (R1\|\|K) is sent to tag.

4 Tag receives request message from Reader and encrypted data block AES (R1\|\|K), and then executes decipherment algorithm AES-1 (R1\|\|K) to get R1. The result data R1 and ID operate the Hash function getting H(R1\|\|ID), which is sent back to the reader by the tag.

5 Sever extracts the high-order bits of all the PID, and checks the list for matching PID that has the same bits as ID. If there is a match, Database will recognize Tag as a legal device. Afterwards, Database searches for the stored ID in its memory and then hashes ID and R to gain H(R1\|\|PID). After that, DB compares the hash value that weather H(R1\|\|PID) and H(R1\|\|ID) is equal. If at least one of the two verifications does not pass, the protocol will be aborted and the Database responds a FALSE code. If the two values are equal, Database may consider that the reader R is legal, then executes AES encryption algorithm, i.e. AES(PID\|\|K) and goes to next step.

6 Reader obtains random number R2 by pseudo random number generator, and executes AES

encryption algorithm to acquire encryption module AES (R2\|\|K), which is sent to tag with the previous data AES (PID\|\|K).

7 Tag identify the authentication of Reader, i.e. after the tag receives the encryption module AES (PID\|\|K), then executes the AES decipherment algorithm AES-1(PID\|\|K) to obtain the PID, comparing with itself ID. If ID=PID, the authentication is successful. If two values are not equal, the protocol will be aborted and the Database responds a FALSE code.

8 After the Tag identifies the legal Reader, it executes AES decipherment algorithm AES-1(R2\|\|K) to get R2. At the same time refreshes itself ID, ID=H(ID\|\|R2).

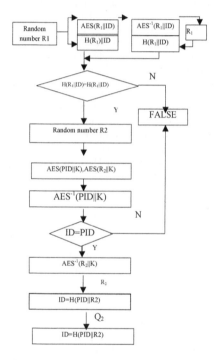

Figure 3. The state flow chart of our proposed identification.

9 Tag transmits trigger signal q to reader. The Reader refresher its ID, i.e. ID=H(ID\|\|R2)

4 ANALYSIS

4.1 Security analysis

The user's privacy mainly results from the leakage of location information or tag information of the tag's owner. The messages sent and received between the tag and the reader are transmitted as different messages each time during all the authentication procedures. It is impossible to track the tag's location through the previous unchanged messages, because different messages are exchanged each time due to the sending of the randomly selected PID. However, the tag's location can be estimated even though different messages are sent each time, in case of recognition.

In the proposed protocol, the secrecy of the messages sent and received during the authentication procedure is guaranteed, by concealing and sending the sent message (PID) through the AES computation with random numbers. That is, the PID of the sent tag can be calculated, only if the random number and its own PID information are known. It is safe against the message eavesdropping attack, because it is impossible to calculate the tag's overall ID even though the PID is exposed.

There are two kinds of attack; resend attacks disguised as a reader and as a tag. In case of disguising as a reader, the attacker eavesdrops on the message sent from the reader to the tag and resends it. In the proposed protocol, the resend attack is prevented by establishing the pseudo random numbers R1, R2.

Hash function is one-way irreversible. And then, AES algorithm is more secure than the protocol described in paper [8], which applies bits computation instead of AES. What is more, the crashes of AES algorithm 128bit need considerable amounts of time. The continuous refreshing ID has successfully avoided the attacker to acquire information of importance. The security performance comparison is illustrated in Table 3, according to the analysis of related references.

4.2 The algorithm complexity analysis

An encryption of a plaintext block works as follows. Before encryption is started the plaintext block has to be loaded into the RAM of the AES module. In the RFID tag application, the plaintext block is the 128-bit challenge which was received from the reader. Every received byte can be stored in the AES module. No intermediate memory is necessary. The cryptographic key is obtained in a similar way from the tag's EEPROM. Now the AES algorithm can be executed. It starts with a modification of the State by an Add Round Key operation using the unaltered cipher key. Only the last round lacks the MixColumns operation. After emerging it calculates one fourth of the full MixColumn operations.

After getting R tag decryption, the decryption of R with its own identity on $H(R||ID)$. Certification protocol at the end of the process, tag also decryption operation needs to execute $AES-1(ID||K)$. Hash algorithm is a one-way function, which determines the complexity of the algorithm, and this function can be different encryption algorithms, when using AES.

For the convenience of analysis and comparison, this paper sets the parameters of the different protocols. The length of data ID, R1, R2, k, PID is '1' bits. The tag, reader, database execute Hash function is represented by the 'H', AES encryption and decryption algorithm are represented by AES(). R means necessary random number in the different protocols and devices. 'ø' means not existence. The number of tags saved in the database is 'N'. For example, performing a distributed RFID ask – response authentication protocol, tag needs two Hash operations and generate a random number. Then fill in the corresponding cell 'H: 2, R:1'; Reader needs to generate a random number, so fill in the corresponding cell 'R:1'; Sever needs N/2 Hash operation, so fill in the corresponding cell 'H:N/2', and so on.

Table 2. Security performance comparison between related protocols.

Authentication protocol	1	2	3	4	5	6	7	8	9	10	11
Hash-Lock protocol	×	×	×	×	×	√	√	×	×	Read only	[1]
Randomized-hash-lock protocol	×	×	√	√	×	√	√	×	√	Read only	[2]
Hash-chain protocol	√	×	×	×	√	√	√	√	√	Read &write	[3]
Based on the Hash ID change protocol	√	√	√	√	√	×	×	√	√	Read &write	[4]
LCAP RFID protocol	√	√	√	√	√	×	×	√	×	Read &write	[5]
Distributed inquiry reply protocol	√	√	√	√	√	√	√	√	×	Read only	[11]
David digital library RFID protocol	√	√	√	√	√	√	√	√	×	Read &write	[6]
AES-Hash	√	√	√	√	√	√	√	√	√	Read &write	

1 privacy protect; 2 replay attack ; 3 tag spoofing attack; 4 reader spoofing attack; 5 location analysis; 6 data synchronism; 7 DOS attack; 8 the forward security; 9 changeable ID; 10 tag type

Table 3. The algorithm complexity analysis about related protocols.

related protocols	Operations			Tag storage requirement			Hardware circuit of tags
	Tag	Reader	Sever	Tag	Reader	Sever	
Hash-Lock protocol	H:1	∅	H:1	2l	∅	3l	Hash function
Randomized-hash-lock protocol	H:1 R:1	∅	H:N/2	1	∅	1	Random number generator, Hash function
Hash-Chain protocol	H:2	∅	H:N/2	1	∅	2l	Hash functions
Based on the Hash ID change protocol	H:3	∅	H:3 R:1	4l	∅	5l	Hash function
LCAP RFID protocol	H:2	R:1	H:2	2	1	8l	Hash function
Distributed inquiry reply protocol	H:2	R:1	H:N/2	1	1	1	Random number generator, Hash function
David digital library RFID protocol	H:2 R:2	R:2	H:N/2	4l	1	2l	Random number generator
AES-Hash	H:3 AES():2	R:1	H:3 AES():2	1	1	6l	Random number generator, hash functions

5 CONCLUSIONS

This paper presented a strong-security protocol for passive RFID tags which achieves a relatively strong intensity of cryptographic authentication. This is because AES arithmetic has high quality in encryption and high rate of execution, but has low demand for configuration of hardware. Compared with the David digital library RFID protocol, this scheme costs lower in tag hardware, but more complicated in sever design. Furthermore, AES arithmetic belongs to the category of symmetric arithmetic and is very suitable to run in RFID system because it can be easily executed in the semiconductor devises. The improvement of the A E S arithmetic to enhance execution efficiency of program is studied. This challenge-response authentication protocol was proposed which was integrated into the existing ISO 09798-2 standard. With this strong-security protocol, we basically achieve the demand of the current RFID system security.

Finally, the protocol security has carried on the concrete analysis, shows that the protocol cannot only ensure the security of RFID system efficient, but can also provide better openness, and can adapt to the complex application environment, and thus lay the foundations for the further wide application of RFID system.

REFERENCES

Sarma S. E., Weis S. A., Engels D. W. Radio-frequency identification: secure risks and challenges. *RSA Laboratories Cryptobytes*, 2003, 6(1):2–9.

Weis, S.A., Sarma, S., Rivest, R., Engels, D. Security and privacy aspects of low-cost radio frequency identification systems. In: Hutter, D., Müller, G., Stephan,W., Ullmann, M. (eds.) Security in Pervasive Computing. LNCS, vol. 2802, pp. 201–212. Springer, Heidelberg (2004).

Ohkubo, M., Suzuki, K., Kinoshita, S. A Cryptographic Approach to 'Privacy Friendly' tag, RFID Privacy Workshop (November 2003).

Henrici D.,Muller P. Hash-based enhancement of location privacy for radio-frequency identification devices using varying identifiers. In: Proceedings of the 2nd IEEE Annual Conference on Pervasive Computing and Communications Workshops (PERCOMW 04), Washington, DC, USA, 2004, 149–153.

Lee S. M. Hwang Y. J. Lee D. H., Lim J. L Efficient authentication for low cost RFID systems. In: Gervasi O. eds. Proceedings of the International Conference on Computational Science and Its Applications (ICCSA 2005). Lectures Notes in Computer Science Berlin: Springer-Verlag, 2005, 619–627.

Molnar D, Wagner D. Privacy and security in library RFID; Issues, practices, and architectures. In Proceedings of the 11th ACM Conference on Computer and Communications Security(CCS' 04), Washington, DC, USA, 2004, 210–219.

Rhee K., Kwak J., Kim S., Won D. Challenges response based RFID authentication protocol for distributed database environment. In; Hutter D, eds. Proceedings of the 2nd International Conference on Security in Pervasive Cotrrputing(SPC 2005). Lectures Notes in Computer Science 3450. Berlin: Springer Verlag, 2005, 70–84.

Strong Authentication for RFID Systems Using the AES Algorithm Martin Feldhofer, Sandra Dominikus, and Johannes Wolkerstorfe.

Raghavan NS.AES croptography *Advances into the Future Java World* 2000, 12(4):47–51.

Martin Feldhofer and Christian Rechberger A Case Against Currently Used Hash Functions in RFID Protocols In Embedded and Ubiquitous Computing – EUC 2005 Workshops, volume 3823 of LNCS, pages 945–954. Springer, December 2005.

T. Dimitriou. A Lightweight RFID Protocol to protect against Traceability and Cloning attacks. In First International Conference on Security and Privacy for Emerging Areas in Communications Networks (SecureComm 2005), pages 59–66, Athens, Greece, September 2005, IEEE Computer Society.

ZHOU Yong-Bin FENG Deng-Guo. Design and Analysis of Cryptographic Protocols for RFID. *Chinese Journal of Computers*, 2006, 29(4):581–589.

M.Feldhofer and S.Dominikus Strong Authentication for RFID Systems Using the AES Algorithm. In CHES2004, volume 3156 of Lecture Notes in Computer Science, pages 357–370. Springer, 2004.

Computational Intelligence in Industrial Application – Ling (ed.)
© *2015 Taylor & Francis Group, London, ISBN: 978-1-138-02818-0*

An optimized nonlinear grey Bernoulli model and its application for epidemic prediction

L. P. Zhang, Y. L. Zheng & Y. J. Zheng*
School of Public Health, Xinjiang Medical University, Urumqi, China

K. Wang & X. L. Zhang
Department of Medical Engineering and Technology, Xinjiang Medical University, Urumqi, China

ABSTRACT: Being able to forecast an epidemic tendency accurately has been quite a popular subject for researchers both in the past and at present. This paper presents an adaptive nonlinear grey Bernoulli model named ONGBM(1,1) by using genetic algorithms to solve the optimal parameter estimation problem, to predict the epidemic trend of hepatitis B (HB) in Xinjiang, China. Three grey models, traditional GM(1,1), Grey Verhulst Model (GVM) and traditional NGBM(1,1) model are also established for comparison. Numerical results illustrate that the proposed model is more accurate than the traditional GM(1,1), GVM and NGBM(1,1) models. The results also show that the optimum mechanisms indeed improve the grey model for prediction accuracy by using genetic algorithms approach and the grey system theory is a feasible analysis method for epidemic forecasting.

KEYWORDS: Nonlinear grey Bernoulli model; Genetic algorithm; Grey model.

1 INTRODUCTION

Epidemic forecasting plays an important role in epidemic prevention, treatment and health policy research. A number of methods have been proposed for epidemic forecasting, such as statistical methods (Ellner et al., 1998; Zhang et al., 2013; Liu et al., 2011; Chen et al., 2012), neural network models (Bai & Jin, 2005; Guan et al., 2004), the differential equation models (Bhadra et al., 2011; Goodreau et al., 2012; Andrews & Basu, 2011), grey prediction models (Shen, et al., 2013; Xu & Zhang, 2012), fuzzy forecasting models (Huarng et al., 2012; Hsieh and Tsaur, 2013; Jones et al., 2010) and the geographic information system models (Eisen & Eisen, 2011; Wei et al., 2011; Nykiforuk & Flaman (2011)). To obtain a reliable forecast, certain laws governing the phenomena of system development must be discovered on the basis of either natural principles or real observation (Cui et al., 2013). The traditional approach often needs to meet with large number of samples, normal distribution or on the assumption of realizing the structure of the system to be forecasted. However, in practical life, system information is always poor and searching the natural principles is very hard. Forecasting system development directly from actual data characteristics has become an alternative method.

The grey forecasting model was fist proposed by Deng (1982), which is well-known for limited observed data modeling. The theory does not rely on a statistical method to deal with a grey quantity, but deals indirectly with original observations, and searches its intrinsic regularity. Grey forecasting model has been successfully applied in various fields and has demonstrated satisfactory results. In recent years, grey modelling has been used increasingly as an epidemic forecasting tool (Lei et al., 2010; Shen et al., 2013; Xu & Zhang, 2012).

Although traditional grey models are easy to understand and simple to calculate, with satisfactory accuracy, but it also lack flexibility to adjust the model to acquire higher forecasting precision. Literatures show performance of traditional grey models still could be improved. The nonlinear grey Bernoulli model NGBM(1,1) proposed by Chen et al. (2008) is a nonlinear differential equation with power index γ. It is a simple modification of GM(1,1) combining with Bernoulli differential equation. By adjusting power exponent, the curvature of the solution curve could be adjusted to fit the result of accumulated generating operation of raw data. To further improve the fitness of NGBM model, many researchers have tried to improve the prediction accuracy of the grey forecasting model by optimizing the selection of model

* Corresponding author: J. Zheng E-mail: 41430622@qq.com

parameters and power exponent (Chen et al., 2010; Hsu, 2010).

The current paper aims to develop an approach to increase the predictive precision of the NGBM(1,1) by optimizing the initial condition, interpolated coefficient in the background value and the power exponent. By using genetic algorithm (GA) to solve the optimization problem, we presented an optimized based nonlinear grey Bernoulli model abbreviated as ONGBM(1,1). To verify the efficiency and rationality of the proposed model, we apply the optimized model to forecast the incidence of HB with nonlinear small sample characteristics in Xinjiang, China. To determine the optimal model parameters, we select Matlab, which is a GA based optimization solver, to find the optimal values of model parameters.

The organization of this survey is as follows: the routine modelling method for NGBM(1,1) is described and its disadvantages are pointed out in Section 2. The optimized NGBM(1,1) based on GA are also presented and discussed in Section 2. In Section 3, an empirical analysis on incidence of HB is applied to verify the feasibility and effectiveness of the optimized NGBM(1,1) model proposed here. Finally, conclusions are made in Section 4.

2 METHODS

2.1 The original nonlinear grey Bernoulli model NGBM(1,1)

The nonlinear grey Bernoulli model NGBM(1,1) proposed by Chen et al. (2008) is a first order single-variable nonlinear differential equation combining both the traditional GM(1,1) model and the Bernoulli equation. For predications involving nonlinear small sample time series, its performance is better than that of the traditional grey forecasting models. The NGBM(1,1) model constructing process is described below:

Step 1: Let $X^{(0)} = (x^{(0)}(1), x^{(0)}(2), \cdots, x^{(0)}(n))$, $(n \geq 4)$ be an original non-negative time series. Its 1-AGO (accumulated generating operator) sequence $X^{(1)}$ is

$$X^{(1)} = (x^{(1)}(1), x^{(1)}(2), \cdots, x^{(1)}(n)),$$

where

$$x^{(1)}(k) = \sum_{i=1}^{k} x^{(0)}(i); k = 1, 2, \cdots, n. \tag{1}$$

Step 2: The grey differential equation of NGBM(1,1) is defined as

$$x^{(0)}(k) + az^{(1)}(k) = b(z^{(1)}(k))^{\gamma}; k = 2, 3, \cdots, n, \gamma \neq 1. \tag{2}$$

where $x^{(0)}(k)$ is called a grey derive, and $z^{(1)}(k) = \alpha x^{(1)}(k) + (1 - \alpha) x^{(1)}(k - 1))$ is referred to the background value of the grey derivative. When α is equal to 0.5, the above equation is called the nonlinear grey Bernoulli model, which is abbreviated as NGBM(1,1) (Chen et al., 2008). When α is an indefinite value in the interval [0,1], the model is called the Nash nonlinear grey Bernoulli model, which is abbreviated as NNGBM(1,1) (Chen et al., 2010).

When $\gamma = 0$, the Eq. (2) reduced to the traditional GM(1,1) model,

$$x^{(0)}(k) + az^{(1)}(k) = b. \tag{3}$$

When $\gamma = 2$, the Eq. (2) reduced to the grey Verhulst model (GVM),

$$x^{(0)}(k) + az^{(1)}(k) = b(z^{(1)}(k))^2. \tag{4}$$

Step 3: The whitenization differential equation of the NNGBM(1,1) is a first-order differential equation, which is defined by

$$\frac{dx^{(1)}}{dt} + ax^{(1)} = b(x^{(1)})^{\gamma}. \tag{5}$$

Step 4: The structural parameters a and b can be estimated by using the least square method,

$$A = [a, b]^T = (B^T B)^{-1} B^T Y \tag{6}$$

Where

$$Y = \begin{bmatrix} x^{(0)}(2) \\ x^{(0)}(3) \\ \vdots \\ x^{(0)}(n) \end{bmatrix}, B = \begin{bmatrix} -z^{(1)}(2) & [z^{(1)}(2)]^{\gamma} \\ -z^{(1)}(3) & [z^{(1)}(3)]^{\gamma} \\ \vdots & \vdots \\ -z^{(1)}(n) & [z^{(1)}(n)]^{\gamma} \end{bmatrix}. \tag{7}$$

Step 5: Set the initial value as $\hat{x}^{(1)}(1) = x^{(0)}(1)$, we can get the solution of whitenization equation as

$$\hat{x}^{(1)}(k+1) = \{[(x^{(0)}(1))^{(1-\gamma)} - \frac{b}{a}]e^{-a(1-\gamma)k} + \frac{b}{a}\}^{1/(1-\gamma)}; \gamma \neq 1. \tag{8}$$

Step 6: Apply the first-order inverse accumulated generation operation (1-IAGO) to $\hat{x}^{(1)}(k+1)$, we can obtain the simulation and forecasting function of $x^{(0)}$ as

$$\hat{x}^{(0)}(k+1) = \hat{x}^{(1)}(k+1) - \hat{x}^{(1)}(k); k = 1, 2, \cdots. \tag{9}$$

Step 7: Modeling error analysis.

For the purpose of evaluating the out-of-sample forecast capability, the forecasting accuracy is examined by calculating two different evaluation statistics: the root mean square error (RMSE) and the mean absolute percentage error (MAPE). These are expressed as follows:

$$\text{RMSE} = \sqrt{\frac{\sum_{k=1}^{n}[(x^{(0)}(k) - \hat{x}^{(0)}(k))]^2}{n}}, \tag{10}$$

$$\text{MAPE} = (\frac{1}{n}\sum_{k=1}^{n}\left|\frac{x^{(0)}(k) - \hat{x}^{(0)}(k)}{x^{(0)}(k)}\right|) \times 100\%, \tag{11}$$

where $x^{(0)}(k)$ is the actual value at time k, $\hat{x}^{(0)}(k)$ is its fitting value and n is the number of data used for prediction.

2.2 The optimized nonlinear grey Bernoulli model ONGBM (1,1)

2.2.1 The genetic algorithm

The genetic algorithm (GA), developed by Holland (1970), can be understood as an "intelligent" probabilistic search algorithm which can be applied to a variety of combinatorial optimization problems. With the characteristics of easier application, great robustness, and better parallel processing than most classical methods of optimization, GA has been widely used for combinatorial optimization, production scheduling, automatic control, image processing, machine learning, data mining, etc.

A GA consists of a string representation ("genes") of the nodes in the search space, a set of genetic operators for generating new search nodes, a fitness function to evaluate the search nodes, and a stochastic assignment to control the genetic operators. Typically, a GA consists of the following steps (Hou et al., 1994):

Step 1: Randomly generate an initial population of the search nodes.

Step 2: Evaluate fitness of individuals in the population according to the fitness function (objective function).

Step 3: Genetic operations-new search nodes are generated randomly by examining the fitness value of the search nodes and applying the genetic operators to the search nodes.

Step 4: Repeat Steps 2 and 3 until the algorithm converges.

Genetic algorithm based on natural evolutionism is a good method to resolve complicated nonlinear programming problem. Many scholars tried to estimate the model parameters by using GA and achieve good results (Hsu, 2009, Hsu, 2011). In this article,

we take genetic operators as a parameter search technique which utilize the genetic operators to find near optimal solutions to construct the ONGBM (1,1).

2.3 The GA based nonlinear grey Bernoulli model ONGBM (1,1)

It has been shown from the prediction function of the NGBM (1, 1) is that, the prediction accuracy is determined by the initial conditions, the background value, structural parameters a, b, and the power exponent γ.

As the structural parameters can be replaced by the original data, α and γ (Eq. (6)), only the α in background value, the initial value and exponent power γ are required to be optimized. Forecasting can be performed after obtaining the parameters a, b, initial value and γ in Eq. (8).

In view of the above analysis, we can see that some disadvantages exist in routine NGBM (1, 1) model:

1 Based on the hypothesis that the first datum of sequence is invariable, the original NGBM (1,1) model takes the first datum in observed data sequence as the initial condition. For observation errors always exist, especially when great deviation from the real value at the first point exists, the constant c defined with the above method becomes incorrect inevitably. Actually, the first datum is the oldest information in a sequence. So, the hypothesis violates with the fact. The existing literature show that the first datum has no relation to the simulative value $\hat{x}^{(1)}(k)$ (Deng, 2002; Liu, 1997).

2 For the original NGBM (1,1) model, the generating coefficient α in the background value is set to be 0.5 as tradition. However, the fixed α value is not the optimal selection for some series in reality (Hsu, 2009). If α is adjusted according to the characteristics of original series, it will provide a better prediction performance.

3 The power γ is used to be the adjustable parameter. Chen et al. (2010) and Wang (2013) solved the value of γ by Nash equilibrium theory. Because of the complex design calculation and manufacturing process, the promotion and application were hindered.

To improve the forecasting accuracy of the grey model, Dang et al. (2005) used $x^{(1)}(n)$ as the initial condition. However, considering the complexity and uncertainty involved, it cannot be proved theoretically that this method is any better than the traditional methods. Cui et al. (2013) used $x^{(1)}(1) + c$ as the initial condition and solve the value of c by least square method. To use all the information in the sample data to predict the behavior of an actual system, we use the weighted sum of $x^{(1)}(1)$ and $x^{(1)}(n)$ as the initial condition (Wang et al., 2010) and the weighted sum of $x^{(1)}(k)$ and $x^{(1)}(k+1)$ as the background value in the NGBM(1,1) model. Then we use GA to optimize the weighted coefficient and exponent power γ. We construct an optimized NGBM model abbreviated as ONGBM as follows.

$$
\begin{cases}
\hat{x}^{(1)}(k) = \{[(\hat{x}^{(1)}(1))^{(1-\gamma)} - \dfrac{b}{a}]e^{-a(1-\gamma)(k-1)} + \dfrac{b}{a}\}^{1/(1-\gamma)}, \\
\hat{x}^{(0)}(k) = \hat{x}^{(1)}(k) - \hat{x}^{(1)}(k-1), \\
\hat{x}^{(1)}(1) = \beta x^{(1)}(0) + (1-\beta)x^{(1)}(n), \\
z^{(1)}(k+1) = \alpha x^{(1)}(k+1) + (1-\alpha)x^{(1)}(k), \\
0 \le \alpha \le 1, \\
k = 2, 3, \cdots, n.
\end{cases} \quad (12)
$$

Noting that a and b are the structure parameters, which can be estimated by using the least square method (Eq. (6)). Thus, the ONGBM(1,1) grey model only has three parameters α, β and γ. The modelling procedure can be regarded as optimization in a three dimensional space. GA is very effective for optimizing in multidimensional space. To optimize the NGBM(1,1) prediction model by GA, the MAPE, Eq. (12), is used as the objective function for a nonlinear integer programming with parameters and variable as decisions variables.

To find the optimal α, β and γ by using GA, define the fitness function as

$$
\min_{\alpha, \beta, \gamma} \text{MAPE} = (\frac{1}{n}\sum_{k=1}^{n}\left|\frac{x^{(0)}(k) - \hat{x}^{(0)}(k)}{x^{(0)}(k)}\right|) \times 100\%. \quad (13)
$$

The software Matlab 7.1 is used in this study to find the optimal value of parameters $\hat{\alpha}$, $\hat{\beta}$ and power exponent $\hat{\gamma}$ that correspond to the smallest in-sample forecasting error. Forecasting can be performed after obtaining the parameters.

Regarding the settings of GA parameter, a number of articles have already established the concept and setting values (Glodberg, 1989; Pan, 2002). In this article, the Genetic Algorithm Toolkit (GATOOL) of Matlab was used to realize optimization algorithm. The default crossover rate, mutation rate, and population size are set to 0.8, 0.01, and 100, respectively. The generations and stall generations are set to 500. All other settings are left at their default values. After setting the above-mentioned conditions, one can set the GAT of Matlab to start optimizing. When the stopping criteria is met, Matlab stop the optimization process, and output the minimum in-sample MAPE and optimal value of α, β and γ.

The follow numerical examples in this research will show the ONGBM is powerful in improving the model precision.

3 EMPIRICAL RESULTS

In this section, to verify its effectiveness and feasibility in epidemic prediction, the ONGBM(1,1) is used to forecast the incidence of HB in Xinjiang, China.

3.1 Data

Viral hepatitis is a serious public health problem affecting billions of people globally. The World Health Organization (WHO) has estimated that there are 360 million chronically hepatitis B virus (HBV) infected people and 5.7 million HBV-related cases worldwide. HBV accounts for an estimated 500,000–700,000 annual deaths worldwide (Clements et al., 2006; Lok & McMahon, 2007). HB, caused by infection with HBV, is a common liver disease in China. According to the latest national HB seroepidemiological survey, approximately 20 million suffer from HB, and almost 300,000 die annually from consequences of HBV infection in China. Both liver cancer and cirrhosis are among the 10 most common causes of mortality in China, for both, HBV causes the majority of deaths (He et al., 2005; Liang et al., 2009).

HBV infection is highly endemic in Xinjiang Uygur Autonomous Region, China, with transmission occurring in childhood and adulthood. A group data of Xinjiang CDC shows that the incidence of HB has been increasing gradually year by year, which rose from $101/10^5$ in 2004 to $210/10^5$ in 2012. HBV infection results in a heavy disease burden not only in premature death but also in health impairment. HB forecasting can be an invaluable tool for HB control and elimination efforts.

For the purpose of forecasting the incidence of HB in Xinjiang, we collected the monthly data set from 2011 to 2012. Epidemiological incidence data was obtained from the report of Xinjiang Center for Disease Control and Prevention (Xinjiang CDC). The 18 observations from January 2011 to June 2012 are used for model building, and the remaining observations from July 2012 to December 2012 are considered to evaluate the out of sample forecasting performance of the model.

3.2 Simulation analysis

Endemic HBV infection is a serious health problem in Xinjiang, causing a substantial burden of acute and chronic liver disease. As limited information is available on this issue in Xinjiang, traditional statistic forecasting methods always cannot obtain good performance. The spread of an epidemic is a random, changeable, unstable, and complicated nonlinear system influenced by a lot of uncertain factors. Therefore, it can be regarded as a grey system.

The real incidence of HB in Xinjiang from 2011 to 2012 is shown in Table 1 and Fig. 1. Fig. 1 demonstrates that there are some nonlinear fluctuations in the original data. In order to grasp the trends of HB, we constructed the ONGBM(1,1) forecasting model. Three grey models GM(1,1), GVM, the traditional NGBM(1,1) are also established for comparison.

For the sake of convenience, the detailed calculation and modeling process are omitted here. Only the original data, simulation, forecasting results and the corresponding errors are shown in Table 1 and Fig. 1.

Table 1. Forecasting performance evaluation and comparison.

Time	Original value	GM(1,1)	GVM	NGBM(1,1) $\alpha = 0.5, \gamma = 0.2019$	ONGBM(1,1) $\alpha = 0.0082, \beta = 1.0493, \gamma = 0.1006$
201101	16.79	16.79	16.79	16.79	16.77
201102	15.29	17.67	3.89	15.38	16.35
201103	20.81	17.67	4.73	16.62	17.03
201104	17.01	17.67	5.72	17.38	17.43
201105	18.59	17.68	6.87	17.86	17.68
201106	16.96	17.68	8.18	18.17	17.84
201107	18.56	17.68	9.66	18.34	17.94
201108	18.42	17.68	11.28	18.42	17.99
201109	16.83	17.69	13.00	18.42	18.01
201110	16.00	17.69	14.77	18.36	18.00
201111	16.46	17.69	16.50	18.26	17.96
201112	17.41	17.69	18.09	18.11	17.92
201201	16.28	17.69	19.42	17.94	17.85
201202	21.45	17.70	20.40	17.74	17.77
201203	20.12	17.70	20.94	17.53	17.69
201204	15.59	17.70	20.99	17.29	17.59
201205	18.32	17.70	20.55	17.04	17.49
201206	16.59	17.71	19.65	16.78	17.38
MAPE(%)		8.18	32.22	7.87	7.43
RMSE		1.75	7.41	1.87	1.71
201207	17.58	17.71	18.37	16.51	17.27
201208	17.09	17.71	16.83	16.24	17.15
201209	16.28	17.71	15.12	15.96	17.02
201210	13.32	17.72	13.35	15.67	16.90
201211	19.08	17.72	11.62	15.38	16.77
201212	17.89	17.72	9.97	15.09	16.63
MAPE(%)		9.04	16.14	10.96	8.79
RMSE		1.99	4.48	2.20	1.84

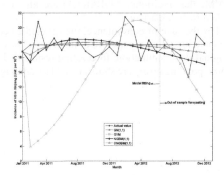

Figure 1. The comparison of the original and forecast values for four models.

The optimization stopping criterion was that the change of objective value in last 1000 valid trials is less than 0.1%. The default generations and stall generations are set to 1000. After setting the above-mentioned conditions, one can set the GAT of Matlab to start optimizing. When the stopping criteria is met, Matlab stop the optimization process, and output the minimum in-sample MAPE and optimal value of γ. And in this article, the optimization stopping criterion was that the change of objective value in last 1000 valid trials is less than 0.1%. By taking the above settings, we obtain the minimum MAPE (7.43%) and optimal parameter value $\hat{\alpha} = 0.0082$, $\hat{\beta} = 1.0493$ and $\hat{\gamma} = 0.1006$.

Table 1 shows that the in-sample MAPE of these four forecasting models are 8.18%, 32.22%, 7.87% and 7.43% respectively, and the corresponding model RMSE are 1.75, 7.41, 1.87 and 1.71. Comparing the out-of sample forecast results from June 2012 to December 2012, we can see that the MAPE (8.79%) and RMSE values (1.84) of ONGBM(1,1) model are also all smaller than the traditional GM(1,1) model, GVM model and NGBM(1,1) model. According to the results described above, the improved NGBM(1,1) model on the optimal parameter estimation by GA obtains the lowest out-of sample forecasting errors among these models. This implies ONGBM(1,1) modeling results are reliable and it has better prediction ability than the other three models.

4 DISCUSSION AND CONCLUSION

Accurate disease predictions and early-warning signals of increased disease burden can provide public health and clinical health services with the information needed to strategically implement prevention and control measures. Due to the lack of more additional epidemiological data published by Xinjiang CDC at present, few methods can be applied to predict the epidemic tendency in Xinjiang except grey model.

Although the traditional grey models have been successfully applied in various fields, Literature shows that if the original data hold with high degree of nonlinearity, the model precision will be lower than linear cases (Chen et al., 2010). The NGBM(1,1) proposed by Chen (2008) is a recently developed grey forecasting model which has a power exponent γ that can effectively manifest the nonlinear characteristics of real systems and flexibly determine the shape of the model's curve.

In this article, by optimizing the initial condition, interpolated coefficient in the background value and power exponent of the NGBM model, we presented an improved NGBM model termed ONGBM and solved the parameters based on GA. To verify the proposed model's validity and reliability for epidemic forecasting, the ONGBM was used to forecast the incidence of HB in Xinjiang, with nonlinear small sample characteristic. Numerical results illustrate that the optimized model can enhance the prediction accuracy. Moreover, the proposed optimized method is effective for use in the epidemic prediction and the optimum mechanisms indeed improve the grey model of prediction accuracy by using genetic algorithms approach.

ACKNOWLEDGMENTS

This research was supported by the Xinjiang Postgraduate Scientific Research and Innovation Pprojects [XJGRI2014101] and the National Natural Science Foundation of China [11401512, 11461073, 11401512, 81260410].

REFERENCES

Andrews, J. R., & Basu, S. 2011. Transmission dynamics and control of cholera in Haiti: an epidemic model. *The Lancet* 377(9773):1248–1255.

Bai, Y., & Jin, Z. 2005. Prediction of SARS epidemic by BP neural networks with online prediction strategy. *Chaos, Solitons & Fractals* 26(2):559–569.

Bhadra, A., Ionides, E. L., Laneri, K., Pascual, M., Bouma, M., & Dhiman, R. C. 2011. Malaria in Northwest India: Data analysis via partially observed stochastic differential equation models driven by Lévy noise. *Journal of the American Statistical Association* 106(494):440–451.

Chen, C. I., Chen, H. L., & Chen, S. P. 2008. Forecasting of foreign exchange rates of Taiwans major trading partners by novel nonlinear Grey Bernoulli model NGBM(1,1). *Communications in Nonlinear Science and Numerical Simulation* 13(6):1194–1204.

Chen, C. I., Hsin, P. H., & Wu, C. S. 2010. Forecasting Taiwans major stock indices by the Nash nonlinear grey Bernoulli model. *Expert Systems with Applications* 37(12): 7557–7562.

Chen, Y., Wu, A., Fan, H., & Wang, C. 2012. Forecasting hepatitis epidemic situation by applying the time series model. *International Journal of Simulation and Process Modelling* 7(1): 42–49.

Clements, C. J., Baoping, Y., Crouch, A., Hipgrave, D., Mansoor, O., Nelson, C. B., Treleaven, S., van Konkelenberg, R., & Wiersma, S. 2006. Progress in the control of hepatitis B infection in the Western Pacific Region. *Vaccine* 24(12): 1975–1982.

Cui, J., Liu, S. F., Zeng, B., & Xie, N. M. 2013. A novel grey forecasting model and its optimization. *Applied Mathematical Modelling* 37(6): 4399–4406.

Dang, Y. G., Liu, S. F., & Liu, B. 2005. The GM models that x(1)(n) be taken as initial value. *Chinese Journal of Management Science* 13(1): 132–135.

Deng, J. 2002. Grey prediction and grey decision. *Press of Huazhong University of Science & Technology, Wuhan,* 100–190.

Deng, J. L. 1982. Control problems of grey systems. *Systems & Control Letters* 1(5): 288–294.

Eisen, L., & Eisen, R. J. 2011. Using geographic information systems and decision support systems for the prediction, prevention, and control of vector-borne diseases. *Annual review of entomology* 56: 41–61.

Ellner, S., Bailey, B., Bobashev, G., Gallant, A., Grenfell, B., & Nychka, D. 1998. Noise and nonlinearity in measles epidemics: combining mechanistic and statistical approaches to population modeling. *The American Naturalist* 151(5): 425–440.

Glodberg, D. E., & Holland, J. H. 1989. Genetic algorithms and machine learning. *Machine learning* 3(2): 95–99.

Goodreau, S. M., Cassels, S., Kasprzyk, D., Montaño, D. E., Greek, A., & Morris, M. 2012. Concurrent partnerships, acute infection and HIV epidemic dynamics among young adults in Zimbabwe. *AIDS and Behavior* 16(2): 312–322.

Guan, P., Huang, D. S., & Zhou, B. S. 2004. Forecasting model for the incidence of hepatitis A based on artificial neural network. *World J Gastroentero* 10(24): 3579–3582.

He, J., Gu, D., Wu, X., Reynolds, K., Duan, X., Yao, C., Wang, J., Chen, C. S., Chen, J., Wildman, R. P. et al. 2005. Major causes of death among men and women in China. *New England Journal of Medicine* 353(11): 1124–1134.

Holland, J. H. 1970. Robust algorithms for adaptation set in a general formal framework. *In Adaptive Processes (9th) Decision and Control, 1970. 1970 IEEE Symposium on* (Vol. 9, pp. 175–175). *IEEE.*

Hou, E. S., Ansari, N., & Ren, H. 1994. A genetic algorithm for multiprocessor scheduling. *Parallel and Distributed Systems, IEEE Transactions on* 5(2): 113–120.

Hsieh, W. Y., & Tsaur, R. C. 2013. Epidemic forecasting with a new fuzzy regression equation. *Quality & Quantity* 47(6): 3411–3422.

Hsu, L. C. 2009. Forecasting the output of integrated circuit industry using genetic algorithm based multivariable grey optimization models. *Expert systems with applications* 37(6): 7898–7903.

Hsu, L. C. 2010. A genetic algorithm based nonlinear grey Bernoulli model for output forecasting in integrated circuit industry. *Expert Systems with Applications* 37(6): 4318–4323.

Hsu, L. C. 2011. Using improved grey forecasting models to forecast the output of opto-electronics industry. *Expert Systems with Applications* 38(11): 13879–13885.

Huarng, K. H., Yu, T. H. K., Moutinho, L., & Wang, Y. C. 2012. Forecasting tourism demand by fuzzy time series models. International Journal of Culture, *Tourism and Hospitality Research* 6(4): 377–388.

Jones, R. A., Salam, M. U., Maling, T. J., Diggle, A. J., & Thackray, D. J. 2010. Principles of predicting plant virus disease epidemics. *Annual review of phytopathology* 48: 179–203.

Lei, Z., Yafei, Y., Zhihong, L., & Weiguo, X. 2010. Application of Grey Artificial Neural Network Analysis on Forecasting Epidemic of the Rice Blast. *Chinese Agricultural Science Bulletin* 12: 053.

Liang, X., Bi, S., Yang, W., Wang, L., Cui, G., Cui, F., Zhang, Y., Liu, J., Gong, X., Chen, Y. et al. 2009. Epidemiological serosurvey of hepatitis B in Chinadeclining HBV prevalence due to hepatitis B vaccination. *Vaccine* 27(47): 6550–6557.

Liu, Q., Liu, X., Jiang, B., & Yang, W. 2011. Forecasting incidence of hemorrhagic fever with renal syndrome in China using ARIMA model. *BMC infectious diseases* 11(1): 218.

Liu, S. F. 1997. The trap in the prediction of a shock disturbed system and the buffer operator. *Journal-Huazhong University of Science and Technology Chinese Edition* 25: 25–27.

Lok, A. S., & McMahon, B. J. 2007. Chronic hepatitis B. *Hepatology,* 45(2): 507–539.

Nykiforuk, C. I., & Flaman, L. M. 2011. Geographic information systems (GIS) for health promotion and public health: a review. *Health Promotion Practice* 12(1): 63–73.

Pan, Z. 2002. Genetic algorithm for routing and wavelength assignment problem in all-optical networks. *Depatrment of Electrical and Computer Engineering, University of California, Davis, Tech. Rep.*

Shen, X., Ou, L., Chen, X., Zhang, X., & Tan, X. 2013. The Application of the Grey Disaster Model to Forecast Epidemic Peaks of Typhoid and Paratyphoid Fever in China. *PloS one* 8(4): e60601.

Wang, Y., Dang, Y., Li, Y., & Liu, S. 2010. An approach to increase prediction precision of GM(1,1) model based on optimization of the initial condition. *Expert Systems with Applications* 37(8): 5640–5644.

Wang, Z. X. 2013. An optimized Nash nonlinear grey Bernoulli model for forecasting the main economic indices of high technology enterprises in china. *Computers & Industrial Engineering* 64(3): 780–787.

Wei, L., Qian, Q., Wang, Z. Q., Glass, G. E., Song, S. X., Zhang, W. Y., Li, X. J., Yang, H., Wang, X. J., Fang, L. Q. et al. 2011. Using geographic information system-based ecologic niche models to forecast the risk of hantavirus infection in Shandong Province, China. *The American journal of tropical medicine and hygiene* 84(3): 497–503.

Xu, F., & Zhang, S. 2012. Measles Trends Dynamic Forecasting Model Based on Grey System Theory. *Management Science and Engineering* 6: 71–74.

Zhang, X., Liu, Y., Yang, M., Zhang, T., Young, A. A., & Li, X. 2013. Comparative Study of Four Time Series Methods in Forecasting Typhoid Fever Incidence in China. *PloS one* 8(5), e63116.

225

Computational Intelligence in Industrial Application – Ling (ed.)
© 2015 Taylor & Francis Group, London, ISBN: 978-1-138-02818-0

Study on communication interface solutions between FPGA and DSP

Bo Song & YongShan Liu
School of Aerospace Engineering, Beijing Institute of Technology, Beijing, China

Lei Feng & ZhaoYu Zhang
China North Industries Corporation, Beijing, China

ABSTRACT: Aiming at the communication problem between DSP and FPGA in an embedded system, this paper made an in-depth analysis from the circuit, communication rate, logic control, resource occupation, hardware cost, applicable situation and other aspects. By combining practical application, it summarized the application characteristics of each program, and gave the recommendation on the system design scheme selection, thus providing a reference for future embedded system design.

KEYWORDS: Data-transmission, DSP, FPGA.

1 INTRODUCTION

Along with the Digital Signal Processor (DSP) and Field Programmable Gate Array (FPGA) development, the use of DSP+FPGA embedded hardware system is showing its superiority. Data communication performance between DSP and FPGA processor is one of the major problems affecting the whole performance of embedded systems, thus communication between FPGA and DSP has become the key of the whole system design. At present, there are a variety of schemes for implementation of communication between FPGA and the DSP, including external memory interface such as EMIF scheme, host interface such as HPI scheme, serial type interface such as McBSP scheme, and special interface such as RapidIO scheme. This paper will focus on the above common implementation scheme, take TI company's DSP processor and Xilinx company's FPGA processor as an example, make an analysis from the circuit, the communication rate, logic control, resource occupation, hardware cost, applicable situation and other aspects, gave the recommendation on the system design scheme selection.

2 COMMUNICATION DESIGN OF HPI INTERFACE BASED ON DSP

2.1 HPI introduction

HPI (Host Port Interface) is the Interface of main equipment or main processor with DSP, in TMS320C67xx series, Host Interface is parallel Port, can realize the parallel high-speed data transmission, the Host can directly access to all the DSP storage space and its peripherals on chip storage mapping.

2.2 The HPI interface and working mechanism

Based on HPI interface, DSP and FPGA communication scheme and interfaces are defined as shown in Fig. 1, Table 1.

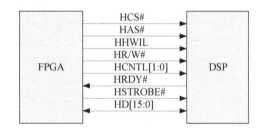

Figure 1. DSP and FPGA communication scheme based on HPI interface.

Table 1. HPI interface signal.

Pin	Function
HD[15:0]	Data bus
HCNTL[1:0]	Register Controller
HHWIL	Host half-word select
HAS	Address latch
HR/W	Read/Write
HCS	Host chip select
HRDY	Host ready
HSTROBE	Host select

The host (FPGA) access to the HPI is divided into the following 3 steps:

1 Initialize HPI control registers (HPIC);
2 Initialize HPI address register (HPIA);
3 Write data to HPI data register (HPID) or read data from the HPID register;

The host in the communication process with the HPI through HCNTL (1-0) to select which register for operation. For 16 bit HPI, any HPI register host access requires the HPI bus by two half words; HHWIL low represent first half word, and second half word to a high level.

HPI transmission of data can be read operation without address increment, read operation with address auto increment, write operation without address increment and write operation with address auto increment, mode selection be controlled by HCNTL[1:0] signal. Take the read and write operation without the address increment as an example, analysis the data transmission efficiency of FPGA and DSP through the HPI interface, read sequence as shown in Fig. 2.

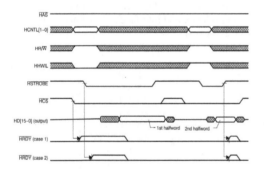

Figure 2.   Read sequence of HPI.

If the clock cycle for T, read and write a 32bit data needs time for:

$$T_B = T_1 + T_2 + T_3 + T_4 + T_5 \qquad (1)$$

where T_1 for period number when the HSTROBE# signal keep high before access the first half-word, generally 1.5T;

T_2 for period number as HPI in a busy state, when reading the on-chip peripherals, this parameter can be 15ns;

T_3 for period number, when the HRDY# signal effective until HSTROBE# rise, usually 3T; T_4 for period number when the HSTROBE# signal keep high before access the second half-word, generally 1.5T;

T_5 for time of data duration when transfer the second half-word , usually 1T.

According to the above data, assuming that the clock frequency is 100 MHz, Then, the transmission rate is V:

$$V = \frac{D}{T_B} \qquad (2)$$

where D is the number of byte in whole data transmission; T_B is FPGA read and write a word (32 bit) from the DSP requires time; so the transmission speed V is 47.1 MB/s.

3   COMMUNICATION DESIGN BASED ON THE SRIO INTERFACE

As an open standard interconnection technology with the characteristics of new type high performance, low pin count, high stability, RapidIO can widely meet the demand of the application of embedded system. RapidIO supports series and parallel interconnection form, with same programming mode, transaction and addressing mechanism, are respectively suitable for different applications. Parallel interface is suitable for high-performance, high-bandwidth, transmission distance is short; serial interface for serial backplane, DSP and serial control plane. As Parallel RapidIO with more signal lines, it is difficult to be widely used, while the 1×/4× serial RapidIO only have 4 or 16 signal lines, gradually prevalent, so this paper only introduces serial RapidIO.

SRIO (Serial Rapid I/O) based on SerDes (Serialize Deserialize) technology which has been widely used in backplane interconnection. The standard 1×/4× SRIO interfaces support four ports, with a total of 16 signals lines. SRIO take advantage of 8B/10B coding strategy embedding clock signal in the data signal, to support the three service frequency: 1.25 GHz, 2.5 GHz and 3.125 GHz.

3.1   *The working principle of SRIO*

The RapidIO package is basic communication unit between system endpoint devices; each operation is based on the request and response pockets to complete the transaction. Initiating device or the master device generates a request transaction; this transaction is sent to the target device. Then the target device generates a response transaction back to initiate device to complete the operation. The transmission operation process of SRIO shown in Fig. 3.

3.2   *SRIO interface and implementation*

DSP and FPGA with a 4×SRIO interface to realize the interconnection scheme as shown in Fig. 4.

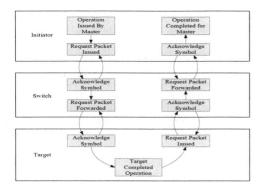

Figure 3. SRIO's transmission operation process.

Figure 4. DSP and FPGA communication scheme based on SRIO.

C6455 is TI company's high-performance fixed-point DSP, based on the 3rd generation, frequency up to 1.2 GHz, integrated four standards 1×/4× SRIO interface. These four interfaces can be used as four independent 1×interfaces, transfer different data; also be combined together as a 4× interface in order to improve the throughput of a single interface.

Xilinx's SRIO IP core can easily achieve SRIO interface protocol on FPGA devices, and Virtex-6 series devices support SRIO IP core can achieve single-channel data transmission up to 6.25 Gbps, supports 1×, 2× and 4× interface mode.

SRIO's transmit and receive ports all need to be initialized, including enabling port, configure the port operation modes, settings and enable PLL module, set the device ID and data transfer rate. When sender end DSP though write operation sends a frame data, then sends a doorbell data packet, doorbell data packet at the receiving end FPGA generates an system interrupt to inform the data arrives effective, FPGA then sent doorbell packet back to the sender DSP, alike generates an interrupt to the sender DSP, the sender DSP receives interrupt and then continue sending the next data frame, and so move in circles to achieve high-speed data transmission.

As SRIO data packets using 8B/10B code, theoretically at 3.125 GHz work rate, the effective data transmission rate of X4 mode of SRIO is:

$$V = 3.125GHz \times 4 \times 8 / 10 = 1250MB / s.$$

4 COMMUNICATION DESIGN OF EMIF INTERFACE BASED ON DSP

EMIF is the external memory expansion interface, it is integrated of DSP processor, supports several of seamless connection with external parallel interface device. EMIF is connected with FPGA; let the FPGA acts as a coprocessor system to realize high speed data transmission system. EMIF bus width of C67xx series DSP processor is 32 bits, with a total of 4 storage space, addressing capability is 512M bytes, with independent synchronous clock, support 8/16/32 data width, support any type of storage in the CE1 space segment, control signal can be selected, also support ROM/ FLASH, asynchronous memory I/O and synchronous SDRAM.

4.1 The EMIF interface signal

The EMIF interface signal line have five types, including enhanced data memory and controller signal, interface external control signals, asynchronous control signal, response signal and the internal interconnect bus. Among them, the interface external control signal has clock input signal (ECLKIN), clock output signal (ECLKOUT), data bus(ED), the address bus (EA), chip enable (CE#) and byte enable (BE#); asynchronous control signal is mainly output enable (AOE#), the read signal (ARE#) the write signals (AWE#), and the ready signal; the response signal is mainly HOLD# and HOLDA# and the bus request signal BUSREQ.

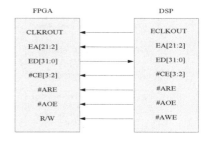

Figure 5. DSP communicate connection with FPGA though EMIF interface.

4.2 The EMIF communication mode access FPGA register directly

In the communication system of DSP and FPGA, there are four CE spaces inside the DSP that can be used in FPGA, FPGA adopts the register access mode in the process of cooperating with DSP, mapping the register settings to a CE space of DSP, DSP though access this CE space complete reading and writing FPGA data, FPGA as coprocessor complete the cooperative work with the main processor DSP by simple EMIF sequential logic control.

4.3 EMIF communication mode though the dual port RAM and FPGA

During data sharing communication when the DSP and FPGA are all the host, by means of the dual port RAM as a data relay station to achieve real-time efficient transmission between two pieces of asynchronous chip. Dual port RAM with the characteristics of high communication speed, strong real-time, simple interface, both sides of CPU can be active for data transmission. Dual port RAM can be an external independent chip and integrated in the FPGA, and the plug-in chip performance is similar to follow 4, this chapter mainly studies dual port RAM communication embedded in FPGA .

In general, one end of the dual port RAM connected EMIF, another end in the internal FPGA and connect to the FPGA logic control module, to avoid competition phenomenon when FPGA and DSP simultaneously write with the same internal storage unit of dual port RAM, usually using two pieces of dual port RAM, of which write the first piece, read the first one after fully write the second block, alternate implementation realize Ping Pong mechanism, improve the data transmission rate. The main function of FPGA in this architecture is to provide sequential logic control. Specific process is shown in Fig. 6.

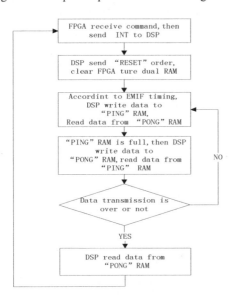

Figure 6.   EMIF read and write dual port RAM.

Each read and write cycle of EMIF asynchronous interface consist of 3 stages: establish, trigger and maintain, setup time starts from the register access

cycle (optional, address effective) to read and write trigger effectively, triggering time is read and write register from effectiveness to invalid, hold time begin with read and write signal invalid to the end of access period. Read and write operations can independently set up above 3 stages in the CEXCTL register, the minimum time of establishment and holding can be set 1, the minimum time of holding can be set 0, after read and write the last data, CE immediately high after the hold time.

Transmission rate: for the C67 series chip, the fastest time to read and write clock of establish, trigger and maintain is 2 clock cycles, if maximum of the clock is 100 MHz, and transfer 32 bits data, the transmission rate:

$$V = \frac{D}{T_B} = \frac{32/8}{20ns} = 200MB/\text{s}.$$

5   SHARED SDRAM ACHIEVE THE COMMUNICATION OF DSP AND FPGA

5.1 SDRAM introduction

SDRAM (Synchronous Dynamic Random Access Memory) is used to meet the needs of the cache and the large amount of data communication, according to the actual need to select or configuration the storage depth; can be external independent chip and integrated in the FPGA, this chapter mainly studies the way of external connection with independent chip.

Taking the model of MT48LC4M32B2 SDRAM as an example, this paper introduces SDRAM application in the interconnection between DSP and FPGA. MT48LC4M32B2 adopts synchronous interface mode, all the signals are recorded on the rising edge of system clock; with the internal pipeline operation; programmable burst length of 1, 2, 4, 8, or full page.

5.2 SDRAM interface signal and working principle

Communication scheme of DSP and FPGA based on the SDRAM is shown in Fig. 7.

The basic signal of SDRAM can be divided into the following categories:

1 Control signal: including chip select (CEX), synchronous clock (CLK), read and write select (#WE), data valid signal (DQM3-DQM0);
2 The address selection signal: including the row address select (#RAS), the column address select (#CAS), row/column address lines (A11-A0) multiplexing at different time, bank block address lines (BA1-BA0);
3 The data signal: bidirectional data port (DQ31-DQ0);

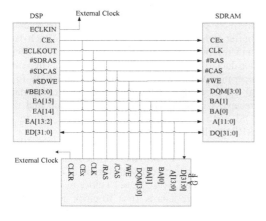

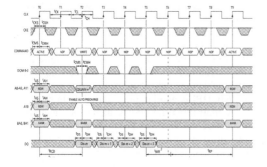

Figure 7. Connection diagram based on the SDRAM communication scheme.

DSP chip's access to external chip memory is through the EMIF, the C6713 provides 4 independent external memory interface (CEx), all the CEx space support directly interface to SDRAM except CE1, and there exist specialized SDRAM control register (SDTCL) and sequential control register (SDTIM) used to control.

DSP through the EMIF to initialize the SDRAM, write data to SDRAM, then sent the write end command to FPGA; later FPGA can read data from SDRAM.

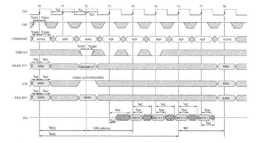

Figure 8. SDRAM write time of 4 words burst.

It is used to be 4 words burst (auto pre-charge) mode when DSP access to SDRAM, CAS delay select 1 clock cycle is 10ns, and data size is 16 byte. The basic write operation of SDRAM is needed to match the control line and address line, then sent a series of commands to complete: first DSP issue a BANK activation command, and latches the corresponding BANK address (BA0, BA1 row address (A0-A11) and given). After BANK activate command, it is must wait for more than T_{RCD} time (delay index of RAS to CAS), sent the write command word; write commands can be written immediately, during the T_{DATA} time (write time of 4 words burst), write data sequentially to DQ

(data bus). Delayed T_{WR} after the last data is written (the time of the last data output to the next command), issued a pre-charge command, close the activated page. Waiting for T_{RP}(after PRECHARGE command, time apart, can once again access the row), it is able to start the next operation, and then DSP sent FPGA write end signals. If the working clock is 100 MHz, then the time of DSP write 16byte to SDRAM is:

$$T_W = T_{RCD} + T_{DATA} + T_{WR} + T_{RP} = 30 + 30 + 10 + 20 = 90ns$$

Figure 9. Read sequential of SDRAM.

FPGA's basic read operation on SDRAM is needed to match the control line and address line, then issuing a series of commands to complete, setting FPGA recognize DSP write end signal during T_{CON}time, FPGA read the data in the SDRAM, the operation steps similar to the writing process. The total time consumption from the activation to preprocessor is T_{RAS}, at the end of the read operation, send the PRECHARGE command to SDRAM, to close the activated page. After Waiting for T_{RP}, it can start the next read operation, and then sends a read end signal to the DSP. The time of FPAG read DRAM is:

$$T_R = T_{CON} + T_{RAS} + T_{RP} = 10 + 70 + 20 = 100ns$$

It can be drawn from the above calculation, the transmission rate between DSP and FPGA is:

$$V = \frac{D}{T_B} = \frac{D}{T_R + T_W} = \frac{16}{100 + 90} = 84.2MB / s.$$

6 COMMUNICATION DESIGN OF MCBSP INTERFACE BASED ON DSP

McBSP (Multichannel Buffered Serial Port) is provided by TI Company's DSP processor chip, which is generally used to connect the serial interface and implement communication with other DSP or FPGA. McBSP has a double buffered data register, it provides the bit synchronous clock which sending and receiving independently, with multi-channel receiving and serial word length is optional; the generation of the internal clock and frame synchronization pulse can also be programmed, with great flexibility.

Hardware connection of DSP and FPGA communication which is based on the McBSP interface as shown in Fig. 10:

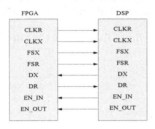

Figure 10. Hardware connection graph based on the McBSP interface.

McBSP is composed of external equipment's data channel and control channel. Data via data sending pin(DX) and data receiving pin (DR)communicating with a device which connected to the McBSP, control information (timing and frame synchronization) realize the pin communication by sending/receiving clock (CLKX/CLKR) and frame sync signal (FSX/FSR).

When the DSP and FPGA build communication, CLKR, CLKX, FSR, FSX signals that needed for McBSP are all generated by the FPGA. The EN_IN, EN_OUT on the Diagram are enable signal when DSP controlling the McBSP interface of FPGA. When EN_IN is high, the FPGA receiving the DSP data; when EN_OUT is high, FPGA began to send DSP data.

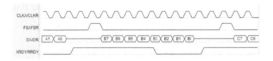

Figure 11. Read/Write sequence of McBSP.

When FPGA detected FSX signal is effective, after a certain delay data (set in the data time delay control register XDATDLY), the data which be sent to the transmission shift register (XSR) began to shift output to the DX pins in turn. At the end of each data unit reception and the rising edge of CLKX clock, if the data send register (DXR) was ready for new data, new data in DXR will be automatically copied to XSR. Copy operations of DXR to XSR will active the sending status bit (XRDY) in the next falling edge of CLKX, which means it can write new data to DXR, after FPGA write data, XRD will be invalid automatically.

Once DSP detecting the FSR signal is effective, the effective state will be detected in the drop edge of the first receive clock CLKR, and then after a certain data delay, the data on DR pin will shift into the receive shift register (RSR) in turn. If the receiver buffer register (RBR) is empty, at the end of each data unit reception and the rising edge of CLKX clock, the data in RSR will be copied to the RBR. This copy operation will trigger receiving state bit RRDY to 1 at the drop edge of the next clock, indicating data receive register (DDR) has been prepared and the DSP controller can read data. When the data is read, RRDY will be invalid automatically.

Assuming that McBSP is set to single frame mode, each frames for 1 word, the word length of data unit is 8 bits, 1 bits data delay, then time of FPAG transmit data to the DSP is Tx = 10 clock cycle, the time of DSP receive data is Tr = 10 clock cycles. If the working clock is 100 MHz, transfer 8byte data, the transmission speed is:

$$V = \frac{D}{T_B} = \frac{D}{T_R + T_W} = \frac{8}{20ns} = 40MB \, / \, \text{s}.$$

7 A BRIEF COMPARISON OF THE CHARACTERISTICS OF COMMUNICATION SCHEMES

Different applications require different processing capacity, the requirement of the data traffic and delay between DSP and FPGA is also different, so it is needs different communication interface way. The McBSP interface is simple, less PCB signal line, easy wiring, while implementing the communication between DSP and FPGA, the required sending/receiving signals are generated by the FPGA, with simple logic, easy to realize, suitable for low-speed embedded system which the request to the transmission speed is not high. HPI is good at the external main processor FPGA controlling the DSP, suitable for client/server architecture system. EMIF data transmission rate is high, suitable for high speed real-time data acquisition system. SDRAM's read and write operation with strict timing requirements, and when FPGA is controlling SDRAM, it is necessary to handle DSP's request operation to bus, the programming requirements to bus arbitration are higher, and a large number of signal lines increased PCB wiring difficulties. The storage depth of external storage chip compared to the embedded dual port RAM, it can select or configure according to the actual requirement; for FPGA, the higher demands to embedded dual port RAM are more expensive; while for PCB layout, the more complex is more expensive. SRIO with high speed transmission, less signal lines, high reliability, while as it is the forefront technology, research and development cost is high, time-consuming, SRIO's switching chip manufacturer is little, the price of the terminal products that based on this technology is high.

Table 2. A brief comparison of the characteristics of the 5 communication schemes.

Solution	Transmission rate (MByte/s)	Communication mode	Resource consumption	Hardware complexity	Logic control	Economic	Application
McBSP	40	Serial communication	DSP specific control module, FPGA general IO interface	Simple	Simple	Cheap	Small amounts of data serial communication
SDRAM	84.2	Data Sharing	DSP and FPGA general IO interface	Complex	Complex	Expensive	High data rate and depth of storage sharing occasions
EMIF	200	Host DSP, parallel communication	DSP specific control module, FPGA general IO interface	Complex	Medium	Medium	Master-slave parallel architecture, DSP host, high data rate applications
HPI	47.1	Host FPGA, parallel communication	DSP specific control module, FPGA general IO interface	Medium	Sophisticated	Medium	Master-slave parallel architecture, DSP slave, Medium rate
SRIO	1250	Serial Communication	DSP and FPGA specific control module	Simple	Complex	Exorbitant	Ultra-high-speed serial real-time communication

8 CONCLUSION

This paper aiming at communication problem between DSP and FPGA in embedded system, 5 kinds of present common schemes are introduced, aiming at the circuit, communication rate, logic control, resource occupation, hardware cost, applicable situation and other aspects make an in-depth analysis, By combining practical application summarized the application characteristics of each program, and gave the recommendation on the system design scheme selection, thus providing a reference for future embedded system design.

REFERENCES

Zhu Shanying. Study of the control system of linear motor pumping based on DSP and ARM. *Huazhong University of Science and Technology*, 2007, 27–28.

Texas Instruments Incorporated. 2003. TMS320C6000 DSP Host Port Interface (HPI) Reference Guide: 16–21.

Sun Xiujuan. 2004. Design of airborne information processing system with double CPU based on HPI interconnection. Systems Engineering and Electronics.

Zhao Longbo. 2009. RapidIO interconnection technology research and model validation. Aeronautical Computer Technology.

Huang Kewu. 2008. Design of high-speed SRIO interface based TMS320C6455. Electronic Measurement Technology.

Texas Instruments Incorporated. 2003. TMS320C6000 DSP External Memory Interface (EMIF) Reference Guide: 74–81.

Cypress Semiconductor Corporation. 2005. CY7C026A CY7C036A 16K×16/18 Dual-Port Static RAM data sheet. September 6: 5–7.

Micron Technology Inc. 2002. SYNCHRONOUS DRAM MT48LC4M32B2 data sheet: 7–9.

Wei Jinchen & Li Gang & Wang Chen-ye. 2012. TMS320C6000 DSP System Design Principles and Applications, 116–128.

Computational Intelligence in Industrial Application – Ling (ed.)
© *2015 Taylor & Francis Group, London, ISBN: 978-1-138-02818-0*

Three novel chaotic polynomials for image encryption using three different MOD operators

H.B. Kekre & P. Halarnkar
MPSTME, NMIMS University, Mumbai, India

Tanuja Sarode
TSEC, Mumbai University, Mumbai, India

ABSTRACT: Nowadays data storage is not limited to standalone machines but it has to be transmitted across the network. Due to advancement in technology, novel techniques need to be devised to protect data. Data includes multimedia information like images. To protect images one of the methods used is image encryption. In this paper three novel chaotic polynomials are proposed along with three different variations of MOD operators. Experimental results prove that the method is suitable for encryption.

1 INTRODUCTION

Data security is of high concern due to increasing thefts and misuse of digital data. Authentication is one aspect of providing security with the help of which one can confirm its ownership. Another very important aspect is making the image data unreadable to the users, which can be achieved by image encryption. Many traditional methods and approaches are available to secure digital images, a discussion of few of them are presented here.

A novel approach for image encryption is proposed by Abd El-Latif et al (2013). The proposed system makes use of quantum chaotic system and color spaces. YCbCr color space is used and only Y component is scrambled which is a low frequency subband. Experimental results prove the performance of the method to be good.

Bibhudendra Acharya et al. (2011) proposed image encryption using index based chaotic sequence, M-sequence and Gold sequence. The proposed method was tested on grayscale images. In chaotic based sequence, two 1D maps are used logistic and tent map. Execution time and Histogram are used as the measures for experimental results.

Quantum logistic map is used for image encryption, the method is proposed by Akhshani et al. (2012). The proposed approach makes use of a quantum Logistic map. This sequence is XORed with the image pixel values to obtain the encrypted image.

Armand et al. (2014) proposed an image encryption technique based on fast chaotic block cipher. This paper uses solutions of the Linear Diophantine Equation (LDE) whose coefficients are integers.

These coefficients can be generated from any chaotic system. This results in low computational complexity and gives high security.

Azoug et al. (2013) proposed double image encryption scheme using chaotic maps and Reciprocal Orthogonal Parametric transform. The encryption keys used are the parameters of ROP transform and the initialization of chaotic sequences. Experimental analysis is carried out by brute force attack, blind decryption.

Behnia et al. (2013) presented a novel image encryption based on Jacobian elliptic maps. In the method initial condition, control parameters are considered as keys. Jacobian Elliptic map is iterated 1000 times and the sequence is XORed with the image pixel values to encrypt them. The experimental results show that the method is practical and can be used for video encryption also.

A new hyperchaotic map is proposed by Boriga et al. (2014) and its use for image encryption is proved. The hyperhcaotic map finds its base from the parametric equations of the serpentine curve. Statistical analysis proves the method is efficient.

An image encryption technique based on cellular automata and SCAN is proposed by Chen et al. (2005). Based on the pattern generated by the SCAN technology the pixels of the image are permuted. Then these pixels are replaced by cellular automata substitution.

A batch image encryption technique combining vector quantization and index compression is proposed by Chen et al. (2010). The proposed technique makes use of vector quantization for the same. An additional index compression is also been done.

Experimental results show the performance of compression rate and computational cost.

An analysis and improvement of Hash based image encryption scheme is proposed by Deng et al. (2011). The existing Hash based technique's diffusion process was too weak to resist the chosen plain text attack. Experimental parameters like NPCR, UACI give good results which prove the proposed scheme is suitable for image encryption.

Dong (2014) proposed an image encryption technique which uses one time keys and coupled chaotic systems. Experimental results show that the proposed method is reliable and can be used for image encryption and secured communications.

Ginting et al. (2013) proposed a color image encryption scheme based on RC4 stream cipher and logistic map. Experimental results show that the image is visually not identifiable, correlation in ciphered image is eliminated, it is a lossless image encryption scheme which is verified by hash value of the original image and decrypted image.

Kadir et al. (2014) proposed color image encryption using skew tent map and hyper chaotic system of 6th order. The method is resistant against common attacks.

Kanso et al. (2012) proposed an image encryption scheme based on 3D chaotic map. The scheme has three phases, in the first phase the pixels are shuffled using a search rule of 3D chaotic map, in the second and third phase the chaotic sequence is used to scramble the shuffled pixels through mixing and masking process.

A chaos based keyed hash function is proposed by Kanso et al. (2013). The proposed method makes use of single, 4D chaotic cat map. The method is fast and efficient; it takes an arbitrary length of input but gives a fixed length output (n).

A fast image encryption technique based on chaotic maps is proposed by Kwok (2007). Chaotic sequence is used as a stream cipher based on skewed tent map; this sequence undergoes a mixer operation which is done by m-dimensional cat map. The random sequence generated is also tested under NIST which is a statistical test suite.

An image encryption scheme based on AES (Advanced Encryption Standard) whose keys are generated using 2D Henon map and 2D Cheyshev map is proposed by Li et al. (2013). Experimental results prove that the proposed method is secured against cryptanalysis.

Lima et al. (2013) proposed an image encryption scheme based on finite field cosine transform. The proposed scheme has the advantage of low computational complexity.

A key stream generated by chaos and plain images is used for image encryption process proposed by Liu et al. (2013). The method is a triple color image encryption scheme, in which three color images are taken, they are firstly converted to grayscale images, and these grayscale images are combined to form a color image. Experimental results show that the proposed method is suitable for encrypting images in batches.

A family of new chaotic system is proposed by Liu et al. (2013), which is based on complex number. Two maps of the family are selected for the purpose of image encryption. Two entropy coding methods are employed; Run length coding and Huffman coding are used to reduce the correlation between the signals. NIST tests were used to check the randomness.

An image encryption scheme using Arnold transform and discrete fractional angular transform is proposed by Liu et al. (2012). The discrete fractional angular transform is applied over the changed complex function. This process is repeated many number of times. The amplitude is regarded as the encrypted image and its phase as the key.

Liu et al. (2013) proposed an image encryption technique based on fractional Fourier transform and pixel scrambling operation based on double random phase encoding. Numerical simulations prove the validity of the proposed method.

An image encryption scheme is proposed which is a combination of wavelet and time domain is used by Luo et al. (2014). Experimental results shows that the proposed technique is secured, robust and efficient.

A fast color image encryption technique is proposed by Mohammad et al. (2012). The proposed algorithm is based on coupled two dimensional piecewise chaotic map (CTPNCM). Apart from its sensitivity to initial condition and other parameters of chaotic maps, the method also provides a fast and highly secured way for image encryption.

A gray image encryption scheme is proposed by Pareek et al. (2013). The visual degradation of the gray image is achieved by mixing the pixel values. The resulting image is then passed through a diffusion process which makes use of the 128 bit key.

2 POLYNOMIAL MAP

In this paper three novel chaotic polynomial maps are introduced. The equations followed by the map and bifurcation diagram are given below.

2.1 Polynomial 1

$$x = 2 * \left(x - x^2 \right) * a \tag{1}$$

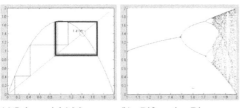

(a) Polynomial 1 Map   (b) Bifurcation Diagram

Figure 1.   Polynomial 1.

2.2 *Polynomial 2*

$$x = 2 * x - \left(0.95 * x^2\right) - \left(0.05 * x^3\right) * a \qquad (2)$$

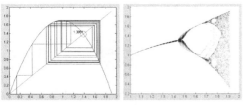

(a) Polynomial 2 Map   (b) Bifurcation Diagram

Figure 2.   Polynomial 2.

2.3 *Polynomial 3*

$$x = 2 * x - \left(0.95 * x^2\right) - \left(0.05 * x^4\right) * a \qquad (3)$$

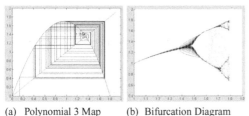

(a) Polynomial 3 Map   (b) Bifurcation Diagram

Figure 3.   Polynomial 3.

3 PROPOSED APPROACH

In this paper, three new chaotic polynomials are pro-posed. All the three polynomials have chaotic nature, which can be seen from the experimental results above. All the three were thoroughly tested and then used for image encryption.

The steps followed for image encryption

1   Read 24 bit color image
2   Separate the R G and B components
3   Generate three different chaotic sequences using either Polynomial 1, Polynomial 2 or Polynomial 3.
4   Use either MOD 2 bit operator, MOD 4 bit Operator or MOD 8 bit operator, apply this oper-ator separately on R, G and B component and the chaotic sequences.
5   The resulted image is encrypted

The decryption process involves generating the same chaotic sequences using the required Polynomial and applying the appropriate MOD operator on the encrypted values.

3.1 *MOD operator*

In this paper, a variation over the MOD operator is proposed. They are termed as MOD 2 bit operator, MOD 4 bit operator and MOD 8 bit operator.

Note: For chaotic sequence generated, three digits after the decimal points are been made use for the encryption purpose for e.g. 0.5678, then 567 is used for the same.

The details of how the MOD operator plays its role in encryption process are given below.

Let A be the original image pixel value e.g. 80, Let B be the chaotic sequence value for e.g. 0.4566, follow-ing is the procedure how the MOD operator is applied.

1   Convert A and B to binary form.
2   For MOD 2 or 4 or 8 bit operator, 2 or 4 or 8 bits of A are considered and converted to decimal, so are the 2 or 4or 8 bits of B are considered and con-verted to decimal
3   Apply the following on A1 and B1 obtained in step 3
    C= (A1+B1) MOD 2^(2 or 4 or 8)
4   The result C obtained in step no 3 is converted back to binary 2 or 4 or 8 bits and stored in tem-porary array
5   If the bits of A and B are not over go to step 2 else go to step 6
6   Convert the binary 8 bits stored in temporary array to decimal equivalent. This is the encrypted value

4 EXPERIMENTAL RESULTS

For experimental purpose, 24-bit color images of size 256×256 were used. The method was tested on a number of images out of which the results for 5 images is presented in the paper. The proposed approach applies three different polynomials, and every polynomial with three different variations of MOD operator. The results obtained are given below.

4.1 Polynomial 1 Map

4.1.1 MOD 2 bit

(a) Original image (b) Encrypted image (c) Decrypted image

(d) Histogram (e) Histogram (f) Histogram

Figure 4. Polynomial 1 MOD 2bit operator.

4.2 Polynomial 2 Map

4.2.1 MOD 4 bit

Figs. 4–6 show the original, encrypted and decrypted images for Polynomial 1 and MOD 2 bit,

(a) Original image (b) Encrypted image (c) Decrypted image

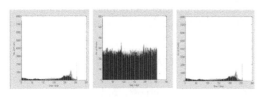

(d) Histogram (e) Histogram (f) Histogram

Figure 5. Polynomial 2 MOD 4bit operator.

Polynomial 2 & MOD 4 bit and Polynomial 3 and MOD 8 bit.

4.3 Polynomial 3 Map

4.3.1 4.3.1 MOD 8 bit

(a) Original image (b) Encrypted image (c) Decrypted image

(d) Histogram (e) Histogram (f) Histogram

Figure 6. Polynomial 3 MOD 4bit operator.

Table 1. Values of intersection point, Min and Max value for period doubling and range for Polynomial 1, Polynomial 2 and Polynomial 3.

Type of Map	Intersection point	Min Value (a)	Max value (a)	Range
Polynomial 1 Map	1.418	1.5	2	0.5
Polynomial 2 Map	1.385	1.45	1.9	0.45
Polynomial 3 Map	1.355	1.35	1.8	0.45

Table 1 shows the values obtained for all three Polynomials for period doubling.

Table 2. Shows the values used for encryption purpose for all three Polynomials in R, G, and B Planes.

4.4 Polynomial 1

Following are the experimental results obtained for Polynomial 1 for parameters Entropy, row and column correlation and PAFCPV

Table 2. Values used for encryption and decryption process in Polynomial 1, Polynomial 2 and Polynomial 3 in R, G and B planes.

Polynomial 1	Initial Condition	Factor a
R-Plane	0.545	1.851
G-Plane	0.645	1.951
B-Plane	0.145	2.000
Polynomial 2		
R-Plane	1.425	1.823
G-Plane	1.435	1.899
B-Plane	1.445	1.856
Polynomial 3		
R-Plane	1.425	1.723
G-Plane	1.426	1.699
B-Plane	1.545	1.710

Table 3. Value of entropy in original and encrypted images for Polynomial 1 using MOD 2 bit, MOD 4 bit and MOD 8 bit operator.

| Image name | Original image | Encrypted images | | |
		MOD 2 Operator	MOD 4 Operator	MOD 8 Operator
Lena	7.3433	7.9854	7.9819	7.9781
Tiger	7.3129	7.9566	7.9576	7.9556
Micky	4.7265	7.4702	7.4752	7.4697
Rainbow Birds	7.5662	7.9805	7.9799	7.9776
Fruits	7.2263	7.9901	7.9888	7.9886

Table 4. Value of row correlation in original and encrypted images for Polynomial 1 using MOD 2 bit, MOD 4 bit and MOD 8 bit operator.

| Image name | Original image | Encrypted images | | |
		MOD 2 Operator	MOD 4 Operator	MOD 8 Operator
Lena	0.8454	0.2476	0.2361	0.2326
Tiger	0.5264	0.2063	0.2011	0.1982
Micky	0.5164	0.2297	0.2168	0.2138
Rainbow Birds	0.5005	0.2014	0.1973	0.1964
Fruits	0.5452	0.2205	0.2110	0.2072

Table 5. Value of column correlation in original and encrypted images for Polynomial 1 using MOD 2 bit, MOD 4 bit and MOD 8 bit operator.

| Image name | Original image | Encrypted images | | |
		MOD 2 Operator	MOD 4 Operator	MOD 8 Operator
Lena	0.7005	0.2778	0.2645	0.2638
Tiger	0.3580	0.2060	0.2000	0.1977
Micky	0.5468	0.2470	0.2241	0.2169
Rainbow Birds	0.3085	0.2160	0.2102	0.2058
Fruits	0.5127	0.2459	0.2339	0.2315

Table 6. Value of PAFCPV in encrypted images for Polynomial 1 using MOD 2 bit, MOD 4 bit and MOD 8 bit operator.

| Image Name | Encrypted image | | |
	MOD 2 Operator	MOD 4 Operator	MOD 8 Operator
Lena	0.3135	0.3336	0.3365
Tiger	0.3721	0.3894	0.3941
Micky	0.4741	0.4869	0.4775
Rainbow birds	0.3588	0.3753	0.3794
Fruits	0.3711	0.3914	0.3958

4.5 Polynomial 2

Following are the experimental results obtained for Polynomial 2 for parameters entropy, row and column correlation and PAFCPV.

Table 7. Value of entropy in original and encrypted images for Polynomial 2 using MOD 2 bit, MOD 4 bit and MOD 8 bit operator.

| Image name | Original image | Encrypted images | | |
		MOD 2 Operator	MOD 4 Operator	MOD 8 Operator
Lena	7.3433	7.9912	7.9898	7.9894
Tiger	7.3129	7.9910	7.9910	7.9909
Micky	4.7265	7.9722	7.9697	7.9668
Rainbow birds	7.5662	7.9944	7.9946	7.9945
Fruits	7.2263	7.9901	7.9890	7.9893

Table 8. Value of row correlation in original and encrypted images for Polynomial 2 using MOD 2 bit , MOD 4 bit and MOD 8 bit operator.

Image name	Original image	Encrypted images		
		MOD 2 Operator	MOD 4 Operator	MOD 8 Operator
Lena	0.8454	0.1800	0.1811	0.1808
Tiger	0.5264	0.1814	0.1801	0.1814
Micky	0.5164	0.1815	0.1811	0.1810
Rainbow Birds	0.5005	0.1801	0.1824	0.1806
Fruits	0.5452	0.1810	0.1805	0.1803

Table 9. Value of column correlation in original and encrypted images for Polynomial 2 using MOD 2 bit , MOD 4 bit and MOD 8 bit operator.

Image name	Original image	Encrypted image		
		MOD 2 Operator	MOD 4 Operator	MOD 8 Operator
Lena	0.7005	0.1793	0.1826	0.1811
Tiger	0.3580	0.1816	0.1808	0.1803
Micky	0.5468	0.1807	0.1804	0.1802
Rainbow Birds	0.3085	0.1789	0.1809	0.1800
Fruits	0.5127	0.1801	0.1817	0.1811

Table 10. Value of PAFCPV in encrypted images for Polynomial 2 using MOD 2 bit, MOD 4 bit and MOD 8 bit operator.

Image Name	Encrypted images		
	MOD 2 Operator	MOD 4 Operator	MOD 8 Operator
Lena	0.2835	0.2946	0.3001
Tiger	0.3422	0.3526	0.3555
Micky	0.4385	0.4713	0.4742
Rainbow birds	0.3309	0.3447	0.3490
Fruits	0.3346	0.3448	0.3486

4.6 Polynomial 3

Following are the experimental results obtained for Polynomial 3 for parameters Entropy, row and column correlation and PAFCPV

Table 11. Value of entropy in original and encrypted images for Polynomial 3 using MOD 2 bit , MOD 4 bit and MOD 8 bit operator.

Image name	Original image	Encrypted images		
		MOD 2 Operator	MOD 4 Operator	MOD 8 Operator
Lena	7.3433	7.9668	7.9638	7.9557
Tiger	7.3129	7.9415	7.9442	7.9401
Micky	4.7265	7.4094	7.4161	7.4031
Rainbow birds	7.5662	7.9698	7.9700	7.9661
Fruits	7.2263	7.9839	7.9841	7.9834

Table 12. Value of row correlation in original and encrypted images for Polynomial 3 using MOD 2 bit , MOD 4 bit and MOD 8 bit operator.

Image name	Original image	Encrypted images		
		MOD 2 Operator	MOD 4 Operator	MOD 8 Operator
Lena	0.8454	0.2714	0.2553	0.2638
Tiger	0.5264	0.2126	0.2096	0.2065
Micky	0.5164	0.2396	0.2228	0.2216
Rainbow birds	0.5005	0.2014	0.1984	0.1998
Fruits	0.5452	0.2354	0.2261	0.2257

Table 13. Value of column correlation in original and encrypted images for Polynomial 3 using MOD 2 bit , MOD 4 bit and MOD 8 bit operator.

Image name	Original image	Encrypted image		
		MOD 2 operator	MOD 4 operator	MOD 8 operator
Lena	0.7005	0.1895	0.1948	0.2106
Tiger	0.3580	0.1840	0.1845	0.1865
Micky	0.5468	0.1949	0.2117	0.2211
Rainbow birds	0.3085	0.1832	0.1847	0.1877
Fruits	0.5127	0.1817	0.1865	0.1918

5 OBSERVATIONS AND CONCLUSION

From experimental results it can be observed that in Table 1 the maximum range is obtained for Polynomial 1 of 0.5. In Polynomial 1 for MOD 4 operator, Micky

Table 14. Value of PAFCPV in encrypted images for Polynomial 3 using MOD 2 bit, MOD 4 bit and MOD 8 bit operator.

| Image name | Encrypted image | | |
	MOD 2 Operator	MOD 4 Operator	MOD 8 Operator
Lena	0.3067	0.3204	0.3151
Tiger	0.3945	0.4007	0.3986
Micky	0.4701	0.4560	0.4404
Rainbow birds	0.3554	0.3626	0.3560
Fruits	0.3657	0.3747	0.3719

image gives a rise of 58% in entropy, PAFCPV is the highest, A reduction of 62% in row correlation is obtained in Lena for MOD 8 operator and column correlation by 72%. In Polynomial 2 micky image gives the highest entropy of 68% with MOD 2 operator, Row correlation is obtained in lena for MOD 2 operator with a reduction of 74% and column correlation of 78% reduction, the highest value for PAFCPV is obtained in MOD 8 bit operator for Micky image.

In Polynomial 3, highest entropy is obtained for Micky image with MOD 4 operator with 56% rise, row correlation is obtained in lena image with MOD 2 operator with a reduction of 73%, column correlation with 67% reduction, the highest value of PAFCPV is obtained in MOD 2 operator in Micky image.

From the experimental analysis it can concluded that Polynomial 2 gives the best performance for entropy, row correlation and column correlation. As per NPCR all the polynomials are giving equally excellent results and hence not included in experimental results

REFERENCES

Abd El-Latif, A. A., Li, L., Wang, N., Han, Q., & Niu, X.A new approach to chaotic image encryption based on quantum chaotic system, exploiting color spaces. *Signal Processing*, 2013, 93(11): 2986–3000.

Acharya, B., Sunder, S. S., Thiruvenkatam, M., & Sajan, A. K. 2011. Image encryption using index based chaotic sequence, M sequence and gold sequence. In: *Proceedings of the 2011 International Conference on Communication, Computing & Security*, 541–544.

Akhshani, A., Akhavan, A., Lim, S. C., & Hassan, Z. An image encryption scheme based on quantum logistic map. *Communications in Nonlinear Science and Numerical Simulation*, 2012, 17(12): 4653–4661.

Armand Eyebe Fouda, J. S., Yves Effa, J., Sabat, S. L., & Ali, M. A fast chaotic block cipher for image encryption. *Communications in Nonlinear Science and Numerical Simulation*, 2014, 19(3): 578–588.

Azoug, S. E., & Bouguezel, S. 2013. Double image encryption based on the reciprocal-orthogonal parametric transform and chaotic maps. In: *8th International Workshop on Systems, Signal Processing and their Applications (WoSSPA)*: 156–161.

Behnia, S., Akhavan, A., Akhshani, A., & Samsudin, A. Image encryption based on the Jacobian elliptic maps. *Journal of Systems and Software*, 2013, 86(9): 2429–2438.

Boriga, R., Dăscălescu, A. C., & Priescu, I. A new hyper-chaotic map and its application in an image encryption scheme. *Signal Processing, Image Communication*.

Chen, R. J., Lu, W. K., & Lai, J. L. 2005. Image encryption using progressive cellular automata substitution and SCAN. . *IEEE International Symposium on Circuits and Systems. ISCAS 2005*, 2014, 1690–1693.

Chen, T. H., & Wu, C. S. Compression-unimpaired batch-image encryption combining vector quantization and index compression. *Information Sciences*, 2010, 180(9):1690–1701.

Deng, S., Zhan, Y., Xiao, D., & Li, Y. Analysis and improvement of a hash-based image encryption algorithm. *Communications in Nonlinear Science and Numerical Simulation*, 2011, 16(8): 3269–3278.

Dong, C. E. Color image encryption using one-time keys and coupled chaotic systems. *Signal Processing: Image Communication*, 2014, 29(5): 628–640.

Ginting, R. U., & Dillak, R. Y. Digital color image encryption using RC4 stream cipher and chaotic logistic map. *International Conference on Information Technology and Electrical Engineering (ICITEE)*, 2013, 101–105.

Kadir, A., Hamdulla, A., & Guo, W. Q. Color image encryption using skew tent map and hyper chaotic system of 6th-order CNN. *Optik-International Journal for Light and Electron Optics*, 2014, 125(5): 1671–1675.

Kanso, A., & Ghebleh, M. A novel image encryption algorithm based on a 3D chaotic map. *Communications in Nonlinear Science and Numerical Simulation*, 17(7): 2943–2959.

Kanso, A., & Ghebleh, M. 2013. A fast and efficient chaos-based keyed hash function. *Communications in Nonlinear Science and Numerical Simulation*,18(1):109–123.

Kwok, H. S., & Tang, W. K. 2007. A fast image encryption system based on chaotic maps with finite precision representation. *Chaos, solitons & fractals*,32(4):1518–1529.

Li, J., & Liu, H. 2013. Colour image encryption based on advanced encryption standard algorithm with two-dimensional chaotic map. *IET Information Security*,7(4):265–270.

Lima, J. B., Lima, E. A. O., & Madeiro, F. 2013. Image encryption based on the finite field cosine transform. *Signal Processing: Image Communication*,28(10):1537–1547.

Liu, H., & Wang, X. 2013. Triple-image encryption scheme based on one-time key stream generated by chaos and plain images. *Journal of Systems and Software*, 86(3):826–834.

Liu, Y., Tong, X., & Hu, S. 2013. A family of new complex number chaotic maps based image encryption algorithm. *Signal Processing: Image Communication*, 2012, 28(10):1548–1559.

Liu, Z., Gong, M., Dou, Y., Liu, F., Lin, S., Ashfaq Ahmad, M., & Liu, S. Double image encryption by using Arnold transform and discrete fractional angular transform. *Optics and Lasers in Engineering*, 2012, 50(2):248–255.

Liu, Z., Li, S., Liu, W., Wang, Y., & Liu, S. 2013. Image encryption algorithm by using fractional Fourier transform and pixel scrambling operation based on double random phase encoding. *Optics and Lasers in Engineering*, 51(1):8–14.

Luo, Y., Du, M., & Liu, J. A symmetrical image encryption scheme in wavelet and time domain. *Communications in Nonlinear Science and Numerical Simulation*. 2014.

Mohammad Seyedzadeh, S., & Mirzakuchaki, S. A fast color image encryption algorithm based on coupled two-dimensional piecewise chaotic map. *Signal Processing*, 2012, 92(5):1202–1215.

Pareek, N. K., Patidar, V., & Sud, K. K. Diffusion–substitution based gray image encryption scheme. *Digital Signal Processing*, 2013, 23(3):894–901.

Computational Intelligence in Industrial Application – Ling (ed.)
© 2015 Taylor & Francis Group, London, ISBN: 978-1-138-02818-0

Energy consumption performance of TMN based on data packet set

JinSha Yuan & HongYin Xiang

Department of Electronic and Telecommunication Engineering, North China Electric Power University, Baoding, Hebei, China

ABSTRACT: Based on the IEEE 802.15.4 standard, Zigbee Wireless Sensor Networks (WSNs) has been widely used in environmental monitoring, assembly line production, mining development, smart home and civilian industry. In this paper, we proposed a electric switchgear cabinet Temperature Monitoring Network (TMN) deployed in a single substation wireless sensors based on Zigbee protocol, and carried out simulations for energy consumption. According to the results of the experiment, a conclusion was concluded that the influence on energy consumption of the network data packet length size is linear. The longer the data packet, the greater the energy consumption. The greater the network size, the more obvious the effect.

KEYWORDS: Energy consumption; Switchgear cabinet; Wireless sensor networks; Zigbee.

1 INTRODUCTION AND RELATED WORK

1.1 Introduction

Switchgear usually works in the state of high voltage and high current. Circuit breaker in the switchgear cabinet and some inside areas, such as the connections between plugs will be bad for manufacturing, transportation, installation and aging causing contact resistance too large and fever [Yanwei, Chu.& Bo, Yang.& Heng, Liu. 2013]. If these parts cannot get timely maintenance, they will produce abnormal temperature rise, leading to local spark, even arc discharge, and resulting in electrical equipment damage [Xiying, Qian.& Shichang, N.i& Jie, Zheng. et al. 2012]. The key equipments in switchgear cabinet, for instance, circuit breaker, isolating switch and cable joint, need effective temperature monitoring for fire accident protection.

1.2 Related work

Switchgear cabinet temperature monitoring is preserved by waxy temperature measurement, optical fiber temperature measurement, infrared temperature measurement, surface acoustic wave test, and wireless temperature measuring methods. According to [McMullin, J.N.& Eastman, C.D.& Pulikkaseril, C.& Adler, D. 1999] and [Takamatsu, T. 1991], Waxy is simple, low cost and easy to implement. But it can only be qualitatively reflect the change of the temperature with low precision and poor real-time performance. [Xiupeng, Zhang.& Handong, Li.& Hao, Yu. 2013] analyzed the optical fiber temperature measurement method, which can be real-time online temperature monitoring, having strong resistance to electromagnetic interference, good corrosion resistance and large temperature measurement range. [Gang, Chen.& Jiaomin, Liu.& Xuejie, Wei.& Jinmei, Liu. 2007] pointed out that there were some restrictive factors constraining its extensive use, such as high equipment cost, narrow interior space, difficult arrangement of wire, line ageing, not easy for installation and remove.[Jinsoo, Han.& Changsic, Choi.& Ilwoo, Lee. 2011] designed and analyzed infrared thermometry, and [Ling, Hong.& Chengmao, Cao.& Quan, Liu. 2010] studied the performance of a practical item. But [Xiying, Qian.& Shichang, N.i& Jie, Zheng. et al. 2012] stated that some shortages still could not be ignored, just as expensive equipments, high artificial costs and unable to realize real-time online monitoring temperature. Meanwhile, strong magnetic field environment would cause certain damage to personnel health of staff working long hours monitoring. [Hinrichsen, V.& Scholl, G.& Schubert, M.& Ostertag, T. 1999] discussed a novel scheme, which employed a typical kind of surface acoustic wave temperature sensor with antenna capturing electromagnetic waves, providing energy for the sensor chip. Yet, it is a pity that there is no large-scale application for this approach, maybe some lack of maturity.

Wireless technology has been accepted more and more to measure temperature, in which temperature measuring devices were installed in high voltage area, directly contacting with the testing objects. Then, typical wireless modules transmit temperature information to the external concentrator. So, temperature data collection in high voltage area turns into a simple thing. Thus, this way is indeed a very good choice to switchgear temperature measuring practice for no need wired layout, high precision, real time transition, low cost and low energy consumption. [Monchau, Shie.& Pochen, Lin.& Tsungmao, Su.& Poki, Chen.& Hutahaean, A. 2014] mainly adopted the Zigbee wireless sensor temperature measuring network technology.

From above, it can be seen that the wireless temperature measuring is an ideal option for switchgear cabinet temperature measurement method. This article mainly analyzes temperature measurement performance in a wireless sensor network based on Zigbee technology.

2 TEMPERATURE MONITORING SCHEME

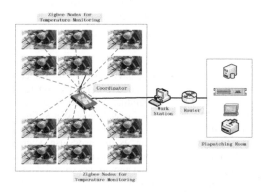

Figure 1. A temperature measurement network deployed in switch cabinet based on the technology of Zigbee.

As shown in Fig. 1, the high voltage switchgear cabinet intelligent wireless temperature measurement system is mainly composed of temperature measuring node, the coordinator, monitoring station of three parts. In general, there are 8–20 switchgear cabinets in a switchgear room, and usually 3–12 points in a single monitoring tank. Thus, a range of 24–240 points need to be monitored in a single substation. What is more, a star topology network in Zigbee protocol can handle up to 65535 temperature nodes by a coordinator. And so, the capacity is enough for substation temperature monitoring [Chunliang, Hu.& Shengyuan, Yang.& Tengyaw, Hu.& Weibin, Wu. 2010]. Among them, the coordinator node is played the part by fully functional nodes (Full Function Device, FFD), which can communicate with any other devices in the network. While, temperature measurement nodes are usually acted by simple functional nodes (Reduced Function Device, RFD), which can only communicate with the coordinator node.

In Fig. 1, temperature measurement nodes binding on the circuit breakers form a Zigbee network.

Monitoring nodes measure, read and report contact temperatures up to the network coordinator at regular intervals [Hainan, Long.& Leyang, Zhang.& Jiao, Pang.& Caixia, Li.& Tierui, Song. 2010]. Coordinator is generally installed in the substation interior walls, and interacts with dispatching server through cable and monitoring workstation. Here, coordinator is used for node management and data forwarding [Gang, Zhang.& Shuguang, Liu. 2010]. Because of relatively harsh environment and limited battery power supply, the set of scientific power application scheme becomes the most restriction for wireless sensor nodes requiring long survival time for months or even years. Therefore, this article mainly analyzes the temperature monitoring performance of energy consumption of the Zigbee network.

3 ENERGY CONSUMPTION OF SENSOR NODES

3.1 Node structure

Wireless sensor node is different from general sensors, which usually contains processing module, communication module, perception module and power supply module. The fore two modules consume the most energy.

Processing module is the node control and data processing center, consisted by an embedded system, including the microprocessor, memory, operating system, etc., responsible for sensor control algorithm, communication protocol and sensory data for complex processing. The general microprocessor supports a variety of work modes, like sleep, idle and work.

Communication module, including baseband part and RF part, is responsible for receiving and sending data between nodes. Energy consumption of a wireless communication system is influenced by different modulation mode, data rate, transmission power and operating frequency [Kwang-il Hwang. 2011].

3.2 Energy consumption of microprocessor

Microprocessor has three states, and they are running state, idle state and the sleep state. Running state declares the processor is executing instructions at the moment, so producing benefits. When only some part of the modules is in running condition, saving of energy consumption, processor turns into idle state. Further on, when most modules hangs up on work, sleep state starts. Now, sleep provides the lowest energy consumption, and the availability of the system resources reaches the least [Cagnetti, M.; Leccese, F.; Proietti, A. 2012].

When a task needs to be handled, CPU is in running state. On the contrary, when any task is in absence for processing or lack of energy, operating system would change to sleep. At this very moment, CPU and internal modules stop working, besides the necessary clock and power management modules, meaning minimum energy consumption [August, Betzler.; Carle, Gomez.; Ilker, Demirkol.; Josep, Paradells. 2014]. That is, compared with the running state and idle state, sleep state consumes less and save energy.

When an exception occurs, CPU will be awakened to return to running mode. Yet, this state switch will spend longer time consuming more resources [August, Betzler.; Carle, Gomez.; Ilker, Demirkol.; Josep, Paradells. 2014]. Once tasks have been completed, operating system can choose to enter idle mode, with moderate energy consumption between running and sleep state. If in a period of certain time, no message queues, system predicts no events will generate recently, and sleep mode may be another optional selection. Surely, if a new task arrives, idle state shall translate into running mode [Sadik, K.Gharhan.; Rosdiadee, Nordin.; Mahamod, Ismail. 2014].

Power in different modes can be got by checking the reference manual, and then the energy consumption of processor comes into being in the premise of statistics of the duration of corresponding states.

This system employs CC2420 chip of TI company, with power supply voltage is 3.3V, send current 17.4 mA, receives current 18.8 mA, sleep current 20 mA. Obviously, energy consumption is more than 900 times between send/receive mode and sleep mode. If the whole network works in a rational mechanism less run mode and more sleep mode under the condition of guaranteed delivery demand, better energy saving would be achieved efficiently.

As mentioned above, sensor nodes switching from one state to another state will also cost a certain amount of time and energy consumption, and so the number of state transitions must be controlled in a reasonable range. Table 1 illustrates key data of CC2420 chip in different working modes and mode switches.

Table 1. Working currents and mode switch time of CC2420 chip.

Mode	Current	Mode Switch	Time
Ioff	$0.02\mu A$	Toff-idle	$1ms$
Itx	$17.4mA$	Tcca/ed-idle	$2\mu s$
Irx	$18.8mA$	Tidle-cca/ed	$192\mu s$
Iidle	$426\mu A$	Tsleep-idle	$0.6ms$
Icca/ed	$17.4mA$	Tidle-sleep	$192\mu s$
Isleep	$20\mu A$	Ttx-idle	$2\mu s$
–	–	Tidle-tx	$192\mu s$
–	–	Trx-idle	$2\mu s$
–	–	Tidle-rx	$192\mu s$

3.3 *Energy consumption of communication module*

For CC2420, communication module has six states, and they are TX, RX, OFF, IDLE, SLEEP and CCA/ED. TX names transmitting status, launching packets from one node to other nodes. RX names receiving state, accepting packets from other nodes; OFF means system closing; IDLE stands for events waiting. SLEEP indicates only individual modules working with low energy consumption. CCA/ED expresses the action of Clear channel assessment (CCA) and channel energy detection (ED).

After launch, communication module comes into IDLE state from OFF. It is observed that IDLE is a core state approach to other modes. Case data emitting, state changes into TX. Case packet receiving, state comes into RX. Case channel assessment, states turns into CCA/ED. When tasks are completed, it will return to the IDLE mode. Similar to the processor module, switch between IDLE and SLEEP mode carries out under the same conditions.

4 PERFORMANCE SIMULATION

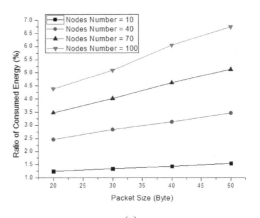

(a)

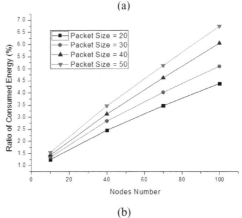

(b)

Figure 2. Influence on energy consumption of data packet length.

(a) $PacketSize = \{20, 30, 40, 50\}$

(b) $N = \{10, 40, 70, 100\}$

Fig. 2(a) shows the energy consumption performance with different data packet length under various network scales. In this figure, the horizontal axis indicates packet length, and the vertical axis demonstrates the average energy consumption ratio of all nodes in this network, represented by E_{r-mean} shown in formula (1).

$$E_{r-mean} = \frac{1}{N} \sum_{n=1}^{N} \left(1 - \frac{E_{nr}}{E_{ni}}\right) \cdot 100\% \qquad (1)$$

Among them, N makes sensor node number in this network, E_{nr} indicates the residual energy of each node in a certain state, E_{ni} stands for the initial energy of each node.

From Fig. 2(a), two conclusions can be drawn as follows:

1 In the same scale of network, the network energy consumption increases in a one-way linear style with data packet length increasing.
2 Along with larger the network scale, faster and faster the network energy consumption increases. For specific performance, the slope of curve increases continually. More than that the linear changes little up and down. On the whole, the greater the network size, the greater the fluctuation of average energy consumption. Amplitude of variations is shown in Table 2.

Table 2. In different network scale, the maximum average energy consumption contrast when $PacketSize = \{20,30,40,50\}$.

N	max E_{r-mean}	min E_{r-mean}	max $DiffE_{r-mean}$
10	1.54	1.24	0.30
40	3.46	2.46	1.00
70	5.13	3.47	1.66
100	6.74	4.38	2.36

Table 2 shows that corresponding to the number of sensor nodes $N = \{10, 40, 70, 100\}$, there are max $DiffE_{r-mean} = \{0.30, 1.00, 1.66, 2.36\}$. This demonstrates that the larger the network size, the greater the impact on energy consumption of the network from data packet length. Moreover, the longer the data packet, the higher the energy consumption.

Fig. 2(b) shows the energy consumption performance with different network scale under various data packet length. In this figure, the horizontal axis indicates the number of sensor nodes, and the vertical axis demonstrates the average energy consumption ratio of all nodes in this network, represented by E_{r-mean} shown in formula (1).

From Fig. 2(b), two conclusions can be drawn as follows:

1 In the same data packet length, the network energy consumption increases in a one-way linear style with network size increasing.
2 Along with longer the data packet length, the network energy consumption increases up almost in linear style. Amplitude of average energy consumption variations are shown in Table 3.

Table 3. In different packet size, the maximum average energy consumption contrast when $PacketSize = \{20,30,40,50\}$.

$PacketSize$	max E_{r-mean}	min E_{r-mean}	max $DiffE_{r-mean}$
20	4.38	1.24	3.14
30	5.09	1.35	3.74
40	6.05	1.43	4.62
50	6.74	1.54	5.20

Table 3. shows that corresponding to the data packet length

$PacketSize = \{20, 30, 40, 50\}$,

there exists

max $DiffE_{r-mean} = \{3.14, 3.74, 4.62, 5.20\}$.

This demonstrates that the larger the packet size, the greater the impact on energy consumption of the network from network scale. Moreover, the more the sensor node, the higher the energy consumption.

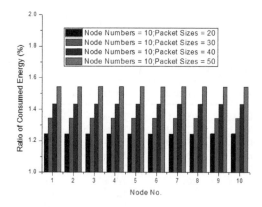

Figure 3. Each sensor node energy consumption in different packet size when $N = 10$.

Fig. 3 shows the each sensor node energy consumption under different packet size. It can be seen that densities both in horizontal and vertical axis are basically the same performance. In other words, the energy consumption of this system shows very good linear overall.

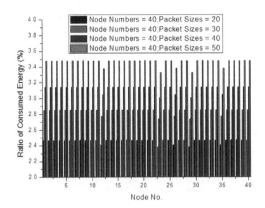

Figure 4.   Each sensor node energy consumption in different packet size when $N = 40$.

Fig. 4 shows the each sensor node energy consumption under different packet size. It can be seen that densities in horizontal axis direction are basically the same performance, but densities in vertical axis direction are changes up and down to some extent. Of course, the energy consumption of this system is still linear in general. That is the cause accounting for the different details of the curve in Fig. 2(a) from (b).

5   CONCLUSIONS

To sum up, the influence of data packet size on energy consumption of the network is linear. In the same network scale, larger packet size leads to higher energy consumption. Moreover, bigger network scale gets greater influence. Similarly, bigger network scale gets greater influence also. According to the results of above simulations, some corresponding network parameters may be appropriately adjusted so as to get the best performance of energy consumption, which will make it more suitable for power switchgear cabinet wireless temperature monitoring.

ACKNOWLEDGMENTS

This work was supported by National Natural Science Foundation of China (No. 61302106), the Fundamental Research Funds for the Central Universities (No. 2014MS105) and (No. 13MS66), and Hebei Province Natural Science Foundation of China, Youth Science Fund (No. E2013502267). Here, we are obliged to above foundations, their responsible persons and all our lab associates.

REFERENCES

August, Betzler.; Carle, Gomez.; Ilker, Demirkol.; Josep, Paradells. 2014. A Holistic Approach To Zigbee Performance Enhancement For Home Automation Networks. *Sensors14:*14932–14970.

Chunliang, Hu.& Shengyuan, Yang.& Tengyaw, Hu.& Weibin, Wu. 2010. Practical Design of Active & Intelligent Energy-Saving System With Modules Strategy. *In the Proceedings of 2010 International Conference on Machine Learning and Cybernetics, Qingdao, China, 11–14 July 2010.*

Cagnetti, M.; Leccese, F.; Proietti, A. 2012. Energy Saving Project For Heating System With Zigbee Wireless Control Network. *In the Proceedings of 2012 11th International Conference on Environment and Electrical Engineering, Roma, Italy, 18–25 May 2012.*

Gang, Chen.& Jiaomin, Liu.& Xuejie, Wei.& Jinmei, Liu. 2007. The Testing and Analysis Technology For Temperature Field of Low-Voltage Circuit Breaker. *In the Proceedings of 2007 International Conference on Machine Learning and Cybernetics, Hongkong, China, 19–22 August 2007.*

Gang, Zhang.& Shuguang, Liu. 2010. Study on Electrical Switching Device Junction Temperature Monitoring System Based on Zigbee Technology. *In the Proceedings of 2010 International Conference on Computer Application and System Modeling, Taiyuan, China, 22–24 October 2010.*

Hainan, Long.& Leyang, Zhang.& Jiao, Pang.& Caixia, Li.& Tierui, Song. 2010. Design of Substation Temperature Monitoring System Based on Wireless Sensor Networks. *In the Proceedings of 2010 2nd International Conference on Advanced Computer Control, Shenyang, China, 27–29 March 2010.*

Hinrichsen, V.& Scholl, G.& Schubert, M.& Ostertag, T. 1999. Online Monitoring of High-Voltage Metal-Oxide Surge Arresters By Wireless Passive Surface Acoustic Wave (SAW) Temperature Sensors. *In the Proceedings of 1999 Eleventh International Symposium on High Voltage Engineering, London, England, 23–27 August 1999.*

Jinsoo, Han.& Changsic, Choi.& Ilwoo, Lee. 2011. More Efficient Home Energy Management System Based on Zigbee Communication and Infrared Remote Controls. *In the Proceedings of 2011 IEEE International Conference on Consumer Electronics, Melbourne, Australia, 19-20 November 2011.*

Jukk, Suhonen.& Timo, D.Hamalainen. & Marko, Hannikainen. Availability and end-to-end reliability in low duty cycle multi hop wireless sensor networks. *Sensors* 2009, (9):2088–2116.

Kwang-il Hwang. 2011. Energy Efficient Channel Agility Utilizing Dynamic Multi-Channel CCA For Zigbee RF4CE. *IEEE Transactions on Consumer Electronics1:*113–119.

Ling, Hong.& Chengmao, Cao.& Quan, Liu. Real-Time Temperature Monitoring System For High Voltage Switchgear Based on Infrared Wireless Transmission. *In the Proceedings of 2010 2nd International Conference on Information Science and Engineering, Hangzhou, China, 3–5 December 2010.*

Monchau, Shie.& Pochen, Lin.& Tsungmao, Su.& Poki, Chen.& Hutahaean, A. 2014. Intelligent Energy Monitoring System Based on Zigbee-Equipped Smart Socketsj. *In the Proceedings of 2014 International Conference on Intelligent Green Building and Smart Grid, Taipei, China, 23–25 April 2014.*

McMullin, J.N.& Eastman, C.D.& Pulikkaseril, C.& Adler, D. 1999. Measurement of The Wax Appearance Temperature In Crude Oil By Laser Scattering. *Electrical And Computer Engineering* 3:1755–1758.

Pollin, S. & Ergen, M. & Ergen, S. & Bougard, B. & Der Perre, L. & Moerman, I. & Bahai, A. & Varaiya, P. & Catthoor, F. Performance Analysis of Slotted Carrier Sense IEEE 802.15.4 Medium Access Layer. *IEEE Transactions on Wireless Communications* 2008, (9):3359–3371.

Saad A.Khan.& Fahad A.Khan. 2009. Performance Analysis of A Zigbee Beacon Enabled Cluster Tree Network. *In the Proceedings of 2009 ICEE 3rd International Conference on Electrical Engineering, Tianjin, China, 3–5 November 2009.*

Sadik, K.Gharhan.; Rosdiadee, Nordin.; Mahamod, Ismail. 2014. Energy-Efficient Zigbee-Based Wireless Sensor Network For Track Bicycle Performance Monitoring. *Sensors14:15573–15592.*

Takamatsu, T. Life time of thermal electrets of carnauba wax, esters, fatty acids and alcohols. *Electrets* 1991, (1): 106–110.

Xiying, Qian. & Shichang, N.i & Jie, Zheng. et al. High voltage switchgear on-line infrared temperature measurement based on zigbee technology. *Mechanical and Electronic Information* 2012, (36): 136–137.

Xiupeng, Zhang.& Handong, Li.& Hao, Yu. Application Research On High Voltage Switch Cabinet Based On Fiber Bragg Gratings Temperature Measurement System. *In the Proceedings of 2013 5th International Conference On Intelligent Human-Machine Systems and Cybernetics (IHMSC), Hangzhou, China , 26–27 August 2013.*

Yanwei, Chu.& Bo, Yang.& Heng, Liu. Wireless Intelligent Temperature-Measuring System for High-Voltage Switchgear Based on Zigbee Technology. *Electrician and Electrical 2013 (6): 23–28.*

Computational Intelligence in Industrial Application – Ling (ed.)
© 2015 Taylor & Francis Group, London, ISBN: 978-1-138-02818-0

Study on synthetic aperture sonar imaging algorithm of UUV

NaiQiang Fan, YingMin Wang & YanNi Gou
Northwestern Polytechnical University, Xi'an, China

ABSTRACT: The traditional synthetic aperture algorithm is mainly used in narrow-band and narrow-beam signal imaging. When used for synthetic aperture imaging in UUV (Unmanned Underwater Vehicle), the traditional algorithm generates poor imaging effect due to slower sonar platform, larger signal bandwidth and broader beamwidth. On the basis of the principle of classical Range-Doppler algorithm, the traditional algorithm is improved in accordance with characteristics of UUV synthetic aperture imaging. Fresnel assumption under narrow bandwidth signal is abandoned, more accurate range hyperbolic curve model is used to derive the algorithm, and arbitrary point targets in the imaging area are simulated on the basis of the derived result. Simulation results show that the improved algorithm has a higher resolution and moderate amount of computation and is thereby more suitable for UUV synthetic aperture imaging.

KEYWORDS: UUV; Synthetic Aperture Sonar; Range-Doppler Algorithm; Broad Bandwidth Signal.

1 INTRODUCTION

UUV (Unmanned Underwater Vehicle) is an indispensable tool for ocean exploitation by human beings. In terms of deep-sea geological survey, seabed topographical mapping and submarine resources survey, the use of UUV requires the technology of synthetic aperture imaging to perform high-resolution imaging to submarine targets. Synthetic aperture refers to an imaging technology that uses mini-size array of linear movement, processes target echo signals in different positions, obtains equivalent virtual large aperture array through synthesis and thereby realizes high azimuth resolution[1]. Currently, synthetic aperture is the only technology to increase the resolution of side scan sonar azimuth by an order of magnitude[2]. Due to the limited volume of small UUV, synthetic aperture sonar is limited by location and space during the installation. In addition, the sonar signal has large bandwidth and wide beam, and puts forward higher requirements for synthetic aperture imaging algorithm.

Synthetic aperture imaging is mainly divided by point-by-point imaging and line-by-line imaging[3][4]. Point-by-point imaging algorithm (e.g. time domain correlation) is more accurate, yet less used due to huge amount of computation. The line-by-line imaging algorithm such as Range-Doppler, Chirp Scaling (CS), Frequency Scaling (FS) and ωk are all extensively applied to Synthetic Aperture Radar (SAR). However, the synthetic aperture imaging algorithm must be improved when used in Synthetic Aperture Sonar (SAS) since Fresnel hypothesis fails to stand all the time due to its broad bandwidth and broad-beam signal.

In the synthetic aperture algorithm, the most important relationship is the instantaneous range from the point target to the sonar array. The range determines the phase characteristics of the signal, and behavior analysis for phase modulation works through the synthetic aperture imaging algorithm. The range of sonar array for receiving echo wave signals will cover a plurality of range cells with time, and the range variation is called range cell migration. To obtain ideal image focus effect, range migration correction turns to be the very problem for each algorithm. The core idea of synthetic aperture imaging algorithm is based on matched filtering for echo wave in the directions of range and azimuth. Modulation with upward range is determined by phase code of transmitted pulse, while modulation with upward azimuth comes from motion of sonar platform. The basic idea of Rang-Doppler algorithm is to use elimination of range migration to decouple a two dimensional signal from the range and azimuth into two one-dimensional signals which then undergo matched filtering (pulse compression), greatly reducing the amount of computation. CS algorithm avoids interpolation operation in range migration correction, and can perform minute non-linear scale variation under linear frequency modulation hypothesis of signals. The algorithm can perform range migration correction with range variation in an efficient and accurate manner through simple phase multiplication. The ωk algorithm processes coupling for range and azimuth in two-dimensional frequency domain through reference function multiplication and Stolt interpolation, and can process data with wide apertures or large squint angles. In this thesis, the Range-Doppler algorithm is taken as an example to improve the traditional algorithm to be suitable for UUV synthetic aperture imaging.

In [5][6] of the references, improved algorithms are all proposed for narrow-beam signals, and generate poor imaging effect for wide beam signals in UUV. In the application of UUV (synthetic aperture imaging), higher range resolution and azimuth resolution are always expected. The range resolution requires a large bandwidth of signal to obtain; azimuth resolution mainly relies on enough length of synthetic aperture and meanwhile requires wide wave beams of signals. Therefore, the synthetic aperture imaging algorithm must have both characteristics of broad bandwidth and wide beam to meet the requirements of imaging in UUV. On the basis of analyzing the principle of classical Range-Doppler algorithm, Fresnel assumption is abandoned, a more accurate range hyperbolic curve model is established, the improved Range-Doppler algorithm which can satisfy both broad band and wide beam conditions under the positive side-looking and small side-looking mode is derived to be more suitable for broad bandwidth and wide beam signal imaging in UUV.

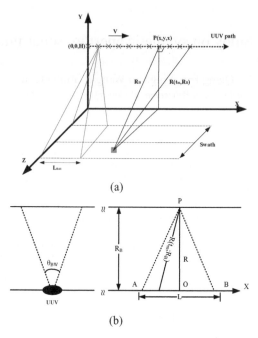

(a)

(b)

Figure 1. (a) Synthetic aperture imaging geometry, (b)Side-looking imaging mode by synthetic aperture technique.

2 SYNTHETIC APERTURE IMAGING MODEL

In a strip-map imaging system, UUV transmits beams perpendicular to the direction of motion while passing through the aperture and irradiates a target area, as shown in Figure 1(a). Here we need to make the following assumptions to the data model: First, the model ignores influences from media disturbance, refraction and multi-path, the speed of sound remains constant, the signal is transmitted along a straight line, so the propagation delay is proportional to the distance between the target and the platform. Second, the model assumes that the complex index of reflection of each target is stable and does not change with the difference of squint angles. Finally, the model assumes that the UUV is stationary when transmitting and receiving signals. This is the stop-start-stop mode commonly used in an SAS data model.

As shown in Figure 1(b), the beam width is θ_{BW}. When the UUV arrives at Point A, the leading edge of the wave velocity touches any point target P in the imaging scene.

When the UUV arrives at Point B, the back edge leaves Point P, and the length from A to B is the effective synthetic aperture L. For synthetic aperture imaging, while analyzing echo wave of point target P, assume the vertical range from this point target to the UUV route to be R_B, the slow time t_m of the node

of the vertical range line and the route to be zero time (origin), and the slant range from the phase center of the sonar array to P at any time t_m to be $R(t_m; R_B)$.

Assume the sonar transmitting signal to be linear frequency modulation signal $s(\hat{t}) = a_r(\hat{t}) \exp(j\pi\gamma\hat{t}^2)$, thus baseband signal of the echo of the point target in the fast range time slow azimuth time domain ($\hat{t} - t_m$ domain) has the model as follows:

$$S_0(\hat{t}, t_m; R_B) = A_0 w_r[\hat{t} - 2R(t_m; R_B)/c]$$
$$\times w_a \ (t_m - t_c)$$
$$\times \exp\{j\pi K_r[\hat{t} - \frac{2R(t_m; R_B)}{c}]^2\} \quad (1)$$
$$\times \exp\{-j\frac{4\pi}{\lambda}R(t_m; R_B)\}$$

Where A_0 is an arbitrary complex constant, $\hat{t}$ is the fast range time, t_m is the slow azimuth time for the motion of UUV, t_c is the deviation time of the beam center from the point target, $w_r(\hat{t})$ is the range envelope, rectangular window function $w_a(t_m)$ is azimuth envelope, λ is wave length of transmitting signal, K_r is linear frequency modulation frequency, and $R(t_m; R_B)$ is instantaneous slant range of UUV to target.

3 TRADITIONAL SYNTHETIC APERTURE IMAGING ALGORITHM

In synthetic aperture imaging processing, two-dimensional range and azimuth is generally simplified to two one-dimensions. Typically, imaging algorithm aims at narrow-band narrow-beam signals, i.e. the signal satisfies the following conditions:

1 Narrow-band condition

$$B = |f - f_c| << f_c \qquad (2)$$

Where B is signal bandwidth, and f_c is carrier frequency of transmitting signal. It is generally agreed that when B meets the condition of the above formula, it is the narrow band signal. Usually when $B / f_c < 0.1$, it can be regarded as the narrow band signal.

2 Narrow-beam condition

$$Vt_m << R_B \qquad (3)$$

Where V is the velocity of UUV. When the length of the wave beam sweeping over the synthesized array of the point target is much smaller than the range from the UUV's trace to the imaging central line, it is generally regarded as narrow beam.

3.1 Original Range-Doppler algorithms

It can be derived from Formula (1) that the systematic matching function of the matched filtering for the range is:

$$s_r(\hat{t}) = s_t^*(-\hat{t}) = a_r(\hat{t}) \exp(j\pi\gamma\hat{t}^2) \qquad (4)$$

In order to reduce the amount of calculation, matched filtering for the range and azimuth are respectively completed in the frequency domain. For the range, Fast Fourier Transform (FFT) and inverse Fourier Transform (IFFT) are used for conversion between the time domain and the frequency domain to obtain the output of the matched filtering as follows:

$$s(\hat{t}, t_m; R_B) = IFFT_{f_r}\{FFT_{\hat{t}}[s(\hat{t}, t_m; R_B)] \cdot FFT_{\hat{t}}[s_r(\hat{t})]\} \qquad (5)$$

Assume the range to be a rectangular window, after receiving signal of Formula (1) is processed as stated above, it can be derived:

$$s(\hat{t}, t_m; R_B) = A\sin c\{\Delta f_r[\hat{t} - \frac{2R(t_m; R_B)}{c}]\} \times a_a(t_m)\exp[-j\frac{4\pi}{\lambda}R(t_m; R_B)] \qquad (6)$$

Where A is echo signal amplitude after range compression, and Δf_r is bandwidth of the linear frequency modulation signal.

After range compression is completed, the next step should be azimuth compression. For point target P at the nearest range R_B, the length of wave beam sweeping over integrated array of point target is much smaller than R_B, i.e. when $Vt_m << R_B$, the narrow-beam condition is satisfied, the relationship between R_B and t_m can be expressed through Fresnel assumption as:

$$R(t_m; R_B) = \sqrt{R_B^2 + (Vt_m)^2} \approx R_B + \frac{(Vt_m)^2}{2R_B} \qquad (7)$$

Where item 2 refers to the value of range migration. As a result of $Vt_m << R_B$, $R_B + \frac{(Vt_m)^2}{2R_B} \approx R_B$.

Fast range time slow azimuth time domain signal may be expressed as:

$$s(\hat{t}, t_m; R_B) = A\sin c\{\Delta f_r[\hat{t} - \frac{2R_B}{c}]\} \times a_a(t_m)\exp\{-j\frac{4\pi}{\lambda}[R_B + \frac{(Vt_m)^2}{2R_B}]\} \qquad (8)$$

Echo wave envelope is a straight line in the two-dimensional surface, i.e. coupling of the range and the azimuth is released, so that matched filtering of the azimuth is simplified. Here, the matched filtering function of azimuth is:

$$S_a(t_m; R_B) = a_a(t_m)\exp(-j\pi\gamma_m(R_B)t_m^2) \qquad (9)$$

Where Doppler frequency modulation rate is $\gamma_m(R_B) = -\frac{2V^2}{\lambda R_B}$. Similar to matched filtering of range, the matched filtering of azimuth can be performed in the Doppler domain as well, and the output of the matched filtering is:

$$s(\hat{t}, t_m; R_B) = IFFT_{f_a}\{FFT_{t_m}[s(\hat{t}, t_m; R_B)] \cdot FFT_{t_m}[s_a(t_m; R_B)]\} \qquad (10)$$

Finally the imaging result of point target is derived:

$$s(\hat{t}, t_m; R_B) = C\sin c\{\Delta f_r[\hat{t} - \frac{2R_B}{c}]\} \cdot \sin c(\Delta f_a t_m) \qquad (11)$$

Where Δf_r is transmitting signal bandwidth, and Δf_a is Doppler bandwidth.

3.2 Analysis on limitation of original Range-Doppler algorithm

Azimuth resolution of synthetic aperture processing may be expressed by the following formula:

$$\rho_a = \frac{K_w \lambda}{4\sin(\theta_a/2)} \tag{12}$$

Where K_w is weighting factor, and θ_a is beam angle of azimuth processing. According to Formula (12), azimuth resolution of the target depends on the beam angle between the wave length and the azimuth. Under determinate wave length, large enough azimuth beamwidth and large synthesized aperture are essential to obtain the required azimuth resolution. In SAS imaging, carrier frequency of sonar transmitting signal is as low as several kHz to scores of kHz, and the signal is broad bandwidth signal, failing to meet the condition of narrow bandwidth signal in Formula (2). Moreover, in the application of UUV, both the azimuth beam angle of transmitting signal is large, the cross range Vt_m of wave beam sweeping over the point target and the length $L_s = R_B\theta_a$ of synthesized array (positive side-looking mode) can be stimulated as compared with R_B, failing to meet the narrow-beam condition in Formula (3). In this case, the wave front of the echo wave is spherical phase modulation, which cannot be assumed by parabola, i.e. Fresnel assumption no longer stands. Therefore, the original Rang-Doppler algorithm must be improved to meet the requirements of UUV synthetic aperture imaging for broad bandwidth wide beam signals.

4 IMPROVED ALGORITHM

In the positive side-looking or small squint angle mode and under the condition of broad bandwidth and wide beam signals, Fresnel assumption is abandoned, and the range between UUV and point target in the imaging domain can be expressed with a more accurate hyperbolic curve model as:

$$R(t_m; R_B) = \sqrt{R_B^2 + (Vt_m)^2} \tag{13}$$

When Formula (13) is substituted into Formula (1):

$$S_0(\hat{t}, t_m; R_B) = A_0 w_r[\hat{t} - 2\sqrt{R_B^2 + (Vt_m)^2}/c]$$

$$\times w_a \ (t_m - t_c)$$

$$\times \exp\{j\pi K_r[\hat{t} - \frac{2\sqrt{R_B^2 + (Vt_m)^2}}{c}]^2\}$$

$$\times \exp\{-j\frac{4\pi}{\lambda}\sqrt{R_B^2 + (Vt_m)^2}\}$$

1 Range Fourier Transform

Range Fourier transform can be expressed as:

$$S_0(f_r, t_m; R_B)$$

$$= \int_{-\infty}^{\infty} s_0(\hat{t}, t_m; R_B)\exp(-j2\pi f_r\hat{t})d\hat{t} \tag{14}$$

According to the principle of stationary phase[5], the result of Formula (14) is:

$$S_0(f_r, t_m; R_B) = A_0 A_1 w_r(f_r)w_a(t_m - t_c)$$

$$\times \exp\{-j\frac{4\pi(f_0 + f_r)\sqrt{R_B^2 + (Vt_m)^2}}{c}\} \tag{15}$$

$$\times \exp\{-j\frac{\pi f_r^2}{K_r}\}$$

Where A_1 is constant, and f_0 is center frequency of sonar transmitting signal.

2 Range compression

According to Formula (15), the range matched filter of $S_0(f_r, t_m; R_B)$ is $H_{rc}(f_r) = \exp\{j\frac{\pi f_r^2}{K_r}\}$. The signal after range compression is:

$$S_{rc}(f_r, t_m; R_B) = S_0(f_r, t_m; R_B) \cdot H_{rc}(f_r)$$

$$= A_0 A_1 w_r(f_r)w_a(t_m - t_c) \tag{16}$$

$$\times \exp\{-j\frac{4\pi(f_0 + f_r)\sqrt{R_B^2 + (Vt_m)^2}}{c}\}$$

3 Azimuth Fourier transform

For Formula (16), use the principle of stationary phase again, and the result of azimuth Fourier transform is:

$$S_{2df}(f_r, f_a)$$

$$= A_0 A_1 A_2 W_r(f_r)W_a(f_a - f_{ac}) \tag{17}$$

$$\times \exp\{j\theta_a(f_r, f_a)\}$$

Where A_2 is the constant, $W_a(f_a - f_{ac})$ is the azimuth spectrum envelope taking Doppler center frequency f_{ac} as the center, and $\theta_a(f_r, f_a)$ is the phase angle after Fourier transform.

4 Range inverse Fourier transform

Range inverse Fourier transform is performed to $S_{2df}(f_r, f_a)$, i.e. $S_{rd}(\hat{t}, f_a) = \int_{-\infty}^{\infty} S_{2df}(f_r, f_a)\exp\{j2\pi f_r\hat{t}\}df_r$. The principle of stationary phase can be used to find the solution. When phase term $\theta_a(f_r, f_a)$ in the synthesis is developed to power series of f_r, it can be obtained:

$$S_{rd}(\hat{t}, f_a) = A_0 A_1 A_2 A_3 w_r \{ \frac{1}{1 - K_r Z} [\hat{t}$$

$$- \frac{2R_B}{cD(f_a, V)}]\}$$

$$\times w_a \ (f_a - f_{ac}) \qquad (18)$$

$$\times \exp\{-j \frac{4\pi R_B f_0 D(f_a, V)}{c}\}$$

$$\times \exp\{j\pi K_m[\hat{t} - \frac{2R_B}{cD(f_a, V)}]^2\}$$

Where K_m is a new range modulation frequency,
$$K_m = \frac{K_r}{1 - K_r Z}, \quad Z = \frac{cR_0 f_a^2}{2V^2 f_0^3 D^3(f_a, V)}.$$

Range envelope w_r is the range migration, the first exponential term is azimuth modulation induced by range migration, and the second exponential term is the range modulation.

5 Range migration correction

Range migration correction may be achieved by the processing method of Sinc interpolation. Sinc core function is truncated and weighted by sharpening window. After being corrected, the signal is formula (19).

Where the range envelope has nothing to do with the azimuth frequency f_a, which indicates that the range migration has been corrected accurately.

$$S_{rd}(\hat{t}, f_a) = A_0 A_1 A_2 A_3 p_r \{ \frac{1}{1 - K_r Z} [\hat{t}$$

$$- \frac{2R_B}{c}]\}$$
$$\times w_a \ (f_a - f_{ac}) \qquad (19)$$

$$\times \exp\{-j \frac{4\pi R_B f_0 D(f_a, V)}{c}\}$$

$$\times \exp\{j\pi K_m[\hat{t} - \frac{2R_B}{cD(f_a, V)}]^2\}$$

6 Azimuth compression

After the range migration correction is completed, azimuth focus can be performed to data through matched filter. The compressed signal is converted to:

$$S_{ac}(\hat{t}, f_a) = A_0 A_1 A_2 A_3 p_r \{ \frac{1}{1 - K_r Z} [\hat{t}$$

$$- \frac{2R_B}{c}]\}$$

$$\times w_a \ (f_a - f_{ac})$$

$$\times \exp\{-j \frac{4\pi R_B f_0 D(f_a, V)}{c}\}$$

7 Azimuth inverse Fourier transform

The principle of stationary phase is used to convert slow time into frequency domain (Doppler domain), and constant phase, constant coefficient and amplitude function are ignored, thus obtaining:

$$S(\hat{t}, t_m; R_B) = \int_{-\infty}^{\infty} S_{ac}(\hat{t}, f_a) \exp\{j2\pi f_a t_m\} df_a$$

$$= A_0 \sin c(\hat{t} - \frac{2R_B}{c}) \sin c(t_m)$$

$$\times \exp\{-j \frac{4\pi f_0 R_B}{c}\}$$

$$\times \exp\{j2\pi f_{ac} t_m\}$$

The above improved algorithm implementation flow chart is shown in Figure 2.

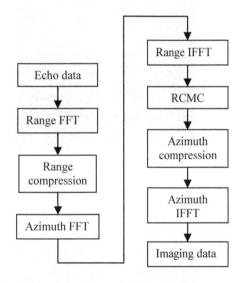

Figure 2.   Improved Range-Doppler algorithm implementation flow chart.

5 ALGORITHM SIMULATION AND ANALYSIS

The traditional Range-Doppler algorithm and the improved algorithm in this thesis are respectively used for imaging point target. Parameters of Simulation is shown in Table 1.

Figure 3 shows the echo wave envelope of single point target under the influence of range migration. Figure 4 shows the echo wave of single point target after range migration is eliminated through the traditional Range-Doppler algorithm.

253

Table 1. Parameters of simulation.

C(m/s)	1500
Carrier f_0	80KHz
Bandwidth B	20KHz
Beamwidth	12°
Slant angle	3°
Antenna D(m)	0.2
Platform V(m/s)	1.5
PRF	10Hz
Stand-off	100m
Swathwidth	200m

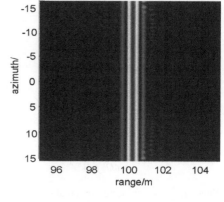

Figure 5. Range migration correction by improved Range-Doppler algorithm of this paper.

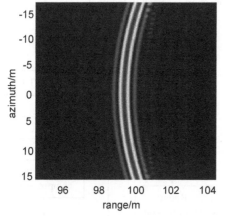

Figure 3. Single point target of echo envelope under the influence of rang migration.

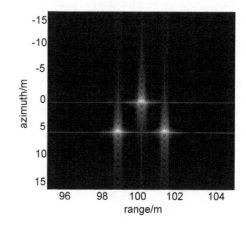

Figure 6. Three point targets imaging by classical Range-Doppler algorithm.

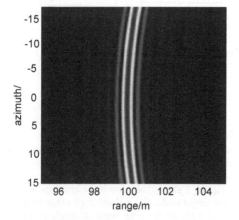

Figure 4. Classical range migration correction base on Fresnel approximation.

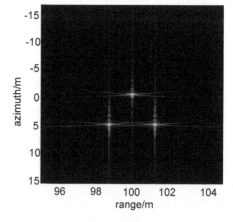

Figure 7. Three point targets imaging by improved Range-Doppler algorithm of this paper.

Figure 5 shows the result of range migration elimination through accurate range hyperbolic curve model in this thesis. Figure 6 shows three point target echo wave imaging through Rang-Doppler algorithm under the condition of traditional Fresnel assumption. Figure 8 shows the result of three point target imaging through algorithm in this thesis.

It is observed from the simulation result that the traditional Rang-Doppler algorithm has certain error since it aims at narrow band narrow-beam signals and adopts Fresnel assumption for eliminating range migration, and this kind of error becomes relatively obvious in the case of broad bandwidth wide beam signals. In the Improved algorithm, Fresnel assumption and narrow-beam assumption are both abandoned while the more accurate range hyperbolic curve model is adopted to eliminate the error of range migration, bringing better imaging effect.

6 CONCLUSION

The following conclusions can be drawn from simulation and interpretation of imaging results:

1 In the UUV synthetic aperture imaging, the instantaneous range from sonar to the target is one of the most important parameters, and the range hyperbolic curve model can bring higher precision than other models when describing instantaneous slant range.
2 Certain assumption is used for the improved Rang-Doppler algorithm in the process of derivation. This assumption would generate certain influence to imaging resolution under the condition of large squint angle, and would generate little influence to imaging focus effect under the condition of positive side-looking or small squint angle.
3 Under both conditions of positive side-looking and small squint angle modes, the improved Rang-Doppler algorithm can meet the requirement of UUV synthetic aperture imaging for broad bandwidth and wide beam signals, and has higher resolution and equivalent amount of computation as compared with the primal algorithm.

REFERENCES

Liu Yongtan, Radar Imaging Technology [M], Harbin: Harbin Institute of Technology Press, 2001.
M.A Pinto, High Resolution Seafloor Imaging With Synthetic Aperture Sonar, IEEE Oceanic Engineering Society Newsletter, Summer 2002:15-20.XIAOYUN QI, JI QI. A robust content-based digital image watermarking scheme [J]. Signal Processing, 87, 2007.1264–1280.
Peter T.Gough, Unified Framework for Modern Synthetic Aperture Imaging Algorithms [J]. John Wiley&Sons, Inc, 1997.
Jacques Chatillon, Alam E, Michael A et al. SAMI: A low-frequency prototype mapping and imaging of the seabed by means of synthetic aperture [J]. IEEE Journal of Oceanic Engineering, 1999, 24(1):4–15.
Bao Zheng, radar imaging technology [M], Beijing: Publishing House of Electronics Industry, 2005.
Wang Xuyan, algorithmic study on synthetic aperture sonar imaging and motion compensation [D]. Doctoral Dissertation of Northwestern Polytechnical University, 2008.

Computational Intelligence in Industrial Application – Ling (ed.)
© 2015 Taylor & Francis Group, London, ISBN: 978-1-138-02818-0

Evolutionary games on a spatial network model

Y.X. Li, Y.H. Chen & C.G. Huang
School of Information, Zhejiang University of Finance and Economics, Hangzhou, P.R .China

ABSTRACT: Different macroscopic or microscopic patterns of complex networks influence the outcomes of evolutionary games substantially. Though a plethora of studies have been devoted to spatial networks, i.e., complex networks embedded in a space, evolutionary games on these network types have seldom been considered. We study evolutionary games on a spatial network model (Louf et al. 2013. PNAS 110: 8824-8829). Two types of pairwise games, i.e., Prisoner's Dilemma game and Snowdrift game, are adopted. We performed numerical simulations to study the evolution of cooperation at different parameter settings of the network model. Varying the parameter which weighs the relative importance of the cost with regard to the benefits can generate the artificial network from the star graph to the minimum spanning tree. In the intermediate regime which the empirical estimates for some railway networks fall in, evolutionary game dynamics exhibits the best performance of cooperation for both types of games. This result might indicate that the structural evolution of real-world spatial networks is also beneficial to the overall cooperation if their nodes have fictitious or real mutual interplays.

KEYWORDS: Evolution of cooperation; Prisoner's Dilemma game; Snowdrift game.

1 INTRODUCTION

Our daily activities firmly rely on various networks, i.e., social networks, economic networks, transportation networks, etc. Studying the structural characteristics of as well as the dynamical processes on these networks can help us design better mechanisms to guide the relevant social or economic activities toward higher efficiency (Newman 2003, Chen et al. 2012). When modeling mutual interactions between two agents, people often resort to games. Game theory has developed as a unified paradigm behind many scientific disciplines (Weibull 1995). In a simple game model, agents can have two strategic choices, i.e., cooperation and defection. An agent can increase others' incomes by cooperating with them, whereas he can contribute nothing with a defection strategy. Notwithstanding an increase of the overall benefit via cooperation, defection is a rational choice for an agent himself. In an evolutionary system where the interacting agents can update their strategies, the evolution of cooperation becomes a puzzle.

When the game-playing agents have fixed partners, the interaction or partnership structure of them can be described as a network or graph. Nowak and May first introduced a regular lattice as the interaction structure of agents into evolutionary game models (Nowak & May 1992). Their model resembles a cellular automaton and the cooperation level as a function of the benefit-to-cost parameter exhibits a first-order phase transition. Recent years has witnessed the rise of research interests in the study of complex networks. The study of evolutionary games on complex networks thereby becomes an interest. Abramson & Kuperman (2001) studied the Prisoner's Dilemma game on W-S small-world networks. Pacheco and Santos studied evolutionary games on B-A scale-free networks (Santos & Pacheco 2005). Surprisingly, for both Prisoner's Dilemma game and Snowdrift game, the equilibrium frequency of cooperation keeps a high level on the whole parameter range. Later, it is further shown that scale-free networks promote cooperation in an analytic framework by using public goods game as a metaphor (Santos et al. 2008). The wide existence of scale-free structures in nature and human societies might indicate that many systems have evolved a structure to favor the evolution cooperation.

Rong et al. showed that assortative mixing, i.e., large-degree vertices (hubs) tend to interconnection to each other closely, degrades cooperation levels in scale-free networks (Rong et al. 2007). An empirical study showed that an online social network exhibits disassortative mixing pattern besides small-world and scale-free features (Fu et al. 2007). They studied evolutionary games on such a real network and found that the enhancement and sustainment of cooperative behaviors are attributable to the underlying network topological organization. Perc studied evolutionary games on scale-free networks that are subjected to intentional and random removal of vertices (Perc 2009). Similar to the structure resilience, evolutionary game dynamics exhibits a robust cooperation against random deletion of vertices. However, cooperation is fragile to intentional removal targeted at hubs.

Besides small-world and scale-free features, complex networks exhibits various macroscopic and microscopic patterns which influences evolutionary game dynamics substantially (Szabó & Fáth 2007). Ohtsuki et al. (2006) concluded a simple rule that is a good approximation for general networks: if the benefit of the altruistic act, b, divided by the cost, c, exceeds the average number of neighbors, k, cooperation can be favored. Recently, Wang et al. (2013) analyzed the time course of cooperation evolution on networks. They found a typical evolution process includes a period that cooperators try to endure defectors' invasion and a period that perfect C clusters fast expand their area. The equilibrium frequency of cooperation at the final stage depends on two important factors: the formation of the perfect C cluster at the end of the first period and the expanding fashion of the perfect C cluster during the second period.

Most complex networks are embedded in a space, e.g., transportation networks, power supply networks, Internet, brain networks, etc. Even individuals in a social network have their geographical positions. These large-scale networks which are characterized by the embedding of their nodes in space are commonly known as spatial networks. A network model which merely portrays the topology cannot satisfy our need to characterize these networks. Therefore, researchers have been trying to invent various network models to mix space and topology. Rozenfeld et al. (2002) proposed a simple method to construct a scale-free network on a lattice. Recently, Louf et al. (2013) introduced a generic model for the growth of spatial networks based on the general concept of cost-benefit analysis. Their model can produce a family of networks that range from the star graph to the minimum spanning tree.

Our interest in spatial networks focuses on how their unique features influence the outcome of evolutionary games. Power (2009) proposed a spatial agent-based model of N-person Prisoner's Dilemma game in a socio-geographic community. Buesser & Tomassini (2012) studied evolutionary dynamics of pairwise games on networks embedded in a Euclidean two-dimensional space with different kinds of structures and degree distributions. They found that spatial scale-free networks are still good for cooperation but to a lesser degree. In this work, we study evolutionary games on top of the artificial networks generated by Louf et al.'s algorithm. Interestingly, in the parameter regime which fits some real spatial networks, evolutionary game dynamics achieves the best performance of cooperation for both Prisoner's Dilemma game and Snowdrift game.

2 MODEL

2.1 Evolutionary game dynamics as a testing framework

A number of agents play pairwise games and update their strategies. Each agent is located in one node of a spatial network which is embedded on a two-dimensional plane and their mutual interactions are determined by the edges. Agents' strategies are either cooperation (C) or defection (D). At each time step of the evolution, an agent plays a game with each of its neighbors and obtains a payoff. An agent's fitness is the sum of all its obtained payoffs. A payoff matrix A determines the payoff for an agent when it plays a game with another one. C strategy is represented as $[1, 0]^T$, while D strategy is $[0, 1]^T$. Therefore, we can write the fitness of agent i as the following equation:

$$f_i = \sum_{j \in \mathcal{N}_i} s_i^T \mathbf{A} s_j, \tag{1}$$

where $\mathcal{N}_i$ is the set of neighbors of agent i.

The payoff matrix is crucial to understand the interplay between two agents. Usually, such a matrix has the following form:

$$\mathbf{A} = \begin{bmatrix} R & S \\ T & P \end{bmatrix}. \tag{2}$$

In an evolutionary step, an agent is randomly selected to update its strategy after all agents have finished game playing. An agent chooses a random neighbor as its model and imitates its strategy with a probability (Szabó & Toke 1998)

$$\omega(s_i \leftarrow s_j) = \frac{1}{1 + exp(\frac{-(f(j) - f(i))}{\theta})}, \tag{3}$$

where θ is the noise extent of imitation. Under this function, agents imitate evolutionarily-advantaged ones smoothly.

The fraction of cooperators to all agents, ρ_C, measures the level of cooperation. The evolution experiences a number of time steps until there is a relatively-steady fraction of cooperators in the population. In our simulations, we usually let the evolution run for 10^5 time steps. At the steady regime of an evolution, the averaged fraction of cooperators was computed. Data points in the following figures were further averaged over 100 independent simulation runs; namely, the equilibrium frequencies of cooperation $\langle \rho_C \rangle$ are presented.

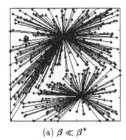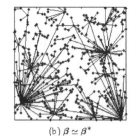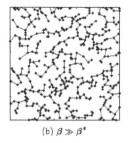

| (a) $\beta \ll \beta^*$ | (b) $\beta \simeq \beta^*$ | (b) $\beta \gg \beta^*$ |

Figure 1. A visualization of generated networks with three different regimes of β. (a) $\beta/\beta^* = 0.01$; (b) $\beta/\beta^* \simeq 1$; (c) $\beta/\beta^* = 100$. Other parameters are $\mu = 1.1$, $L = 100$, $N = 400$, $k = 10$ and $a = 1.1$. The appearances of these networks are quite different.

2.2 Louf et al.'s spatial network model

Before simulating the evolution, a spatial network is generated using the algorithm proposed by Louf et al. (2013). This model is based on a cost-benefit analysis mechanism. First, the nodes are distributed uniformly in the two-dimensional plane. For the simulations, this plane is simulated by a $L \times L$ lattice. The weight M of each node is distributed according to the power law

$$P_M(x) = \frac{\mu}{x^{\mu+1}}. \qquad (4)$$

Such a weight conveys a certain socioeconomic index. For example, it can be interpreted as the size of the population resided in the node if the artificial network represents a railway network. The network is generated from a root node which is selected randomly. Then, edges are added recursively by connecting a new node to one that is already connected to the network. At each time step, the nodes belonging to the graph is defined as "inactive" nodes and the other, not yet connected nodes are "active". An edge is generated by connecting an active node to an inactive node, such that

$$R'_{ij} = k \frac{M_i M_j}{d_{ij}^{a-1}} - \beta d_{ij} \qquad (5)$$

is maximum. Here, d_{ij} is the Euclidean distance between i and j. Obviously, $a \geq 1$, $k > 0$, $\beta > 0$. In this function, the quantity β weighs the relative importance of the cost with regard to the benefits, which is the most influential parameter to shape the generated network. A criterion to set such a parameter is to determine an average magnitude $\beta^* = k\bar{M}^2 \rho^{a/2}$. $\bar{M}$ is the average weight of all nodes, $\rho = N/L^2$ denotes the node density on the given plane. There are three regimes of β, i.e., $\beta \ll \beta^*$, $\beta \simeq \beta^*$ and $\beta \gg \beta^*$. As is shown in their article [15], the variation of this parameter produces a family of networks that ranges from the star graph to

the minimum spanning tree. The empirical estimates of β/β^* for some real railway networks in different countries usually fall in the intermediate regime.

3 RESULTS

3.1 Results using the payoff matrix of Prisoner's dilemma game

We implemented the simulation framework in C using igraph library (Csardi & Nepusz 2006). Figure 1 gives an illustration of three typical generated networks. Parameters are the same for them except different settings of β. In Figure 1a, vertices connects to several hub ones. In Figure 1c, the generated network approaches a minimum spanning tree, as is discussed in their article (Louf et al. 2013). In the following subsections, the equilibrium levels of cooperation as a function of the game parameter at different network parameter settings are presented. In the initial stage of a simulation, each agent is assigned with a random strategy. The fraction of cooperation is measured at the steady regime of the evolution. A node's weight is sampled from [1, 50); namely, we set up a maximal limit.

Prisoner's Dilemma game is the most famous pairwise game. In simulations, the following form of the payoff matrix with only one parameter is frequently adopted (Nowak & May 1992):

Table 1. The payoff matrix for prisoner's dilemma game.

	C	D
C	1	0
D	b	0

Here, $b \in (1, 2]$ is a usual setting. The payoff matrix tells us that an agent will obtain a higher pay by playing D strategy than C strategy when it meets a cooperator. However, an agent will obtain nothing when it meets a defector. Hence, the strategic choice for a rational agent is defection. The simulation results are shown in Figure 2. $\langle \rho_C \rangle$ as a function of b is presented with different magnitudes of β. The legend specifies the order of magnitude for β/β^*. It can be seen that $\beta/\beta^* \simeq 1$ leads to the highest level of cooperation, compared with other β/β^* values. In Figure 3, we further show $\langle \rho_C \rangle$ as a function of b with some specific values in the intermediate regime $\beta/\beta^* \simeq 1$. The parameter range $0.1 < \beta/\beta^* < 1$ facilitates the evolution of cooperation the most. As is shown in their article (Louf et al. 2013), the generated networks possess the property of a spatial hierarchy in the intermediate regime $\beta/\beta^* \simeq 1$. Interestingly, the estimate β/β^* for some real-world networks falls in this regime.

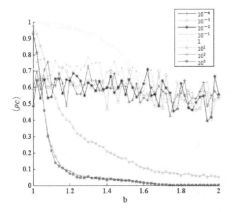

Figure 2. The equilibrium frequency of cooperation as a function of the parameter of the PDG for different magnitudes of β. In the regime $\beta/\beta^* \simeq 1$, the best performance of cooperation is yielded. Figures 2-4 have the same parameter values with Figure 1 except β.

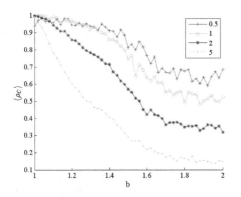

Figure 3. The equilibrium frequency of cooperation as a function of the parameter of the PDG with some specific values within the intermediate regime $\beta/\beta^* \simeq 1$.

3.2 Results using the payoff matrix of snowdrift game

Besides Prisoner's Dilemma game, the Snowdrift game (also known as Chicken or Hawk/Dove game) was also adopted for testing the optimal parameter range of β. Let us look at the payoff matrix in Table 2.

Table 2. The payoff matrix for snowdrift game.

	C	D
C	1/(2r)	1/(2r) − 1/2
D	1/(2r) + 1/2	0

Here, $0 < r \leq 1$. Obviously, Snowdrift game is different from Prisoner's Dilemma game. An agent faces the same circumstance when its co-player is a cooperator. However, an agent will obtain a higher pay by playing C strategy than D strategy when it meets a defector. In this dilemma, a cooperative behavior does not merely contribute to others. Meanwhile, it brings about some benefit to the subject.

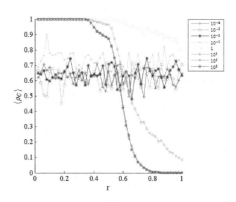

Figure 4. The equilibrium frequency of cooperation as a function of the parameter of the SDG for different magnitudes of β. The qualitative result is the same with that in the situation of PDG.

In Figure 4, $\langle \rho_C \rangle$ as a function of b is shown for different magnitudes of β in the case of Snowdrift game. Figures 4 and 2 can be compared as the generated networks have the same parameter setting. Here, we can obtain the same result that the parameter range $0.1 < \beta/\beta^* < 1$ is the most optimal for the evolution of cooperation. A comparison between Figure 4 and Figure 2 further shows that Snowdrift game has a remarkably better performance of cooperation than Prisoner's Dilemma game in the regime $\beta \gg \beta^*$. This is attributed to the benefit-cost sharing mechanism of Snowdrift game. Therefore, for both types of games, the optimal parameter regime of β/β^* for cooperation coincides with the parameter regime of β/β^* which gives the best fit to some real-world spatial networks.

3.3 The influences of some other factors

Since the network model has several tuning parameters, we further investigate whether some other factors influence the qualitative results of the evolutionary game dynamics. Each simulation started with a newly-generated network. Therefore, the simulation results are based upon "averaged" networks. Without specifying the values, network parameters are set the same with those of Figure 1. Figure 5 compares the results for two different network sizes. Notice that $L = 100$. Therefore, the network with $N = 40$ has a more scattered distribution of nodes on the two-dimensional plane. At $\beta/\beta^* = 1$, $N = 40$ promotes cooperation to a lesser degree than $N = 400$. However, given $N = 40$, the case of $\beta/\beta^* = 1$ outperforms the cases of $\beta/\beta^* = 0.01$ and $\beta/\beta^* = 100$ in increasing the levels of cooperation on average.

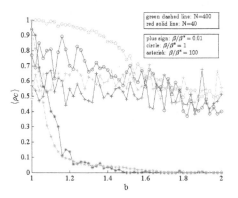

Figure 5. The influence of the network size over the outcome of the evolutionary dynamics. The equilibrium frequencies as a function of b with $N = 400$ and $N = 40$ are presented. For both cases, the qualitative results are alike.

μ is set at 1.1 in Louf et al. (2013)'s simulations, which is motivated by empirical results on city populations. In our simulations, the default setting of μ is 1.1. The influence of the parameter μ in Equation 4 is reported in Figure 6. The equilibrium frequencies of cooperation with $\mu = 1.5$ and $\mu = 1.1$ are at the same level when other parameters are the same.

The rate with the process of link formation is controlled by the parameter k in Equation 5. In Figure 7, $\langle \rho_C \rangle$ as a function of b is shown with varied k. For two different settings of this parameter, the quantitative levels of cooperation are almost equal when other parameters are the same. These simulations with varied network parameters indicate that the intermediate

regime of β leads to the best performance of cooperation and such a result is robust to the variation of some other factors.

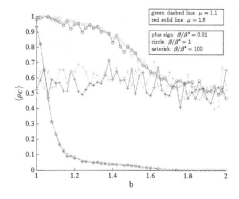

Figure 6. The simulation results with varied μ in Equation 4. The equilibrium frequencies as a function of b with $\mu = 1.1$ and $\mu = 1.5$ are presented. The two cases yield quite similar results.

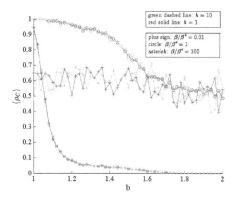

Figure 7. The influence of the parameter k in Equation 5 over the outcome of the evolutionary dynamics. The equilibrium frequencies as a function of b with $k = 1$ and $k = 10$ are presented.

4 CONCLUSIONS

A continuing effort in the complex network field is inventing more accurate models to capture the structural features of real-world networks. Based on these models, the selection of their parameters is a further step towards a good description or representation of a particular network data set. Humans heavily rely on various man-made spatial networks, e.g., railway networks. Louf et al. proposed a theoretical model to characterize these spatial networks based on a cost-benefit mechanism. Depending on the parameter β which represents the relative importance of the cost with regard to the benefits, the model can generate a family of networks. In this work, we have studied evolutionary games on these spatial networks. It is interesting that equilibrium frequencies of cooperation achieve the best level in the intermediate regime of the cost-benefit parameter β. It is shown that the empirical estimates of this parameter for railway networks in several countries fall in this intermediate regime. Therefore, such a parameter regime leads to an optimal network structure. From another point of view, evolutionary game dynamics might help designing robust spatial networks.

ACKNOWLEDGEMENTS

Y.L. would like to acknowledge supports from the NSFC (11347201), the Zhejiang Provincial Natural Science Foundation (LQ13F030004) and the MOE Project of Humanities and Social Sciences (13YJC630084).

REFERENCES

Abramson, G. & Kuperman, M. 2001. Social games in a social network. *Physical Review E* 63(3): 030901.

Buesser, P. & Tomassini, M. 2012. Evolution of cooperation on spatially embedded networks. *Physical Review E* 86(6): 066107.

Chen, S., Huang, W., Cattani, C. & Altieri, G. 2012. Traffic Dynamics on Complex Networks: A Survey. *Mathematical Problems in Engineering* 2012: 732698.

Csardi, G. & Nepusz, T. 2006. The igraph software package for complex network research. Inter. J. Complex Syst. 2006: 1695.

Fu, F., Chen, X. Liu, L. & Wang, L. 2007. Social dilemmas in an online social network: the structure and evolution of cooperation. *Physics Letters A* 371(1-2): 58–64.

Louf, R., Jensen, P. & Barthelemy, M. 2013. Emergence of hierarchy in cost-driven growth of spatial networks. *Proceedings of the National Academy of Sciences USA* 110: 8824–8829.

Newman, M.E.J. 2003. The Structure and function of Complex Networks. *SIAM Review* 45(2): 167–256.

Nowak, M.A. & May, R.M. 1992. Evolutionary Games and Spatial Chaos. *Nature* 359(6398): 826–829.

Ohtsuki, H., Hauert, C., Lieberman, E. & Nowak, M.A. 2006. A simple rule for the evolution of cooperation on graphs and social networks. *Nature*, 441(7092): 502–505.

Perc, M. 2009. Evolution of cooperation on scale-free networks subject to error and attack. *New Journal of Physics* 11(3): 033027.

Power, C. 2009. A Spatial Agent-Based Model of N-Person Prisoner's Dilemma Cooperation in a Socio-Geographic Community. *Journal of Artificial Societies and Social Simulation* 12(1): 8.

Rong, Z., Li, X. & Wang, X. 2007. Roles of mixing patterns in cooperation on a scale-free networked game. *Physical Review E* 76(2): 027101.

Rozenfeld, A.F., Cohen, R., ben-Avraham, D. & Havlin, S. 2002. Scale-Free Networks on Lattices. *Physical Review Letters* 89: 218701.

Santos, F. C. & Pacheco, J. M. 2005. Scale-free networks provide a unifying framework for the emergence of cooperation. *Physical Review Letters* 95(9): 098104.

Santos, F.C., Santos, M.D. & Pacheco, J.M. 2008. Social diversity promotes the emergence of cooperation in public goods games. *Nature* 454(7201): 213–216.

Szabó G. & Fáth, G. 2007. Evolutionary games on graphs. *Physics Reports* 446(4-6): 97–216.

Szabo, G. & Toke, C. 1998. Evolutionary prisoner's dilemma on a square lattice. *Physical Review E* 58: 69.

Wang, Z., Kokubo, S., Tanimoto, J., Fukuda, E., & Shigaki, K. 2013. Insight into the so-called spatial reciprocity. *Physical Review E* 88(4): 042145.

Weibull, J.W. 1995. *Evolutionary Game Theory*. Cambridge, MA: MIT Press.

Computational Intelligence in Industrial Application – Ling (ed.)
© 2015 Taylor & Francis Group, London, ISBN: 978-1-138-02818-0

Adaptive road-restricted vehicle detection on aerial image sequence

S.Y. Zhang, Y.S. Li, J.W Guo & Z. Li
Southwest Jiaotong University, Chengdu, Sichuan, China

ABSTRACT: A new adaptive method of detecting vehicles from aerial image sequences was proposed in this paper, with the restriction of road obtained from Geographic Information System (GIS). Vehicle detection was executed by virtue of an optimal space searching algorithm in a multi-dimensional space which includes scale, angle and 2D image plane dimension. This paper adopted a typical, commonly used Histogram of oriented Gradients (HoG)-based Support Vector Machine (SVM) classifier, and then gained the best responses of the locations and directions of vehicles through specially designed multi-dimensional searching and removal of repetitive responses. This new approach was tested with an aerial image sequence with dense traffic conditions. The result reveals that an average quality of more than 76 percent can be achieved with road information assistance, and of more than 72 percent without road information guidance. Although most efforts have been paid on the development of advanced classifier, the result demonstrates that the detector could also have played a considerable role in vehicle detection.

1 INTRODUCTION

Traffic information is significant for the traffic management and planning. Current traffic data can be gathered in various ways. Traditionally, widely used method is induction loops (Davidson & Valentine 2001), which is a low-priced solution to gather traffic data continuously; another widely used technology is stationary traffic video cameras (Bischof et al. 2010); and floating car method can also give information about the traffic flow (Kerner et al. 2005). All these can only detect cars in some points or segments of the road, there are some limitations for wide range traffic monitoring. Today, remote sensing enables gathering geo-information from a distance (Hinz et al. 2006). Especially, satellites allow mapping of very large areas (Leitloff et al. 2010, Salehi et al. 2012), but often have a low repetition time and a low resolution. More flexible are airborne sensors operating on helicopters (Karimi Nejadasl et al. 2006), UAVs (unmanned aerial vehicle) (Moranduzzo & Melgani 2014) or other aircrafts.

SAR, LiDAR and optical sensors are three primarily used airborne remote sensors in traffic monitoring. For SAR, an advantage is that it is independent of weather condition (Hinz et al. 2007, Suchandt et al. 2010). Velocities can also be derived with a device of moving target indication (Baumgartner & Krieger 2012). In terms of LiDAR, car detection and velocity deduction are also possible (Yao & Stilla 2011), and LiDAR is somewhat suitable for nadir view in urban areas having a side looking geometry. Both SAR and LiDAR, however, cannot provide discernable color

information about the images. In contrast, optical sensors make color discrimination of different objects in the images possible. IR cameras, a type of optical sensors for instance, can distinguish cars with different activity states (e.g. moving or stationary) because warmer parts such as an engine appear to be brighter in the images (Hinz & Stilla 2006). But this discrimination is only to a limited degree. That is to say, only warm objects can be detected, while cold targets will be neglected. Although having a high image acquisition frequency, IR cameras possess only a small pixel and therefore a low spatial resolution which may significantly lower the detection efficiency, and its higher price relative to other optical sensors also makes it not a favorable choice for vehicle detection.

Currently aerial video and high frequency aerial image sequences are applied to vehicle detection, Reinartz et al. (2006) studied serial aerial images based traffic monitoring, and their result indicate that the frequency of aerial images sequence used in traffic vehicle detection should be greater than 3Hz. Airborne vehicle detection basically includes automatic extraction and manual extraction, while the former is the mainstream.

Mtir et al. (2012) studied the image difference based moving vehicles detection. Firstly, calculate the transformation matrix, then rectify adjacent frames, and make difference between adjacent frames, finally moving vehicles are extracted with some postprocessing (thresholding, morphologic processing etc.). Difference detection approach reliance upon temporal change is very reliable, and only detects moving vehicles. For aerial video, the usual approach is optical flow based moving vehicles tracing (Karimi Nejadasl et al. 2006), which is similar to difference method that cannot distinguish the stationary vehicles.

A method is manually constructing a car model based on empirical knowledge for car detection. Moon et al. (2012) constructed a vehicle rectangular model which consists of four edge detectors having the size and the shape of an average car, then used Canny algorithm extracting edges from the aerial image, and matched edge map with the car model to distinguish cars. The method's main compute cost is edge detection with small price, but the model is mainly dependent on the vehicle edges, which is insufficient for vehicle detection to eliminate false positives (like the vegetation area with the similar structure to vehicles).

Currently, the researchers tend to utilize off-line or on-line machine learning methods using training samples. Kembhavi et al. (2011) calculates three feature classes (Histogram of Gradients (HoG), Color Probability Maps (CPM) and Pairs of Pixels (PoP)) which are concatenated in a feature vector of approximately 70000 elements for a sample, and describe the vehicle detection as a regression problem. Finally, solve the problem by utilizing the partial least squares (PLS).

One of the implicit methods makes use of the Histogram of oriented Gradients (HoG), Haar-like features and Local Binary Patterns (LBP) (Grabner et al. 2008). These features are passed to an on-line Boosting algorithm to generate the strong classifier. The focus is on the on-line ability of the machine learning algorithm. Almost all of the related publications try to improve the detection of vehicles by introducing new features of applying different training of machine learning methods (Tuermer et al. 2013).

The complexity of vehicle postures makes them not easy to be detected from an aerial image. A successful detection so far may rely on a good arrangement of vehicles. For instance, the detection work by Cao et al. (2011) was based on accordant direction of cars which can be seen from the traffic video. This one-direction traffic trend made the detection process simple. Similarly, vehicles detected by Tuermer et al. (2013) also showed a fixed direction in the image data. However, complicated and various vehicle postures may need to be handled during traffic detection of large areas, and thus a more sophisticated and advanced detection method is requited to cover a more complicated traffic area.

Auxiliary data has been introduced to improve vehicle detection. Wei & Zhou (2008) used GIS road data GIS to constrain their detection area. Moranduzzo & Melgani (2014) employed a classification stage to speed up vehicle detection process. Tuermer et al. (2013) adopted image segmentation pretreatment and disparity map to constrain search area, by which they improved their vehicle detection in spite of a small number of samples.

In addition, Tuermer et al. (2013) demonstrated that a better detector would help to improve vehicle detection. Therefore, one may expect that combining of designing an efficient detector and making full use of GIS road information will promote the detection efficiency significantly.

In this study we proposed a new adaptive method of detecting vehicles from aerial image sequences, with road information obtained from GIS. The following contain two sections: Section 2 describes the framework and key technology of our method and Section 3 is the experimental results and analysis.

2 METHODOLOGY

This study uses machine learning technology to train a classifier to recognize vehicles from aerial images. Two main parts of the method involve a typical classifier (light grey) and a special detector (dark grey) with high performance (Fig.1).

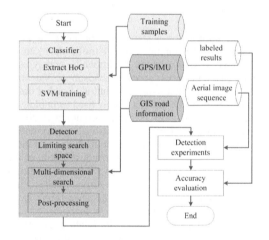

Figure 1. The framework of our method.

2.1 Feature extraction and machine learning

HoG feature, which was firstly introduced by Dalal & Triggs (2005), was chosen as the descriptive element of the cars. A sliding window based approach where the car sample size is 32×64 was selected. We set the size of the HoG's block as 16×16, cell as 8×8, bin number as 9, and slide overlapping subwindows with sizes of 8×8 over the whole window area, and then we calculated features for every subwindow. The final result would be a 756 features vector. A detailed description of how to create features can be found in Dalal & Triggs (2005), and thus will not be iterated here. The HoG of a car is shown in Figure 2.

A number of positive samples were extracted manually (Fig. 2, left), before which the directions of these cars had been corrected to the same orientation. Negative samples were randomly generated, as shown in Fig. 2 (right). The HoG features of these positive and negative samples were utilized as SVM's input, and a linear-SVM was employed to train the classifier and test its accuracy.

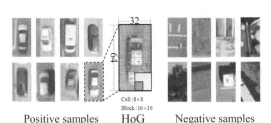

Positive samples   HoG   Negative samples

Figure 2.   Training samples and HoG feature.

2.2 *Adaptive road-restricted detection*

First of all, we introduced the detection model, and based on this, the following subsections describe in detail our vehicle detection method (Fig. 1), which mainly contains three parts: (1) limiting search space, (2) multi-dimensional searching, and (3) post-processing.

2.2.1 *Detection model*
The detector can be described as a hypothesis $H(X)$ which can discriminates between vehicles and background (tree, road, and building, etc.). Detection is a process which searching the maximum response of $H(X)$ from a 4-D aerial image space (i.e. scale, angle, and 2-D image plane) which can be defined as:

$$(x, y, \sigma, \theta) = \arg\max\left(H(X)\right) \tag{1}$$

Where x, y is the plane coordinates, σ is scale factor and θ is angle. $X = f(x, y, \sigma, \theta)$ is a function of plane coordinates, scale and angle. $H(X)$ is the response of H for X, and in here H is SVM classifier.

If one can determine the domain of parameters above, and limit search spaces as Tuermer et al. (2013) did, then vehicle detection in a multi-scale and multi-angle space can be accomplished.

2.2.2 *GIS-based detection limit searching space*
GIS provides the road information which can be used as the prior knowledge (i.e. the road plane position and orientation) of our adaptive detector to constrain search spaces. The GPS and IMU of aircrafts can be used simultaneously to rectify aerial images, and then approximate image plane search spaces can be achieved.

Scale Space: The key to determine search scale is calculating image spatial resolution, which will be affected by the flight height of the airborne platform and also status of the camera. Therefore, the scale space can be described with the following equitation:

$$\sigma = g(h, r, \varphi) \tag{2}$$

Where h is flight height, r is posture parameters, and φ is camera elements. The h, r are given by GPS and IMU respectively, φ is provided by camera calibration. Then, the scale σ can be obtained.

Image Plane: Based on the road information provided with GIS, and combined with scale space parameters and the prior knowledge regarding vehicles (e.g. road width, vehicle sizes and traffic rules, etc.), it is easy to extract a rough traffic region (S in Fig. 3, left) from an image area with GIS spatial analysis technology. Most commonly, using road centerlines (l in Fig. 3, left) to do extraction is sufficient. Traffic regions can be divided into two types based on vehicle moving behaviors: one is non-intersection traffic area (S_1 in Fig. 3, right) and the other is intersection area (S_2 in Fig. 3, right).

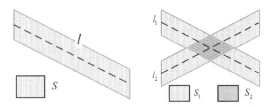

Figure 3.   Two types of image plane.

The moving direction of vehicles in S_1 area is relatively fixed, that is, consistent with the road orientation and parallel to the l_1 or l_2 line. In S_2 area, however, the vehicle direction is variable and not easy to determine due to the complexity of traffic rules in the intersection region and driver's destination.

Angle Space: this paper employed k-nearest neighbor-based method to initialize the angle search parameter. A road l can be regarded as composed of n points $\{p_1 \dots p_n\}$. If a car V is on the road l, the angle search parameter then needs to be initialized for S_1 and S_2 area separately, based on the previous discussion of different features of the two areas.

In S_1 area where the road's orientation and the vehicle's direction are roughly parallel, as shown in Figure 4, the V's initial direction θ was estimated by searching the two road points $\{p_i, p_{i+1}\}$ nearest to V using k-NN algorithm (where $k=2$), and then $\pm\Delta=5°$ was selected as V's angle search range.

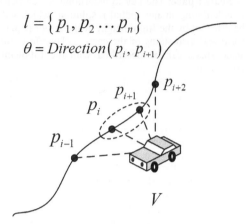

$$l = \{p_1, p_2 \cdots p_n\}$$
$$\theta = Direction(p_i, p_{i+1})$$

Figure 4.   k-NN angle initialization (k=2).

In S_2, however, the flexibility of vehicle's direction makes $\pm\Delta=90°$ (i.e. 0°–180°) the necessary and reasonable range of search angle.

2.2.3   Multi-dimensional searching
To avoid false responses, a threshold δ was set to filter out samples with weak positive responses which are the distances from the HoG feature vectors of samples to the classification hyper-plane in SVM. The

---

Algorithm 1: Multi-dimensional searching

Input:
source image **I**: response threshold δ
Initialization:
initial angle θ_0 (default is 0)←k-NN-based method
initial scale σ_0 (default is 1)←GPS/IMU
I's response matrix $\mathbf{R} \leftarrow \mathbf{0}$
I's orientation matrix $\mathbf{O} \leftarrow \mathbf{0}$
Iteration:
for $\mathbf{I}_{ij}$ in **I** do
   $r_{temp} = 0; o_{temp} = 0;$
   if *region type* $== S_1$ then
      for θ in $[\theta_0 - 10°, \theta_0 + 10°]$ do
         for σ in $[\sigma_0-0.1, \sigma_0, \sigma_0+0.1]$ do
            if SVM$(\theta, \sigma) > r_{temp}$ then
               $r_{temp} = $ SVM(θ, σ); $o_{temp} = \theta$;
   else if *region type* $== S_2$ then
      for θ in $[0°, 180°]$ do
         for σ in $[\sigma_0-0.1, \sigma_0, \sigma_0+0.1]$ do
            if SVM$(\theta, \sigma) > r_{temp}$ then
               $r_{temp} = $ SVM(θ, σ); $o_{temp} = \theta$;
if $r_{temp} > \delta$ then
   $\mathbf{R}_{ij} = r_{temp}$; $\mathbf{O}_{ij} = o_{temp}$;
Output:
   Response matrix $\mathbf{R}$ and orientation matrix $\mathbf{O}$ of I;

---

threshold was calibrated with training dataset and test dataset. To be more specific, the threshold to remove weak responses was set at a distance where no less than 95% vehicles could be detected successfully (Sec. 3.2 and Fig. 7).

The discussion above has revealed that GPS/IMU and road information together can limit search spaces to a great extant; In particular road information can define both the image plane search space and the angle search space very well. Algorithm 1 is the pseudo-code of multi-dimensional searching that was used for limited search spaces (Fig. 5).

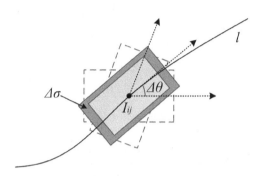

Figure 5.   Diagram of multi-dimensional searching.

2.2.4   Post-processing
It is obvious that the sliding-window detection can lead to the detection of one car more than once, that is, multiple false positives for the same car. Therefore, only when the center of a window has the local maximum response relative to neighbors can the center value be kept, otherwise the center will be recorded as 0 (Fig. 6).

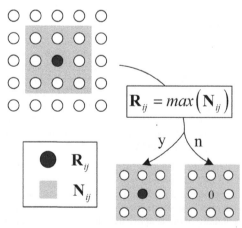

$$\mathbf{R}_{ij} = max(\mathbf{N}_{ij})$$

* $\mathbf{N}_{ij}$ is the $k \times k$ neighborhood of $\mathbf{R}_{ij}$, here $k = 3$.

Figure 6.   Local non-maximal suppression.

3 EXPERIMENTS

The experimental results are evaluated using the following assessment standards. Correctness is TP/TP+FP, completeness is TP/TP+FN and quality is TP/TP+FP+FN; where TP is the sum of true positives, FP is the sum of false positives, and FN is the sum of false negatives.

The accuracy of position and direction of detected vehicles is evaluated by the root mean square error (RMS) between detection results and manual-labeled results.

3.1 Dataset

Experimental dataset: The performance of the proposed approach was evaluated on the KIT AIS aerial imagery dataset (Schmidt 2012),which was acquired by the German Aerospace Center (DLR), and processed by the Karlsruhe Institute of Technology (KIT) Institute of Photogrammetry and Remote Sensing (IPF) to be used for traffic monitoring and pedestrian tracking. Nine image sequences with high spatial resolution of about 15cm comprise the dataset, and all of them are geo3-referenced by GPS/IMU.

Training samples: 420 vehicle samples were manually extracted for training, and 100 samples for testing, then 900 negative samples were randomly generated from the background of aerial images. Some samples are shown in Figure 2.

3.2 Training classifier

HoG features were extracted for all positive and negative samples, and linear-SVM was employed for training the classifier, detailed training process was similar to Tuermer et al. (2013). In our training, the HoG parameters are consistent with descriptions of section 2.1, and the value of parameter C in linear-SVM is 0.01 which is determined using cross validation. Test results showed that our SVM classifier achieved 98% detection accuracy.

Using the method described above in Section 2.3, all of the vehicle samples (420 training vehicle samples and 100 test vehicle samples) have been used to calibrate the threshold value δ of SVM detection. Figure 7 (left) shows the statistical histogram of the SVM classifier's responses and the fitting curve with normal probability distribution, and Figure 7 (right) shows the accumulation normal distribution. 95% correct classification rate of calibrating datasets required the threshold value δ to be 0.48.

After getting the classifier and the threshold value, MunichStreet04 area of KIT AIS with dense vehicles was chosen to do experiment.

During the experiment, the road-restricted detection method (favored in this paper) was compared with unrestraint detection to test its validity. All detections used the same parameter setting. We padded all image boundaries with 8 pixels for detecting vehicles located at the edge of images, and set the size of sliding windows as 3×3; for the angle varying step size, $[\theta_0-10°, \theta_0+10°]$ in the increment of 2° was used as the range of angle θ for detection when the initial angle $\theta 0$ could be initialized (usually in region S_1); otherwise the range was $[0°, 180°]$ in step of 10° (usually in region S_2). Scale factor σ was taken $[\sigma_0-0.1, \sigma_0, \sigma_0+0.1]$) during detection, where the initial scale factor σ_0 was computed using GPS/IMU. The size of local non-maximizing window was set as 3×3 in post-processing.

3.3 Results and discussion

Results of the experiment are shown in Figure 8 where Figure 8a is the result of unrestraint detection, while Figure 8b gives the result of road-restricted detection.

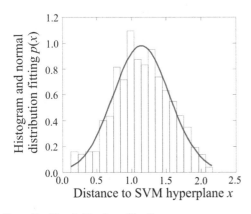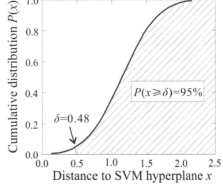

Figure 7.   Threshold value calibration.

Figure 8a-1 is the hand-labeled vehicles (the ground truth used in detections a and b), Figure 8a-2 displays the SVM response map of unrestraint detection, and Figure 8a-2 is the detection results after post-processing, in which × represents a false positive, → is a true positive (the starting points of the arrows indicate the locations of the detected vehicles and the directions of the arrows characterize vehicle orientations), and ○ is a false negative. Figure 8b-1 shows the center lines and the buffer zone of road-restricted detection. Since the spatial resolution of MunichStreet04 sequence was 0.2m, the buffering distance of 50 pixels was adopted in this experiment. Symbols in 2 and 3 of Figure 8b have the same meaning as that of Figure 8a, except for that Figure 8b is road-restricted detection.

For the convenience of comparison, the detecting directions of true positives were corrected to be consistent with the labeled directions. It should be noted that detected vehicles with location errors or orientation errors were all considered as false positives.

The effectiveness of this road-restricted detection method can also be expressed quantitively, by doing the average of each parameter (correctness, completeness and quality) of all frames in the experiment, which has been shown in Table 1.

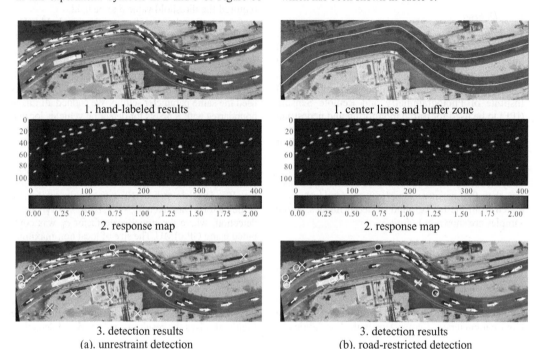

1. hand-labeled results

2. response map

3. detection results
(a). unrestraint detection

1. center lines and buffer zone

2. response map

3. detection results
(b). road-restricted detection

Figure 8. MunichStreet04 detection results: Both unrestraint detection (a) and road-restricted detection (b) employed the frame 1 of MunichStreet04 dataset. In (a), correctness=75.8%, completeness=88.7%, quality=69.1%; in (b), correctness=85.7%, completeness=90.6%, quality=78.7%. From the result, the improvement is obvious for road-restricted detection, and the searching space of (b) is much smaller relative to (a)'s which means faster detection.

Table 1. Detection evaluation.

Average Value	Normal detection			Without consideration of truck		
	Correctness	Completeness	Quality	Correctness	Completeness	Quality
Unrestraint detection	81.9%	86.8%	72.8%	88.6%	88.0%	79.1%
Road-restricted detection	84.5%	89.0%	76.4%	92.3%	89.5%	83.3%
Comparison	+2.6%	+2.2%	+3.6%	-	-	-

Both Figure 8 and Table 1 show that the road-restricted method also performs quite well in vehicle-populated regions, the correctness, completeness and quality have been improved by 2.6%, 2.2% and 3.6% respectively relative to the unrestraint detection. An interesting problem was that most false positives occur on trucks; this was probably due to the lack of truck-like cars in our training samples. Therefore, the effectiveness of this method should be evaluated without including trucks. Exclusion of trucks in the experiment, also related false positives, true positives and false negatives, led to a significant increase in the correctness, completeness and quality.

Other aspects that needed to be evaluated were the location and orientation accuracy of the method. The average RMSs were thus calculated by comparing both location (x and y) and orientation (angle) between the road-restricted and the manual-labeled detection. Statistical results are shown in Table 2.

In conclusion, compared with the unrestraint detection, the number of false positives of road-re-

Table 2. Location and orientation accuracy.

	Average RMS-x (pixel)	Average RMS-y (pixel)	Average RMS-$angle$ (rad)
Average Value	2.25	1.48	0.059

stricted detection was reduced (Fig. 8 and Tab.1). This can be readily explained by the fact that the road-restricted detection has constrained some parts of a full 4-D space (image plane, scale, and angle) which has been searched by the unrestraint detection. Clearly, an increase of the number of false positives and also of the detection time will occur if the size of the search space is increased.

Proved through the experiment, road-restricted detection can greatly limits the searching space, indirectly reduces the false positives and false negatives number, and then improves the detection results. At the same time, the location and orientation accuracy of all experiments is well.

4 CONCLUSION

This paper has presented an adaptive vehicle detection method that can accurately detect the location and direction of the vehicle in aerial images. On KIT AIS datasets, we have achieved good detection results using SVM with a small number of training samples.

The proposed approach doesn't restrict the vehicle's orientation, and using the GIS road information improves detecting speed and detection results. It

is also important generating negative samples from multi-dimensional space to avoid obvious false negatives. On the other hand, our detection results can provide initialization for the vehicle track, and further are used to estimate the velocity of the vehicle etc.

Our results show that the detector is as important as the classifier, and it is still interesting to design more sophisticated ways to improving the detector. One possible way is to combine a number of simple and different types of detectors, like house detector and road detector, to obtain better detection. Also, the detection of our current pipeline is somewhat slow, and it is interesting to exploit the detection algorithm to speed it up.

ACKNOWLEDGMENTS

This work was supported by "the Fundamental Research Funds for the Central Universities", Special Fund by Mapping Technology Plan in 2014 "Using UAV images to Identify and Extract Disaster information" and Open Research Fund by Sichuan Engineering Research Center for Emergency Mapping & Disaster Reduction.

REFERENCES

Bischof, H., Godec, M., Leistner, C., Rinner, B. & Starzacher, A. 2010. Autonomous audio-supported learning of visual classifiers for traffic monitoring. *IEEE Intelligent System* 25(3):15–23.

Baumgartner, S. V. & Krieger, G. 2012. Fast GMTI algorithm for traffic monitoring based on a priori knowledge. *IEEE Transactions on Geoscience and Remote Sensing* 50(11):4626–4641.

Cao, X., Wu, C., Lan, J., Yan, P. & Li, X. 2011. Vehicle detection and motion analysis in low-altitude airborne video under urban environment. *IEEE Transactions on Circuits and Systems for Video Technology* 21(10):1522–1533.

Dalal, N. & Triggs, B. 2005. Histograms of oriented gradients for human detection; *proceedings of the Computer Vision and Pattern Recognition*. 2005 IEEE Computer Society Conference on Computer Vision and Pattern Recognition Workshops (CVPR 2005), pp. 886–893.

Grabner, H., Nguyen, T. T., Gruber, B. & Bischof, H. 2008. On-line boosting-based car detection from aerial images. *ISPRS Journal of Photogrammetry and Remote Sensing* 63(3): 382–396.

Hinz, S., Bamler, R. & Stilla, U. 2006. Editorial theme issue: Airborne and spaceborne traffic monitoring. *ISPRS Journal Photogrammetric Remote Sensing* 61(3–4):135–136.

Hinz, S., Meyer, F., Eineder, M., & Bamler, R. 2007. Traffic monitoring with spaceborne SAR—Theory, simulations, and experiments. *Computer Vision and Image Understanding* 106(2):231–244.

Hinz, S. & Stilla, U. 2006. Car detection in aerial thermal images by local and global evidence accumulation. *Pattern recognition letters* 27(4):308–315.

Kerner, B. S., Demir, C., Herrtwich, R. G., Klenov, S.L., Rehborn, H., Aleksic, M. & Haug, A. 2005. Traffic state detection with floating car data in road networks. *IEEE 2005 proceedings of the Intelligent Transportation Systems*, IEEE.

Karimi Nejadasl, F., Gorte, B. G., & Hoogendoorn, S. P. 2006. Optical flow based vehicle tracking strengthened by statistical decision. *ISPRS journal of photogrammetry and remote sensing* 61(3): 159–169.

Kembhavi, A., Harwood, D. & Davis, L. S. 2011. Vehicle detection using partial least squares," *IEEE Transactions on Pattern Analysis and Machine Intelligence* 33(6):1250–1256.

Leitloff, J., Hinz, S. & Stilla, U. 2010. Vehicle detection in very high resolution satellite images of city areas. *IEEE Transactions on Geoscience and Remote Sensing* 49(7):2795–2806.

Moranduzzo, T., & Melgani, F. 2014. Automatic Car Counting Method for Unmanned Aerial Vehicle Images. *IEEE Transactions on Geoscience and Remote Sensing* 52(3), 1635–1647.

Mtir, I. H., Kaaniche, K., Chtourou, M. & Vasseur, P. 2012. Aerial sequence registration for vehicle detection. *Proceedings of the 9th International Multi-Conference on Systems, Signals and Devices (SSD '12)*, pp. 1–6.

Moon, H., Chellappa, R. & Rosenfeld, A. 2012. Performance analysis of a simple vehicle detection algorithm. *Image and Vision Computing* 20(1): 1–13.

Reinartz, P., Lachaise, M., Schmeer, E., Krauss, T. & Runge, H. 2006. Traffic monitoring with serial images from airborne cameras. *ISPRS Journal of Photogrammetry and Remote Sensing* 61(3):149–158.

Salehi, B., Zhang, Y. & Zhong, M. 2012. Automatic moving vehicles information extraction from single-pass WorldView-2 imagery. *IEEE Journal of Selected Topics in Applied Earth Observations and Remote Sensing* 5(1): 135–145.

Suchandt, S., Runge, H., Breit, H., Steinbrecher, U., Kotenkov, A., & Balss, U. 2010. Automatic extraction of traffic flows using TerraSAR-X along-track interferometry. *IEEE Transactions on Geoscience and Remote Sensing* 48(2):807–819.

Schmidt, F. 2012. *Tracking Vehicles in Aerial Image Sequences, media release*, 12 November, Karlsruher Institute Of Technology (KIT), Institut für Photogrammetrieund Fernerkundung (IPF), viewed 30 January 2013, <http://www.ipf.kit.edu/english/downloads_707.php>.

Tuermer, S., Kurz, F., Reinartz, P. & Stilla, U. 2013. Airborne vehicle detection in dense urban areas using HoG features and disparity maps. *IEEE Journal of Selected Topics in Applied Earth Observations and Remote Sensing* 6(6):2327–2337.

Wei, D. Y. & Zhou, G. Q. 2008. Traffic Spatial Measures and Interpretation of Road Network Using Aerial Remotely Sensed Data. *IEEE International Geoscience and Remote Sensing Symposium, 2008(IGARSS 2008)* 3:1319–1322.

Yao, W. & Stilla, U. 2011. Comparison of two methods for vehicle extraction from airborne lidar data toward motion analysis. *IEEE Geoscience and Remote Sensing Letters* 8(4): 607–611.

Computational Intelligence in Industrial Application – Ling (ed.)
© 2015 Taylor & Francis Group, London, ISBN: 978-1-138-02818-0

The exploration and practice of ERP and E-business application technical talents training

ChaoYi Chen, Jian Yong Zhang & Jun Li Liu
Tianjin Institute of Software Engineering, Tianjin, China

ABSTRACT: With the rapid development of information technology, mankind has entered the information age. The demand of information industry development for talents is increasing, while the dependence of enterprises and consumers for ERP and E-business is growing. Considering the characters of ERP and E-business, the talents training in this area should be focussed on the practicability and innovation. By means of available administration, enterprises and schools can have a further and deeper cooperation in course, resource of education, training and practice and entrepreneurship. Establishing a comprehensive system of theory, practice, innovation and entrepreneurship is necessary and beneficial for improving the practice and innovation ability of students.

KEYWORDS: ERP; E-business; School-enterprise cooperation; Application talents training.

1 INTRODUCTION

The concept of ERP (Enterprise Resource Planning) was presented by Gartner Group company in 1990. As the next generation of Manufacturing Resource Planning (MRP II), ERP increases quality management, distribution and transportation management, human resources management these new contents are very important in nowadays (Scott & Vessey, 2000).

For the purpose of making profits, enterprises have to integrate the resources of human, equipment and material so that the effective product organization and control can be came true by achieving lowest cost and shortest product period. With the development of science and technology, scientific management is increasingly relying on information technology. In the early, enterprises used management information system (MIS) to record large amount of initial data and carried out querying and summarizing data; and then, for dealing with the abnormal variations of materials and inventory, some enterprises used the ROP (Re-Order Point) to purchase according to bulk purchase rule when the available inventory decrease to reorder point; in the mid-1960s, IBM proposed material requirements planning (MRP) which considers all the products, raw materials and components as materials which can be fall into independent demand and dependent demand. The concept of MRP is decreasing the inventory overstock by getting materials in right time at right place and following the right quantity; in the end 1970s and early 1980s, manufacturing resources planning (MRP II) developed fast. The core concept of (MRP II) is integrating the production,

sale, purchase, finance and engineering and share the data in order to become a mode of integrated production management; ERP developed in the base of MRP II and not only expand the scope of enterprise management but also focus on the transregional and transnational cooperation management in order to deal with the coordination problems between different organization departments or different enterprises (Huang et al., 2003); Nowadays, with the rapid development of E-business, ERP is facing a new challenge which is how ERP adjusts its content and function to be fit for the application of E-business. More and more enterprises are getting to involve E-business, so it is very important to focus on the integration between ERP and E-business.

2 INTEGRATION BETWEEN THEORY AND PRACTICE EDUCATION

2.1 *School-enterprise cooperation*

School-enterprise cooperation is a collaboration mode between schools and enterprises they have complementary advantages. Schools and enterprises provide their advantage resources to contain programs and make great efforts to cooperation so that they can make new profits more than they make by themselves (You & Zhou, 2004). Talents training need the support of school-enterprise cooperation, especially for the practical talents training.

There are few kinds of school-enterprise cooperation mode. The first is the mode of students enter companies, under this mode students can gain more

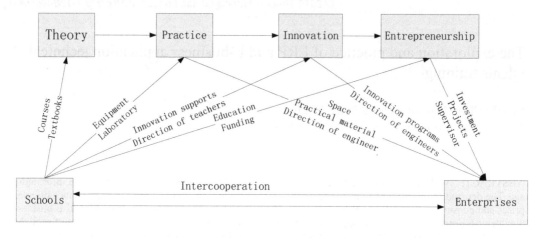

Figure 1.　The talents training system of ERP and E-business.

practical experience and be familiar with real company environment; the second is the mode of companies come into campus, which mode can help companies to choose outstanding students in advance and students can also join the project team during school hours; the third is interaction between schools and enterprise, under this mode, enterprises provide more internship opportunities, equipments, engineers and materials while schools offer more supports of space, teachers and facilities. Both sides carry out deeper cooperation in teacher training, science study, talents cultivation, technology transformation and, etc.

The industry of ERP and E-business had a very fast development in the last decade, which needs the support of efficient management and research and development capabilities. The staffs in this industry should have higher practical ability and professional knowledge that makes a higher request to staff training. For providing enough qualified staffs, it needs to strengthen the task of practical and creative talents training through the School-enterprise cooperation (Jin & Zhang, 2012). So improve the effect of school-enterprise cooperation can make lots benefits for both of schools and enterprises.

2.2 Course system design

The talents training of ERP and E-business have to focus on interdisciplinarity which includes management, economics, computer science, software engineering and etc. The course design should pay attention to the character of ERP and E-business so as to integrate the advantages of them. The structure of course cannot just be designed with basic courses but also should be included in professional basic courses, practical courses and innovative courses, which can

come into a courses system in order to direct students more comprehensively.

The theory foundation courses include higher mathematics, university physics, linear algebra, probability theory, English and etc. The professional basic courses include computer basic theory, management, economics, computer network, object oriented programming, data structure, Java and etc. Specialized courses include ERP and supply chain management, electronic commerce law, e-commerce payment and settlement, logistics and storage, security management, e-commerce network marketing, website development and so on; the practical and innovation courses include practical training suite, entrepreneurship training, E-business development, case studies and etc.

The design of course system should pay attention to the development of ERP and E-business. There are more and more E-business application be reflected in ERP and E-business enterprises are coming to attach importance to supply chain management and customer relationship management which is the important part of ERP system.

3 TALENTS TRAINING SYSTEM BUILDING OF ERP AND E-BUSINESS

The training of ERP and E-business talents should focus on systemic education with theory, practice and innovation rather than carry out single point. The theory is the basic of practice, which can guide the practice in right direction. In other hand, practice can check out if theory is right or wrong. Innovation comes into being while practice accumulated. Innovation should be transformed to industry so that can produce its effect (Yu & Feng, 2012).

The development of ERP and E-business industry has to rely on technology research and application innovation so as to talents is the key. It is not easy to handle well on specialized course teaching, practical training and innovation for neither school nor enterprise. So school-enterprise cooperation is a good way to integrate more resources in order to improve every partner's advantages. Schools can provide teachers and textbook compilation for theory, laboratory place and equipment for practice, education and funding for students' entrepreneurship. Enterprises can provide practice place, engineer and practice textbook for practice education, project and instruction for innovation, investment, program and supervisor for entrepreneurship (Li et al. , 2007) .

Schools and enterprises should establish an ecosystem by carrying out available cooperation, which means schools and enterprises not only pay lots of efforts and offer advantage resource but also they can get benefit from the cooperation which is bigger than they do it by themselves. The economic profit is the best benefit of enterprise and the improvement of talents training is the best benefit of schools. If we want to balance both of their benefit, it should rely on building a perfect talents training system so that the school-enterprise cooperation system will have a sustainable development.

4 CASE STUDIES

This paper uses Tianjin Institute of Software Engineering (TJISE) as a case to have real case analysis. On September 2010, 300 acres, 165,000 square meters TJISE that was near to the third higher education district was built and operated. The duty of TJISE is to build domestically top-ranking and internationally advanced software. Meanwhile, this institute will be a service base that provides software technical support and social service.

TJISE has an outstanding performance in school-enterprise cooperation, it has already cooperated with many domestic or international leading manufacturers, software companies and education training institutions which include SAP, IBM, Oracle, EMC2, Autodesk, Adobe, Isoft stone, China Soft International IT Company, AVN&IS, CSDN, China Orient. It involves outsourcing of ERP, E-business, service, database, mosaic and digital media. TJISE has also invited industry organization to lead the development of the institute, such as Tianjin Software Industry Association, Tianjin Service outsourcing Association, Tianjin Integrated System Association and so on. The schools including Tianjin Polytechnic University, Tianjin University of Technology, Tianjin Normal University more than 10 schools has a deep cooperation with TJISE. Under the guide of

government TJISE established "schools union" and "enterprises union". With the supports of Tianjin Binhai Hi-tech Zone Software Park and Tianjin E-business Association, TJISE promoted the cooperation of "schools union" and "enterprises union" in the talents training of ERP and E-business. The employment rate and employment quality of TJISE are both very excellent compared to same level schools.

In the field of building course system, TJISE established E-business and Internet marking professional direction facing on the practical talents training of ERP and E-business. In the cross-border E-commerce professional direction, the training department of IBM designed course scheme involves supply chain management, ERP application, policies and regulations, inventory management and investment these contents. Teachers come from schools are in charge of the theory and professional basic courses teaching while engineers come from enterprises are responsible for the practical training. Meanwhile they promote together to the innovation practice in order to offer innovation and entrepreneurship opportunities to students.

The performance of TJISE in the practical technology talents training of ERP and E-business, school-enterprise cooperation plays a very important role in order to integrate every partner's resources. Meanwhile, building a comprehensive course system includes theory, practice, innovation is also crucial to ERP and E-business talents training.

REFERENCES

Judy E. Scott, Iris Vessey. Implementing Enterprise Resource Planning Systems: The Role of Learning from Failure. *Information Systems Frontiers*. 2007, 213–218.

Ying Huang, Jinlong Zhang, Shuqin Cai, Shuangyuan Shi. 2003. Study of the Integration of ERP, SCM and CRM in E-Commerce. Journal of Wuhan University Technology, 25(1): 122–125.

Wenming You, Sheng Zhou. The research on mechanism optimization of integration of production, education and research, *Science of Science and Management*, 2004, 9–12.

Zhen Jin, Jilan Zhang. 2012. Research on the Practice Mode Based ERP. Information Engineering Research Institute,USA. Proceedings of 2012 International Conference on Education Reform and Management Innovation (ERMI 2012) Volume 2. Information Engineering Research Institute, USA.

Jianguo Yu, Meilin Feng, Dinghua Xiao, Pengpeng Huang. 2012. Research and Practice of the Open Experiment Teaching for ERP Course Group. Information Engineering Research Institute, USA.Selected Papers from 2012 2nd International Conference on Education and Education Management (EEM 2012) Volume 3[C]. Information Engineering Research Institute, USA.

Qi Li, Li Zhang, Lifang Peng. Thought on Cultivating Patterns for Personnel with Diversified Electronic Commerce. *Economic Management*, 2007, 29(14): 58–63.

Computational Intelligence in Industrial Application – Ling (ed.)
© 2015 Taylor & Francis Group, London, ISBN: 978-1-138-02818-0

Dynamic linkages between exchange rates and stock prices

Hao Guo

Robotics and Microsystems Center, Soochow University, Suzhou, China

Fei Qi*

School of Mechanical and Electric Engineering, Soochow University, Suzhou, China
Collaborative Innovation Center of Suzhou Nano Science and Technology, Soochow University, Suzhou, China

BoPing Tian

Department of Mathematics, Harbin Institute of Technology, Harbin, China

ABSTRACT: Dynamic linkages among USD/RMB exchange rate, the Shanghai Stock Exchange (SSE) Composite Index and the S&P 500 Index are investigated in this paper by using daily data between Jan. 5, 2004 and Sep. 25, 2009. We divide the sample period into four subintervals by wavelet transform method. Based on the subintervals divided, co-integration test, Granger causality test, and variance decomposition are used to investigate these linkages. The results indicate that USD/RMB exchange rate leads the SSE Composite Index with a negative correlation after Chinese exchange rate reform. The SSE Composite Index leads USD/RMB exchange rate after the explosion of "global market crash" in Jan. 2008. The S&P 500 Index is the Granger causality of USD/RMB exchange rate after the US subprime crisis. The US subprime crisis has a temporary effect on the linkages between exchange rates and stock prices.

1 INTRODUCTION

Dynamic linkages between exchange rates and stock prices have been the hot issues after the Asian Financial Crisis in 1997 (Granger et al. 2000, Nagayasu 2001, Chang 2002, Shamsuddin & Kim 2003, Phylaktis & Ravazzolo 2005, Yau & Nieh 2006, Pan et al. 2007). Compared with the Asian Financial Crisis, US subprime crisis had caused a financial tsunami in global financial markets, such as bankruptcy of financial institutions, closedown of investment funds, and fluctuation of stock markets.

Previous studies show that correlations among asset prices might be significant confronting the crisis (Granger et al. 2000, Nagayasu 2001, Phylaktis & Ravazzolo 2005, Baig & Goldfajn 1999). Baig & Goldfajn (1999) indicate that the correlation coefficients of exchange rates, stock prices, interest rates, and bond prices are significantly larger during a crisis in Korea, Malaysia, Indonesia, Thailand and Philippines. Granger et al. (2000) find that exchange rates lead stock prices in South Korea but show an opposite relation in Philippines, and data from Hong Kong, Malaysia, Singapore, Thailand, and Taiwan indicate strong feedback relations. Yet, Shamsuddin & Kim (2003) believe that stable long-run relationships among Australian, US, and Japanese markets exist before Asian financial crisis, but these relationship disappear in the post-Asian crisis period. Nagayasu (2001) indicates some banking and financial sectors seem to have caused upward pressure on exchange rates. What is more, Pan et al. (2007) state that these linkages are also respect to the exchange rate regime, the trade size, the degree of capital control, and the size of equity market.

However, studies above put emphasis on the Asian Financial Crisis and countries with freely floating exchange rate system. On Jul. 21 2005, a managed floating exchange rate regime based on market supply was established in China. From then on, RMB was no longer pegged to US dollars alone but a basket of currencies. The exchange rate decreased by 7.4% from 8.11 USD/RMB to 7.51 USD/RMB from Jul.21, 2005 to Oct. 16, 2007, meanwhile, the SSE Composite Index increased by 500% from 1020.63 to 6124.04 points. Whether stock prices are affected by the fluctuations of exchange rates, or vice versa? Deng & Yang (2008) and Ni et al. (2008) both indicate that USD/RMB exchange rate is a unidirectional Granger causality of the SSE Composite Index. Yet, relationship that stock prices lead exchange rates has not been previously established.

The aim of the present study is to investigate the linkages among USD/RMB exchange rate, the SSE Composite Index and the S&P 500 Index under financial events, such as exchange rate reform and US subprime crisis, using Granger causality test, wavelet transform method, and variance decomposition.

*corresponding author

2 METHODS

2.1 *Johansen co-integration test*

The co-integration theory was created by Engle and Granger in 1987. If each component of vector time series $\{X_t\}$ is $I(1)$ process, and some linear combination $\alpha' X_t$ of these components is stationary, then the components of vector series $\{X_t\}$ are said to be co-integration, where α denotes a co-integration vector. This theory laid foundation for finding balanced relationships among non-stationary series and establishing VEC models. Johansen (1995) used maximum likelihood estimate for testing the co-integration relationships among variables.

2.2 *Granger causality test*

When it comes to "*x* is the Granger causality of *y*", it means *x* is conducive to forecast *y*. In addition, research series should be stationary or have a co-integration relationship, then a VAR or VEC model can be established to test the relationship. If series do not satisfied these conditions, a VAR model can also be established by making series stationary after the first or second order difference. Two models of Granger causality test are listed here (Granger et al., 2000):

If a co-integration relationship does not exist, it is need to establish a VAR model

$$\Delta y_t = c + \sum_{i=1}^{k} \alpha_i \Delta y_{t-i} + \sum_{i=1}^{k} \beta_i \Delta x_{t-i} + \varepsilon_t \qquad (1)$$

Where y_t and x_t indicate economic variables, ε is the white noise, c is a constant term. x_t is said to be a Granger causality of y_t by rejecting the hypothesis: $H_0 : \beta_1 = \beta_2 = \cdots = \beta_k = 0$.

If a co-integration relationship exists among variables, a VEC model with the error correct term ECT_{t-1} is established

$$\Delta y_t = c + \delta ECT_{t-1} + \sum_{i=1}^{k} \alpha_i \Delta y_{t-i} + \sum_{i=1}^{k} \beta_i \Delta x_{t-i} + \varepsilon_t \quad (2)$$

Where δ denotes the speed of correcting errors. If x_t stands for exchange rate, y_t is stock price in the VEC model, then exchange rate is not Granger causality of stock price by accepting the hypothesis: $H_0 : \beta_1 = \beta_2 = \cdots = \beta_k = 0$ and $\delta = 0$.

3 DATA AND EMPIRICAL RESULTS

3.1 *Data and breakpoints test*

This paper analyzes exchange rate of RMB against 100 US dollars (ER), the closing prices of Shanghai Stock Exchange Composite Index (SSE) and the S&P 500 Index (SPX) by using daily data from Jan. 5, 2004 to Sep. 25, 2009.

Before a stationary test, the sample period would be divided into several subintervals. Since there are many financial events including in the sample, and ADF test was suspect when a sample period includes some major events, such as great depression, oil shocks. "Failure to consider it properly could lead to erroneous conclusions when the null is not rejected" (Granger et al. (2000)). What is more, divide the sample into subintervals would be attributed to compare differences before and after financial events (Shamsuddin and Kim, 2003; Granger et al., 2000; Phylaktis and Ravazzolo, 2005; Pan et al., 2007). We will employ wavelet transform to estimate the structural breakpoints of ER, SSE and SPX, then divide the sample by these selected breakpoints.

Db3 (Daubechies 3) wavelet with four levers wavelet transform is applied to deal with ER, SSE and SPX. Points which have the larger amplitude may be breakpoints. We use the first level which include high-frequency information and can describe tremendous changes of series to test breakpoints. Pick out the maximum modulus of the amplitude (*Max*) in the first level. Then, determine a threshold (*H*) based on *Max*. The threshold is defined by $H = p \cdot Max$, where $0 < p < 1$ as the threshold coefficient. We demarcate points that have larger modulus than the threshold degressively. Breakpoints should not be selected too much to ensure sample capacity in every subinterval. Hence, ER, SSE, and SPX are taken to wavelet transform, and we choose $p = 1/4$, $2/3$ and $2/3$ as their thresholds coefficient respectively. Demarcated breakpoints are illustrated in Figure 1(a), Figure 2(a), and Figure 3(a).

The breakpoints and corresponding events are listed in Table 1. For every series (ER, SSE, and SPX), the former date listed have larger modulus than the latter, because we demarcate points degressively. Firstly, we pick out the events which have great influence on exchange and stock markets. Then, we must make sure there are enough sample points in every subinterval.

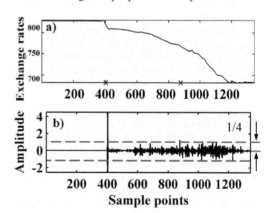

Figure 1. a) Demarcate breakpoints of ER by wavelet transformation with b) the first level.

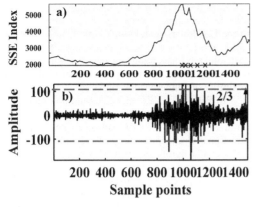

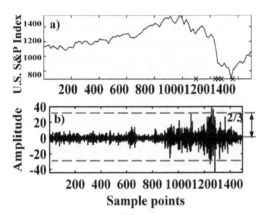

Figure 2. a) Demarcate breakpoints of SSE by wavelet transformation with b) the first level.

Figure 3. a) Demarcate breakpoints of SPX by wavelet transformation with b) the first level.

Table 1. Breakpoints and financial events.

Series	Breakpoints	Events and corresponding date
ER	**2005.07.20**[*] 2007.05.15	Exchange rate reform (2005.07.21).
SSE	2007.11.29 **2008.01.18**[*] 2007.10.24 2008.03.28 2008.06.18	"Global market crash" with many primary stock markets in the world declined (2008.01.21).
SPX	2008.10.28 2008.03.18 2008.09.26 2008.11.28 **2009.03.23**[*]	G20 Summit was hold in London for funding and reforming the international financial institutions to overcome the subprime crisis (2009.04.02).

Note: [*] denotes the selected breakpoint.

Therefore, we select three points and divide daily data (5 days a week) in number of 1495 into four subintervals. Then, ER, SSE and SPX are transformed to natural logarithm forms and named LER, LSSE and LSPX. The logarithm values and the divided four subintervals (Period1~Period4) are shown in Figure 4.

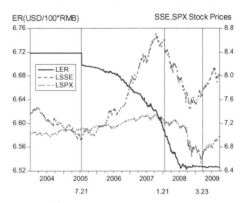

Figure 4. Exchange rates and stock prices (in logarithm) with four subintervals: Period1 (2004.1.5~2005.7.20): Before exchange rate reform; Period2 (2005.7.21~2008.1.18): After exchange rate reform; Period3 (2008.1.21~2009.3.23): In US subprime crisis; Period4 (2009.3.24~2009.9.25): After US subprime crisis.

3.2 Panel unit root and co-integration test

IPS panel unit root test is used to discuss stationarity of LER, LSSE and LSPX. The results are summarized in Table 2, which show that the null of non-stationarity is accepted by Period1, and rejected by Period2 to Period4. That is to say, in Period1, a VAR model can be established to investigate causality between exchange rates and stock prices. For Period2 to Period4, the null is rejected at 1% significance level in the first-difference, which satisfies $I(1)$ process, and Johansen cointegration test is employed to test whether it can be established a VEC model for checking linkages among variables.

Table 2. IPS panel unit root test.

	Level[b]		1st diff.[a]	
	IPS	t-statistics	IPS	t-statistics
Period1	−2.794	0.003 ***	–	–
Period2	2.744	0.997	−24.633	0.000 ***
Period3	0.684	0.753	21.976	0.000 ***
Period4	−0.204	0.419	−9.7445	0.000 ***

Notes: ***denote significance at 1% levels.[a]: Individual intercept,[b]: Individual intercept and trend (Johansen, 1995).

The results of Johansen co-integration test are shown in Table 3. It has a co-integration relationship in Period2 by rejecting there are no co-integration equations among variables. According to Period3 and Period4, it is not found any co-integration relationships between exchange rate and stock prices. According to the lack of the co-integration relationship, Pan et al. (2005) summarize two reasons: 1) it seemed to be a common phenomenon for countries using a managed floating exchange rate regime; 2) this may relate to the noise contained in daily data. Hence the results in Period3 and Period4 may presumably attribute to the noise of data which caused by the US subprime crisis.

For Period2, the co-integration equation is listed as follow:

$$LER = -0.049\, LSSE + 0.127\, LSPX \qquad (3)$$

Equation (3) illustrates that exchanger rate (LER) has a negative correlation with the SSE Composite Index (LSSE), and a positive correlation with the S&P 500 Index (LSPX) in Period2. In addition, it seems that LSPX have a lager influence on LER owing to the absolute value of coefficient $|0.127| > |-0.049|$.

Table 3. Johansen co-integration test.

	H_0	Trace	Critical	Prob.
Period2	$r = 0$	38.917	29.797	0.003***
	$r \leq 1$	11.565	15.495	0.179
	$r \leq 2$	1.262	3.841	0.261
Period3	$r = 0$	30.871	42.915	0.451
	$r \leq 1$	15.347	25.872	0.546
	$r \leq 2$	4.112	12.518	0.726
Period4	$r = 0$	9.685	29.797	0.985
	$r \leq 1$	4.153	15.495	0.891
	$r \leq 2$	0.626	3.841	0.429

Note: ***denote significance at 1% levels; r denotes the number of significant vectors. Lag intervals in the first difference are used 1-4.

3.3 Granger causality test

For Period1, LER, LSSE and LSPX are stationary and should be created a VAR model (Equation (1)). However, USD/RMB exchange rate has almost no changes before Chinese exchange rate reform, which caused too many zeroes in the first-order difference sequence. So, a VAR model is invalid to test the linkages between LER and LSSE. The same situation also appears between LER and LSPX. Figure 4 also reveals there are no linkages between exchange rates and stock prices in Period1. This is presumable attributed to the Chinese exchange rate regime pegged to the US dollars merely. In addition, from Table 4, LSSE and LSPX has no Granger causality between each other by accepting the hypothesis (0.160 and 0.496) in Period1. This may be respect to the weak flexibility of Chinese stock markets.

In Period2, a VEC model should be established based on the results of Johansen co-integration tests. From Table 4, USD/RMB exchange rate is the unidirectional Granger causality of the SSE Composite Index at 1% significant level. This result is accorded with Deng and Yang (2008). We think, expectation of RMB appreciation causes vast international hot money entering into Chinese foreign exchange and stock markets. This will make the excess liquidity of macroeconomy and overvalued stock prices. Based on this perspective, exchange rates are Granger causality of the SSE Composite Index. The S&P 500 Index is Granger causality of USD/RMB exchange rate and the SSE Composite Index in Period2. The volatility of the US stock market, which has abundant capitalization and large trading volume, may affect investors' decisions. But, we still believe RMB exchange rate reform is the key factor on fluctuation of exchange rates and stock prices in Period2. Maybe the three variables have a co-movement in this period.

According to Period3 and Period4, there are no co-integration relationships among variables, so two VAR models are established to test Granger causality based on the first-order difference of LER, LSSE and LSPX. In Period3, the SSE Composite Index is Granger causality of exchange rate at 10% significant level. It is indicates that the crash of Chinese stock prices puts pressure on RMB revaluation in long-run dynamic linkages. The S&P 500 Index is Granger causality of the SSE Composite Index in Period3 at 5% significant level. This is presumably attributed to the US subprime crisis and a large amount of hot money flowing out of Chinese stock markets.

The S&P 500 Index is the Granger causality of USD/RMB exchange rate in Period4. We believe the US subprime crisis has a temporary effect on the linkages between exchange rates and stock prices, which is in accordance with Ravazzolo (2005). What is more, if there are no financial events, we also can not exclude the possibility of no linkages between exchange rates and stock prices.

3.4 Variance decomposition

Comparing with the co-integration and Granger causality, which are long-run co-movement tests, variance decomposition can be employed to test the short-run dynamic linkages among variables, by providing information about relative importance of each random innovation in affecting the variables.

Table 4. Granger causality/ block exogeneity wald tests.

	Period1 (VAR), Sample size: 399
LSPX $-/\rightarrow$ LSSE	3.381 (0.496)
LSSE $-/\rightarrow$ LSPX	6.572 (0.160)

	Period2	Period3	Period4
	(VEC) 650	(VAR) 301	(VAR) 130
ΔLSSE $-/\rightarrow \Delta$LER	2.741 (0.602)	8.479 (0.076)*	2.475 (0.649)
ΔLSPX $-/\rightarrow \Delta$LER	25.614 (0.000)***	5.942 (0.204)	34.320 (0.000)***
ΔLER $-/\rightarrow \Delta$LSSE	17.269 (0.002)***	5.822 (0.213)	2.035 (0.729)
ΔLSPX $-/\rightarrow \Delta$LSSE	17.517 (0.002)***	11.843 (0.019)**	4.215 (0.378)
ΔLER $-/\rightarrow \Delta$LSPX	2.762 (0.598)	8.349 (0.080)*	1.832 (0.767)
ΔLSSE $-/\rightarrow \Delta$LSPX	3.593 (0.464)	4.076 (0.396)	4.472 (0.346)

Notes: ***, ** and * denote significance at 1%, 5% and 10% levels; Number in this table is F-statistic and in the parentheses indicates probabilities. $x-/\rightarrow y$: x is not Granger causality of y.

From Table 5, both exchange rate (LER) and SSE Composite Index (LSSE) are affected by the S&P 500 Index (LSPX) with relative importance of 7.378% and 4.212% in Period2. It means the US stock prices lead Chinese exchange rates and stock prices with short-run dynamic linkages in Period2. In Period3, the relative importance of the S&P 500 Index (ΔLSPX) on the SSE Composite Index (ΔLSSE) is 4.016%. That is to say, after the "global market crash", SSE Composite Index was affected by the S&P 500 Index. The result is presumably attributed to the US subprime crisis. Comparing with Period3, the relative importance of exchange rate (ΔLER) on the SSE Composite Index (ΔLSSE) and the S&P 500 Index (ΔLSPX) increased from 2.628% to 3.703% and 2.663% to 15.107% respectively in Period4. Namely stock prices are short-run causality of exchange rates after the US subprime crisis. In addition, the SSE Composite Index has influence on the S&P 500 Index in the short term with relative importance of 4.893% in Period4 (they have an opposite lead-lag relationship in the long-run linkage in Table 4). It means, exchange rates and stock price have significant differences for long-run and short-run causality relationship after the subprime crisis.

4 CONCLUSIONS

This paper analyzes linkages among USD/RMB exchange rate, the SSE Composite Index and the S&P 500 Index. The results are listed as follows:

- Before exchange reform (Period1), there is no significant relationship among the three variables.
- USD/RMB exchange rates lead the SSE Composite Index with a negative correlation in Period2. We believe RMB exchange rate reform is the key factor on fluctuation of exchange rates and stock prices in this period.
- The SSE Composite Index lead exchange rates after the "global stock market crash" (Period3). US subprime crisis is considered to be the key event which reverses this linkage in Period3.
- The S&P 500 Index is the Granger causality of USD/RMB exchange rate in Period4. The US subprime crisis has a temporary effect on the linkages between exchange rates and stock prices. We do not exclude the possibility of no linkage among the three variables when great financial events are absence.

Table 5. Variances decomposition for exchange rate and stock prices innovations (10 days).

	LER explained by			LSSE explained by			LSPX explained by		
	LER	LSSE	LSPX	LER	LSSE	LSPX	LER	LSSE	LSPX
Period2	92.537	0.085	**7.378**	1.849	93.939	**4.212**	0.730	1.285	97.985

	ΔLER explained by			ΔLSSE explained by			ΔLSPX explained by		
	ΔLER	ΔLSSE	ΔLSPX	ΔLER	ΔLSSE	ΔLSPX	ΔLER	ΔLSSE	ΔLSPX
Period3	94.709	**2.628**	**2.663**	1.679	94.305	**4.016**	3.248	1.796	94.956
Period4	81.190	**3.703**	**15.107**	2.248	95.211	2.541	1.413	**4.893**	93.694

ACKNOWLEDGEMENTS

The authors of this paper gratefully acknowledge the financial supports from the National Natural Science Foundation of China (71350005), China Postdoctoral Science Foundation (2014M551655), and Natural Science Foundation of Jiangsu Province of China (BK20130325).

REFERENCES

Baig, T. & Goldfajn, I., 1999. Financial market contagion in the Asian crisis. IMF Staff Paper. *International Monetary Fund* 46, 167–195.

Deng, S. & Yang, C., 2008. An empirical study on the relationship between stock price and exchange rate in China. *Journal of Financial Research* 1, 29–41.

Granger, C.W.J. et al., 2000. A bivariate causality between stock prices and exchange rates: evidence from recent Asian flu. *The Quarterly Review of Economics and Finance* 40, 337–354.

Johansen, S., 1995. *Likelihood-based inference in cointegrated vector autoregressive models*. Oxford: Oxford University Press.

Nagayasu, J., 2001. Currency crisis and contagion: evidence from exchange rates and sectoral stock indices of the Philippines and Thailand. *Journal of Asian Economics* 12, 529–546.

Ni, K. & Ni, Q., 2008. An empirical study of the influence of international stock market and exchange rate impact to the stock price of China. *Financial Theory and Practice* 9, 65–68.

Pan, M.S. et al., 2007. Dynamic linkages between exchange rates and stock prices: evidence from East Asian markets. *International Review of Economics and Finance* 6, 503–520.

Phylaktis, K. & Ravazzolo, F., 2005. Stock prices and exchange rate dynamics. *Journal of International Money and Finance* 24, 1031–1053.

Shamsuddin, A.F.M. & Kim, J.H., 2003. Integration and interdependence of stock and foreign exchange markets: an Australian perspective. *International Financial Markets, Institutions and Money* 13, 237–254.

Application of MRTD on novel ultra-wideband bandpass filter

MuJie Fan

Department of Computer Science and Engineering, Changchun Normal University, China

ABSTRACT: In this paper, a novel Asymmetrical Co-Planar Waveguide (ACPW) ultra-wideband bandpass filter is proposed, the Multi-Resolution Time-Domain formulation (MRTD) is proposed and detailed illustration of this method is given. The UWB bandpass filter is analyzed with MRTD method. The numerical and measured results demonstrate the advantages of MRTD over conventional FDTD method. The correctness and effectiveness of MRTD method are verified by the results.

KEYWORDS: Ultra-wideband; Filter; MRTD.

1 INTRODUCTION

Since C. P. Wen advanced CPW [1], coplanar waveguides (CPWs) have predominantly been used in microwave and millimeter-wave integrated circuits. The most of them are given only for symmetric CPW, there are few literatures published about the asymmetric coplanar waveguides (ACPWs). The asymmetric coplanar waveguide (ACPW) is flexible and general compared with CPW, and the symmetric lines can be considered as the special cases of asymmetric lines, and the popular methods are used to compute ACPW such as mapping technique, FDTD method and wavelet method [2-4].

The finite difference time domain method (FDTD) is a very effective method for solving electromagnetic problems. However, the mesh size must satisfy the stability conditions in order to avoid numerical dispersion. Computer memory and calculation time will be required when physical size to be computed is large. In recent years, the wavelet analysis methods used in the numerical calculation of the magnetic field obtained are developed quickly. In 1996, MRTD method was used to solve Maxwell equation by Krumpholz, It has not only good dispersion characteristics, but also the calculation memory and computing speed are obtained. The mesh numbers needed by MRTD algorithm are less than that needed by traditional FDTD algorithm.

Electromagnetic will occupy the infinite space in the calculation of electromagnetic radiation and scattering. The effective boundary condition must be set at the cutting position in order to avoid reflection wave with absorption by boundary. The setting of absorbing boundary conditions is relatively complex because of the complexity of wavelet. At present, the popular absorption boundary conditions are used such as PEC, Bayliss-Trukel, Engquist-Majda, MUR, and PML in the calculation of electromagnetic field, the application of these boundary conditions in MRTD algorithm has been the developed [5-7]. This paper focuses on application to the PML boundary condition used in the MRTD algorithm combined with MUR and PEC boundary conditions, and different boundary condition is set according to the calculation model and the characteristics of the distribution.

The formula of ADI-MRTD algorithm and PML absorbing boundary condition is given by this paper. A novel structure filter with asymmetric coplanar waveguide is presented and analyzed with MRTD method compared with traditional FDTD algorithm. The computation time and precision are better by using MRTD method compared with FDTD method. Boundary condition is simpler with MRTD method.

2 MRTD METHOD

According to Maxwell's equations:

$$\nabla \times \vec{H} = \varepsilon \frac{\partial \vec{E}}{\partial t} \tag{1}$$

$$\nabla \times \vec{E} = -\mu \frac{\partial \vec{H}}{\partial t} \tag{2}$$

The electric and magnetic fields are expanded using MRTD method as follows:

$$
\begin{aligned}
{}_{k+1}E^{\phi z}_{l,m,n+1/2} = {}_k E^{\phi z}_{l,m,n+1/2} &+ \frac{\Delta t}{\varepsilon \Delta x} \sum_{i=L}^{L-1} a(i)_{k+1/2} H^{\phi y}_{l+1/2+i,m,n+1/2} \\
&- \frac{\Delta t}{\varepsilon \Delta y} \sum_{i=L}^{L-1} a(i)_{k+1/2} H^{\phi x}_{l,m+i+1/2,n+1/2}
\end{aligned}
\tag{3}
$$

$$_{k+1}H^{\phi z}_{l+1/2,m+1/2,n} =_k H^{\phi z}_{l+1/2,m+1/2,n} - \frac{\Delta t}{\mu\Delta x}\sum_{i=L}^{L-1}a(i)_{k+1/2}E^{\phi y}_{l+i,m+1/2,n}$$

$$+\frac{\Delta t}{\mu\Delta y}\sum_{i=L}^{L-1}a(i)_{k+1/2}E^{\phi x}_{l+1/2,m+i,n} \quad (4)$$

3 ABSORBING BOUNDARY CONDITION

In 1994, the perfectly matched layer (PML) absorbing boundary condition was proposed by Berenger, the reflection wave is reduced when electric and magnetic field components are devised in absorb boundary region. For example, the component and the Maxwell equation are split into: $E_x = E_{xy} + E_{xz}$ difference equation with PML absorbing boundary condition is:

$$_{k+1/2}E^{\phi xy}_{l+1/2,m,n} = \frac{2\varepsilon-\sigma_y\Delta t}{2\varepsilon+\sigma_y\Delta t}\,_k E^{\phi xy}_{l+1/2,m,n} + \frac{\Delta t}{2\varepsilon+\sigma_y\Delta t}\frac{\Delta t}{\varepsilon\Delta y}\sum_{i=L}^{L-1}a(i)_k H^{\phi z}_{l+1/2,m+i+1/2,n}$$

$$_{k+1/2}E^{\phi xz}_{l+1/2,m,n} = \frac{2\varepsilon-\sigma_z\Delta t}{2\varepsilon+\sigma_z\Delta t}\,_k E^{\phi xz}_{l+1/2,m,n} - \frac{\Delta t}{2\varepsilon+\sigma_z\Delta t}\frac{\Delta t}{\varepsilon\Delta z}\sum_{i=L}^{L-1}a(i)_{k+1/2}H^{\phi y}_{l+1/2,m,n+i+1/2} \quad (5)$$

Because the MRTD algorithm is based on Daubechies functions wavelet, the conductivity distribution function is $\sigma_i(r)=\int_{m-8}^{m+8}\sigma_{max}(1-\frac{r}{\delta})^2\phi(r)dr$. δ is thickness of PML medium . The max conductivity distribution is $\sigma_{max}=\frac{m+1}{150\pi\Delta\sqrt{\varepsilon_r}}$, and m is order of conductivity distribution. Outside medium is ideal conductor. Other electric and magnetic field component can be obtained by using similar method.

4 NUMERICAL ANALYASIS

The MRTD algorithm and the PML boundary conditions are applied to the asymmetric coplanar waveguide (ACPW) and calculation of relevant parameters of coplanar waveguide filter to verify the correctness and superiority of the algorithm comparing with the traditional FDTD algorithm. The ACPW structure is shown in Fig. 1, in which, for s_1 is 0.6 mm, s_2 is 0.3 mm, w is 0.4 mm, and thickness of medium is H. The characteristic impedance of ACPW was calculated by FDTD and ADI-MRTD, PML boundary is 4 layers, boundary wall is PEC boundary, the excitation source with pulse excitation as $\psi = \exp[-(n\Delta t - 200\Delta t)/(100\Delta t)]^2$, the computing parameters are shown in Table 1.

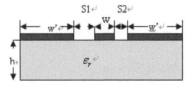

Figure1. Structure of ACPW.

Table 1. Compute parameters.

Method	Mesh number	Steps	Computer time
MRTD	40*40*70	1200	13 minutes
FDTD	100*100*120	1200	6 minutes

Computing memory and computing speed are obtained with MRTD method, and the calculation results are similar compared with traditional FDTD method. In order to further validate the effectiveness of the MRTD and PML boundary conditions, a new structure of ACPW broadband filter as shown in Fig. 2 is calculated and analyzed.

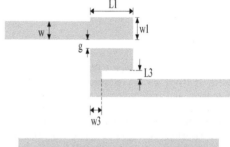

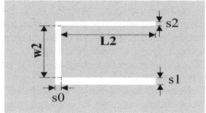

Figure 2. ACPW filter.

A slot line on both sides of the center of the conduction band width is different in ACPW.

In this paper, the design of ultra wideband ACPW filter is obtained by adjusting the slot width. The specific size of optimized ultra wideband ACPW filter is shown as follows: w = 1.23 mm, g=0.4 mm, w1=1.7 mm, L1=5.7mm, w2=2 mm, w1=g=3.8 mm, L2=6.8 mm, w3=1 mm, L3=1.8 mm, s0=0.1 mm, s1=0.3 mm, s2=0.6 mm. The 4 filter structure is shown in Fig. 3. The simulation calculation and test results of filter are shown in Fig. 3.

As can be seen from the calculations, deviation exist between measurement values and numerical values because the frequency is too high, but the overall trend is consistent. The correctness and effectiveness of MRTD method are verified by the results.

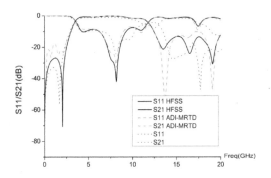

Figure 3. Calculation and test results.

5 CONCLUSIONS

The computational grid numbers are significantly reduced with MRTD algorithm compared to traditional FDTD algorithm by the calculation of ACPW. MRTD algorithm with PML boundary condition is effective to solve complex size model to compute electromagnetic field. The MRTD method (ANN-MRTD) is proposed, and detailed illustration of this method is given. A novel UWB band pass filter based on asymmetrical coplanar waveguide (ACPW) is analyzed using MRTD method. The correctness and effectiveness of MRTD method are verified by the results compared to traditional FDTD method.

ACKNOWLEDGMENT

This work was supported by the National Natural Science Foundation of Changchun Normal University.

REFERENCES

Wen, C.P.Coplanar waveguide: a surface strip transmission line suitable for nonreciprocal gyromagnetic device applications. *IEEE Transactions on Microwave Theory and Techniques* 1969 (17), 1087–1090.

Fang, S.J., Wang, B.S. 1999. Analysis of asymmetric coplanar waveguide with conductor backing. *IEEE Transactions on Microwave Theory and Techniques* 1999, (47): 238–240.

Frumpholz, M., Katehi, L.P.B. MRTD: new time-domain schemes based on multi resolution analysis. *IEEE Transactions on Microwave Theory and Techniques* 1996, 44, 555–571.

Chen, P., Fang, S.J. Calculation and analysis of dispersion characteristic of ACPW using FDTD method. *Acta Electronica Sinica* 2006, 34, 1610–1612.

Emmanouil M. Tentzeris, Robert L. Robertson. PML Absorbing Boundary Conditions for the Characterization of Open Microwave Circuit Components Using Multiresolution Time-Domain Techniques (MRTD). *Transactions on Antennas and Propagation*, 1999, 47(11): 1709–1715.

Fang Shao-Jun,WANG Bai-suo. A CAD-Oriented Model for Asymmetric Ally Shielded Multilayered CPW [J]. *Acta Electronica Sinica*, 2002, 30(6):804–807.

Xiaoming Li, Shaojun Fang. 2009. Ultra-wideband Bandpass Filter Using Hybrid Microstrip/ACPW Structures [M]. MAPE 2009, 1147–1149.

Computational Intelligence in Industrial Application – Ling (ed.)
© 2015 Taylor & Francis Group, London, ISBN: 978-1-138-02818-0

Wavelet process neural network for power load forecasting

Y. Li, D. Wang, N. Yu & F. Wang
State Grid Information & Telecommunication Company of SEPC, Taiyuan, Shanxi, P.R. China

ABSTRACT: In this paper, the Wavelet Process Neural Network (WPNN) model is proposed based on the wavelet theory and the Process Neural Network (PNN) model. WPNN incorporates the neural network in learning from processes and the time-frequency localization property of wavelet. Moreover, the network can deal with continuous input signals, which make it facilitates in tackling dynamics of complex processes. The corresponding learning algorithm is given and the network is used to solve the problems of power load forecasting. The simulation test results indicate that the WPNN has a faster convergence speed and higher accuracy than the same scale PNN. This provided an effective way for the problems of power load forecasting.

KEYWORDS: Process neuron; Wavelet Process Neural Network (WPNN); Learning algorithm; Power load forecasting.

1 INTRODUCTION

In the past years, instead of using common sigmoid activation functions, the wavelet neural network employing nonlinear wavelet basis functions, which readily reveals properties of the function in localized regions of the joint time-frequency space, has been widely studied for their outstanding capability of fitting nonlinear models in time-series prediction, signal processing, and many other fields [1-5]. In addition, wavelet neural network combines the capability of artificial neural network in learning and the capability of wavelet transform, which offers several advantages over traditional method. However, in the complex case, the inputs of many complicated systems are continuous time-varied functions or procedures. Unfortunately, the inputs of traditional artificial neural networks are discrete values. To overcome this limitation, He and Liang proposed a process neuron model in 2000[6,7]. From the point view of architecture, the major characteristics that distinguish process neuron from traditional artificial neuron are that inputs and corresponding connection weights of process neuron can be time-varied functions.

Process neural network (PNN) architecture is adopted as an alternative approach to solve nonlinear problems. But the structure and the parameters of process neural network are selected at random. The existing PNN method lacked for an efficient constructive model. All of these defects cause great effect in influencing practical application. In this paper, wavelet process neural network (WPNN) model is proposed based on the wavelet theory and process neural network. There are several advantages of combining wavelets and process neural network, for instance, the wavelet transformation provides an effect way for structure selection and parameters confirmation. The network is more suitable in learning functions with local variations. The inputs of the WPNN are time-varied functions. The activation function of hidden layer is wavelet-based function. The threshold of hidden layer is replaced by dilation and translation parameters of wavelet. WPNN introduces the wavelet analysis property into process neural network. As a result, WPNN shows surprising effectiveness in solving the nonlinear problem of poor convergence or even divergence encountered in other kinds of neural networks. It can markedly increase the convergence speed and provide better function approximation ability.

The rest of this paper is organized as follows. In Section 2, the process neuron model and the wavelet transform is reviewed. In Section 3, the topological architecture of the WPNN is described. In Section 4, a corresponding learning algorithm for the WPNN is developed. In Section 5, the effectiveness of the WPNN and its learning algorithm is proved by power load forecasting. The comparative simulation test results highlight the property of the WPNN. Conclusions are given in Section 6.

2 PROCESS NEURAL NEURON AND WAVELET ANALYSIS

2.1 Process neuron

The process neuron model is composed of three sections: process inputs signal weighting, time, space two dimensional aggregate and threshold active output. The inputs and connection weights of process neuron are continuous time-varied functions. An aggregation operator on time is added to the process neuron, which provides the process neural network with the capability of handling simultaneously two dimension information of time and space. The schematic architecture of the process neuron is depicted in Fig. 1.

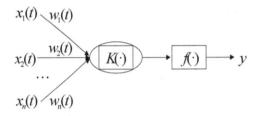

Figure 1. The sketch diagram of process neuron model.

The output of the process neuron model can be expressed as

$$y = f((W(t) \oplus X(t)) \otimes K(\cdot) - \theta) \qquad (1)$$

where $X(t) = (x_1(t), x_2(t), \cdots, x_n(t))$ is the input function. $W(t) = (w_1(t), w_2(t), \cdots, w_n(t))$ is the weight function. "$\oplus$" denotes a certain space aggregation operator. It can be defined as $A(t) = W(t) \oplus X(t) = \sum_{i=1}^{n} x_i(t)w_i(t)$, "$\otimes$" denotes a certain time aggregation operator. It can be defined as $A(t) \otimes K(\cdot) = \int_0^T A(t)K(\cdot)dt$. θ is the threshold. $f(\cdot)$ is the activation function.

2.2 Wavelet analysis

Wavelet transformation is a mathematical theory developed in recent years. It is considered as a great breakthrough following Fourier analysis, and has been widely used in a lot of research fields. The outstanding characteristic of the wavelet transform is that wavelet transform has good localization in both time and frequency space. This property is very useful for the analysis of non-stationary signals and the learning of the nonlinear function. Suppose $\psi(x) \in L^2(R)$ satisfies the admissibility condition

$$C_\psi = \int_R \frac{|\hat{\psi}(\omega)|^2}{|\omega|} d\omega < \infty \qquad (2)$$

where $\hat{\psi}(\omega)$ is the Fourier transform. The $\psi(x)$ as a mother wavelet is a single fixed function from which all basis function are generated a family of functions by the following operation of dilation and translation [8]. Where $\psi_{a,b}(x)$ can be expressed as

$$\psi_{a,b}(x) = \frac{1}{\sqrt{a}} \psi(\frac{x-b}{a}) \qquad (3)$$

where the parameters a and b are the dilation and translation parameters expressed in real number R, respectively. The basic idea behind the wavelet transform is to represent arbitrary function $f(x)$ as a superposition of wavelets. The continuous wavelet transform of $f(x)$ is given by:

$$W_f(a,b) = (f, \psi_{a,b}) = \frac{1}{\sqrt{|a|}} \int_{-\infty}^{+\infty} \psi_{a,b}(x) f(x) dx \qquad (4)$$

In terms of wavelet transformation theory, the inverse continuous wavelet transform can be obtained through the following formula:

$$\begin{aligned}
f(x) &= \frac{1}{C_\psi} \int_0^{+\infty} \frac{da}{a^2} \int_{-\infty}^{+\infty} W_f(a,b) \psi_{a,b}(t) db \\
&= \frac{1}{C_\psi} \int_0^{+\infty} \frac{da}{a^2} \int_{-\infty}^{+\infty} W_f(a,b) \frac{1}{\sqrt{a}} \psi(\frac{t-b}{a}) db
\end{aligned} \qquad (5)$$

3 WAVELET PROCESS NEURAL NETWORK

The proposed WPNN consist of three layers. The first layer is input layer, which is composed of n nodes. The second layer is the hidden layer, which is composed of m wavelet process neurons. The last layer is the output layer. The topological architecture of the WPNN model is depicted in Fig. 2.

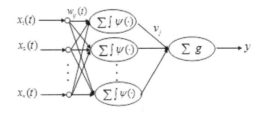

Figure 2. The topological structure of the WPNN model.

The input function is $X(t) = (x_1(t), \cdots, x_n(t))\ t \in [0, T]$. The outputs can be expressed as

$$y = g(\sum_{j=1}^{m} v_j \psi(\frac{\int_0^T \sum_{i=1}^n x_i(t) w_{ij}(t) dt - b_j}{a_j})) \qquad (6)$$

where $w_{ij}(t)$ is the link weight function between the jth process neuron in the hidden layer and the ith unit in the input layer. v_j is the link weight function between the jth process neuron in the hidden layer and the output layer. $\psi(\cdot)$ is the wavelet basis function of the process neurons in the hidden layer. $g(\cdot)$ is the activation function of the neuron in the output layer. y is the output of the network.

4 LEARNING ALGORITHM

4.1 Learning algorithm based on orthogonal basis functions

It is obvious that (6) is very difficult to be calculated by the computer directly. In order to raise the efficiency of the computation and the adaptability to practical problems resolving of the WPNN, a group of appropriate orthogonal basis functions are introduced into the input space.

Supposing that $b_1(t), b_2(t), \cdots, b_k(t), \cdots$ is a set of standard orthogonal basis functions, $x_i(t), w_{ij}(t)$ can be express by the same set of standard orthogonal basis functions as

$$x_i(t) = \sum_{k=1}^{\infty} c_i^k b_k(t), \qquad c_i^k \in R \qquad (7)$$

$$w_{ij}(t) = \sum_{k=1}^{\infty} w_{ij}^k b_k(t), \qquad w_{ij}^k \in R \qquad (8)$$

By mathematics analysis theorem[9], if any $\varepsilon > 0$ is given, then there exists L_i such that

$$\underset{0 \le t \le T}{Sup} \left| x_i(t) - \sum_{l=0}^{L} c_i^l b_l(t) \right| \le \varepsilon \text{ for all } t \in [0, T]. \text{ If } L =$$

$max\{L_1, L_2, \ldots, L_n\}$, then

$$x_i(t) = \sum_{k=1}^{L} c_i^k b_k(t), \qquad c_i^k \in R \qquad (9)$$

$$w_{ij}(t) = \sum_{l=1}^{L} w_{ij}^l b_l(t), \qquad w_{ij}^l \in R \qquad (10)$$

According to the property of orthogonal function
$$\int_0^T b_k(t) b_l(t) dt = \begin{cases} 1, k = l \\ 0, k \ne l \end{cases}, \text{ Eq. (6) can be expressed as}$$

$$y = g(\sum_{j=1}^{m} v_j \psi(\frac{\sum_{i=1}^n \sum_{k=1}^L c_i^k w_{ij}^k - b_j}{a_j})) \qquad (11)$$

4.2 Learning procedure

Given P number of learning functions: $(x_{p1}(t), x_{p2}(t), \cdots, x_{pn}(t), d_p), p = 1, 2, \cdots, P$, where the first subscript of $x_{pi}(t)$ denotes the serial number of learning functions. The second subscript denotes the serial number of subvector. d_p denotes the desired output of the inputs $(x_{p1}(t), x_{p2}(t), \cdots, x_{pn}(t))$

Supposing that y_p is the corresponding actual output function of the WPNN model. Thus, the mean square error of the WPNN model can be defined below

$$E = \frac{1}{2} \sum_{p=1}^{P} (y_p - d_p)^2$$

$$= \frac{1}{2} \sum_{p=1}^{P} (g(\sum_{j=1}^{m} v_j \psi(\frac{\sum_{i=1}^n \sum_{k=1}^L c_{pi}^k w_{ij}^k - b_j}{a_j})) - d_p)^2 \qquad (12)$$

Applying gradient descent algorithm, w_{ij}^k, v_j, a_j, b_j should be adjusted to train the WPNN. We have

$$\begin{cases} w_{ij}^k(s+1) = w_{ij}^k(s) + \alpha \Delta w_{ij}^k \\ v_j(s+1) = v_j(s) + \beta \Delta v_j \\ a_j(s+1) = a_j(s) + \gamma \Delta a_j \\ b_j(s+1) = b_j(s) + \eta \Delta b_j \end{cases} \qquad (13)$$

where, $\alpha, \beta, \gamma, \eta$ are learning rate. s is the iteration.

For the convenience of analysis, u is defined as

$$u = \frac{\sum_{i=1}^n \sum_{k=1}^L c_{pi}^k w_{ij}^k - b_j}{a_j}, \text{ Applying gradient descent on this}$$

estimate of the mean square error, we also have

$$\begin{cases} \Delta w_{ij}^k = -\frac{\partial E}{\partial w_{ij}^k} = -(y-d)(\sum_{j=1}^m v_j \psi'(u)) \sum_{i=1}^n \sum_{k=1}^L \frac{c_{pi}^k}{a_j} \\ \Delta v_j = -\frac{\partial E}{\partial v_j} = -(y-d)\psi(u) \\ \Delta a_j = -\frac{\partial E}{\partial a_j} = -(y-d)(\sum_{j=1}^m v_j \psi'(u)(-\frac{u}{a_j})) \\ \Delta b_j = -\frac{\partial E}{\partial b_j} = -(y-d)(\sum_{j=1}^m v_j \psi'(u)(-\frac{1}{a_j})) \end{cases} \qquad (14)$$

The usually used basis functions include Gaussian radial basis functions, B-spline basis functions, wavelet basis functions and some neurofuzzy basis functions. In this study, we choose Morlet wavelet function as the activation function. The Morlet wavelet is a cosine modulated Gauss wave. When x goes larger or smaller, it decays very rapidly. The wavelet exhibit good locality features in both time and frequency. The Morlet wavelet function can be expressed as

$$\psi(x) = \cos(1.75 \frac{x-b}{a}) \exp(-(\frac{x-b}{a})^2 / 2) \quad (15)$$

The complete procedure of the learning algorithm is summarized in Fig. 3.

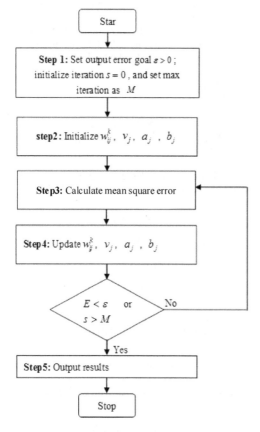

Figure 3. The learning algorithm of WPNN.

5 SIMULATION TEST

5.1 *Power load forecasting by the WPNN*

Load forecasting is an integral part of electric power system operations. With the worldwide deregulation of the power system industry, load forecasting is now becoming even more important not only for system operators, but also for market operators, transmission owners, and any other market participants so that adequate energy transactions can be scheduled and appropriate operational plans and bidding strategies can be established [10-11]. The forecast ranging from 5 to 20 years is termed as long term forecasting while the forecast ranging from few months to 5 years is termed as medium term forecasting. If the duration of the forecast varies from few hours to weeks, it is called as Short-Term Load Forecasting. The long and medium term forecasting are used to determine the capacity of generation, transmission or distribution system additions and the type of facilities required in transmission expansion planning, annual hydro thermal maintenance scheduling, etc. The short-term load forecasting is needed for control and scheduling of power system and also as inputs to load flow study or contingency analysis. The purpose of very short-term load forecasting (ranging from minutes to hours) is for real time control and security evaluation.

With power systems growth and the increase in their complexity, many factors have become influential to the electric power generation and consumption Therefore, the forecasting process has become even more complex, and more accurate forecasts are needed. The relationship between the load and its exogenous factors is complex and non-linear, making it quite difficult to model through conventional techniques, such as time series and linear regression analysis. Besides not giving the required precision, most of the traditional techniques are not robust enough.

The power load data used in this paper was taken from one city of China from January 1990 to December 1998, and the sampling interval is about a month. We get a power load time series with 108 discrete points such as $\{LOAD_j\}_{j=1}^{108}$. $(LOAD_i, LOAD_{i+1}, \ldots, LOAD_{i+5})$ can be used to generate an input function IF_i by nonlinear least-squares method, $LOAD_{i+6}$ is the output OF_i of the network, where $i=1, \ldots, 103$. Thus, we can get 103 samples such as $\{IF_i, OF_i\}_{i=1}^{103}$. The samples $\{IF_i, OF_i\}_{i=1}^{91}$ are selected to train the WPNN. The WPNN used in this section is composed of three layers, the first layer is the input layer with one unit, the second layer is consisted of 24 wavelet process neurons, the last layer is output layer with one neuron. The orthogonal Legendre basis functions are selected to expand input functions and weight functions. According to [12], initialize the dilation and translation parameters of WPNN. The error goal is set to 0.0001, and the initialize learning rate is set to 0.01, the max iteration number is set to 5000. In order to have an idea about the success of the model, the same scale PNN model is formed. The PNN is trained and tested by the same training samples. The learning error curves are depicted in Fig. 4.

After 428 iterations, WPNN has converged, and after 874 iterations, PNN model has converged. As we can see from this figure the WPNN has good convergence behaviors than that of the PNN.

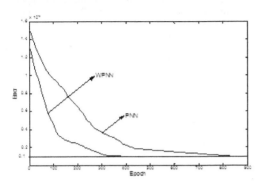

Figure 4.   Learning error curve.

The samples $\{IF_i, OF_i\}_{i=92}^{103}$ are selected to test the WPNN. The test results as shown in Table 1.

Table 1.   Simulation test results of WPNN.

Date	Desired value (MW)	Actual value (MW)	Absolute error(MW)	Relative error(%)
January, 2013	75672	74983	689	0.91
February, 2013	69867	69356	511	0.73
March, 2013	87785	89039	1554	1.43
April, 2013	83817	83407	410	0.49
May, 2013	86835	86346	489	0.56
June, 2013	84529	84531	2	0.00
July, 2013	101733	101062	671	0.66
August, 2013	105179	102823	2356	2.24
September, 2013	86984	85672	1312	1.51
October, 2013	85989	86489	480	0.56
November, 2013	85457	87145	1688	1.97
December, 2013	94483	94415	68	0.07

The PNN is trained and tested by the same training samples. The test results are described in Table 2.

Table 2.   Simulation test results of PNN.

Date	Desired value (MW)	Actual value (MW)	Absolute error(MW)	Relative error(%)
Jan, 2013	75672	74098	1574	2.08
Feb, 2013	69867	70887	1020	1.46
Mar,2013	87785	90462	2677	3.05
Apr,2013	83817	85386	1569	1.87
May,2013	86835	85668	1167	1.34
Jun,2013	84529	85212	685	0.81
Jul,2013	101733	97693	4040	3.97
Aug,2013	105179	98037	7142	6.79
Sep, 2013	86984	91264	4280	4.92
Oct,2013	85989	88775	2786	3.24
Nov,2013	85457	83201	2256	2.64
Dec,2013	94483	95995	1512	1.60

As seen in Tables 1 and 2, the WPNN is much superior to the PNN. The average relative error of WPNN is 0.93% compared with 2.81% of PNN. The learning ability of WPNN is better than that of PNN. The simulation test results indicate that the WPNN has a higher accuracy than the PNN. The WPNN seems to perform well and appears suitable for using as a predictive tool for the power load forecasting.

6   CONCLUSIONS

In order to solve the problem of medium term power load forecasting, Wavelet process neural network model is proposed in this paper. WPNN, which combine the wavelet theory and process neural network, utilize wavelets as the basis function to construct the networks. Wavelet function has the spatial-spectral zooming property, and therefore, it can represent the local spatial-spectral characteristics of the function. Moreover the network can deal with continuous input signals directly. A corresponding learning algorithm is developed. In consideration of the complexity of the aggregation operation of time in the WPNN, a group of appropriate orthogonal basis functions in the input function space of the WPNN is selected, and then the input functions and the network weight functions are represented as expansion of the same orthogonal basis functions. This application shows that the learning algorithm based on the expansion of the orthogonal basis functions simplifies the computing complexity of the WPNN. The effectiveness of the WPNN and the corresponding learning algorithm is proved by forecast medium term power load. The simulation test results indicate that the proposed network architecture and the associated learning algorithm are quite effective in modeling the dynamics of complex processes and perform accurate forecasting.

REFERENCES

Zhang Q.H, Benveniste A. Wavelet networks. *IEEE Transactions on Neural Networks*, 1992, 3(6): 889–898.

Zhang Q.H. Using Wavelet network in nonparametric estimation, *IEEE Transactions on Neural Networks*, 1997, 8 (2): 227–236.

Zhang J, Gilbert W.G, Miao Y.B, et al. Wavelet neural networks for function learning. IEEE *Transactions on Signal Processing*, 1995, 43(6):1485–1497.

Chen Y.H, Yang B, Dong J.W. Time-series prediction using a local linear wavelet neural network. *Neurocomputing*, 2006, 69: 449~465.

M. Ulagammaia, P. Venkatesh, P.S Kannan, et al. Application of bacterial foraging technique trained artificial and wavelet neural networks in load forecasting. *Neurocomputing*, 2007, 70: 2659–2667.

He X.G, Liang J.Z. Process Neural Networks. Proceedings of the Conference on Intelligent Information Processing. Publishing Houses of Electronics Industry, Beijing, 2000, 143–146.

He X.G, Liang J.Z. Some theoretical issues on process neural networks. *Engineering Science*, 2, 2000, 40–44. (in Chinese with English abstract)

Daubechies I. The wavelet transform, time-frequency localization and signal analysis. *IEEE Transactions on Information Theory*, 1990, 136(5):961–1005.

Jeffreys, H., Jeffreys, B.S. Methods of Mathematical Physics. 3rd edn. Cambridge University Press, Cambridge, 1988, 446–448.

Gwo C.L, Ta P.T. Application of fuzzy neural networks and artificial intelligence for load forecasting. *Electric Power Systems Research*, 2004, 70: 237–244.

T. Yalcinoz, U. Eminoglu. Short term and medium term power distribution load forecasting by neural networks. *Energy Conversion and Management*, 2005, 46: 1393–1405.

Zhao X.Z, Zhou C.H, Chen T.J, et al.: A research on the initialization of parameters of wavelet neural networks, *Journal of South China University of Technology*, 2003, 31(2):77–79. (in Chinese with English abstract).

Section 3: Industrial engineering, product design and manufacturing

Study on the influence of the spindle system joints on the dynamic characteristics of a spindle system

YongSheng Zhao & YanHu Li
College of Mechanical Engineering and Applied Electronics Technology, Beijing University of Technology, Beijing, China

ABSTRACT: The high speed motorized spindle mainly includes the spindle-bearing, spindle-the handle, and the handle-tool joints. These joints have a significant impact on the dynamic characteristics of the spindle system. In this paper, a complete finite element model of the spindle system is established based on consideration of the characteristics of bearing-spindle, spindle-the handle, and the handle-tool joints. The influence of the joints on the dynamic characteristic of the spindle system and the harmonic response analysis is carried out to get the frequency response function of the tip point, which will provide the basis for predicting the cutting stability of the spindle system.

KEYWORDS: Spindle system; Joints; Finite element method; FRF.

1 INTRODUCTION

Machine tool spindle system is one of the core components of the NC machine tools. The processing performance of the machine tool is directly affected by the spindle system dynamics. Therefore, it is necessary to study the dynamic characteristics of spindle system.

The high speed motorized spindle, mainly includes the spindle-bearing, spindle-the handle, the handle-tool joints, is a relatively independent unit. As the vulnerable spot of the spindle system, these joints have a significant impact on the dynamic characteristics of spindle system. At present, study on the motorized spindle dynamics modeling mainly focuses on the modeling of angular contact ball bearing, which usually ignores the influence of the joints. Reference [2] established the dynamic model of the bearing-spindle system and got the bearing stiffness, natural frequency and frequency response function by using quasi statics and finite element modeling method. The quality of the handle and the tool was introduced in reference [3], but the effect of joints was not taken into consideration. Literature [4] came up with a test method to identify joint stiffness by using frequency response function, which is still an effective way to get the joint stiffness at present. In this paper, a complete finite element model of the spindle system is established based on consideration of the characteristics of bearing-spindle, spindle-the handle, the handle-tool joints. The influence of the joints on the dynamic characteristic of the spindle system and the harmonic response analysis is carried out to get the frequency response function of the tip

point, which will provide the basis for predicting the cutting stability of spindle system.

2 THE FINITE ELEMENT MODELING OF MOTORIZED SPINDLE SYSTEM

In this paper, the finite element model of motorized spindle system is established by using ANSYS. There are generally two ways for the finite element modeling of motorized spindle system: simplified model and solid model. Considering the convenient of subsequent modeling of joints, a simplified model is applied here.

The motorized spindle system finite element modeling process is as follows.

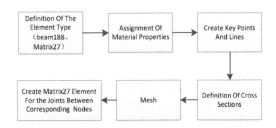

2.1 The finite element modeling of spindle

The dimension of each part in the spindle system is shown in Tables 1–3.

Assuming that each part of the spindle system has the same material properties: elastic modulus is

Table 1. The spindle unit size.

lIne number	1	2	3	4	5	
Line length	90	288	45	216	54	
Inside radius	20.75	20.75	17.25	14.6	14.6	23.1
Outside radius	27.5	30	30	35	40	40
Element number	10	32	5	24	6	

Table 2. The handle unit size.

Line number	1	2	3	4	5	6	
Line length	54	9	9	9	36	27	
Inside radius	12.5	12.5	12.5	12.5	10	10	10
Outside-radius	14.6	23.1	31.5	26.5	31.5	27.5	20
Element number	6	1	1	1	4	3	

Table 3. Cutter unit size.

Line number	1	2	3
Line length	72	30	8
Inside radius	0	0	0
Outside radius	10	10	10
Element number	8	3	1

2.06E11, Poisson's ratio is 0.3, the density is 7900 kg/m³. BEAM188 unit is selected in the modeling of spindle system. The influence of axial, torsion, bending and shear deformation is included in the unit, so it can better describe the dynamic characteristics of the spindle system. The first step of modeling is creating key points, and then the line segment is created by connecting key points. Because the cross section characteristics of BEAM188 unit cannot be defined through real constant, and must be defined by the section number, it is necessary to define the geometry size of different cross sections and assign the size to each line.

2.2 The modeling of spindle system joints

The model of spindle-bearing, spindle-the handle, hangdle-tool joints will be set up respectively. The bearing and the joints stiffness will be simulated by Matrix27 unit. There are two bearings on the front and one bearing on the back of the spindle. Front bearings

center are located in node 65, 65 while after bearing center is located in node 21. The stiffness of joints can be identified by the frequency response function method. The stiffness of the parts is shown in Table 4.

Table 4. Stiffness of joints.

Joints	Radial stiffness (N/m)	Angular rigidity (rad/N.m)
Bearing joint	2.1×10^8	2.9×10^5
The handle joint	6.31×10^7	4.28×10^5
Tool joint	5.97×10^7	3.96×10^5

Use the distributed spring model to simulate the spindle system joints. Matrix27 unit connect to the corresponding two nodes of the joints. When define key word KEYOPT (2) = 0, the matrix is symmetrical as follows:

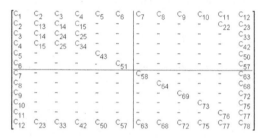

The joint stiffness value is assigned to the corresponding matrix elements.

The established finite element model in ansys is shown in Fig. 1.

Figure 1. Finite element model for the spindle system.

2.3 Experimental verification

The spindle-handle system is subjected to a hammering test in order to obtain its modal parameters, and then the test data is compared with the results of simulation to verify the reliability of the model. The single-point excitation, multi-point pickup method is adopted in the experiment. The vibration signal is received by acceleration sensors, and a 24-channels of data collector are used for the vibration signal acquisition. Experimental devices are shown in Fig. 2.

The experimental data is processed by LMS signal processing system to obtain the frequency response

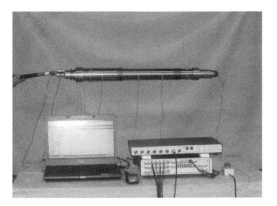

Figure 2. Experimental devices.

function. The frequency response function of each measuring point is shown in Fig. 3.

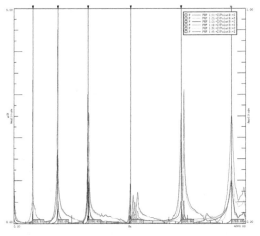

Figure 3. The frequency response function of each measuring point.

Matrix27 element is used to simulate the characteristics for the connection, and to establish the finite element model of the spindle system. The natural frequencies of the spindle system are obtained by simulation and compared with the experimental value, as shown in Table 5.

Table 5. The comparison of experimental and simulation data.

Order	1	2	3	4
experimental	323.51	1280.83	2021.22	2893.66
ANSYS simulation	344.2	1327.3	2086.6	2936.7
error	6.4%	3.6%	3.2%	1.5%

As Fig. 3, Table 5 shows the maximum error between experimental value and simulation value is 6.4%, while the minimum error is only 1.5%. The data was consistent, and this shows that the modeling method is reliable.

3 DYNAMIC CHARACTERISTICS ANALYSIS OF SPINDLE SYSTEM

3.1 Modal analysis of the spindle system

By solving the completed finite element model, the natural frequencies and the mode of vibration are obtained. Three different models will be established for the comparison of influence of different joints on the dynamic characteristics: no joint, only the bearing joint, and all of the joints. If a joint is ignored, it will be viewed as rigid connection. Table 6 shows the first four order natural frequency in the three cases.

Table 6. The natural frequency of spindle system(Hz).

Order	1	2	3	4
Not to consider all of the joints	543.97	2055.9	2936.5	4023.75
Only considering the bearing joint	500.9 7.9%	1530.6 25.6%	2656.1 9.5%	3978.45 1.1%
Considering all of the joints	211.3 57.8%	643.71 57.9%	739.6 72.2%	1906.0 52.1%

Fig. 4 shows the mode of vibration for the condition of no joints. Fig. 5 shows the mode of vibration for the condition of only considering the bearing joints. Fig. 6 shows the mode of vibration for the condition of considering all of the joints.

Figure 4. Considering no joint.

Figure 5. Only considering the bearing joints.

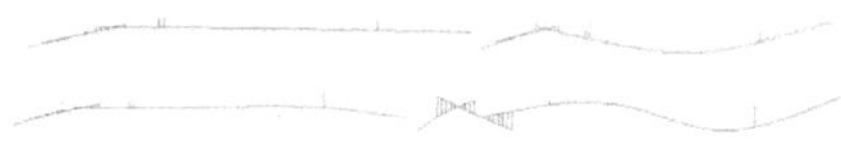

Figure 6. Considering all of the joints.

295

3.2 Influence of joints on the dynamic characteristics of spindle system

The results in Table 6 show that the natural frequencies of the spindle system are reduced after considering the effects of the junction. If only the bearing joint is considered, the first four order natural frequencies are reduced in different degree. The second order natural frequency has the maximum decline of 25.6%. For the condition of the handle and the tool joints, each order of the natural frequency is significantly lower; the fourth order has the minimum decline of 52.1%. Therefore, each order of the natural frequency of the spindle system is decreased obviously when taking the influence of the spindle system joints into consideration. Especially for the condition of considering the handle and the tool joints, the natural frequency and the mode of vibration has the most obvious change. This also proves the importance of the handle and the tool joints in the spindle system modeling.

4 FREQUENCY RESPONSE FUNCTION OF THE TOOL TIP

4.1 Harmonic response analysis of the spindle system

Harmonic response analysis is a method to determine the steady-state response of a structure under sine wave load. Given a frequency range, harmonic response analysis can be used to calculate the response of structure versus frequency and draw out the amplitude–frequency response curve of the structure. Steps of harmonic response analysis by ansys are as follows:

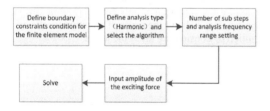

The frequency range for analysis is set between 0 and 4000Hz, the amplitude of excitation force is 200 N. The amplitude–frequency response curve of the tip point is obtained through complete method as shown in Fig. 7.

4.2 Getting the frequency response function of the tool tip

The natural frequency and the dynamic stiffness (the reciprocal of the frequency response function) of the spindle system can be obtained by frequency response function. By substituting the frequency

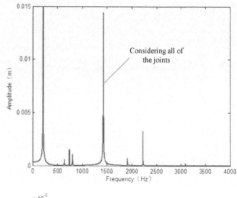

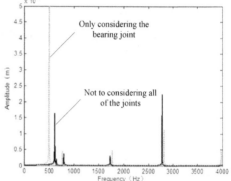

Figure 7. Amplitude-frequency response curve of the tool tip.

response function into the tool-work piece cutting model, we can also predict the cutting stability of spindle system according to the stability condition. So the frequency response function of the tip point has important significance for the analysis of dynamic characteristics and predicting the cutting stability of spindle system.

According to the definition of the frequency response function: $\{X\} = \{H\}\{F\}$

The frequency response function can be obtained by dividing the amplitude–frequency response of tip point by the amplitude of excitation force. Fig–.8 shows the frequency response functions.

4.3 Influence of joints on the frequency response functions

Figs. 7 and 8 show that each order of the natural frequency decreases after considering the stiffness of bearing joint amplitude–frequency curves and frequency response functions shift to the left. When continue to consider the stiffness of the handle and the tool joints, each order natural frequency is largely reduced, and the vibration amplitude have increased substantially. At the same time, the result of harmonic

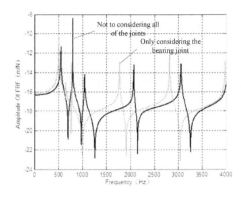

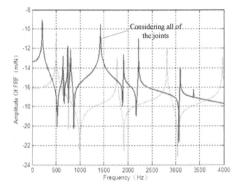

Figure 8. The frequency response functions of tip point.

response analysis is generally consistent with that of the modal analysis. This suggests that the joints of a spindle system, especially the handle and the tool joints, have important influence on the frequency response function of tip point, and affect the cutting performance of the system. So the joints cannot be ignored in the spindle system modeling.

5 CONCLUSION

The stiffness of joints is included in constructing the finite element model of the spindle system, and the influence of joints on the dynamic characteristics of the spindle system has been analyzed. The conclusions are as follows:

1 After considering the influence of joints, the natural frequency of spindle system has a significantly reduction. The handle and tool joints have a larger influence on the natural frequency and the mode of vibration.
2 The amplitude–frequency response shows that the response amplitude of the tip point increase obviously when considering all of the joints.
3 Getting the frequency response function of the tip point further validates the modal analysis results, and also provides the basis for forecasting the cutting stability of the system at the same time.

ACKNOWLEDGEMENT

The authors are most grateful to the Beijing Nature Science and Technology (No. 3132004) for supporting the research presented in this paper.

REFERENCES

Hongwei Zhang, XiangshengGao, Yuqing Zhang. ANSYS nonlinear finite element analysis method and example applications [M].Beijing: China WaterPower Press, 2013.
Yuzhong Cao, Yusuf Altintas. A General Method for the Modeling of Spindle-Bearing Systems. *Journal of Mechanical Design*, 2004,126(11):1089–1104.
DehaoMeng. The dynamic characteristics ofhigh speed motorized spindle and its effect on the stability of processing. Shanghai: Shanghai Jiao Tong University, 2012.
GuopingAn, Bingbing Zhang, YongshengZhao, LigangCai. Effect of bearing configuration on spindle dynamic characteristics.*Advanced Mechanics and Materials*. 2013,437:76–80.
D. M. Shamine, S. W. Hong, Y. C. Shin. Experimental identification of dynamic parameters of rolling element bearings in machine tools. *Journal of Dynamic Systems, Measurement, and Control*, MARCH 2000, 122:95–101.
Li, H. Q., Yung, C. Shin. Analysis of bearing configuration effects on high spindles using integrated dynamic thermo-mechanical spindle model. *International Journal of Machine Tools &Manufacture*, 2004, 44(4):347–364.
LONG, X, H., Balachandran, B. Stability analysis for milling process. *Nonlinear Dynamic*.2007, 3:349–359.

Computational Intelligence in Industrial Application – Ling (ed.)
© *2015 Taylor & Francis Group, London, ISBN: 978-1-138-02818-0*

The study of static and dynamic characteristics of hydrostatic rotary tables

ChengPeng Zhan, WenDi Pan, ZhiFeng Liu & YongSheng Zhao
College of Mechanical Engineering and Applied Electronics Technology, Beijing University of Technology, Beijing, China

ABSTRACT: Static and dynamic performance are the important properties of hydrostatic rotary tables, and in-depth study of the stiffness and damping coefficients is useful for designers and users. In this paper stochastic Reynolds equations are established based on Christensen roughness model. Then, load-carrying capacity, stiffness and damping coefficients of the supporting oil pads and preloaded oil pads are calculated by solving stochastic Reynolds equations. Next the stiffness and damping coefficients of the turntable under the tilt state can be obtained. Finally, this paper analyzes the impact of film thickness, roughness and the tilting angle.

KEYWORDS: Reynolds equations; Hydrostatic rotary table; Tilting; Surface roughness.

1   INTRODUCTION

Hydrostatic bearings have many advantages, like low coefficient of friction, high load capacity and long life. So they are widely used in industrial practice, such as machine tools, aircraft, ship and telescopes. Because of its importance, many researchers have been studied for hydrostatic bearing.

In the aspect of dynamics, Garratt [1] provided a compressible flow model for air bearings that uses a modified Reynolds equation formulation incorporating the effect of centrifugal inertia for high-speed operation and presented a steady-state and unsteady dynamics analysis of this model. Chen [2] presented a simplified model of a coupled spindle shaft and hydrostatic bearings. Analyze the effects of the eccentricity of the spindle system and the load on the deformation of the spindle system. Yang [3] investigated the influences of constant compensations that are produced by capillary or constant flow pump on the dynamic characteristics of a circular worktable supported by a closed-type hydrostatic thrust bearing.

In the aspect of surface roughness, Burton [4] modeled roughness by a Fourier series type approximation. Since, the surface roughness distribution is random in nature; a stochastic approach has to be adopted. Thus, Christensen [5] used a polynomial to approximate Gaussian distribution density function and based on stochastic theory he developed a stochastic theory for the study of rough surfaces in hydrodynamic lubrication. Since then, many researchers have adopted and used this approach to study roughness effect on bearing surfaces extensively. For example, Lin [6] investigates the effects of surface roughness and centrifugal inertia for dynamic stiffness and damping characteristics of compensated hydrostatic thrust bearing. Ahmad [7] studied load capacity, volume flow rate, friction factor, power factor and stiffness factor of hydrostatic thrust spherical bearings which consider the influence of surface roughness and centrifugal inertia. Naduvinamani [8] considered surface roughness and MHD on couple stresses squeeze film in hydrostatic thrust bearing and calculated the mean squeeze film pressure, load carrying capacity and mean squeeze film time of the bearings.

Although many researchers have studied surface roughness effects and various characteristics of hydrostatic bearing, they seldom care about the tilt state. This work will calculate the stiffness and damping coefficients and analyze the influence of film thickness, tilting angle and surface roughness.

2   FORMULATIONS

Basic structure of hydrostatic rotary table is shown in Fig. 1, its supporting system consists of supporting oil pad, preloaded oil pad and radial bearing. The supporting oil pad is circular step pad and rotary table has n supporting pads in total. They are evenly arranged in one supporting circle and the radius of this circle is R_L. The preloaded oil pad is annular recess pad and provides a pre-pressure that can enhance the stiffness of the turntable. The radial bearing is mounted on the center of the turntable, it makes more stable rotation. All hydrostatic oil in the pads are supplied by a constant flow pump.

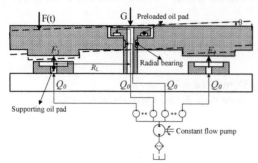

Figure 1.   The simple hydrostatic rotary table.

2.1 Establishment of Reynolds equation

In this study, it is assumed that thin film lubrication theory is applicable; the flow in bearing is isothermal, laminar and axial symmetry. So the N-S equations can be simplified as:

$$\frac{1}{r}\frac{\partial(ru_r)}{\partial r}+\frac{\partial(v_z)}{\partial z}=0 \tag{1}$$

$$\frac{\partial p}{\partial r}=\eta\frac{\partial^2 u_r}{\partial z^2} \tag{2}$$

$$\frac{\partial p}{\partial z}=0 \tag{3}$$

Film pressure has no change in z direction according to Eq. (3) above, so integrating both sides of Eq. (2) with boundary conditions: $z=h_T, u_r=0$; $z=0, u_r=0$ at the surface of the pad, then the tangential velocity is found to be:

$$u_r=\frac{z(z-h_T)}{2\eta}\frac{\partial p}{\partial r} \tag{4}$$

Substituting Eq. (4) into Eq. (1) and integrating it with boundary conditions: $z=h_T, u_r=0, v_z=\frac{\partial h}{\partial t}$; $z=0, u_r=0, v_z=0$, then the Reynolds equation is obtained as:

$$\frac{1}{r}\frac{\partial}{\partial r}(\frac{rh_T^3}{12\eta}\frac{\partial p}{\partial r})=\frac{\partial h_T}{\partial t} \tag{5}$$

Where the film thickness h_T in Eq. (5) can be considered to be made up of two parts:

$$h_T=\delta+h(t) \tag{6}$$

In which δ represents the random part and h represents the nominal smooth part of the film geometry.

2.2 Stochastic Reynolds equation

According to Christensen's stochastic approach of rough surfaces [2], take the expectation of the Reynolds equation (7) and obtained stochastic Reynolds equation.

For radial one-dimensional roughness $h_T=\delta(\theta,\xi)+h(t)$ and the Reynolds-type equation is:

$$\frac{1}{r}\frac{\partial}{\partial r}(\frac{rE(h_T^3)}{12\eta}\frac{\partial p}{\partial r})=\frac{\partial E(h_T)}{\partial t} \tag{7}$$

For circumferential one-dimensional roughness $h_T=\delta(r,\xi)+h(t)$ and the Reynolds-type equation is:

$$\frac{1}{r}\frac{\partial}{\partial r}(\frac{r}{12\eta E(h_T^{-3})}\frac{\partial p}{\partial r})=\frac{\partial E(h_T)}{\partial t} \tag{8}$$

Since most of the engineering rough surfaces are Gaussian in nature, a polynomial form is chosen to approximate a Gaussian distribution [5–7].

$$f(\delta)=\begin{cases}\frac{35}{32c^7}(c^2-\delta^2)^3 & -c\le\delta\le c\\0 & \text{elsewhere}\end{cases} \tag{9}$$

In which c is the half total range of random film thickness variable and function terminates at $c=\pm3\sigma$, as a result, one has:

$$E(h_T)=h, E(h_T^3)=h^3+\frac{1}{3}hc^2, \frac{1}{E(h_T^{-3})}=h^3-\frac{2}{3}hc^2 \tag{10}$$

So the stochastic Reynolds equation can be written as:

$$\frac{1}{r}\frac{\partial}{\partial r}(rE\frac{\partial p}{\partial r})=12\eta T \tag{11}$$

In which

$$E=\begin{cases}h^3+\frac{1}{3}hc^2 & radial\ roughness\\h^3-\frac{2}{3}hc^2 & circumferential\ roughness\end{cases} \tag{12}$$

$$T=\frac{\partial E(h_T)}{\partial t} \tag{13}$$

Also the flow rate of the oil pad can be calculated by integrating Eq. (4):

$$Q(r)=-\frac{\pi rE}{6\eta}\frac{\partial p}{\partial r} \tag{14}$$

2.3 Calculation of supporting oil pad

There are boundary conditions for quantitative compensated circular oil pad is:

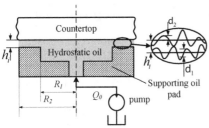

Figure 2.   The structure of supporting oil pad.

$$\begin{cases} r=R_1, p=p_0; r=R_2, p=0 \\ Q(R_1)=Q_0; h=h_{si} \end{cases} \quad (15)$$

Solving the Eqs. (11) and (14) under the boundary conditions (17), one can get the film pressure $p_i(r)$ and oil cavity pressureP_{0i}:

$$p_{0i}=\frac{6\eta\ln(\frac{R_2}{R_1})}{\pi E}(Q_0+\pi R_1^2 T-\frac{\pi(R_2^2-R_1^2)T}{2\ln(\frac{R_2}{R_1})}) \quad (16)$$

$$p_i(r)=\frac{3\eta r^2 T}{E}-\frac{(p_0\ln(\frac{r}{R_2})+\frac{3\eta T(R_2^2\ln(\frac{r}{R_1})-R_1^2\ln(\frac{r}{R_2}))}{E})}{\ln(\frac{R_2}{R_1})} \quad (17)$$

So the load-carrying capacity of circular oil pad is calculated as:

$$F_i=\pi R_1^2 p_{0i}+2\pi\int_{R_1}^{R_2} rp_i(r)\,dr \quad (18)$$

F_i can be divided into two parts one is static load-carrying capacity and anther is dynamic load-carrying capacity , the static part take derivate for film thickness h_{si} and order $\frac{\partial h_{si}}{\partial t}=1$ for the dynamic part, then the stiffness and damping coefficients can be obtained:

$$\begin{cases} K_s=\frac{3\eta Q_0(R_1^2-R_2^2)}{E^2}\frac{\partial E}{\partial h_{si}} \\ C_s=\frac{3\eta\pi(R_1^2-R_2^2)^2}{2E} \end{cases} \quad (19)$$

2.4 Calculation of annular recess oil pad

The boundary conditions for quantitative compensated annular recess oil pad are:

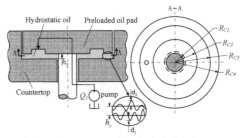

Figure 3.   The structure of preloaded oil pad.

$$\begin{cases} r=R_{C1}, p=0; r=R_{C2}, p=p_{0y} \\ r=R_{C3}, p=p_{0y}; r=R_{C4}, p=0 \\ Q_1=-Q(R_{C2})+Q(R_{C3}); h=h_y \end{cases} \quad (20)$$

Solving the Eqs. (11) and (14) under the boundary conditions (20), one can get the film pressure $p_y(r)$ and oil cavity pressureP_{0y}.

$$p_{0y}=\frac{6\eta\ln(\frac{R_{C4}}{R_{C3}})\ln(\frac{R_{C2}}{R_{C1}})}{\pi E\ln(\frac{R_{C4}R_{C2}}{R_{C3}R_{C1}})}(Q_0+\pi T(R_{c3}^2-R_{c2}^2)$$

$$-\frac{\pi T(\ln(\frac{R_{C2}}{R_{C1}})(R_{c4}^2-R_{c3}^2)+\ln(\frac{R_{C4}}{R_{C3}})(R_{c1}^2-R_{c2}^2))}{2\ln(\frac{R_{C2}}{R_{C1}})\ln(\frac{R_{C4}}{R_{C3}})}) \quad (21)$$

When $r\in(R_{C1},R_{C2})$the film pressure is:

$$p_1(r)=\frac{3\eta r^2 T}{E}-\frac{(p_0\ln(\frac{r}{R_{C1}})+\frac{3\eta T(R_{C1}^2\ln(\frac{r}{R_{C2}})-R_{C2}^2\ln(\frac{r}{R_{C1}}))}{E})}{\ln(\frac{R_{C1}}{R_{C2}})} \quad (22)$$

When $r\in(R_{C3},R_{C4})$the film pressure is:

$$p_2(r)=\frac{3\eta r^2 T}{E}-\frac{(p_0\ln(\frac{r}{R_{C4}})+\frac{3\eta T(R_{C4}^2\ln(\frac{r}{R_{C3}})-R_{C3}^2\ln(\frac{r}{R_{C4}}))}{E})}{\ln(\frac{R_{C4}}{R_{C3}})} \quad (23)$$

So the load-carrying capacity of preloaded oil pad can be solved according to Eqs. (21)–(23):

$$F_y=\pi(R_{C3}^2-R_{C2}^2)p_0+2\pi\int_{R_{c1}}^{R_{c2}} rp_1(r)\,dr+2\pi\int_{R_{c3}}^{R_{c3}} rp_2(r)\,dr \quad (24)$$

Then the stiffness and damping coefficients can be obtained:

$$\begin{cases} K_y = \dfrac{3Q_i\eta((R_{C4}^2-R_{C3}^2)\ln(\frac{R_{C1}}{R_{C2}})+(R_{C1}^2-R_{C2}^2)\ln(\frac{R_{C3}}{R_{C4}}))}{E^2\ln(\frac{R_{C1}R_{C3}}{R_{C2}R_{C4}})}\dfrac{\partial E}{\partial h_y} \\[4mm] C_y = \dfrac{-1}{E\ln(\frac{R_{C1}R_{C3}}{R_{C2}R_{C4}})}[\dfrac{3\pi\eta}{2}(R_{C1}^2-R_{C2}^2+R_{C3}^2-R_{C4}^2)((R_{C1}^2+R_{C2}^2-R_{C3}^2+R_{C4}^2) \\[4mm] \quad \ln(\frac{R_{C1}}{R_{C2}})+(R_{C1}^2-R_{C2}^2+R_{C3}^2+R_{C4}^2)\ln(\frac{R_{C3}}{R_{C4}})-(R_{C1}^2-R_{C2}^2+R_{C3}^2-R_{C4}^2))] \end{cases} \quad (25)$$

2.5 Calculation of the turntable

Figure 4 shows forces exerted on the turntable, all the supporting oil pads provide support force and preloaded oil pad provides preload force. h_s is mean film thickness of all the supporting oil pad, θ_x and θ_y are the tilting angle, so the total force and the total torque is:

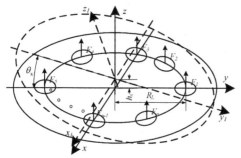

Figure 4.   Forces on the turntable.

$$\begin{cases} F_z = \displaystyle\sum_{i=1}^{n} F_i - F_y - Mg \\[4mm] M = \displaystyle\sum_{i=1}^{n} F_i R_L \sin(\varphi_i) \end{cases} \quad (26)$$

In which:

$$\begin{cases} h_{si} = h_s + R_L \sin(\varphi_i)\tan(\theta_x) + R_L \cos(\varphi_i)\tan(\theta_y) \\ h_y = h_{y0} + h_{s0} - h_s \end{cases} \quad (27)$$

Similarly the stiffness and damping coefficients of the turntable can be obtained:

$$\begin{cases} K_Z = \displaystyle\sum_{i=1}^{n1} K_{si} + K_y, \; K_{\theta_x} = \displaystyle\sum_{i=1}^{n} K_{si}(\dfrac{R_L\sin(\varphi_i)}{\cos(\theta_x)})^2 \\[4mm] C_Z = \displaystyle\sum_{i=1}^{n} C_{si} - C_y, \; C_{\theta_x} = \displaystyle\sum_{i=1}^{n} C_{si}(\dfrac{R_L\sin(\varphi_i)}{\cos(\theta_x)})^2 \end{cases} \quad (28)$$

Stiffness of turntable includes axial stiffness and tilting stiffness, the same axial damping and tilting damping are also included in damping of turntable.

Table 1.   Parameters of the turntable.

Parameters	Value	Parameters	Value
R_1(mm)	160	h_{s0}(mm)	0.1
R_2(mm)	175	h_{y0}(mm)	0.1
R_{C1}(mm)	185	$Q_0\,Q_i$(m³/s)	1e-4
R_{C2}(mm)	220	n (Pa.s)	0.091
R_{C3}(mm)	255	h_M(Kg)	30827
R_{C4}(mm)	290	R_L(mm)	1500

3   SIMULATION RESULTS AND DISCUSSION

To get further insight into the coupled effects of unbalanced load and surface roughness, an example is presented as follows. First the parameters of a turntable is illustrated as follows:

3.1   The influence of mean film thickness

Figure 5 shows the changing curve of stiffness and damping coefficients When mean film thickness h_s decreases, there is no doubt that the stiffness and damping coefficients increase along with the decreases

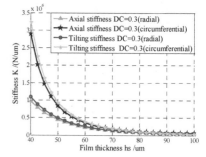

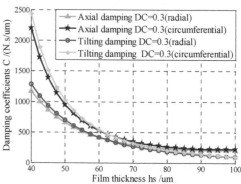

Figure 5.   Stiffness and damping coefficients at versus hs.

in mean film thickness h_s. At the same time, DC is surface roughness parameter and $DC = \frac{c}{h_0}$. For the radial roughness, its stiffness and damping increase amount are less than the circumferential roughness.

3.2 *The influence of surface roughness*

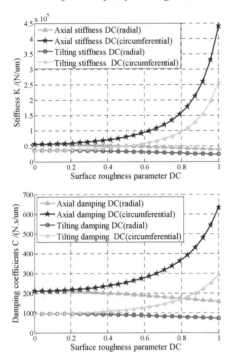

Figure 6.   Stiffness and damping coefficients at versus DC.

Just as shown in Fig. 6, when the value of surface roughness parameters DC increases, for circumferential roughness there is a slight increase in stiffness and damp coefficients at beginning, where, substantial increase when the roughness parameters DC>0.6 Stiffness and damping characteristics slightly decrease along with the increase of DC value for radial roughness.

3.3 *The influence of tilting angle*

Tilting angle of turntable also has significant influence for its stiffness and the damping, just as shown in Fig. 7. For both circumferential roughness and radial roughness its stiffness and the damping increase with tilting angle. The number of its increase is less at the begining, but the the increase becomes larger and larger along with the increase of inclination.

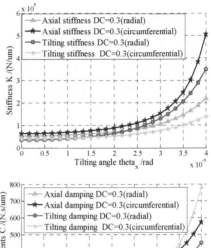

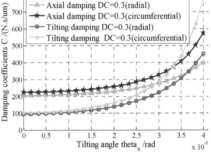

Figure 7.   Stiffness and damping coefficients at versus tilting angle.

4   CONCLUSIONS

This paper deduced stiffness and damping of turntable based on Christensen roughness model and in this calculation considers the impact of roughness and tilting angle. Analyzed the impact of film thickness, surface roughness and tilting of the turntable. The results show that turntable stiffness and damping coefficients increase with the decrease of film thickness and increase with the increase of tilting angle. For circumferential roughness, stiffness and damping increase with the increase of the roughness, but for radial roughness its low just opposite.

ACKNOWLEDGEMENT

The authors are most grateful to the National Science and Technology Major Project of China (No. 2013ZX04013-011).

REFERENCES

JE Garratt, S Hibberd, et al. Centrifugal inertia effects in high-speed hydrostatic air thrust bearings. *Engineering Mathematics*, 2012,,76:59–80.

Dongju Chen, Jinwei Fana. Dynamic and static characteristics of a hydrostatic spindle for machine tools. Manufacturing Systems, 2012,31(1):26–33.

Y Kang, HC Chou, et al. Dynamic behaviors of a circular worktable mounted on closed-type hydrostatic thrust bearing compensated by constant compensations. *Mechanics*, 2013, 29(3):297–308.

RA Burton, Effect of two-dimensional sinusoidal roughness on the load support characteristics of a lubricant film, *J. Basic Eng. Trans. ASME.* 85 (1963)246.

H Christensen, Stochastic model for hydrodynamic lubrication of rough surfaces, Proc. Inst. Mech. Eng. Part-I. 184 (1969) 1013.

Lin JR.Surface roughness effect on the dynamic stiffness and damping characteristics of compensated hydrostatic thrust bearings.International *Journal of Machine Tools & Manufacture*, 2000, 40:1671–1689.

Ahmad W Yacout, Ashraf S Ismaeel, Sadek Z Kassab.The combined effects of the centripetal inertia and the surface roughness on the hydrostatic thrust spherical bearings performance. *Tribology International*, 2007, 40:522–532.

NB Naduvinamani, BN Hanumagowda, Syeda Tasneem Fathima. Combined effects of MHD and surface roughness on couple-stress squeeze film lubrication between porous circular stepped plates. *Tribology International*, 2012, 56:19–29.

Computational Intelligence in Industrial Application – Ling (ed.)
© 2015 Taylor & Francis Group, London, ISBN: 978-1-138-02818-0

Research on the effect of interference fit in the dynamic characteristics of high-speed angular contact ball bearing

TieNeng Guo, XiaoChao Ma, ZhiFeng Liu, Gen Li & WenDi Pan
College of Mechanical Engineering and Applied Electronics Technology, Beijing University of Technology, Beijing, China

ABSTRACT: In the present research, the interference fit of bearing was ignored in the analysis of the dynamical character of bearing. In fact, the interference fit will affect the dynamical character. In order to analyze the interference fit, this paper proposes a general modeling method of high speed angular contact ball bearing. The model is based on the quasi statics analysis and raceway control theory of the bearing. The inner and outer ring radial deformation of the bearing is modeled. Also, the influence of the centrifugal force and gyroscopic moment generated in the rotation process is taken into account. The 7212c bearing was taken as an example. The simulation results display that the axial and radial stiffness which consider the interference fit is larger than that which does not consider the interference fit. According to the increase of interference fit, the axial and radial stiffness increases, especially in the condition of low speed, the radial stiffness considers the interference fit as large as 12.8% than the stiffness without the interference fit.

KEYWORDS: Bearing; Centrifugal force; Gyroscopic moment; Interference fit.

1 INTRODUCTION

High-speed spindle unit is the core of the machine tool, its performance level determines the machining quality of workpiece and machine tool processing efficiency. In the spindle assembly, the installation status of bearing would affect the bearing's positioning, rotating accuracy and bearing's clearance, which in turn affect the contact angle, contact deformation, and the stiffness of bearing. The speed of spindle is relatively high. The spindle and bearing inner ring can produce radial deformation under the centrifugal force. This leads to the change of the interference fit. Therefore, the interference fit is an important influence factor on the stiffness of high-speed bearing stiffness.

The literature[1] showed that in the condition of bearing high speed rotation, the balls were not only subjected to the contact force caused by the interaction between inner and outer rings of the bearing, but also subjected to the centrifugal force and gyroscopic force caused by its rotation. Jones [2] proposed firstly the quasi statics analysis and raceway control theory of the bearing. The concept of bearing stiffness matrix was proposed in the theory. The literature [3] considered the influence of interference fit in the parameter of bearing and the effect of the change of inner ring and outer ring in the bearing's clearance. The scholar from University of science and technology of China

analyzed the influence of interference fit in contact stiffness of bearing[4],his paper used Hertzian contact stiffness mode. But the bearing model can't reflect the relations of global force and displacement of bearing and the bearing contact stiffness can't represent whole bearing stiffness. This paper was based on the quasi statics analysis and raceway control theory and combined the theory of the interference fit with 5 dofs bearing stiffness model .Also, a complete analytical bearing stiffness matrix is expressed. Through this model, the dynamic characteristics of high-speed angular contact ball bearing can be studied in the condition of considering the interference fit.

The modelling method of the bearing interference fit is presented in the first section. In a second section, the 5dofs bearing model is modeled and the bearing stiffness matrix is expressed .Finally, the simulation is shown by studying the model of bearing.

2 MATHEMATICAL MODEL OF ANGULAR CONTACT BALL BEARINGS

2.1 *The change of contact angle and displacement of bearing after installing*

The closely fit problem of the spindle and bearing, bearing and bearing housing can be seen as thick-walled cylinder problem. When the bearing inner ring and spindle in interference fit, the inner ring expands, so that the inner ring raceway diameter increases, Similarly, the outer ring will shrink and the outer ring raceway diameter will decrease. According to elastic mechanics theory, the increasing value of inner ring raceway diameter is δ_F, the decreasing value of the outer ring raceway's diameter is (absolute value) δ_E, the calculation formula:

$$\delta_F = D_i \bullet \Delta f_1 / D_F \tag{1}$$

where D_i is inner diameter of bearing, Δf_1 is the interference fit of inner ring, $D_F = 2r_i$ is diameter of inner raceway curvature center.

$$\delta_E = \cfrac{2\left(\cfrac{D_E}{D_o}\right)\Delta f_2}{\left[1-\left(\cfrac{D_E}{D_o}\right)^2\right]\left\{\left[\cfrac{1+\left(\cfrac{D_E}{D_o}\right)^2}{1-\left(\cfrac{D_E}{D_o}\right)^2}-\mu_b\right]+\cfrac{E_b}{E_h}(1+\mu_h)\right\}} \tag{2}$$

where D_o is outer diameter of bearing, Δf_2 is the interference fit of outer ring, $D_E = 2r_o$ is diameter of outer raceway curvature center, E_b, E_h and u_b, u_h are bearing and bearing housing Modulus of elasticity and Poisson's ratio.

The relation of the initial contact angle with the bearing radial clearance, raceway curvature radius coefficient and ball diameter is

$$\cos\alpha_0 = 1 - \frac{\mu_r}{2(f_o + f_i - 1)D_w} \tag{3}$$

where u_r is radial clearance of bearing, D_w is ball diameter, $f_i = r_i / D_w$ is coefficient of curvature radius of inner ring raceway.

Considering the influence of the inference fit,the initial contact angle becomes the fit angle after installing.

$$\cos\alpha^1 = 1 - \frac{\mu_r - (\delta_F + \delta_E)}{2(f_o + f_i - 1)D_w} \tag{4}$$

After the bearing is installed, axial preload is exerted and the fit contact angle α^1 becomes α.

$$\frac{F_a K_n^{1.5}}{N(BD_w)^{1.5}} = \sin a \left(\frac{\cos a^1}{\cos a} - 1\right)^{1.5} \tag{5}$$

where F_a is axial preload, N is number of balls, $B = f_i + f_o - 1$.

2.2 Centrifugl force and gyroscopic moment

The centrifugal force formula is

$$F_{ck} = \frac{1}{2} D_m \omega^2 \left(\frac{\omega_m}{\omega}\right)_k^2 \tag{6}$$

where $D_m = (D_i + D_o)/2$, ω is rotation speed of bearing, ω_m is revolution speed of ball.

And the gyroscopic moment formula is

$$M_{gk} = J\left(\frac{\omega_b}{\omega}\right)_k \left(\frac{\omega_m}{\omega}\right)_k \omega^2 \sin\beta \tag{7}$$

where J is rotational inertia of the ball, ω_b is rotation speed of ball, β is attitude angle.

The distance of inner and outer ring raceway curvature center increases with the contact deformation under static load. However, the inner and outer ring raceway curvature center and ball center are still on a straight line. The rolling element are affected not only by the contact force but also by the centrifugal force and the gyroscopic moment in the condition of high-speed rotation, which may make the center of the rolling element deviate the connection between the inner and outer raceway curvature centers. This will further lead to the contact angle and the contact force of the inner and outer rings are no longer equal. The connection still can be used as the research object to observe the connection changes and establish the equilibrium equation finally. Considering the outer ring is fixed, the outer ring raceway curvature center will not change under the action of external force. As is shown in Fig. 1.

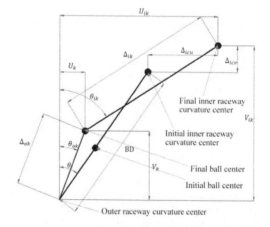

Figure 1.   Displacement of ball and inner ring.

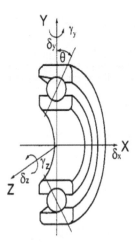

Figure 2. Equilibrium state of the bearing underconsolidated force.

As is shown in Fig. 2, the inner and outer rings of the bearing in each direction of the displacement are: δ_x^o, δ_x^i, δ_y^o, δ_y^i, δ_z^o, δ_z^i, γ_y^o, γ_y^i, γ_z^o, γ_z^i. The contact deformation of rolling element with the inner and outer rings is δ_{ik}, δ_{ok}. Therefore, the relative displacement between the inner and the outer ring of the bearing as is shown in Eq. 8.

$$
\begin{cases}
\Delta\delta_x = \delta_x^o - \delta_x^i \\
\Delta\delta_y = \delta_y^o - \delta_y^i \\
\Delta\delta_z = \delta_z^o - \delta_z^i \\
\Delta\gamma_y = \gamma_y^o - \gamma_y^i \\
\Delta\gamma_z = \gamma_z^o - \gamma_z^i
\end{cases}
\tag{8}
$$

After the ball balanced, the distance of ball center and inner or outer ring raceway curvature centers is

$$
\begin{cases}
\Delta_{ik} = r_i - D_W/2 + \delta_{ik} \\
\Delta_{ok} = r_o - D_W/2 + \delta_{ok}
\end{cases}
\tag{9}
$$

The distance between the inner and outer raceway curvature center after rebalancing is

$$
\begin{cases}
U_{ik} = BD\cos\theta_0 + \Delta\delta_x - \Delta\gamma_z\gamma_{ic}\cos\varphi_k + \Delta\gamma_y\gamma_{ic}\sin\varphi_k \\
V_{ik} = BD\sin\theta_0 + \Delta\delta_y\cos\varphi_k + \Delta\delta_z\sin\varphi_k
\end{cases}
\tag{10}
$$

The relation of contact angle is

$$
\begin{cases}
\sin\theta_{ik} = \dfrac{U_{ik} - U_k}{\Delta_{ik}} \\
\cos\theta_{ik} = \dfrac{V_{ik} - V_k}{\Delta_{ik}} \\
\sin\theta_{ok} = \dfrac{V_k}{\Delta_{ok}} \\
\sin\theta_{ok} = \dfrac{U_k}{\Delta_{ok}}
\end{cases}
\tag{11}
$$

According to the Pythagorean Theorem, the deformation equilibrium equations of the balls are

$$
\begin{cases}
V_k^2 - U_k^2 - \Delta_{ok}^2 = 0 \\
(U_{ik} - U_{ik})^2 + (V_{ik} - V_{ik})^2 - \Delta_{ik}^2 = 0
\end{cases}
\tag{12}
$$

2.3 Force equilibrium equation

When the bearing is processing, due to the high speed rotation itself, the balls are not only affected by contact force, but also by the centrifugal force and the gyroscopic moment.

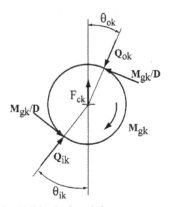

Figure 3. Ball by the force balance.

As is shown in Fig. 3, the force equilibrium equations of the ball is

$$
\begin{cases}
Q_{ok}\sin\theta_{ok} - Q_{ik}\sin\theta_{ik} + M_{gk}/D_w\cos\theta_{ok} \\
-M_{gk}/D_w\cos\theta_{ik} = 0 \\
Q_{ok}\cos\theta_{ok} - Q_{ik}\cos\theta_{ik} + M_{gk}/D_w\sin\theta_{ik} \\
-M_{gk}/D_w\cos\theta_{ok} - F_{ck} = 0
\end{cases}
\tag{13}
$$

Force equilibrium equation of bearing inner ring

$$
\left\{
\begin{aligned}
F_{xi} &= \sum_{k=1}^{N}\left(Q_{ik}\sin\theta_{ik} + \frac{M_{gk}}{D}\cos\theta_{ik}\right) \\
F_{yi} &= \sum_{k=1}^{N}\left(Q_{ik}\cos\theta_{ik} + \frac{M_{gk}}{D}\sin\theta_{ik}\right)\cos\varphi_k \\
F_{zi} &= \sum_{k=1}^{N}\left(Q_{ik}\cos\theta_{ik} + \frac{M_{gk}}{D}\sin\theta_{ik}\right)\sin\varphi_k \\
M_{yi} &= \sum_{k=1}^{N}\left\{r_{ic}\left(Q_{ik}\sin\theta_{ik} + \frac{M_{gk}}{D}\cos\theta_{ik}\right) - f_i M_{gk}\right\}\sin\varphi_k \\
M_{zi} &= \sum_{k=1}^{N}\left\{r_{ic}\left(Q_{ik}\sin\theta_{ik} + \frac{M_{gk}}{D}\cos\theta_{ik}\right) - f_i M_{gk}\right\}\cos\varphi_k
\end{aligned}
\right.
$$

(14)

Assume that pre-tightening force is exerted on the bearing inner ring, the sum of all the reaction between the ball and bearing inner ring and pre-tightening force is equal.

$$
\left\{
\begin{aligned}
&F_a - \sum_{k=1}^{N}\left(Q_{ik}\sin\theta_{ik} + \frac{M_{gk}}{D}\cos\theta_{ik}\right) = 0 \\
&F_{ry} - \sum_{k=1}^{N}\left(Q_{ik}\cos\theta_{ik} - \frac{M_{gk}}{D}\sin\theta_{ik}\right)\cos\varphi_k = 0 \\
&F_{rz} - \sum_{k=1}^{N}\left(Q_{ik}\cos\theta_{ik} - \frac{M_{gk}}{D}\sin\theta_{ik}\right)\sin\varphi_k = 0 \\
&M_{ry} - \sum_{k=1}^{N}\left\{r_{ic}\left(Q_{ik}\sin\theta_{ik} + \frac{M_{gk}}{D}\cos\theta_{ik}\right) - f_i M_{gk}\right\}\sin\varphi_k = 0 \\
&M_{rz} + \sum_{k=1}^{N}\left\{r_{ic}\left(Q_{ik}\sin\theta_{ik} + \frac{M_{gk}}{D}\cos\theta_{ik}\right) - f_i M_{gk}\right\}\cos\varphi_k = 0
\end{aligned}
\right.
$$

(15)

Therefore, assuming that the bearing contains N balls, it can obtain 4N+5 equations and establish a nonlinear equation which includes 4N+5 equations and 4N+5 unknowns. By solving the nonlinear equations, it can get each ball bearing displacement of 5 dof and the inner and outer ring parameters such as contact force, contact deformation and contact angle in the working station. On the basis, the stiffness of bearing can be obtained.

2.4 Stiffness of bearing

Assuming that the outer ring is stationary, the derivative of stress of the inner ring of five degrees of freedom to the corresponding displacement, so the stiffness can be expressed as

$$
K = \frac{\partial F}{\partial \delta}
$$

(16)

Through the above analysis, high-speed angular contact ball bearing model of considering the interference fit can be established. In engineering practice, this article not only provides a reliable guide for the bearing selection, installation and use, but also laid a solid foundation for the research of the spindle system.

3 ANALYSIS OF BEARING CHARACTERISTICS

According to the generic mathematical modeling method of the above established, the 7212c bearing is selected as an object of study. The bearing parameters are shown in the following table [5]:

Table 1. The 7212c bearing parameters.

$\theta[°]$	15
$D_W[mm]$	15.875
$D_i[mm]$	60
$D_o[mm]$	110
$r_i[mm]$	8.176
$r_o[mm]$	8.334
N	14

The bearing characteristic is analyzed in this section. The relations of rotation speed, axial preload and bearing stiffness without considering the interference fit in Fig. 4 and Fig. 5. The simulation of contact angle of considering interference fit can be seen in Fig. 6 and Fig. 7. The comparison of considering interference fit and no interference fit in Fig. 8. Fig. 9 and details in the Table 2. The effect of interference fit in the stiffness of bearing in Fig. 10–Fig. 15.

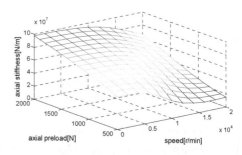

Figure 4. Impact of axial preload and speed on axial stiffness.

Table 2. The comparison of stiffness of bearing.

N/m r/min	No interference fit		Interference fit (0.004mm)		Contrast	
	Axial stiffness	Radial stiffness	Axial stiffness	radial stiffness	Axial	Radial
0	4.99e7	2.73e8	5.60e7	3.08e8	12.2%	12.8%
4000	4.87e7	2.66e8	5.45e7	2.99e8	11.9%	12.4%
8000	3.69e7	1.95e8	4.01e7	2.13e8	8.7%	9.2%
12000	2.02e7	9.24e7	2.10e7	9.63e7	4%	4.2%
16000	1.48e7	5.29e7	1.51e7	5.44e7	2.3%	2.8%
20000	1.45e7	4.29e7	1.47e7	4.40e7	1.4%	2.7%

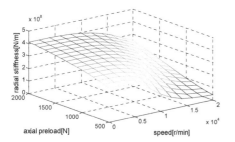

Figure 5. Impact of axial preload and speed on radial stiffness.

As is shown in Fig. 4 and Fig. 5, the rotational speed increases, the axial and radial stiffness decline non-linearly according to increasing speed and increase linearly according to increasing axial preload without considering the interference fit.

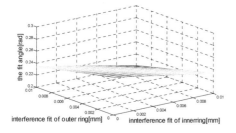

Figure 6. Impact of interference fit of inner and outer ring on the fit angle.

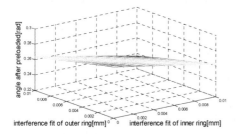

Figure 7. Impact of interference fit of inner and outerring on the preloaded angle.

As is shown in Fig. 6, with the increase of the interference fit of inner and outer ring, the initial angle will become smaller linearly. As is shown in Fig. 7, the fit angle also will become larger after preloaded.

Usually, the choice of the amount of interference fit is on the basis of experience formula. The formula is relevant with the stress and parameter of bearing. So selecting the appropriate amount of interference fit is important for the working of spindle system.

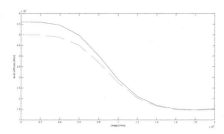

Figure 8. Impact of interference fit and speed on axial stiffness.

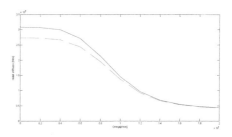

Figure 9. Impact of interference fit and speed on radial stiffness.

As is shown in Fig. 8 and Fig. 9, "b—"represent considering the interference fit. "r-----" represent do not consider the interference. Assume that the axial preload is 500 N and the distance of interference fit is 4e-6m. With the increase in speed, the axial and radial stiffness decline, but the results of considering the interference fit is larger than that do not consider the interference fit.

As can be seen from the above Table 2, while the speed is 0 r/min, the difference of considering the interference fit and no interference fit is greatest. Among them, the difference of axial stiffness is 12.2% and the difference of radial stiffness is 12.8%, with increasing of speed, the difference is narrowed. The results show that the interference fit is more important to improve the stiffness of bearing under the condition of low speed.

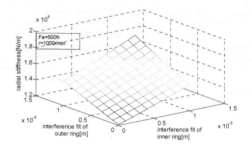

Figure 13. Impact of interference fit of inner and outer ring radial stiffness.

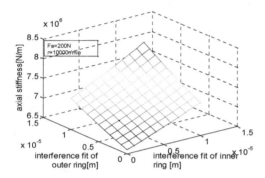

Figure 10. Impact of interference fit of inner and outer ring on axial stiffness.

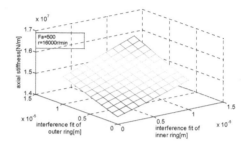

Figure 14. Impact of interference fit of inner and outer ring axial stiffness.

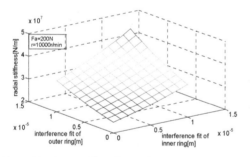

Figure 11. Impact of interference fit of inner and outer ring radial stiffness.

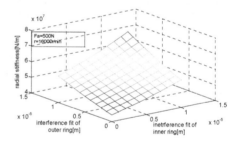

Figure 15. Impact of interference fit of inner and outer ring radial stiffness.

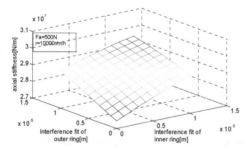

Figure 12. Impact of interference fit of inner and outer ring axial stiffness.

As is shown in Figs. 10 and 11, the axial preload is 200 N and the speed is 10,000 r/min, with the increase of the interference fit of inner and outer ring, the axial and radial stiffness increase nonlinearly. The Figs. 12 and 13 show that the results under the condition of axial preload are 500 N and the speed is 10,000 r/min. Figs. 14 and 15 show that the results under the condition of axial preload is 500 N and the speed is 16,000 r/min. As can be seen from the above results, while the axial preload is from 200 N to 500 N, the axial and radial stiffness increase; while the speed is from 10,000 r/min to 16,000 r/min, the axial and radial stiffness decline.

4 CONCLUSIONS

This paper presents a general method of considering the interference fit of high speed angular contact ball bearing, through this model, we can predict the dynamic characteristics of angular contact ball bearing. The simulation shows that the bearing dynamic characteristic became more complex under the combined effect of the speed, axial force and the interference fit. Through the analysis of the 7212c bearing, we can get the following rule .The stiffness of considering the interference fit is larger than that do not consider the interference fit. Especially in the case of low speed, the interference fit is more important to improve the stiffness of bearing. In addition, with increasing of the interference fit of inner and outer ring, the axial and radial stiffness increase nonlinearly.This also reflects the important role of bearing on the spindle system in practical applications. The model of interference fit should be taken into the whole bearing model.

ACKNOWLEDGEMENTS

This work was supported by the 12th five-year plan 863 Major Project (ss2012AA040702).

The authors are grateful to other participants of the projects for their cooperation.

REFERENCES

Arvid Palmgren: Ball and roller bearing engineering (SKF industries Publications, Sweden 1959).

Jones A B:A general theory for elastically constrained ball and radial roller bearing. *Trans.ASME,J. Basic Eng.*,1960,82:309–320.

Jiwei Luo, Tianyu Luo: Calculation and Application of Rolling Bearings (China Machine Press, China 2009) (In Chinese).

Shuogui Wang, Yuanming Xia: Effect of interference fit and preload for the stiffness of high-speed angular contact ball bearing. *Journal of University of Science and Technology of China*, 2006, 36(12):1314–1320.

Daxian cheng: Mechanical design manual (Chemical industry press,china 2008) (In Chinese).

Tedric A. Harris: Advanced Concepts of Bearing Technology (CRC Press, Inc., United States).

Computational Intelligence in Industrial Application – Ling (ed.)
© 2015 Taylor & Francis Group, London, ISBN: 978-1-138-02818-0

Simulation and optimization of manufacturing enterprise warehousing system based on Witness

DaiJun Pi & QingSong Li

School of Transportation and Automotive Engineering, Xihua University, Chengdu City, Sichuan, China

ABSTRACT: The system simulation, Witness software and enterprise warehousing are first briefly introduced in this paper. Second, the discrete random warehousing system of a manufacturing enterprise is analyzed by using the inventory strategy model, and then Witness simulation software is applied to simulate the warehousing operation system. Finally the system is optimized from the aspect of the inventory strategy, equipment configuration and utilization, the improvement suggestions are put forward, and the feasibility is verified.

KEYWORDS: Witness; Random; Modeling; Simulation; Optimization.

1 INTRODUCTION

For a long time, people have fully realized the superiority of using mathematical model to describe the researched system. However, due to the limitation of mathematical methods, it's very limited to establish and work out mathematical model of complex things. The emergence of computer simulation technology has tremendous advantages; it can be used to solve many complex problems which cannot be solved by mathematical means. In the field of warehousing logistics simulating different target needs to adopt different modeling techniques, in particular, the modeling technology can be divided into two types, i.e., 3D visualization simulation modeling and process simulation modeling. In many cases, simulation modeling is not to pursuit the realistic simulation animation, but to pursuit the process and data fidelity, which is the process simulation modeling technology. Witness software is one of the representatives among them.

Warehousing activities occupy an important position in the enterprise activities. However, there is a big difference between the manufacturing enterprises and logistics enterprises in warehousing management content. The core of manufacturing enterprise warehouse management is to reduce the production cost of products, so the requirements manufacturing enterprise warehouse management control is emphasized on inventory control. Less inventory is more consistent with the goals of manufacturing enterprise warehouse management activities. In the field of manufacturing enterprise warehouse management. Jili Kong & Mengxiao Wang studied inventory management system based on the demand of a large equipment manufacturing enterprise for spare parts inventory management. Ying Zhang used the Hamilton function to solve the optimal control, finally obtained the optimal inventory under the condition of continuous production. Sanyou Ji put forward a new quantitative analysis method about center order information based on the traditional EIQ analysis method and ABC analysis method.

This paper makes a inventory modeling analysis on a manufacturing enterprise through the demand for random variables (s, S) storage model, uses Witness simulation software to simulate the warehousing operation system, finds out the bottleneck in the process. This paper selects the optimal parameter combination based on the analysis of the mathematical model, uses the Witness to simulate again. Under condition of the guarantee process flow to find optimal inventory strategy, improve the equipment configuration and improve equipment utilization, eventually eliminate system bottlenecks, improve the efficiency of system operation, achieve the goal of optimization of the whole system.

2 PROBLEM PRESENTATION

The warehouse operation of a manufacturing enterprise includes the whole process that consists of receiving goods sending goods out according the demands. Warehouse operation process is mainly composed of three stages, i.e. receiving, storage, shipping. Warehouse operation content is shown in Fig. 1.

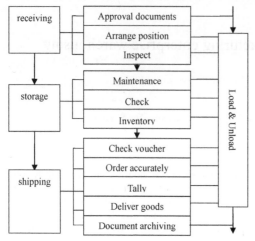

```
receiving ─┬─ Approval documents
           ├─ Arrange position
           └─ Inspect
storage ─┬─ Maintenance
         ├─ Check
         └─ Inventory
shipping ─┬─ Check voucher
          ├─ Order accurately
          ├─ Tally
          ├─ Deliver goods
          └─ Document archiving

Load & Unload
```

Figure 1. Warehouse operation content.

Receiving operation process: There are three different types of goods, goods type uniform distribution. Goods need to be inspected after they were shipped to. The inspection goods are shipped to the staging area waiting for forklift to the corresponding storage.

Shipping operation process: the demand of the goods is random, and it's uniform distribution among the three types. When each kind of goods received random order, the operator will need to make the export goods onto the conveyor belt. After the goods get to the staging area through the conveyor belt, they need to be waiting for the inspection by tester, and then be shipped.

Storage inventory control: With three three-dimensional shelves, three types of goods are stored in the three shelves. Each shelf is set to 10 rows and 10 columns, the maximum capacity is 100, each cargo compartments for a unit of the goods. According to the previous experience, storage strategy is (10, 80).

3 MATHEMATICAL MODELS

3.1 Model description

Inventory strategy model to solve the problem is: how much is the replenishment time, what is the quantity of each replenishment. Inventory strategy model depends on the analysis of cost, impact on the cost in the inventory model is order fee (C_3), storage fee (C_1) and shortage fee (C_2). There are three types of common inventory strategy: t_0 cycle strategy; (s,S) strategy; (t, s, s) mixed strategy.

The storage model include the deterministic model and the random model. The main characteristic of random storage model is demand or ahead of time is random variables, but its distribution law is known. The pros and cons of random storage strategy, usually with the size of the profit expectations as a measure. The warehouse demand is a random variable r, its probability as shown in Table 1 $\left(\sum_{i=0}^{m} P(r) = 1\right)$, each subscription fee is C_3=13 RMB, the unit cost for goods K=0.41 RMB, unit of storage cost is C_1=0.055 RMB, the unit shortage cost is C_2=1.13RMB. Inventory is I, the order quantity is Q, discuss the best (s, S) storage strategy.

Table 1. Demand for the goods.

Demand r(Unit: pcs)	60	70	80	90	100
Probability $P(r)(\sum P(r) = 1)$	0.1	0.2	0.3	0.3	0.1

Order fee:

$$C_3 = KQ = C_3 + K(S- I) \tag{1}$$

Storage fee: failed to sell part need to pay the storage fee when r < S. don't need to pay storage fee when r ≥ S (in order to simplify the model, in this not consider the storage fee of the sold part), so the expectations for the storage fee:

$$\sum_{r \leq S} C_1 (S- r) P(r) \tag{2}$$

Shortage fee: When the demand for r>S, r-S parts need to pay the shortage cost, so the expectations for the shortage fee:

$$\sum_{r > S} C_2 (r- S) P(r) \tag{3}$$

Order fee, storage fee and shortage fee expectations for the sum of:

$$C(S) = K(S- I) + \sum_{r \leq S} C_1 (S- r) P(r) + \sum_{r > S} C_2 (r- S) P(r) \tag{4}$$

3.2 Determine the optimum storage quantity S

The random values of demand r will in order of size $r_i < r_{i+1}$ (i=0,1,..., m-1). when the value of S is r_i, denoted as S_i, Solve S to make C(S) minimum. In order to make the $C(S_i)$ minimum, S_i should satisfy the inequality (3.5) and (3.6).

$$\sum_{r \leq S_i} P(r) \geq \frac{C_2 - K}{C_1 + C_2} \tag{5}$$

$$\sum_{r \le S_{i-1}} P(r) \ge \frac{C_2 - K}{C_1 + C_2} \tag{6}$$

Because $(C_2-K)/(C_1+C_2)$ strictly less than 1, known as the critical value, expressed in N: $N=(C_2-K)/(C_1+C_2)$. Integrated formula (3.5) and (3.6) to get inequalities in which S_i is determined:

$$\sum_{r \le S_{i-1}} P(r) < N = \frac{C_2 - K}{C_1 + C_2} \le \sum_{r \le S_i} P(r) \tag{7}$$

When i=3, S_i=90, meet the formula (3.7), $S=S_3$=90 is the best inventory.

3.3 *Determine the minimum storage S*

If this stage do not order, the total fee of expected value:

$$\sum_{r \le S} C_1(S-r)P(r) + \sum_{r > S} C_2(r-S)P(r) \tag{8}$$

If this stage order, the total fee of expected value:

$$C_3 + K(S-s) + \sum_{r \le S} C_1(S-r)P(r) + \sum_{r > S} C_2(r-S)P(r) \tag{9}$$

If this stage do not order, it can save an order fee C_3. Therefore, we assume the existence of a number of s (s≤S), the formula (3.10) is set up:

$$\sum_{r \le S} C_1(S-r)P(r) + \sum_{r > S} C_2(r-S)P(r) \le$$
$$C_3 + K(S-s) + \sum_{r \le S} C_1(S-r)P(r) + \sum_{r > S} C_2(r-S)P(r) \tag{10}$$

If the type (10) is right, no order is reasonable; otherwise, should order, order quantity is Q=S-I. Finishing type (10), the inequality (11) can be obtained.

$$Ks + \sum_{r \le S} C_1(s-r)P(r) + \sum_{r > S} C_2(r-s)P(r) \le$$
$$C_3 + KS + \sum_{r \le S} C_1(S-r)P(r) + \sum_{r > S} C_2(r-S)P(r) \tag{11}$$

In formula (11), when s=S, because of C_3>0, inequality is clearly established. When s<s, in formula (11) the left side of the expected value of shortage fee is greater than the right side, but the left storage fee in the left is less than the right, so the inequality can still be established. Because of S only from $r_0, r_1, \ldots, r_m$, The minimum r_i ($r_i \le S$), which make the formula (11) is right, must be s.

When S=90, the right side of the formula (3.11) is 51.58; when s=60, the left side of the formula (3.11) is 48.33. Apparently 48.33<51.58, the minimum inventory is s=60.

4 WITNESS MODELING AND SIMULATION

4.1 *Entity object and logic flow chart*

The entity equipments, which are needed by the manufacturing enterprise warehousing operation can be found in corresponding entity library of Witness simulation software, do not need to import the drawing tool to make graphics file. The corresponding relationship between warehousing operation equipments and Witness simulation objects is shown in Table 2.

Table 2. The corresponding object.

Physical equipment	Witness entity object
Goods	Part
Inspection device	Machine
Temporary storage	Buffer
Forklift	Vehicle
Goods shelves	Buffer
Conveyor belt	Conveyor
Receiver	Machine

For the simulation model, commonly used flow diagram to express the discrete time system model, this paper simulated the flow chart of the system simulation is shown in Fig. 2.

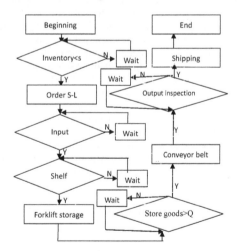

Figure 2. Simulation logic flow chart.

4.2 *Parameters settings*

Goods: there are three different types of goods arrival interval time obey negative exponential distribution

NEGEXP (0.1), and the type of goods in the 1, 2, 3 are uniformly distributed, different types of goods in different colors.

Inspection: incoming inspection are set up in two testers, outbound inspection set only one tester, test time obey negative exponential NEGEXP (0.5).

Forklift: Set up two cars, the maximum carrying capacity is 3; the maximum operating speed is 2 m/s.

Shelves: using three three-dimensional shelves, three types of products are stored in the three shelves, each shelf is set to 10 lines 10 columns, the maximum capacity is 100, each cargo compartments for a unit of the goods. Based on past sales data and taking into account of the cost of storage and safety stock, when the storage shelf is greater than 80, the system will automatically close the input port; when the storage capacity is reduced to 60, the system will automatically open the input port of the shelf to replenishment.

Conveyor belt: each shelf is respectively connected with the one conveyor belt, each belt capacity is 200 units, and the running speed is 1 m/s.

Buffer: set 2 buffers, one for the receiving goods, one for the shipping goods, its capacity is 50 units.

4.3 Visual elements and description

Part elements and Machine elements: in the element selection window respectively select P & Machine element, right-click Display, jump out of the Display dialog box, set the Text, Style, and design the details of the element.

Track elements: in the element selection window select Load element, right-click Display, jump out of the Display dialog box, set the Path. in the element selection window select Unload element, right-click Display, jump out of the Display dialog box, set the Path, and design the details of the element.

Vehicle elements: in the element selection window select Forklift element, right-click Display, jump out of the Display dialog box, set the Style, and design the details of the element.

Buffer element: in the element selection window select Huojia element, right-click Display, jump out of the Display dialog box, set the Text, Icon, count form Part Queue, and design the details of the element. Witness layout as shown in Fig. 3.

5 SIMULATION RUNNING AND DATA COLLECTION

Assumes that the system simulation one unit is equivalent to one hour, set up the simulation time is 365 × 24 = 8760 units, i.e. simulate the manufacturing enterprise actual operation of one year, the simulated production enterprise actual operation of one year. The simulation process is shown in Fig. 4. Repeated

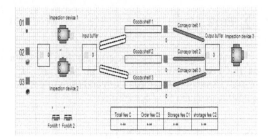

Figure 3. Warehousing operation layout.

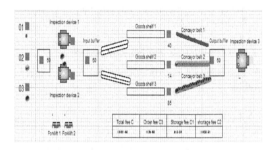

Figure 4. Operation simulation process.

simulation runs 30 times to get the average value, we get important performance indexes of the manufacturing enterprise warehouse operation process, as shown in Tables 3_table 8.

Table 3. Inventory expense.

	Total fee C(RMB)	Order fee C_3(RMB)	Storage fee C_1(RMB)	Shortage C_2(RMB)
Goods	864693	129877	93264	641552

Table 4. Inventory level.

Shelf number	Maximum inventory (box)	Average inventory (box)	Average storage time (h)
1	80	48	11.16
2	80	41	12.03
3	80	37	11.08

Table 5. Buffer.

Buffer	Total in	Total out	Max inventory (box)	Average inventory (box)	Average storage time (h)
Input buffer	86198	86148	50	25.13	2.53
Output buffer	86148	86098	50	49.78	5.00

Table 6. Conveyor belt.

No.	Idle rate (%)	Blocking rate(%)	Busy rate (%)	Line rate (%)	Average number	Average time
1	3.78	2.18	43.60	50.44	17.46	0.51
2	3.07	2.37	43.02	51.54	17.61	0.52
3	4.90	2.94	40.24	51.92	16.77	0.54

Table 7. Inspection device.

No.	Blocking(%)	Busy(%)	Idle(%)
1	3.13	49.68	47.18
2	3.14	49.66	47.20
3	0.00	99.01	0.99

Table 8. Forklift.

No.	Idle(%)	Transport distance ratio (%)	Load& Unload (%)
1	55.52	43.16	1.32
2	51.29	47.35	1.37

6 THE RESULTS ANALYSIS AND SYSTEM OPTIMIZATION

Run the simulation model, observe the simulation process, and analyses the results of the system simulation, we can find some unreasonable places in the operation process of system. Warehousing goods total average fee is 864693 RMB, of which the shortage fee is larger (641552 RMB), that shows the inventory strategy (10,80) from experience is not reasonable; The average storage quantity of the input buffer is 49.78 (the maximum storage capacity is 50), the average storage time is 5 h, suggesting that there is a long time accumulation phenomenon, and the maximum retention amount close to the buffer capacity of 50, this shows that the process was made a system bottleneck in this position, this is due to the speed of the outbound tester that cannot keep up with the outbound goods sorting inspection speed; The idle rate of the two inspection devices in the input process is as high as 47.18% and 47.20%, at the same time, two forklifts idle rates are as high as 55.52% and 51.29%, this reflects the goods arrival rate that cannot keep up with the inspection device and the working efficiency of the forklift, thus causing equipment idle and waste. In addition, each shelf has a capacity of 100, but three shelves inventories only to an average of 41 cases, warehouse utilization rate is low.

Through analyses of the simulation data, we can get that the comprehensive efficiency of the manufacturing enterprise warehousing operation system are closely related with the inventory storage, coordinated operation of equipment and personnel. In order to achieve a smooth flow of the goods in the warehouse operation process, eliminating system bottlenecks, we can use the storage strategy (60, 90), which got from the mathematical model. At the same time, add a delivery inspection device, reduce the quantity of goods in the buffer in the output process; because the two inspection devices and forklifts idle rate is higher, we can reduce one inspection device and one forklift to improve equipment utilization rate.

According to the above strategy, set simulation parameters, operate simulation model again, find that the total average storage fee is reduced to 621832 RMB; three buffers are not long time accumulation and retention of goods, both within the buffer capacity; at the same time, the inspection device and forklift idle rate by before 47.18% and 47.20% are reduced to 26.5% and 26.1%, the utilization rate of equipment is improved greatly; in addition, the average inventory quantity of the three shelves is improved to 61 box, the system is optimized.

7 SUMMARY

Warehousing system is an important part of the logistics system; which use the method of system simulation to modeling, analysis and optimization of the warehousing operation system. In order to improve the efficiency of warehousing operations, hope for give the actual planning, construction and upgrading, transformation a certain theoretical support and reference. Analysis of the warehousing operation system and even the entire logistics system by using Witness logistics simulation software is still at the starting stage in our country, through the analysis and optimization of the simulation software of Witness is applied to the warehousing operation system, Witness simulation software has large potential, if it can be fully applied to the actual, will bring great social and economic benefits.

REFERENCES

Ji Sanyou, Yang Tao. 2012. Analysis and application of EIQ-ABC *method in the iron and steel logistics center planning. Journal of Wuhan University of Technology (information and Management Engineering Edition).*
Kong Jili, Wang Mengxiao. 2011. Large scale equipment manufacturing enterprises of spare parts inventory management system for. *Logistics technology*, 30(11): 109–113.

Kong Jili, Feng Ailan, Jia Guozhu. 2012. *The Enterprise Logistics Management.* Beijing: Peking University press.

Wang Yachao, Ma Hanwu. 2006. *Modeling and Simulation of Production Logistics System.* Beijing: Science Press.

Wu Peng; Lv Youguang. The multi period production optimization. *Operations research and the management of.2014 semi finished goods inventory is considered under stochastic demand.*

Zhang Xiaoping, Shi Wei, Liu Yukun. 2008. *Simulation of Logistics System.* Beijing: Tsinghua University press.

Zhang Ying.2007. Control theory of inventory management of continuous production of industrial engineering based on (06):10–12.

Zhao Ning.2012. *Logistics System Simulation Case.* Beijing: Peking University press.

Zhou Xingjian, Zhang Beiping. 2012. *Modern Warehouse Management and Practice.* Beijing: Peking University press.

Computational Intelligence in Industrial Application – Ling (ed.)
© 2015 Taylor & Francis Group, London, ISBN: 978-1-138-02818-0

The study on isolation measures of heavy NC machine tools

ZhiFeng Liu, EnZhi Yang, Gang Wang & FuPing Li
College of Mechanical Engineering and Applied Electronics Technology, Beijing University of Technology, Beijing, China

ABSTRACT: For the problem of external isolation about heavy NC machine tools, in this paper, we use the ANSYS finite element software, and it is based on a heavy NC machine tool. We use the base surface's vibration displacement of heavy NC machine tool as the research object, and build three-dimensional models. We use the linear isolation trench, and study the passive isolation vibration of a far-field unilateral external source. Through the effect analysis of the isolation and the depth and width of trenches for vibration isolation, we get the effect of the changes in parameter vibration isolation trench for the heavy NC machine tool, and provide a theoretical basis and reference to the vibration isolation for the construction of the external aspects of the Heavy NC machine tool.

KEYWORDS: Heavy NC machine tools; Isolation trench; Isolation.

1 INTRODUCTION

Heavy NC machine tools are urgently needed to develop energy, aerospace, large ships, automotive and other industries in our country; it is a strategic material measure of a country's technology level and the comprehensive national strength. Heavy NC machine tools as machine tools that manufacture large and extra- large parts, it directly determines the precision of machining of large parts. With more and more higher requirements about development of modern industrial technology and high performance technology for the parts precision and surface integrity, the machining accuracy requirements are also getting higher, in order to guarantee the machining accuracy, not only the machine structure itself, but also the influence on the precision of the machine tool that from the external vibration through the foundation and spread to the machine cannot be ignored.

The essence of vibration is vibration wave propagation along the surface, so you can set up barriers to intercept the vibrations so as to achieve the purpose of vibration isolation. Many scholars have done a lot in terms of isolation study. Woods[1] puts forward the both concept of active vibration isolation in the near field and far field of passive vibration, but does not provide quantitative boundary; Lysmer 2] first puts forward the distance from the source 2.5LR as the dividing line; Haupt [3] made 2LR as the near field and far field boundary vibration isolation; Chang Qiu [4] of domestic also analysis boundary value of near field and far field; Woods and Haupt had ever systematically experimental study on air trench isolation barrier ; Chouw and Andersen [5] get the conclusions that trench isolation effect is better than the barrier isolation through the research; Chang Qiu [4] analyzed the main parameters that effect the vibration isolation effect of the barrier, studied the complete wave transmission phenomenon that reduce the barrier isolation efficiency or even failure, and point out that the flexible barrier isolation is easy to appear the phenomenon; Guangyun Gao [6] consider filling groove as different plasmid of elastic exists in half space, they use Rayleigh wave scattering integral equations, and made three dimensional analysis about filled trench far-field passive vibration isolation and they concluded that the vibration isolation effect of the rigid material is better than the flexible material; Adam [7] argued that fill the trench made of flexible material, such as (the extreme empty ditch) works better than rigid vibration isolation materials, as can be seen there is a contradiction. Overall, barrier mechanism of vibration isolation is not deep.

2 THE THEORY AND METHOD

2.1 *The barrier mechanism of isolation trench on the vibration*

When vibration energy propagate into the soil, vibrations propagate in the form of elastic waves. Elastic waves of vibration energy propagate in soil mainly in the form of body waves and surface waves. Body wave can be divided into compression wave and shear wave, surface wave is mainly propagation in the form of Rayleigh wave. The vibration propagations in soils, and energy decreases with increasing distance from the source, whose main reason is the effect of discovering geometry damping and material damping in the transfer process. When elastic waves diffuse along the spherical surface, the energy density will increase as the surface area of a spherical wave increasing, which is increased according to the square value of the radius of the attenuation. The amplitude of vibration is proportional to the square root of the energy density, therefore, geometric damping effect makes the amplitude of elastic wave propagation in soils as the distance increases attenuation. Material damping is attenuation of the amplitude because of wave energy is converted to other forms of energy (such as heat, etc.) and to dissipate when the elastic wave propagation in materials. Considering geometric and material damping case, the experience energy attenuation formula is

$$V_2 = V_1 \sqrt{\frac{r_1}{r_2}} e^{-\alpha(r_2-r_1)} \tag{1}$$

In which: V_1= Vibration amplitude at $r1$ distance; V_2= Vibration amplitude at $r2$ distance; α= Material damping

The energy propagating in the ground, Rayleigh wave accounted for 67%, accounting for 26% of shear wave, longitudinal wave accounted for 7%. When the distance from the source has a certain distance, the amplitude of Rayleigh wave will be greater than the body wave amplitude. So for the far field passive vibration isolation, it is mainly to reduce the Rayleigh wave energy.

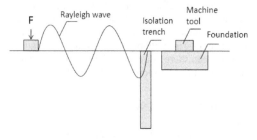

Figure 1. Schematic plan of far-field passive vibration isolation trench isolation.

2.2 Dynamic analysis equation

For external vibration of heavy NC machine tools, its dynamics equation as follows:

$$[M]\{\ddot{u}\} + [C]\{\dot{u}\} + [K]\{u\} = \{F(t)\} \tag{2}$$

In this paper we selected external loads for periodic sinusoidal loads, and that is $F\sin \omega t$. The letter F stands for the amplitude of harmonic force, and the letter of ω stands for the frequency harmonic force.

2.3 Simulation analysis

In this article, it takes ANSYS simulation approach to the problem of heavy NC machine isolation study. When the shock waves in the underground encounter different media; the greater the difference between the media, the more likely the vibration wave reflection and easy interface through the media. The isolation trench in the formation is to prevent the spread of surface waves. In this paper, remote passive vibration isolation analysis is carried out for external local oscillator unilateral, and the external loads using periodic sinusoidal load $F\sin \omega t$,

The direction is vertical downward, the $F = 50$ KN, frequency range of 0–30 Hz, load step 0. 2 Hz. Involves simulating the vibration of the infinite half space soil, the size of the model range is an important problem of dividing grid unit at, if the selection range is too large, more accurate calculation results, but the computing time too long and put forward higher requirements on the machine's memory; if the demarcation of the scope is too small, the calculated results will be affected by the artificial boundary, and large errors will be produced. Xia Shen [8] uses the ratio of horizontal dimension and vertical dimension for reference, when doing numerical simulation analysis, Yongbin Yang [9] proposed in the literature about the environmental vibration caused by the High-speed Rail that 1–1.5 wavelengths both horizontal and vertical size of the numerical model is greater than the soil's shear wave. Based on the above theory, in the analysis, the model selected is relatively larger and the size is 160 * 70 * 40 m ($L* W * D$), base size of 20 * 15 * 5 m ($L * W * D$), the machine tool is simplified as a mass. The following figure shows a three-dimensional simulation model analysis, respectively isolation without trench and isolation with trench.

Various material parameters in the following table:

Table 1. Material parameters.

Materials	Density (kg/m³)	Modulus of elasticity (Pa)	Poisson's ratio
Foundation soil	1960	1.5*108	0.25
Foundation	2400	3*1010	0.2
Machine tool	7200	1.74*1011	0.275

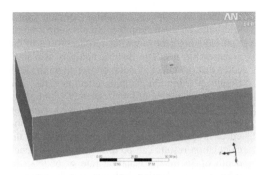

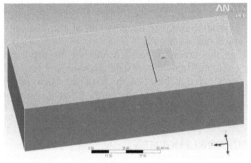

Figure 2. D model of vibration isolation without trench and vibration isolation with trench.

3 SIMULATION RESULTS AND DISCUSSION

3.1 The effect of the isolation trench's depth on the barrier of vibration

The position, width and length of isolation trench invariant, the depth of trenches were obtained from 0, 5, 10, 15, 20, 25, 30, 35 m. Through the simulation calculation and analysis, you can get deep isolation trench isolation effect on the impact of the curve shown in Fig. 3:

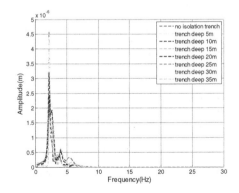

Figure 3. A frequency chart of vertical on the based surface as different isolation trench depth.

Fig. 3 shows the structure of the steady forced vibration condition. Among them: the abscissa is the excitation frequency, and the ordinate is the amplitude of vibration displacement in the vertical direction of the base surface. See (1) when the vibrating at frequencies above 10 Hz, the graph ordinate values is much lower than the ordinate value of below than 10Hz; (2) displacement of the base surface decreased gradually with increasing depth. We picked up the maximum displacement based on vertical surface in Fig. 2, and represented the relationship graphically between the maximum displacement and isolation trench depth, as shown in Fig. 4.

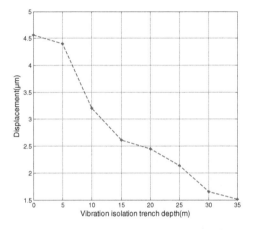

Figure 4. The curve of relationship between an isolation trench depth and the maximum displacement of the vertical direction of the surface of the base.

As can be seen from the graph, with the increase of vibration isolation trench depth, the maximum displacement of the vertical direction based surface gradually decreases, the vibration isolation effect is strengthened gradually, and we can see the impact of isolation trench depth on the isolation effect is relatively big.

3.2 Influence of width of trenches on the vibration block

Fixed location, depth and length of the trenches, trenches width is respectively 1, 2, 3, 4 m. Through calculation and analysis of available isolation trench depth on the vibration isolation effect of the curve as shown in Fig. 5 simulation:

The maximum vertical direction displacement based surface in Fig. 5 was extracted, and the relationship between maximum displacement and isolation trench depth graphed out, as shown in Fig. 6.

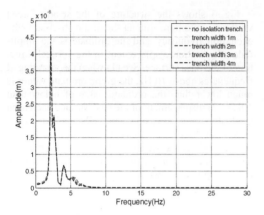

Figure 5. A frequency chart of vertical on the based surface as different isolation trench width.

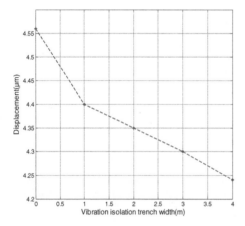

Figure 6. The curve of relationship between an isolation trench width and the maximum displacement of the vertical direction of the surface of the base.

When the groove width increases, the vibration displacement based surface changes little, and this shows that the width of the ditch has no obvious effect on vibration isolation trench damping effect, so the isolation trench width in the engineering can be according to the construction needs to be.

4 CONCLUSION

1 Isolation trench depth is the main factor influencing the damping effect, with the isolation trench depth increases the damping effect is strengthened gradually, to achieve the requirements of Heavy NC machine tool accuracy, in the external conditions, isolation trench depth must meet the demand of vibration isolation; vibration isolation trench width has little effect on the damping effect vibration isolating groove width, can according to the construction needs to decide; analysis also indicates that the effect of isolation trench on the high frequency vibration damping effect is better than that of the damping of low frequency vibration.

2 Keep the other parameters unchanged, isolation trench depth increases, the vibration isolation effect is quite obvious, from the analysis of the calculation results can be drawn, isolation trench depth of 35 m, the isolation rate of 66.9%; the influence of vibration isolation trench width on the vibration isolation effect is relatively small, vibration isolation trench width of 4 m, the isolation rate is only 7%.

3 In this article, it focused only on the depth and width of trenches for data analysis, and for location of trenches, length, trenches filled in the vibration isolation effect of different isolation material pending further study.

ACKNOWLEDGEMENT

The authors are most grateful to the National Science and Technology Major Project of China (No. 2013ZX04013-011).

REFERENCES

Woods R D. Screening of surface waves in soils. *Soil Mech Found Eng Div ASCE*, 1968,94(4): 951–979.

Lysmer J,Waas G. Shear wave in pane infinite structures[J]. *Journal of the Engineering Mechanics Division*, 1972, 98(1):85–105.

Haupt W A. Isolation of vibrations by concrete core walls. Tokyo, Japan :In Proceedings of the Ninth International Conference on Soil Mechanics and Foundation Engineering, 1977,2:251–256.

Qiu Chang. Continuous and discontinuous barrier passive vibration analysis of far field: [Tongji University doctoral dissertation]. Shanghai: Tongji University, 2003.

Chouw N, Le R, Schmid G. Propagation of vibration in a soil layer over bedrock.Engineering Analysis with Boundary Elements, 1991, 125–131.

Gao Guangyun. Discontinuous barrier ground vibration theory and application. Zhejiang University PhD thesis, 1998.

Adam M.Von Estorff O.Reduction of Train-induced Building Vibrations by Using Open and Filled Trenches . *Computers and Structures*, 2005, 83(1):11–24.

Shen Xia. Calculation and analysis of subway vibration in the research on some problems of. Shanghai: Tongji University, 2005.

Yang Yongbin, Shi Rong. Analysis of soil vibration induced by high-speed train. Taiwan consortium of human engineering consultants agency, 1995.

Computational Intelligence in Industrial Application – Ling (ed.)
© 2015 Taylor & Francis Group, London, ISBN: 978-1-138-02818-0

Methodology for the integrated planning of schedule and cost based on WBS and critical chain

Z.J. Jiang, Y.X. Tang, Z.R. Ma & H.Y. Liu
Shandong Jianzhu University, Jinan, China

ABSTRACT: Constantly innovative and economical architectural designs bring big challenges for construction projects, so that in recent years many scholars have begun theoretical research and practical exploration on overall management and full life-cycle on management to meet the challenges. Time and cost are the most closely related targets of project management that should be combined and the existing separate way of management has resulted in the increase of one side and decrease of the other, even decrease of both in practice. Using an underground engineering case, this paper argued the necessity of integration management of schedule and cost and how to make an integrated plan based on WBS and critical chain method, which is the first important stage in integration management.

1 INTRODUCTION

From the systemic point of view, construction project is a system made up of multiple objectives and their inter linkages. Schedule and cost are the most closely associated targets that should be managed in a systematic way of thinking, which is a manifestation and reflection of the dialectical materialism that everything is universally connected and the sum of parts is better than the whole. In addition, due to the specialization and aesthetics of modern architecture presents a lot of challenges for construction, project management area has also led to a revival of the integration management that can solve the low efficiency caused by separate management of the two targets and help managers to shorten the duration and reduce the cost of construction projects as well.

Qi Anbang mentioned in his article that the first major work in integration management was to produce an integrated plan through consolidated consideration of each specific requirement (Qi Anbang, 2002). If we are unable to form a simple plan system, the integrated management equates to talk about stratagems on paper, and cannott apply it in practice. Throughout the academic literature about the integrated method, scholars both here and abroad all paid their attention on how to analyze the system data information synergistically, such as the Earned Value method and various kinds of models based on the mapping relationship between time and cost (Ren Rui, 2008), and they ignored that we should make the plan system first, which is also what this paper calls for. Engineering division and cost structure do not match so that the in-depth study on integrated plan, which is urgently needed, has not achieved satisfied results. Therefore this paper studied on how to integrate the schedule and cost on the basis of WBS and the Critical Chain method to facilitate the next control management.

2 DIALECTIC RELATIONSHIP BETWEEN WBS AND INTEGRATED PLANNING

Work Breakdown Structure (WBS) cuts projects into smaller and easier parts called work package, according to project deliverable, scope or process. In project management, WBS is not a goal but a tool and method, and it is the basic step of management, for example, network planning is carried out on the basis of WBS. On the other hand, integration is making isolated elements together in a certain way to connect the management procedure as an organic system. It is a kind of thinking and idea. Integration is composition while WBS is decomposition, and that seems contradictory, but actually there is a dialectic relationship between them. Favoring each one side will cause management difficulties. Over-stressing on WBS will lead to separation and over- integration will easily go to extremes. Therefore it is necessary for us to discuss the dialectical relationship between them in order to balance their effectiveness.

A reasonable integration of WBS work packages will optimize the overall breakdown structure and get the complementary advantages so that the whole project management can achieve global optimization. Therefore we can say WBS is the basis for integration and integration is an inevitable requirement of

WBS. The integrated plan shows the superiority of WBS, and also make up for its lack of overall benefit to achieve the optimization. Furthermore, we have an evaluation criteria to determine whether they two can work together. If the integration of WBS packages cost is less than the increased effectiveness, we can combine them together (Zhao Xiaochun, 2012).

3 FEASIBILITY OF APPLYING CRITICAL CHAIN METHOD TO INTEGRATED PLANNING

The Critical Chain Method originates from the Theory of Constraints of which the core principle is "The accumulation of local optimum canct achieve the overall optimum". According to such core idea, Goldratt took the factor of resource constraints and human's behavior into consideration when drawing up a schedule plan and proposed that it is the critical chain rather than the critical path determined the project duration. The critical chain is the shortest finish path of a project with consideration of both activity sequence and resource constraints. Goldratt divided activity duration into the shortest duration and the safe time, which was gathered to the buffer to guarantee the stability of critical chain.

Critical Chain Method (CCM), a tremendous development of schedule management in the following of CPM /PERT method, is the latest schedule management method, however, scholars whose researches focused on the determination of working time and buffer management, ignored its potential application in the integration. The critical chain can be a tool for cost management when changing the time for the cost. Han Xiaokang and Lu Mei identified the critical chain in the project cost and used the buffer mechanism to reduce risks that were brought by financing constraints and uncertainty, realizing the applying the critical chain in the cost management (Han Xiaokang & Lu Mei). In this case the CC can be used in both schedule and cost management, we might as well build an integrated plan of time and cost base on the critical chain mode.

4 RESEARCH ON INTEGRATED PLANNING METHOD OF SCHEDULE AND COST

Project departments have their separate plans and their common goal is to complete their task within the overall duration in the contract. But the integrated planning needs each department to balance their different specific program requirement to produce an integrated plan. Based on the above theories, this paper proposes to establish a schedule and cost integrated planning approach by a case study of an underground engineering:

Step 1: Based on the principles and approaches in WBS, we decompose the engineering into the detailed work packages whose duration and budget can be easily calculated. Such breakdown structure calls for coordination of all departments and requires all people from the construction workers to project managers to participate. Taking an underground project for instance, its WBS drawing is shown as Fig. 1.

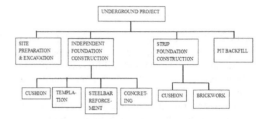

Figure 1. WBS drawing of the underground project.

Step 2: Consult experts and construction workers or investigate to estimate the time of every work package, according to the form and requirement of the critical chain to determine the shortest time and safe time of each activity, then calculate PBt, FBt, their formulas are as follows.

$$PBt = \sqrt{\sum (\Delta Tci)2} \tag{1}$$

ΔTci = safe time of critical activity.

$$FBt = \sqrt{\sum (\Delta Tni)2} \tag{2}$$

ΔTni = safe time of the non-critical activity

Step 3: According to the listing rules, we need to calculate budgets as minimum fees (the most optimistic cost) of each work package, which is used as the lower control limit, and also corresponding to the critical chain, we need to set aside a portion of the preparatory cost as the buffer to reduce the uncertainty of cost estimation. Then we use the preparatory costs to calculate PBc and FBc, their formulas are as follows.

$$PBc = \sqrt{\sum (\Delta Cci)2} \tag{3}$$

ΔCci = preparatory cost of critical activities.

$$FBc = \sqrt{\sum (\Delta Cni)2} \tag{4}$$

ΔCni represents non-critical activities

The underground project's detail data is as shown in Table 1.

Table 1. The underground project's detail data.

Partial Projects	Duration (d)		Cost (yuan)	
	Shortest	Safe	Least	Preparatory
A. Site Preparation	2	2	2000	1000
B. Excavation	3	3	13800	6200
C. Independent Foundation Lime-soil Cusion	1	1	4300	700
D. Independent Foundation Concrete Cushion	3	4	2800	2200
G. Independent Foundation Framework	1	1	810	200
H. Independent Foundation Steel Bar	2	2	8400	1400
I. Independent Foundation Concreting	7	3	7500	4700
E. Strip Foundation Cushion	1	1	5200	800
F. Strip Foundation Brickwork	7	3	14900	6300
J. Backfill	2	3	2300	2000

According the shortest time in the above table, we can draw the network graph of the project case, as shown in figure 2, we can make sure the critical path is A-B-C-D-G-H-I-J (also the critical chain, no detailed calculation here, see Goldratt, 2006), and the non-critical path is E-F (also the non-critical chain, no detailed calculation here, see Goldratt, 2006).

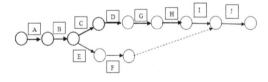

Figure 2. The network graph of the underground project.

By calculation, we can get PBt is rounded to 7d and FBt is rounded to 4d .

Similarly, we find that cost of critical activities is rightly the critical cost of the project; therefore we can use the same critical chain drawing to present the critical chain of both time and cost, as shown in Fig. 3. According to the cost formula expressions, PBc=8500 yuan, FBc=6400 yuan.

A	B	C	D	G	H	I	PB
E	F	FB					

Figure 3. The critical chain drawing (both time and cost) of the underground project.

In summary, building an integrated plan based on WBS and CC is feasible, and in the following control management, we can use the Earned Value Method and buffer warning mechanism to analyze the time and cost data so that an advanced feedback effect will be achieved. The specific procure is not detailed here, just showing in Fig. 4.

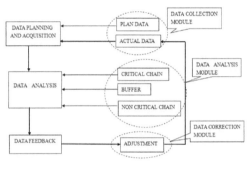

Figure 4. Model of integrated management of schedule and cost.

5 CONCLUSION

The paper firstly discussed the relationship among WBS, the critical chain and integration, then proposed how to make an integrated plan based on WBS and CC, and proved the approach is feasible through an underground project case. Through the establishment of integrated planning, engineering can achieve better integration in the control phase so as to really achieve integration of project management. Furthermore, as BIM (Building Information Model) develops, its influence extends to the area of schedule-cost management, trying to break the information barriers, and its concept is rightly the integration management of construction projects. But how to really accomplish the integration of project management in practice and how to implement sharing of project information and its portability are still needed to research and explore.

REFERENCES

Goldratt, E. 2006. *Critical Chain:*1–235. Beijing: Electronics Industry Press.

Qi Anbang. 2002. On the Multi-factors Integration Management Method for Project Management. *Nankai Business Review.*

Ren Rui. 2008. Application of WBS in Cost-schedule Integration Control. *Science & Technology Progress and Policy.*

Zhao Xiaochun. 2012. *Project Integration Management Research Based on WBS.* Degree thesis, Jiangxi University of Technology.

Han Xiaokang, Lu Mei. 2009. Study on Project Cost Control Based on Critical Chain. *Construction Economy.*

Computational Intelligence in Industrial Application – Ling (ed.)
© 2015 Taylor & Francis Group, London, ISBN: 978-1-138-02818-0

Research on multivariant data integration based on virtual prototyping

E.Z. Wang, M.Q. Luo, P.X. Zhao & H. Liu
Beihang University, Beijing, China

ABSTRACT: This paper describes an overall function and structure design scheme of a multivariant data integration environment based on virtual prototyping. On this basis, a sample project is developed running on a PC system, including the function of integration and interaction of multivariant data such as design parameters, main performance parameters and CAE result data. Also, the experience of a virtual prototype in CAVE environment and mobile platform is explored and tested. According to these studies, this paper discusses the patterns of multivariant data integration based on virtual prototyping and finally puts forward several optimization and improvement scheme of the project and possible development direction in this area.

KEYWORDS: Virtual prototyping; Data integration; CAD/CAE.

1 INTRODUCTION

1.1 Background

This paper involves two important concepts: virtual prototyping and multivariant data integration.

1 Virtual Prototyping (VP) is a new concept gradually raised in the 1990s based on computer technology [1]. The concept of virtual prototyping mentioned in this paper, refers to the realization of 3D aircraft model experience in virtual environment.
2 The objective of multivariant data integration is to integrate CAD light-weighted models, design parameters, performance parameters, CAE analysis results, design feedback data, etc., onto the virtual prototypes of aircraft components, and then to achieve data integration and quick display through interactive techniques such as hotspots.

1.2 Research status home and abroad

A research focus of virtual reality technology is about various interaction methods, including somatosensory, gesture, voice and other natural interaction modes. These interaction methods allow users to more easily interact in the virtual reality environment, improving the authenticity and availability of the virtual reality system.

The virtual prototype can also be used combined with aircraft design and analysis software to compose a collaborative design and simulation environment.

Application of virtual reality technology relies on the support of hardware equipment. One of the applications widely used in aircraft design process, is a multi-projection immersive virtual environment system called CAVE (Cave Automatic Virtual Environment), which is a room-type stereo projection system based on a set of high-end computers, multiple stereoscopic screens (usually consisting of 3-5 large screens) and 3D trackers.

2 DEMONSTRATION OF THE OVERALL SCHEME AND SELECTION OF TOOLS

2.1 Design requirements

The technology of multivariant data integration based on virtual prototyping is a supporting technology for the design evaluation and feedback stage. Figure 1 shows the application of the technology in a typical scenario.

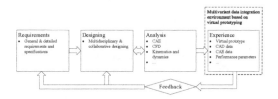

Figure 1. Application in a typical case.

2.2 Design of the system function hierarchy

As is shown in Figure 2, the function of the system is divided into two major parts, namely, virtual prototype display and interaction, and multivariant data integration.

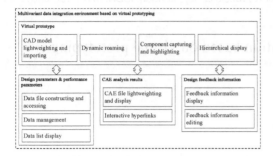

Figure 2.   Application in a typical case.

2.3   Evaluation and selection of the virtual reality platform

By determining the data source format, several interactive graphical development environments were evaluated.

Table 1.   Comparison of virtual reality platforms.

Name	File formats	Platforms	Real-time rendering
Unity 3D	.3DS, .DAE, .FBX, .DXF, .OBJ	PC, WEB, Android, iOS	GOOD
Cult3D	.C3D	PC	FAIR
Blender	.3DS, .DAE, .FBX, .DXF, .OBJ, .x,	PC	FAIR
Ogre	.MESH, .SKELETON	PC	FAIR
G3D Engine	.3DS, .IFS, MD2, .BSP	PC	GOOD

3   SAMPLE IMPLEMENTATION

3.1   Light weighting and importing of CAD model

In this paper, the virtual prototype is established using model import technology based on geometry modeling. The Lightweighting and importing of CAD models is mainly divided into three steps, which are shown in Figure 3.

Figure 3.   CAD data importing and conversion.

In this sample, the generated CATIA file in CGR format is 77.3 MB in size. However, the finally exported FBX file is 8.4 MB in size, which is 89.1% lighter.

3.2   Dynamic roaming

After building the 3D model of virtual prototyping, in order to achieve a higher level of reality experience, basic dynamic roaming and browsing capabilities need to be established, using the design of FPS (First Person Shooting) games.

3.3   Component capturing and highlighting

Component capturing function is one of the key points to achieve multivariant data integration. In order to integrate various types of data associated with each aircraft component, components capturing function in a virtual simulation environment is required, so as to obtain the name of the selected components and other information to get the related parameters and simulation results. As to the using experience, in addition to prompting the name of the captured component, highlighting the component can create better visual effect.

3.4   Hierarchical display of models

In the general design stage of an aircraft, apart from the aerodynamic shape, the 3D model may also include structure, layout, system, etc. In order to complete the experience of aircraft design, a relevance logic hierarchy of the virtual prototype based on the physical logic system needs to be constructed. Hierarchical display of aircraft design information, make it possible for quick view of aircraft internal structure and layout design, so as to reduce the workload of assessment process, and to improve the evaluation efficiency and accuracy.

3.5   Integration of design and performance parameters

There are two parts of contents for design and performance parameters, which are names and values of the parameters. Therefore, similar approaches can be taken for these parameters to be integrated to the virtual prototype. Three aspect of design listed below for this function are majorly considered.

3.5.1   Data file construction and accessing

There are, in fact, several possible approaches for selection of the data structure to be accessed, mainly including local data files, local databases and network databases. The scale of data introduced in the sample of this research is relatively small. Taking operation convenience and format specification into consideration, local data files are used to handle data storage in the sample. Parameter lists will be obtained from a local file and then loaded into memory.

3.5.2 Basic data management function

Basic data management functions include adding, modifying and deleting.

3.5.3 Display of parameter lists

Display function of parameter lists belongs to the interaction characteristics of multivariant data integration based on virtual prototyping. Based on the functions of component capturing and data file accessing, related data are displayed in real time.

3.6 Integration of CAE analysis results

As to the integration of CAE data, file lightweight processing and interactive hyperlink modes are used.

3.6.1 Lightweight and display of CAE files

For CAE simulation and analysis files lightweighting, in order to quickly view the results on ordinary PC, the ultimate goal achieved is: controlling the file size at about 1/10 of the source file, and displaying the files in a regular browser by simply installing a plug-in.

3.6.2 Interactive hyperlinks

CAE simulation and analysis results include three-dimensional grid, images, animation, data and other information. In the virtual prototype simulation environment, the CAE simulation and analysis results of a component are opened through interactive hyperlinks.

3.7 Integration of design feedback information

Design feedback information is integrated to the virtual prototype as one category of the multivariant data, in order to achieve rapid design feedback to improve the efficiency of aircraft design scheme evaluation.

Similar to data management functions, feedback information editing functions allow adding and editing of feedback information.

The difference between feedback information displaying and parameter list displaying lies in the integration of displaying and editing of feedback information, which provides quick creation and real-time modification of feedback information.

3.8 Running test

In the sample, the running test is performed on Windows system. The result shows that the program compatibility is good and interactive operation is quite easy to master.

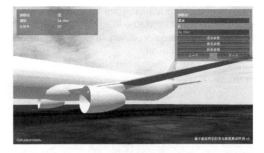

Figure 4. Display and editing of parameters.

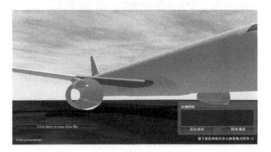

Figure 5. CAE file hyperlink button, feedback information editing and display.

Figure 6. Lightweight CAE document examining.

Figure 7. Aircraft cabin layout demonstration.

With the compiled program accessed to CAVE environment, users are able to experience the virtual prototype and data integration in three dimensional simulative environment.

Figure 8. Components capturing and highlighting, parameter listing in CAVE environment.

4 EXPLORATION OF DATA INTEGRATION PATTERNS

4.1 *Data management*

It is a fundamental aspect to conduct effective data management for data integration, which determines the data exchange standards including sources, formats and storage methods of data. In the sample project design of multivariant data integration, basic requirements for data management have been reflected.

Types of data available for integration into the sample virtual reality platform can be divided into several categories.

1 CAD data, including digital definitions of the aircraft and its components.
2 Design parameters, including weight, general layout parameters.
3 Performance parameters, including flight performance parameters, system and equipment parameters.
4 CAE analysis data, including FEA and CFD data.
5 Design feedback information.
   Other types of data in addition to the above ones in the application of multivariant data integration technology include:
6 General design requirements.

With these types of data and requirements appended, by comparative analysis of design values and predicted or standard values, automatic comparison and initiative warning function can be implemented during the design process.

Multivariant data integration relies on well-established data management systems. Data involved in building virtual prototypes is mainly acquired from BOM (Bill of Material) conversion. BOM of a virtual prototype is basically aiming at display the general design scheme. Taking this into consideration, the BOM tree only needs to be expanded to levels of subsystems or sub-units.

Figure 4.1 shows a typical example of BOM structure.

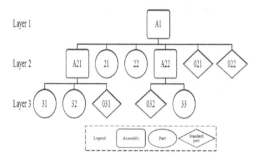

Figure 9. BOM structure diagram.

4.2 *Data conversion*

For data exchange between data management platform and virtual reality platform, data conversion is a key procedure for processing data in large scale or with relatively low compatibility.

1 Lightweighting and importing of 3D digital model files.
   As a basic functional model, it is a significant technique to generate a virtual display environment for CAD technical resources.
   Under the condition of current computing software hardware, it is barely possible to construct large-scale virtual reality models using existing design models. Lightweighting technique is needed to convert these original models to meet the requirements for fast interaction with virtual prototypes in the virtual reality environment.
   3D CAD file formats can be divided into two types: specialized data formats generated by CAD software such as CATProduct and CATPart, and standard data formats such as IGES and STEP. In general, files in standard format are about 1/10 the size of files in specialized format. Conversion from original CAD files into standard files guarantees the model compatibility in different software and implements initial lightweighting at the same time.
   3D models in the virtual reality environment usually require DCC (Digital Content Creation) format files for data exchange. Therefore, 3D digital model files in standard formats have to be lightweighted in further processing.
   Figure 10 shows the two steps for 3D digital model lightweighting.

Figure 10. Lightweighting and importing scheme of 3D digital models.

2 Lightweight processing of CAE analysis results in related disciplines.

Viewing CAE data files usually requires independent operation. Due to the demanding requirements for software and hardware platforms and low efficiency when viewing massive CAE files using CAE software such as ANSYS, it is necessary to apply lightweight processing and visualization to these files.

The lightweighting and visualization process takes an individual scheme as is shown in Figure 11.

Figure 11. Lightweighting and visualization scheme of CAE files.

4.3 *Interactive data experiencing*

Interactive experience of multivariant data makes it possible to browse various types of data in a intuitive way, which requires extra design and development for the virtual reality platform based on the virtual prototype structure.

Interactive experience technology achieved in the sample project mainly involves:

1 List view of design parameters and performance parameters.
2 Visual charts of CAE analysis results connected to the virtual prototype through interactive links.
3 Immediate editing, recording, saving and transmission of design feedback information.

Apart from interactive functions provided in the sample project, other aspects are also possible to be attached including:

4 Replacement and synchronization of analysis result files.
5 Early warning of comparison results of data and its target value.

5 THE DEVELOPMENT PROSPECT

5.1 *Cross-platform distribution and application prospect*

Modern mobile devices, such as smart phones, tablets, are capable of processing complex tasks, and usually can exchange information through technical means such as wireless networks. These characteristics provide hardware and software basics to explore the cross-platform virtual prototype experience and data integration technology.

To test the experience of virtual prototype on a mobile platform, the sample was published to run on Android operating system. The interface is shown in Figure 12.

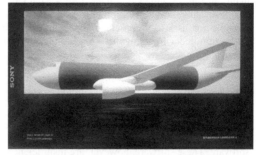

Figure 12. Components capturing and highlighting on mobile devices.

Due to the portability of mobile platforms, virtual prototypes running on these platforms can play a bigger role in the maintenance phase. Data integration in the maintenance phase mainly includes maintenance manual and maintenance log information.

Based on the virtual prototype and its subsidiary functions mentioned above, with part assembling and disassembling functions extended, additional task types which can be applied to maintenance works includes:

1 Viewing, virtual assembling and disassembling of 3D models.
2 Integrated display of component parameters.
3 Integrated display of component maintenance and repair manual information.
4 Viewing, recording and uploading of maintenance log and information.

This designed multivariant data integration platform distributed on mobile devices are supposed to increase the efficiency of maintenance work, and also to speed up the training process by applying these techniques in technical training of maintenance personnel.

5.2 *Enhancing and optimization of virtual experience*

In this topic, the virtual experience in the sample, especially the virtual prototype experience in immersive virtual environment, can be optimized by further enhancing technological means.

1 Somatosensory interaction function

With the somatosensory interaction function applied in virtual prototype experience in the virtual environment, interactive operations can be conducted such as virtual touching, space compatibility testing and intelligent gesture control. This provides more contributive abilities in enhancing virtual experience and assisting the man-machine engineering designing and testing work.

2 Voice interaction function

In the process of experiencing the virtual environment, human-machine voice interaction systems use the most natural way of communication to make it possible for user to complete massive information interaction under light cognitive loads and with less physical resources.

5.3 Uniform design-analysis-simulation-experience system

Design, analysis and evaluation work in the traditional design process are usually based on CAD and CAE software. These software, however, are insufficient in visualization and experience effects. Designers often found defects in design schemes after manufacturing of the prototype, which extended the design cycle of the aircraft.

Therefore, according to the research result of this paper, take airliners for example, there is one possible development direction during the process of designing, manufacturing and operating the aircraft, which is building a multivariant data integration platform based on virtual prototyping.

This can be applied throughout the aircraft development stage and even the whole life cycle of the aircraft. Applications of three key technologies are involved. The first one is rapid modification and display technology, especially for aircraft parameters and layout design. The second one is virtual experiencing technology, including flight environment and task simulation, virtual cabin roaming based on natural man-machine interface, and digital airport building. The third one is supporting technology of maintenance, including virtual assembling/disassembling and maintenance training.

The unified architecture of the D-A-S-E environment provides an integrated platform for supporting comparison and selection of design schemes, visualization of operating environment, and virtual experiencing and evaluation of cabin comfortableness and operational efficiency. It also provides full lifecycle support for multiple aircraft models.

6 CONCLUSION

First of all, this paper elaborated the scheme design, demonstration, technical construction and multi-platform testing of the sample of multivariant data

integration based on virtual prototyping. With the sample designed and built, one of the main research goals of this paper is achieved, which is integration and interaction of multivariant data such as design parameters, main performance parameters and CAE result data in virtual reality environment.

In addition, by analyzing the proposed design and technology in the sample, the pattern of multivariant data integration based on virtual prototyping is explored, especially on three aspects which are data management, data conversion and interactive experiencing technology. Aiming at the aircraft general design stage, application of the multivariant data integration environment based on virtual prototyping can contribute to shortening the development cycle by promoting evaluation efficiency.

Finally, this paper proposes technologies and development prospects in related areas, including the extension of cross-platform versions, the enhanced and optimized virtual simulation and experiencing scheme, and the uniform integrated design-analysis-simulation-experience system.

REFERENCES

Bennett G. R. 1997. Application of virtual prototyping in the development of complex aerospace products. *Aircraft Engineering and Aerospace Technology* 69(1):19–25.

Briggs C. 2010. Process improvement through tool integration in aero-mechanical design. *AIAA*.

Gu X. Y., Fenyes P. A. 2004. Application of the integration framework for architecture development (IFAD) for conceptual vehicle design. *AIAA*.

Hu P., Yu W., Hodges D., et al. 2010. VABS-IDE: VABS-enabled integrated design environment (IDE) for efficient high-fidelity composite rotor blade and wing design. *AIAA*.

Huang J. M. 2005. Study of the pivotal integration technology of CAE, CAD and BOM. *North China Electric Power University*.

Iqbal L. U. 2012. Balanced approach to the aircraft design. *AIAA*.

Iqbal L. U. 2012. Multidisciplinary design and optimization (MDO) methodology for the aircraft conceptual design. *AIAA*.

Liu H., Luo M. Q., Wu Z., et al. 2013. *Exploration and practice on aircraft general design supporting technologies*. Beijing: Beihang University Press.

Wang G. L., Wu Z. 2005. System frame research on aircraft conceptual design platform based on virtual prototyping. *Acta Aeronautica Et Astronautica Sinica* 26(2):162–167.

Whyte J., Bouchlaghem N., Thorpe A., et al. 2000. From CAD to virtual reality: modelling approaches, data exchange and interactive 3D building design tools. Automation in Construction (10):43–55.

Characteristics of flow for three-outlet centrifugal fan based on MRF method and SMM method

J. W. Li*

School of Mechanical Engineering, Wuchang Institute of Technology, Wuhan, China

X. P. Zhong, J. L. Xie & J. J. Ye

Green Fan Manufacturing Collaborative Innovation Center in Hubei Province, Wuhan, China

ABSTRACT: MRF method and SMM method are used for numerical simulation of three-outlet centrifugal fan by comparing the flow field and the corresponding aerodynamic performance parameters. It shows that there is a 5.6% difference between the two air flows and the static pressure rise is consistent. When using SMM method, the flow of three outlets are basically equal. When using MRF method, the difference of the flow of three outlets is relatively large. The maximum difference between the flow of three outlets and the average flow rate is 4.92%. The inconsistent pressure distribution at the volute tongue may be the reason of inconsistent flow by using MRF method. A further study is conducted on the reasons of inconsistent flow of three outlet caused by using MRF method. It presents that changes in the relative positions of the volute and impeller would be the main reason of the outlet flow rate changes.

1 INTRODUCTION

FLUENT proposed three options for studying rotating machinery, namely Multiple Reference Frame(MRF), Mixing Plane Model(MPM) and Sliding Mesh Model(SMM), of which MRF method and SMM method are applied more widely [1-3]. MRF method used multiple reference coordinate system, assuming flow field grid constant and the flow steady. With small amount of calculation and using less time, it was widely used in the simulation of rotating machinery. MRF model has been widely adopted in simulation of single-outlet centrifugal fan to get fairly accurate flow field of centrifugal fan [4,5], but the numerical simulation method of multi-outlet centrifugal fan has rarely been studied. In this paper, the MRF and SMM method are used to simulate the flow field of multi-outlet centrifugal fan. Through comparative analysis of the flow rate, pressure and other parameters as well as local flow field, the differences between the two methods and some theoretical guidance for the numerical simulation of multi-outlet centrifugal fan can be obtained.

2 CALCULATION MODEL AND METHOD

2.1 Calculation model

Three-outlet centrifugal fan used in this paper, as shown in Figure 1, consists of three-exit volute[6],

impeller, collector and other components. Impeller adopts closed impeller, outer diameter D2 = 530mm, inner diameter D1 = 360mm, leaf number Z = 10, the blades are evenly distributed along the circumference. Volute design adopts equal circulation design, using multiple arcs instead of Archimedes spiral, three outlets evenly distributed in the circumferential direction of the outlet, volute width b = 280mm, the volute tongue clearance t = 38mm, which t / D2 = 0.07.

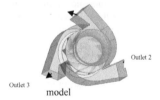

Figure 1. Calculation model.

Figure 2. Computing grid.

*Corresponding Author: Jiawei Li, Email: respaper@163.com

Computational grid was shown in Figure 2. Due to the presence of hook face for instance impeller, the front plate and other surfaces inside three-outlet centrifugal fan, to control the quality of the grid, the whole computational domain was divided into the impeller region, volute region and collector region. In corresponding region, advancing front method was utilized to generate tetrahedral mesh. In the part of the local, refined unstructured tetrahedral mesh of the leading edge of the blade, volute and other parts were generated. Grid number was 1,070,000, of which the impeller area is 550,000, accounting for 50% of the total number of the grid.

2.2 Calculation method

2.2.1 Multi-Reference Frame method (MRF) and Sliding Mesh Method (SMM)

MRF model is to do the steady-state approximate calculation for different rotational velocity or speed of movement of each unit body. It provides a reasonable model for many time-averaged flow field. When the interaction between the rotor and stator is relatively weak, MRF model can be used. Since the influence of rotation was relatively large near the exit area of the impeller, the relative rotation of coordinate system was used to solve the impeller region while absolute coordinate system was adopted in still regions such as volute region and diversion region. The relationship between rotating coordinate system and the absolute coordinate system is as follows:

$$v_r = v - (\omega \times r) \tag{1}$$

Where: v_r is the velocity vector of rotating coordinate system; v is the velocity vector of the absolute coordinate system; ω is the angular velocity of the rotation region; r is the relative position vector.

In the usual momentum equations:

$$\rho \frac{dV}{dt} = \rho R - \nabla \cdot p + \mu \nabla^2 V \tag{2}$$

The formula (1) is substituted into formula (2) to get formula (3):

$$\frac{\partial}{\partial t}(\rho v_r) + \nabla \cdot (\rho v_r v_r) + \rho(2\omega \times v_r + \omega \times \omega \times r + \alpha \times r) \tag{3}$$
$$= \rho R - \nabla \cdot p + \mu \nabla^2 V$$

Where $\alpha = \frac{d\omega}{dt}$, in the formula (2) and (3): V is the velocity vector, R is the force per unit mass; ρ is density; $2\omega \times v_r$ is Coriolis acceleration; $\omega \times \omega \times r$ is centrifugal acceleration; when the rotational speed is constant, $\alpha \times r$ does not exist.

Momentum equation of SMM Law is as shown in equation (4), sliding mesh model is an unsteady calculation method. In rotating machinery, the transient interaction result from the relative motion of stationary and rotating components are generally divided into three kinds of circumstance such as potential interactions (pressure wave interaction), wake interactions and shock wave interactions. Since MRF model ignores the transient interaction, it is limited to the flow of small transient effects. MRF method does not consider the movement of the movement area of the grid relative to the stationary region, grid of computing domain remaining stationary. It is similar to the impeller fixed to a particular position to observe the instantaneous flow of the rotary machine when impeller is at this position. So MRF method is also called frozen rotor method. If transient effects can not be ignored, SMM method could be used to consider the relative motion of stationary and rotating components. Similar to MRF model, when using SSM model the computational domain is also divided into motion part and static part, connected by the non-uniform grid interface connection. Different from MRF model, the grid of each domain of SMM model is a time function which changes over time, making the problem transient. Another difference is that the control equation of SMM model has a new grid form. It dose not use the motion form of coordinate system to get absolute in the static coordinates if there is no additional acceleration of the source term to the momentum. Equations are a special case of general sports/deformation of the grid.

$$\frac{\partial}{\partial t}(\rho v_r) + \nabla \cdot (\rho v_r v_r) + \rho(\omega \times v_r) = \rho R - \nabla \cdot p + \mu \nabla^2 V \tag{4}$$

2.2.2 Numerical solution method

The numerical simulation used three-dimensional Renault average Navier-Stokes equations of conservation and standard $K - \varepsilon$ equations, using the standard wall function near the wall, using SEGREGATED implicit method. Pressure and velocity coupling used SIMPLE algorithm. Turbulence kinetic energy, turbulent dissipation and the momentum equations used second order upwind. The inlet and outlet used pressure boundary conditions with total pressure of the inlet set to 0 Pa and static pressure of the three outlets set to 2500 Pa. When using MRF method, the area of the impeller used rotating coordinate system, setting the rotating wall boundary conditions. Rotating speed of the impeller was 2800 r/min. The volute area and the collector area used the static coordinates. When using SMM method, using MRF method to calculate the steady-state value as the calculation initial value of SMM method to save computation time. The time step t = 5.95238 e-05 (the time which is the impeller rotating 1° used), the rotating speed was still 2800 r/min.

3 CALCULATION RESULTS AND ANALYSIS

3.1 Comparison of calculation time

Table 1. The required resources and computing time of the MRF method and the SMM method to calculate.

	MRF method	SMM method
Computing devices	Eight processes parallel computing	
Computation time /h	About 0.4h	About 8h

MRF method and SMM method used the same grid and the same convergence criteria for numerical calculation. Table 1 shows that the MRF method and SMM method under the same computing devices, it can be seen that the computing time of SMM method is about 20 times of MRF method.

3.2 The contrast of calculation results

Table 2 shows the aerodynamic performance calculation results of three-outlet fan of MRF method and SMM method. Using SMM method simulation results as the benchmark, it can be seen that the distribution of static pressure rise using the two methods is identical while the flow difference is 5.6%. In theory, the flow of three outlets should be basically equal. It can be seen clearly in table 2 that the flow of three outlets using SMM method is basically equal or slightly different. Apparently because of using unstructured tetrahedral mesh, the grid of three outlets is not exactly the same which made the flow of three outlets slightly different. The flow difference of the three outlets using MRF calculation is much bigger with the biggest difference 389 m3/h.

Figure 1 ~ figure 4 (the top outlet is outlet 1,then

Table 2. The fan performance calculation results of two calculation methods.

	air flow rate (m³/h)	Outlet volume 1(m³/h)	Outlet volume 2(m³/h)	Outlet volume 3(m³/h)	Static pressure rise (Pa)
MRF model	14789	5076	5026	4687	3021
SMM model	15678	5263	5180	5234	3085

outlet 2,and 3 in clockwise , the outlets of all figures below are the same) has shown velocity contours and static pressure contours at x = - 70mm (which is a

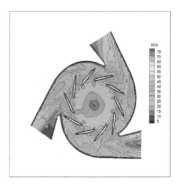

Figure 1. velocity contours at x =-70mm using MRF method.

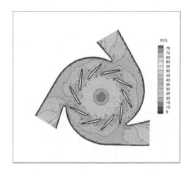

Figure 2. velocity contours at x =- 70mm using SMM method.

Figure 3. The static pressure contours at x=-70mm using MRF method.

split of the impeller) of results respectively using MRF method and SMM method. It can be seen that in the process of air going through the blade channel, the static pressure and speed of air increases due to the work of the impeller. When air goes from the blade channel into the volute impeller, the air inside

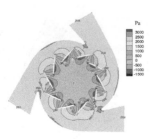

Figure 4. The static pressure contours at x=-70mm using SMM method.

the rotating impeller passage can get gradually transforms the kinetic energy of the gas pressure under the effect of volute constantly expanding pressure. So the kinetic energy of air constantly decreases along the radial direction while the static pressure increases continuously.

According to the comparison of velocity contours, the air velocity distribution of three outlets using SMM method is bigger than it using MRF method, which explains the airflow of simulation using SMM method is bigger than MRF method in table 1. It can be seen that the air velocity of three outlets using SMM method is greater than using MRF method at the same time. Figure 1 clearly shows that the air velocity of three outlets using MRF method is different. The velocity distribution of outlet 1 is the maximum and outlet 3 is the minimum, corresponding to the volume size of three outlets in table 1. In figure 2, the velocity distribution of three outlets is almost the same so the air flow of three outlets using SMM method is the same. According to the comparison of Figure 3 and figure 4, the static pressure distribution of outlet using two methods are basically identical. The volute pressure distribution using SMM method was almost the same while the volute tongue pressure distribution using MRF method is not consistent. This may be the cause of different velocity distribution of three outlets.

3.3 Simulation using MRF method

3.3.1 The calculation result

Inconsistent air flow of three outlets appeared when using MRF method to simulate three-outlet fan. The relative position of the impeller and the volute was changed to further analyze the reasons. The position of volute was not moved while on the basis of the original the impeller rotates clockwise 6°, 12°, 18°, 24°, 30°, 36°(leaf number is 10, take 36° for analysis). The position of original case was recorded as 0°. Numerical analysis of the several models above was

done. Figures 5 and 6 respectively shown the relative position of the impeller and the volute at 0°and 6°. It can be seen that the relative position of the impeller outlet and volute tongue changed.

Table 3 shown the air flow rate of each outlet and pressure rise situation of each angle of fan. Figure 7

Figure 5. The relative position of the impeller and the volute of 0°.

Figure 6. The relative position of the impeller and the volute of 18°.

shown the changes outlet air flow rate of each fan of the various angles. It can be seen from table 3 that in various angles, the air flow of the fan was basically

Table 3. Turbines' air flow rate, each outlet air flow rate and static pressure rise cases under various angles.

	air flow rate (m³/h)	air flow rate of outlet 1 (m³/h)	air flow rate of outlet 2 (m³/h)	air flow rate of outlet 3 (m³/h)	static pressure rise (Pa)
0°	14789	5076	5026	4687	3021
6°	14800	4882	5112	4806	3021
12°	14821	4763	5022	5036	3021
18°	14778	4849	4810	5119	3021
24°	14785	5065	4705	5015	3021
30°	14794	5175	4807	4812	3021
36°	14802	5080	5029	4693	3021

the same, having a small difference with average. The air flow rate of three outlets of fans with different angles was not consistent. It can be seen from table 3 and figure 4 that the aerodynamic performance of the fan of 0°and 36°is basically the same. It shows that with one blade channel angle change, performance of the fan remains unchanged. The air flow rate of single outlet changed with the angle of cyclical changes. In a single cycle the average air flow rate of fan is 14794m³ / h, of which average to each outlet is 4931m³ / h. The air flow of three outlets in the next cycle averaged is respectively 4968m³ / h, 4913m³ / h, 4912m³ / h. The average flow rate difference is small.

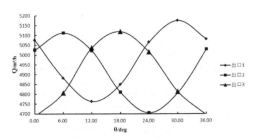

Figure 7. Changes of the outlet air flow rate of each fan of the various angles.

3.3.2 *Analysis of the causes*

In single-outlet centrifugal fan numerical calculation, the interaction between rotor and stator is relatively weak which makes MRF method, the frozen rotor method can accurately simulate the aerodynamic performance parameters of single-outlet centrifugal fan. But it is not possible using MRF method to simulate the pressure fluctuations over time while using SMM Law could make it[7]. Single outlet fan has only one exit thus all the air flow rate have the only exit. For multi-outlet fan using MRF method to simulate three-outlet fan, there are three exits. Since MRF method uses frozen rotor method, the impeller

Figure 8. The partial view of velocity vector of the outlet 3 of 0°.

Figure 9. The partial view of velocity vector of the outlet 3 of 18°.

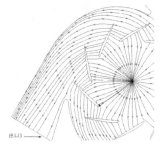

Figure 10. The partial view of motion pattern of the outlet 3 of 0°.

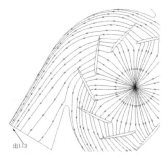

Figure 11. The partial view of motion pattern of the outlet 3 of 18°.

is fixed at one particular position to observe the air flow inside the fan without considering the effect of flow at the former location of the impeller on the current position.

Figures 8 to 11 shows the velocity vector and motion pattern at 0°and 18° of outlet 3 respectively. When the relative angle of volute and impeller changes, since the MRF method adopts to frozen rotor, the airflow velocity vector and the motion pattern direction of the outlet 3 changes. The airstream from near volute tongue to the outlet 3 of 18°is significantly greater than it of 0°. This explains why air flow rate of the outlet 3 of 18°is greater than it of 0°.

4 SUMMARY

MRF method can not be used to simulate pressure fluctuates of three-outlet fan with time since the air flow rate of three outlets is inconsistent. However, due to it using steady calculations, consuming less time, in engineering calculations, we can get the aerodynamic performance parameters of three-outlet fan through the MRF method. If more accurate flow field was needed, the SMM method could be used.

5 CONCLUSION

1 By comparing the simulation results of three-outlet fan using MRF method and SMM Law, it was found that the difference between the two air flows are 5.6% and the static pressure rise is consistent.
2 When using MRF method, the difference of the flow of three outlets is relatively large. The inconsistent pressure distribution at the volute tongue may be the cause of inconsistent flow by using MRF method.
3 It was found that changes in the relative positions of the volute and impeller would be the root cause of the outlet flow rate changes, which is further evidence that the MRF method used frozen rotor law. It could only simulate the aerodynamic performance of the fan when the impeller is at the current state without considering the impact of the last state.
4 In engineering calculations, we can get the aerodynamic performance parameters of three-outlet fan through the MRF method. If more accurate flow field was needed, the SMM method could be used.

ACKNOWLEDGMENTS

This project was supported by the National Natural Science Foundation of China, Grant No. 51106137, the Fundamental Research Funds for the Central Universities, Grant No.2014TS117.

REFERENCES

Deglon D A, Meyer C J. CFD modeling of stirred tanks: Numerical considerations[J]. Minerals Engineering. 2006, 19(10): 1059–1068.
SUN Chendi,WANG Zhenya,YANG Zhigang,etal. Comparison of the Three Methods in the Numerical Simulations of a Transonic Axial-flow Fan[J]. Compressor, Blower & Fan Technology.2011(01):8–12.
HE Wei,MA Jing,WANG Dong,et al.Comparing MRF method with sliding mesh method for automotive front end airflow simulation[J]. Computer Aided Enginerring. 2007(03): 96–100.
Lin S, Tsai M, Shen M, et al. An integrated numerical analysis for a backward-inclined centrifugal fan[C]. 2010.
YOU Bin, E E. Elhadi, XIE Junlong,et al. Three Dimensional Numerical Analysis Of Multi-blade Fan[J]. Journal OF Engineering Thermophysics.2003(03): 419–422.
Eck B. Fans, Beijing China Machine Press.1983:255–257.
Ballesteros-Tajadura R, Velarde-Suarez S, Hurtado-Cruz J P, et al. Numerical calculation of pressure fluctuations in the volute of a centrifugal fan [J]. Journal of Fluids Engineering-Transactions of The ASME. 2006, 128(2): 359–369.

Computational Intelligence in Industrial Application – Ling (ed.)
© *2015 Taylor & Francis Group, London, ISBN: 978-1-138-02818-0*

Dynamic performance of cycloid bevel gear with bionic stripe surface micro-morphology

ZhiFeng Liu, XianFu Wang & ZhiJia Cheng

College of Mechanical Engineering and Applied Electronics Technology, Beijing University of Technology, Beijing China

ABSTRACT: The high-accuracy parameterized cycloid bevel gear model was established by using Pro/E software. Based on the engineering bionics and the finite element theory, the effect of the bionic stripe surface morphology on the improvement of the dynamic performances of the gear was analyzed using ANSYS Workbench software. By comparison of the model analyses of ordinary gear and the gear with bionic stripe surface morphology, the anterior 10 ranks of natural frequencies and mode shapes were calculated. The results show that the maximum amplitude of any rank of the anterior 10 ranks of the gear with bionic surface morphology is smaller than that of the ordinary. The natural frequency of the gear with bionic stripe surface morphology decreases obviously and the trend of the natural frequency is much smoother than that of ordinary gear. The dynamic characteristics and the reliability of the gear have been ameliorated effectively.

KEYWORDS: Engineering bionics; Cycloid bevel gear; Bionic surface morphology; Dynamic performance.

1 INTRODUCTION

Cycloid bevel gear is one of the two significant kinds of spiral bevel gear, and cycloid bevel gear is an important direction of the development of spiral bevel gear. Cycloid bevel gear is being widely used in the heavy-duty and high power transmission field such as truck, aviation, heavy-duty machine tools and so on. With the increasingly development of mechanical transmission system towards high-speed and precise direction, the dynamic characteristics of spiral bevel gear, which is the key transmission components of transmission system, will more significantly affect the transmission performance of transmission system. Researching the dynamic characteristics of cycloid bevel gear is practically valuable and academic significant for designing and manufacturing high quality transmission components with high precision, better durability and low noise characteristics.

It was found that the surface of many organisms characterizes wear resistance owing to their particular non-smooth shape. Long-term living in the soil of soil animals (such as dung beetles, mole crickets, earthworms, pangolins, etc.), after millions of years of evolution, the formation of a biological non-smooth surface, the surface soil contact site showing pits, convex hull, scales, corrugated and other non-smooth shape, it not only has good viscosity reduction, reduction in resistance features, but also has a very high wear resistance. Jilin University researchers after years of research, there have been a variety

of bionic surface pits, wear the gear mesh used in other forms of anti-fatigue properties and research, and have achieved remarkable results [1-10]. In this paper, the cycloid bevel gear with bionic stripe surface micro-morphology used in modal analysis using finite element software ANSYS Workbench, explore the impact of the bionic surface morphology of the dynamic performance of the gear.

2 ESTABLISHMENT OF THREE-DIMENSIONAL MODEL OF CYCLOID BEVEL GEAR

Using quasi-Newton method in Matlab software to meshing equations of tooth surface are dispersed, this

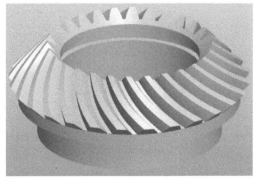

Figure 1. Three-dimensional model of cycloid bevel gear.

paper gets 3d coordinate of gear tooth surface from discrete points. Then the discrete point data import into the Pro/E, and then from points and lines, by the line and plane, from surface to body modeling ideas to build 3d model of gear tooth. Thus it obtained the cycloid bevel gear pair of precise geometric model. Fig. 1 shows the three-dimensional model of cycloid bevel gear.

3 ESTABLISHMENT OF FINITE ELEMENT MODEL

The three-dimensional model of cycloid bevel gear is imported into ANSYS Workbench software. Table 1 shows the basic parameters of cycloid bevel gear pair.

Table 1. The basic parameters of cycloid bevel gear pair.

Parameter name	Code	Driving gear	Driven gear
Midpoint of the normal modulus	m_a	4	4
Big endian modulus	m	5.9	5.9
Number of teeth	z	29	29
Pressure angle	α	20°	20°
Helix angle	β	35°41′	35°41′
Direction of spiral		L	R
Height of tooth	h	9 mm	9 mm
Precision grade	DIN3965/86	6	6
Maximum speed	n_{max}	1200 r/ min	1200 r/ min
Shaft angle	Σ	90°	90°
Midpoint of side gap	j_{nm}	0.14 mm	0.14 mm

0.00   90.00 (mm)
   45.00

Figure 2. The finite element model of cycloid bevel gear.

Set material elastic modulus 2.1×10^5 MPa, Poisson's ratio is 0.3, a density of 7.85×10^3 kg/m³. Geometric model uses second-order tetrahedral element mesh. Fig. 2 shows the finite element model of cycloid bevel gear.

4 THE MODAL ANALYSIS OF ORDINARY GEAR

According to differential equations of motion of elastic mechanics finite element theory, structural model of the system can be obtained

$$[M]\{\ddot{X}\}+[C]\{\dot{X}\}+[K]\{X\}=\{F(t)\} \qquad (1)$$

The formula: $[M]$ is the mass matrix; $[C]$ is the damping matrix; $[K]$ is the stiffness matrix; $[\ddot{x}]$ is the vibration acceleration vector; $[\dot{x}]$ is the velocity vector; $[X]$ is the displacement vector; $\{F(t)\}$ is the exciting force vector.

If it is external force, $\{F(t)\}=\{0\}$, then the free vibration equation of the system. In seeking gear free vibration frequencies and mode shapes (ie seeking gear natural frequencies and mode shapes), the damping has little effect on them, thus damping term can be omitted to obtain the equations of motion for undamped free vibration

$$[M]\{\ddot{X}\}+[K]\{X\}=0 \qquad (2)$$

The corresponding characteristic equation is

$$([K]-\omega_i^2[M])\{x_i\}=0 \qquad (3)$$

ω_i is i rank natural frequency mode. $i=1,2,3...n$.

At this time there is generally a vibration system of n and n main natural frequency vibration mode, each free vibration frequencies and mode shapes representative of a single degree of freedom system, which in the free vibration, the vibration characteristics of the basic model has said modal model. The low-order modes of vibration effects on the structure of the larger model, the dynamic characteristics play decisive role models. Therefore, in the finite element modal analysis, in most cases below a certain frequency, it only considers the low order modes. This paper takes the top 10 natural frequencies and mode shapes have been able to meet the accuracy requirements. Due to limited space, only the first four bands are given modal shape diagram. Fig.3 shows the former 4th modal shape of ordinary gear. Table 2 shows the former 10 ranks of natural frequencies, mode shapes and the maximum amplitude.

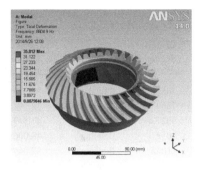

(a) The first-order

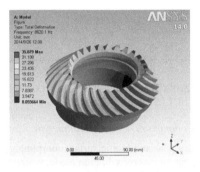

(b) The second-order

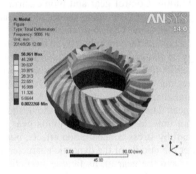

(c) The third-order

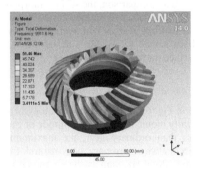

(d) The forth-order

Figure 3.   The former 4th modal shape of ordinary gear.

Table 2.   The former 10 ranks of natural frequencies, mode shapes and the maximum amplitude.

Modal rank	Natural frequencies /Hz	Mode shapes	The maximum amplitude/ mm
1	8604.9	Circumference	35.012
2	8620.1	Radial	35.079
3	9885	Radial	50.961
4	9911.6	Radial	51.46
5	12091	Umbrella	39.745
6	12689	Fold	66.846
7	12709	Fold	64.509
8	12713	Bending	60.174
9	12720	Bending	60.045
10	17252	Bending	65.398

5   THE MODAL ANALYSIS OF CYCLOID BEVEL GEAR WITH BIONIC STRIPE SURFACE MICRO-MORPHOLOGY

The morphologies are similar to surfaces of shells, soft and hard alternated, with micro width and depth (width is 210 um, depth is about 3 um), and arrange regularly. Fig. 4 shows the finite element model with bionic surface morphology and the local structural graph. Modal analysis procedure is identical with ordinary gear. Due to limited space, only the first four bands are given modal shape diagram Fig. 5 shows the former 4th modal shape of the fifth bionic gear. Table 2 shows the former 10 ranks of natural frequencies, mode shapes and the maximum amplitude.

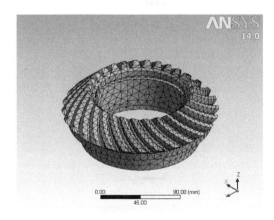

Figure 4.   The finite element model with bionic surface morphology and the local structural graph.

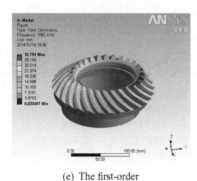

(e) The first-order

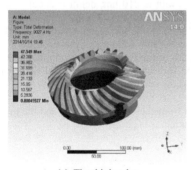

(f) The second-order

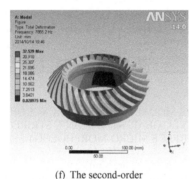

(g) The third-order

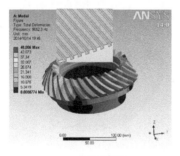

(h) The forth-order

Figure 5. The former 4th modal shape of cycloid bevel gear with bionic stripe surface micro-morphology.

Table 3. The former 10 ranks of natural frequencies, mode shapes and the maximum amplitude.

Modal rank	Natural frequencies /Hz	Mode shapes	The maximum amplitude/mm
1	7855.2	Radial	32.529
2	7869.4	Circumference	32.794
3	9027.4	Radial	47.549
4	9052.3	Bending	48.006
5	11035	Bending	37.069
6	11581	Bending	62.028
7	11594	Bending	52.371
8	11603	Umbrella	57.446
9	11606	Fold	63.191
10	15752	Fold	60.806

6 COMPARISON OF MODAL CALCULATED RESULTS

6.1 Comparison of gear vibration type

Ordinary gear and bionic surface morphology of the front gear 10 diverse vibration mode, vibration mode types are of five kinds, and the main modes are bending vibration. Control gear and bionic surface morphology ordinary gear mode shapes show that both modes are of same type, but, due to the presence of stripes bionic surface morphology of the surface morphology of the gear, so change the order of the two modes differ by vibration type figure shows, when the gear bending vibration and bending reverse vibration, structural changes in the shape of the largest conventional gear. Therefore, to optimize the design of the gear, not only to consider their joints gear and the gear shaft by stress concentration effects, but also to consider the middle of the dynamic characteristics of the gear and the gear on its fatigue strength. Modal analysis by ordinary gear and bionic surface morphology gear comparison, the surface morphology of bionic tooth surface by laser processing stripe-shaped manner, making gear design some of the problems have been solved to provide a design for the future gear surface morphology new ideas.

6.2 Comparison of maximum amplitude

Shown in Fig. 6, the surface morphology of the gear is basically bionic maximum amplitude of each order are less than ordinary gear, the maximum amplitude occurs when ordinary gear natural frequency is around 12689 Hz, namely the six modal vibration mode, while the bionic surface morphology gear the maximum amplitude occurs when the natural frequency of about 11606 Hz, i.e. modal nine vibration

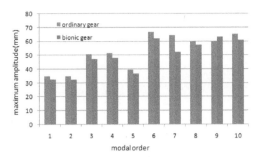

Figure 6. The comparison of the former 10 ranks of maximum amplitude of the ordinary gear and the bionic gear .

mode. Modal analysis results show that the modal distribution of the main gear vibration is folded, while the surface morphology of a stripe shape bionic effectively reduces the maximum amplitude of the gears to help improve the fatigue strength of the gear, gear fatigue life.

6.3 Comparison of gear natural frequency

Shown in Figure 7, and the natural frequency of the conventional gear change of the surface morphology of the gear are bionic increases, the natural frequency of the gear general showed a sharp rise, while the surface morphology of the natural frequency of the gear bionic relatively gentle rising trend can be seen in stripes bionic shaped surface morphology has a certain influence on the gears working conditions.

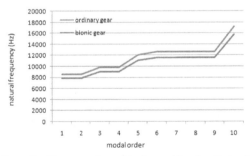

Figure 7. The comparison of the former 10 ranks of natural frequency of the ordinary gear and the bionic gear.

7 CONCLUSION

Modal analysis showed that the surface morphology of bionic produce vibration gear, the maximum amplitude of vibration response parameters such as natural frequency greater impact on the natural frequency range gear smaller, the maximum amplitude decreases, effectively improve the cycloid bevel gear dynamic performance.

REFERENCES

Haidong Xie, Zhaoyao Zhou, Wei Xia, et al. The vibration frequency response analysis of powder metallurgy helical gear transmission system. *Modern Manufacturing Engineering*，2005(2): 26–28.

Luquan Ren, Zhiwu Han, Limei Tian, et al. Characteristics of non-smooth surface morphology of living creatures and its application in agricultural engineering [C] // Design and Nature. Rhodes, Greece: [s. n.]2004: 275–284.

Zhiwu Han, Luquan Ren, Zubin Liu. Investigation on anti-wear ability of bionic non-smooth surfaces made by laser texturing. *Tribology*, 2004, 24(4): 289–293.

Zhiwu Han, Xiaoxia Xu, Luquan Ren Regression analysis of micro-friction and wear on concave non-smooth surface. *Tribology*, 2005, 25(6): 578–582.

Xin Tong, Fuhai Li, Min Liu, et al. Thermal fatigue resistance of non-smooth cast iron treated by laser cladding with different self-fluxing alloys. *Optic, Laser Technology*, 2010, 42(7): 1154–1161.

Zhihui Zhang, Hong Zhou, Luquan Ren, et al. Surface morphology of laser tracks used for forming the non-smooth biomimetic unit of 3Cr2W8V steel under different processing parameters, *Applied Surface Science*, 2008, 254(8): 2548–2555.

Xin Tong, Hong Zhou, Weiwei Chen, et al. Effects of pre-placed coating thickness on thermal fatigue resistance of cast iron with biomimetic non-smooth surface treated by laser alloying. *Optics, Laser Technology*, 2009, 41(6): 671–678.

Hong Zhou, Peng Zhang, Na Sun, et al. Wear properties of compact graphite cast iron with bionic units processed by deep laser cladding WC, *Applied Surface Science*, 2010, 21(15): 6413–6419.

Zhiwu Han, You Lü, Lichun Dong, et al. Modal analysis of gear with bionic surface morphology. *Journal of Jilin University*, 2010, 40(6):1604–1608.

Zhiwu Han, You Lü, Rongfeng Ma, et al. Dynamic performance of gear surface with bionic micro-morphology. *Journal of Beijing University of Technology*, 2010, 37(6):806–810.

Computational Intelligence in Industrial Application – Ling (ed.)
© 2015 Taylor & Francis Group, London, ISBN: 978-1-138-02818-0

Application of ANSYS transient thermal model on reservoir development in deformable media

M.X. Liu, J.H. Li* & D. H. Liu
A.A.Key Laboratory of Exploration Technologies for Oil and Gas Resources, Yangtze University, Wuhan, Hubei, China
Institute of Petroleum Engineering, Yangtze University, Wuhan, Hubei, China

T. Jiang
Department of Sciene and Technology, Yangtze University, Wuhan, Hubei, China

L. Zhang
China Petroleum Pipeline Engineering Corporation, Langfang, Hebei, China

ABSTRACT   Different from the conventional reservoir, deformable media property will change with the overlying formation effective pressure. To date, the finite difference method is mainly used in the numerical simulation studies of reservoir development. Many present papers simulate the application of the conventional method based on the steady thermal analysis of ANSYS. It first derives the similar relationship between pressure sensitive coefficient of the formation and the specific heat of the media in the paper. Then based on the transient thermal conduction model in ANSYS, the simulated result fitted well with the development performance in a certain deformable reservoir. The results show that the transient thermal conduction model on the software of ANSYS simulates the reservoir development. This paper also uses ANSYS finite element software to simulate the production availability of artificial fractured wells. The results show that the ANSYS can well simulate the formation pressure decreases and fractures closure behavior in the production process, it also provides a new approach to research and analyze the effectiveness of fracturing wells.

KEYWORDS: Finite difference; Pressure sensitive; Temperature field; Transient heat conduction; Deformable media.

1 INTRODUCTION

1.1 Finite element theory

Since the research of Courant[1–3] in 1943 and the rapid development that occurred after the 1960s, the finite element method (FEM) has been used in a wide range of applications in the field of engineering, such as in the analysis of objects from elastic materials to plastic, viscoelasticity, and composite materials and in the analysis of solid mechanics to fluid mechanics, heat transfer, and other areas of the seepage flow mechanics of porous media[4]. However, FEM has not been applied to solve seepage problems for a long time. Javandel[1] introduced the basic theory and process of FEM to solve reservoir seepage problems in 1968 and showed the superiority of using FEM in studying seepage problems. Karimi-Fard, RERI, and Firoozabadi[5] used the Galerkin[6] calculus of variations and finite element meshing,

proposed the discretization of the discrete fracture grid model, used line element to discrete fracture, and discretized matrixes by using triangle elements[7, 8].

$$A(u) = \begin{pmatrix} A_1(u) \\ A_2(u) \\ \vdots \end{pmatrix} = 0 \quad (\text{in } \Omega) \tag{1}$$

The domain Ω can be a volume, area, and so on (see Fig. 1). The unknown function u should also satisfy the boundary conditions.

$$B(u) = \begin{pmatrix} B_1(u) \\ B_2(u) \\ \vdots \end{pmatrix} = 0 \quad (\text{in } \Gamma) \tag{2}$$

---

*Corresponding author J.H. Li is an associate professor in Yangtze University.

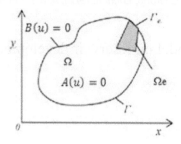

Figure 1. Solving the regional Γ and boundary Ω.

The unknown function u can be a scalar field (e.g., pressure or temperature), and can also be a vector field composed of several variables (e.g., displacement, strain, stress, and so on). A and B represent the differential operator with independent variables (space coordinates, time coordinates, and so on). By using the 2D steady-state heat conduction equation as an example, the control equation and definite condition can be expressed as follows:

$$A(\phi) = \frac{\partial}{\partial x}\left(k\frac{\partial \phi}{\partial x}\right) + \frac{\partial}{\partial y}\left(k\frac{\partial \phi}{\partial y}\right) + q = 0 \,(\text{in } \Omega) \quad (3)$$

$$B(\phi) = \begin{cases} \phi - \bar{\phi} = 0 \\ k\dfrac{\partial \phi}{\partial n} - \bar{q} = 0 \end{cases} \quad (4)$$

where Φ is the temperature (corresponding pressure in the seepage problem), K is the mobility or thermal conductivity (corresponding flow degree k/u in seepage problem), $\bar{\phi}$ and $\bar{q}$ are the given value of temperature and heat flux in the boundary (corresponding pressure and velocity on the side of the boundary in the seepage problem), n is the outward normal direction of the boundary Γ, and q is the source density (corresponding well production in the seepage problem).

In the above problems, if k and q only represent the space position function, the problem is linear; however, if k and q are functions of Φ and its derivatives, the problem is nonlinear.

$$\int_\Omega V^T A(u)\,d\Omega = \int_\Omega (v_1 A_1(u) + v_2 A_2(u) + \cdots)\,d\Omega = 0 \quad (5)$$

$$V = \begin{pmatrix} v_1 \\ v_2 \\ \vdots \end{pmatrix} \quad (6)$$

where V is the vector function, which is a set of arbitrary function, and the numbers are equal to the differential equations.

Eq, (5) and the differential equations (1) are completely equivalent integral form. If the integral equation (5) is established for any V, the differential equation (1) must be satisfied at each point in the domain. Similarly, if the boundary conditions (2) are satisfied at each point in boundary at the same time for a set of arbitrary function $\bar{V}$, the following can be obtained:

$$\int_\Gamma \bar{V}^T B(u)\,d\Gamma = \int_\Gamma (\bar{v}_1 B_1(u) + \bar{v}_2 B_2(u) + \cdots)\,d\Gamma = 0$$

Integral form:

$$\int_\Omega V^T A(u)\,d\Omega + \int_\Gamma \bar{V}^T B(u)\,d\Gamma = 0 \quad (7)$$

Eq. (7) is the equivalent integral form of the differential equation.

1.2 Mechanism of pressure sensitive

Before the reservoir development, the pore pressure in balance with the overlying strata pressure, the reservoir has a certain permeability and porosity. After the reservoir development, especially the first recovery stage, the formation pore pressure drops and the effective stress increases with the fluid produced. The rock skeleton is compressed, leading to the decrease of porosity and permeability. The porosity and permeability decreased with the formation pressure drops, this phenomenon is called the stress sensitivity damage[9,10].

Many experiments have shown that[11-13] the low permeable reservoir's permeability changes with the effective stress obviously, and the changes of reservoir effective stress can be expressed as the function of pore pressure and ground stress, therefore, the reservoir's permeability stress sensitivity can be expressed as:

$$K = K_0 e^{-\alpha(p_i - p)} \quad (8)$$

Type: p, p_i—formation pressure and original-formation pressure, MPa; K— permeability, μm^2;
K_0—original formation permeability, μm^2;
α—pressure sensitive coefficient $1/MPa$.

1.3 Fracturing wells

The hydraulic fracturing has been successfully tested in the United States since 1947. So far, fracturing has become the most effective way to increase production at the oil industry, almost all natural gas and the vast majority of oil wells are fractured[14].

The effect of pressure sensitive effect on the oil field has caused more and more people's attention, In this paper, the first time using the ANSYS transient heat conduction module to simulate the reservoir with the effect of pressure sensitive, take a reservoir in Yu Men oilfield for example, shows that the importance of numerical simulation technology with the effect of pressure sensitive. And using ANSYS to simulate the fracturing horizontal Well. The results show that the ANSYS can well simulate the fractures closure phase in the production process.

2 SIMILAR TO EQUIVALENCE

Current field and seepage field has the similar property; heat flow and seepage fields also have the similar mathematical model[15-18].

2.1 Similar to the differential equation

The differential equation for non-homogeneous elastic anisotropy of single phase media for compressible fluid transient flow:

$$\frac{\partial}{\partial x}\left(K_x\frac{\partial P}{\partial x}\right)+\frac{\partial}{\partial y}\left(K_y\frac{\partial P}{\partial y}\right)+\frac{\partial}{\partial z}\left(K_z\frac{\partial T}{\partial z}\right)=\mu C\frac{\partial T}{\partial t} \quad (9)$$

Type: K_x, K_y, K_z—x, y, z permeability of threedirections; μm^2; C_t—total compressibility, $1/MPa$.

The differential equation for non-homogeneous anisotropic that there is no heat source for transient heat conduction:

$$\frac{\partial}{\partial x}\left(K_{rx}\frac{\partial T}{\partial x}\right)+\frac{\partial}{\partial y}\left(K_{ry}\frac{\partial T}{\partial y}\right)+\frac{\partial}{\partial z}\left(K_{rz}\frac{\partial T}{\partial z}\right)=C\frac{\partial T}{\partial t} \quad (10)$$

Type; K_{rx}, K_{ry}, K_{rz}—x, y, z thermal Conductivity of three directions, $w/(m\cdot°C)$; C—specificheat, $J/(kg\cdot°C)$.

2.2 The initial conditions and boundary conditions

The Table 1 shows the initial conditions and boundary conditions. The Table 2 shows the quantities seepage field and temperature field.

Table 1. The comparison of boundary conditions of temperature field with seepage fields.

	Seepage field	Temperature field
Initial conditions	$P\|_{t=0}$ $= P_0(x, y, z, 0)$	$T\|_{t=0}$ $= T_0(x, y, z, 0)$
The first boundary condition	$P\|_{\Gamma_1}$ $= P(x, y, z, t)$	$T\|_{\Gamma_1}$ $= T(x, y, z, t)$
The second boundary condition	$\frac{K}{\mu}\frac{dP}{dL}\|_{\Gamma_2}$ $= -v(x, y, z, t)$	$K_r\frac{dP}{dn}\|_{\Gamma_2}$ $= -q(x, y, z, t)$

Table 2. The comparison of physical quantities of temperature field with seepage fields.

The seepage field		The temperature field	
Laplace's equation	$\nabla^2 P = 0$	Laplace's equation	$\nabla^2 T = 0$
Darcy's law	$v=-\frac{K}{\mu}\frac{dP}{dL}$	Fourier law	$q=-K_r\frac{dT}{dn}$
The fluid flow	Q	Heat	Q
Mobility	$\frac{K}{\mu}$	Thermal conductivity	K_r
Seepage velocity	v	Heat flow intensity	q
Total compressibility	C_t	Specific heat	C

2.3 Similar constants and similar criteria

1 Similar to the geometric constants

$$C_L=\frac{X}{x}=\frac{Y}{y}=\frac{Z}{z}=\frac{R}{r} \quad (11)$$

2 Similar to the physical constants
Similar to the pressure constant: $C_P = P/T$
Similar to the mobility constant: $C_{K/\mu} = (K/\mu)/K_r$
Similar to the speed constant: $C_v = v/q$
Similar to the flow constant: $C_Q = Q/Q' = C_L^2 C_v$
Combination of the above similar indicators of the similar constants with Darcy's law can be obtained[19]:

$$\frac{C_P C_{K/\mu}}{C_L C_v}=1 \quad (12)$$

2.4 According to the similarity to derive the similar constant of heat capacity

By $C=\frac{Q}{m\cdot\Delta T}$, on both sides of the formula multiply the density, can be obtained:

$$C\cdot\rho=\frac{Q}{V\cdot\Delta T} \quad (13)$$

$$C_t=\frac{\Delta V}{V_b\cdot\phi\cdot\Delta P} \quad (14)$$

$$C_t\phi=\frac{\Delta V}{V_b\cdot\Delta P} \quad (15)$$

Type: V_b—Visual volume of rock, m^3; φ—rock porosity, decimal; C_t—total compressibility, $1/MPa$.

In a unit time, similar to the flow constant:

$$C_Q = \Delta V / Q = C_L^2 C_v \qquad (16)$$

Similar to the volume constant:

$$C_V = V_p / V = C_L^3 \qquad (17)$$

Similar to the specific heat constant:

$$C_C = \frac{C_t \phi}{C_L \cdot C_P} \qquad (18)$$

From all the similar constants have been discussed above can be obtained:

$$C_C \cdot C \cdot \rho = \frac{C_L^2 C_v \cdot Q}{C_L^3 \cdot V \cdot \Delta T \cdot C_P} = \frac{C_v}{C_L \cdot C_P} \cdot \frac{Q}{V \cdot \Delta T} \qquad (19)$$

$$C_C = \frac{C_v}{C_L \cdot C_p} \qquad (20)$$

3 PROJECT CASE

3.1 Modeling and meshing

When meshing the model, and parts of the mesh generation away from are rough, some parts near the wellbore are fine. This mesh can not only guarantee the accuracy of the calculation, but also save the computer storage space, as shown in Fig. 1.

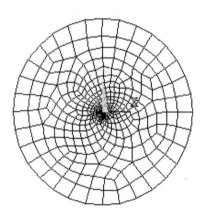

Figure 1. Model grid partitioning.

3.2 The model parameters

According to the similar indicators, the model parameters are set as shown in table 3, the similar constant as follows: $C_v=1.0\times10^{-5}(m{\cdot}s^{-1})/(W{\cdot}m^{-2})$; $C_L=1.0\times10^2$; $C_{K/\mu}=1.0\times10^{-10}m^2(Pa{\cdot}s)^{-1}/(W/m{\cdot}°C)^{-1}$; $C_P=1.0\times10^7Pa/°C$; $C_Q=1.0\times10^{-1}(m{\cdot}s^{-1})/(W{\cdot}m^{-2})$.

Table 3. Parameters of simulated oil field and ANSYS model.

Parameters of simulated oil field		Parameters of ANSYS model	
Supplyradius (m)	100	Supplyradius (m)	1
Well radius (m)	0.1	Well radius (m)	0.001
Permeability ($10^{-3}\mu m^2$)	4.75	Thermal conductivity (W/(m · °C))	0.021
Viscosity (mPa · s)	2.26		
Fluidity (μm^2(Pa · s))	2.10		
Original reservoir pressure (MPa)	44	Original model temperature (°C)	4.4
Fracture and wellbore pressure(MPa)	18.4	Fracture and the wellbore temperature (°C)	1.84
total compressibility (MPa⁻¹)	2.2E4	Specific heat apacity (J · kg−1 · K−1)	0.8876
Porosity	0.1	The density of formation (kg/m³)	2480

Table 4. The relationship between permeability and effective stress.

Effective stress (Mpa)	0	4	9	14	19	24
Pore pressure (Mpa)	44	40	35	30	25	20
K_1 ($10^{-3}\mu m^2$)	4.75	3.89	3.03	2.36	1.84	1.43
K_2 ($10^{-3}\mu m^2$)	4.75	3.82	2.90	2.20	1.67	1.27
K_3 ($10^{-3}\mu m^2$)	4.75	3.74	2.77	2.05	1.52	1.13

K_1—The permeability of the pressure sensitive coefficient is 0.05, 10^{-3} μm². K_2—The permeability of the pressure sensitive coefficient is 0.055, 10^{-3} μm². K_3—The permeability of the pressure sensitive coefficient is 0.06, 10^{-3}μm².

In the ANSYS finite element software, when you define material properties, adding the temperature options in the conductivity isotropic dialog, set the formation's thermal conductivity value at different temperature. According to the similarity principle, this set of data can be seen as the ratio of permeability and viscosity at different pressure. The Table 4 shows the relationship between effective stress and permeability .

3.3 History matching of the vertical well

The Fig. 2(a) shows the temperature distribution around the wellbore of the reservoir with

348

no pressure-sensitive, the Fig. 2(b) shows the temperature distribution around the wellbore with pressure-sensitive coefficient is 0.0551/MPa. We can see from the two pictures: when the reservoir with pressure sensitive, the drainage area around the well-bore is smaller.

The Fig. 3 shows the reservoir's pressure sensitive coefficient is greater,the phenomenon of pressure sensitive is more serious, the single well production capacity's loss is greater. The results also show that it is the most close to actual production performance when the pressure sensitive coefficient is 0.055. This result is close to actual measurement data, this means the ANSYS finite element method can simulate well the reservoir with pressure sensitive.

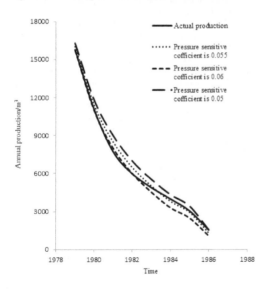

(a)   (b)

Figure 2.   The temperature distribution around the wellbore.

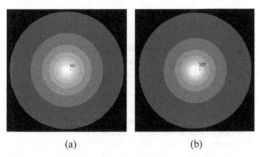

Figure 3.   The production fitting diagram of considering the effect of pressure sensitive.

3.4   *The failure of fracturing vertical wells*

In the low permeability reservoirs generally used fracturing vertical wells, the production

characteristics overall performance for the high production in the initial fracturing operation. As the wells produced, the formation pressure drops, the reservoir permeability decreases, the fractures appear the closure phenomenon, production wells show that the production has fallen sharply, until return to before fracturing, production of fracturing wells changes over time is called the validity period of the fracturing wells. Conventional simulation the effectiveness of fracturing wells is used modify reservoir parameter method in a period of time, in this paper, by using ANSYS transient heat conduction module effectively simulated fractured well all stages of the production process, it has simulated in the process of production, the formation pressure decreases and production declines in process of fractures closure phase.

The Fig. 4 shows the annual production comparison of considering pressure sensitive and does not consider the pressure sensitive. In can be seen from the figure, as the production proceeds, production of fracturing vertical well is falling, the effect of fracturing wells is gradually reduced.

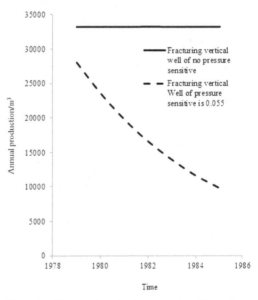

Figure 4.   The annual production comparison figure of considering pressure sensitive and does not consider the pressure sensitive.

4   CONCLUSIONS

1   Because seepage field is similar to temperature field, we can apply ANSYS thermal module to simulate seepage field problems.

349

2 According to the similarity between specific heat androck's integrated compression coefficient, this paper deduces the similar relationship between specific heat and integrated compression coefficient, and using the transient thermal module to simulate the seepage field.

3 Using the ANSYS finite element method can simulate the reservoir with pressure sensitive, the simulation results are close to the actual production performance.

4 ANSYS finite element software can well simulate the formation pressure decreases and fractures closure phase in the production process, it has provides a new approach to research and analyze the effectiveness of fracturing wells.

ACKNOWLEDGEMENTS

Authors thank Yumen oil field for permission to present the paper (2012E-3304) by the support of PetroChina Company Limited. We also wish to acknowledge the support provided by the Scientific Research and Technological Development Projects of China National Petroleum Corporation (2014B-1509).

REFERENCES

Spivak A, Price H S, Settari . 1976.A. Solution of the Equations for Multidimensional, two-phase. Immiscible Flow by Variational Methods [C] *SPEJ* ,Feb.1977:27–41(SPE5723).

McMichael C L, Thomas G W. 1973. Reservoir Simulation by Galerkin'sMethod[J]. SPE 3358.

Javandel I, Witherspoon P A. 1967.Application of the Finite Element Method to Transient Flow in Porous Media [C]. SPEJ, Sept. 1968: 241–252(SPE 2052, Sep).

Jong Gyun Kim, Milind D Deo. 2000.Finite Element, Discrete-Fracture Model for Muliphase Flow in Porous Media. *AICHE Journal*, 46(6):1120–1130.

Karimi-Fard, M. RERI, A. Firoozabadi. 1991.Numerical Simulation of Water Injection in 2D Fractured Media Using Discrete-Fracture Model[J] SPE 71615.

Hoteit, H. Firoozabadi, A. Compositional Modeling by the Combined Discontinuous Galerkin and Mixed Methods[C]. SPEJ, Mar,2006:19–34(SPE 90276,2004).

Geiger, S. Matthai, S. Niessner, J., Helmig, R. 2009. Black-oil simulations for three-component, three-phase flow in fractured porous media[C] SPEJ Jun:338–354(SPE 107485, 2007).

Hoteit, H., Firoozabadi, A. 2005. Multicomponent fluid flow by discontinuous Galerkin and mixed methods in un-fractured and fractured media. Water Resources Research.

FARQUHARRA,SMART B G D. 1993.Stress sensitivity of low- permeability sandstones from the rotliegendes sand stone[R].SPE 26501.

Zhu Shaopeng, Li Mao, Lao Yechun, Liu Shuangqi, Ye Miao. 2012. Studies on Reservoir Numerical Simulation Considering Pressure Sensitive Effect[J]. TUHA OIL& GAS,17(4).

Yang Manping , Li Y un . 2004.The analysis considering about rock stress sensitivity [J] . Natural Gas Geoscience, 15(6) : 601–603.

YuZhongliang,XiongWei.2007.GaoShusheng,etal.Stress-sens-Itivityoftightreservoiranditsinfluenceonoilfieldexploitation[J].ActaPetroleiSinica,28(4):95–98.

Chen Jinhui, Kang Yili, You Lijun, Fang Junwei. Review and Prospect about Study on Stress Sensitivity of Low - Permeability Reservoir. Natura L Gas Geoscience, 2011,22(1).

Liu Fei.2011. Design method of horizontal well fracturing[D].Xi'an Shiyou University.

Li Yaozhu, Lin Gang. 2009. ANSYS-based simulation of temperature field of numerical computation of homogeneous Earth Dam stability of seepage flow. *Port &Waterway Engineering*, 42 (4): 4821.

Jiang Tingxue, Lang Zhaoxin, Shan Wenwen. 2002.Finite element method for post fracturing production performance of wells in low permeability reservoir. *Acta Petrolei Sinica*, 23 (5): 53–58.

LvZhikai, Liu Guangfeng, He Shunli, Dong Kai. 2010. Finite element method of the influence of fracture morphology on horizontal well's productivity. *Science Technology and Engineering*, 10(25).

Su Jianzheng, Huang Zhiwen, Long Qiulian, Liu Changyin. 2012. ANSYS-based simulation of fracturing pressure reducing mechanism. *Oil & Gas Geology*, 33(4): 640–645.

Yang Jinhai,ZhengTianpu.The application of solid-state electric modeling technique in the simulation of and gas flow in an oil (gas) reservoir. Petroleum exploration and development,1994.21(4).

Computational Intelligence in Industrial Application – Ling (ed.)
© *2015 Taylor & Francis Group, London, ISBN: 978-1-138-02818-0*

An approach for the balancing and sequencing problem in the mixed-model U-line

L. Nie, X.G. Wang & K. Liu
School of Mechanical & Electronic Engineering, Shanghai Second Polytechnic University, People's Republic of China

C.T. Pang
Beijing Aviation Key Laboratory of Science and Technology on Precision Manufacturing, People's Republic of China

ABSTRACT: The Mixed-Model Assembly Line (MMAL) balancing and sequencing problem has been widely investigated by many researchers in the operations management field. However, in a practical application, U-shaped assembly line, a special kind of MMAL, is widely used. Due to the particularity of U-shaped assembly line, the methods that are designed for linear assembly line cannot be employed in U-shaped assembly line directly. In order to meet the needs in a practical application, to develop new methods that consider the features of U-shaped assembly line is necessary. The paper mainly studies the integrated approaches for balancing and sequencing problem in a mixed-model U-line with the objective of minimizing ADW (Absolute Deviation of the Workstation). A mathematical programming model has been developed for describing the problem. In addition, a two-stage Multi-Population Genetic Algorithmv (MPGA) is proposed which is incorporated with improved encoding and decoding scheme and crossover operators. Experiments prove the effectiveness and convergence of the algorithm.

1 INTRODUCTION

In mixed-model production, different products or models are produced on the same line with the models varying throughout the production sequence. The U-lines on which mixed model production is performed are called mixed-model U-lines (MMUL). MMUL is a special kind of mixed-model assembly line (MMAL) and is widely used in practical application. The main difference between straight lines and U-lines is that, in U-line the tasks can be assigned from the front to the back and from the back to the front simultaneously, which is impossible for a straight line. Mixed-model U-line production requires the solutions to the following two problems: (1) Balancing problem (BP): How are production tasks assigned to stations on the line? (2)Sequencing problem (SP): In what a sequence are the different models produced on the line? Numerous studies have investigated the BP and the SP separately in a hierarchical manner. They assumed the assembly line is balanced before discussing the product sequencing problem or study how to reach the balance of the assembly line when the models' processing sequence is deterministic. For example, McMullen and Frazier (2000) investigated SP under the assumption that the BP had been solved. Since 2002, more and more researchers

have begun to consider the sequencing and balancing problem in an integrated way. Researches that investigated the BP and SP problems simultaneously have appeared successively. However, they were mainly intended for straight line assembly line. In practice, due to the particularity of U-shaped assembly line, the methods that are designed for linear assembling line cannot be employed in U-shaped assembly line directly. It is important and interesting to consider the two problems simultaneously for MMUL. Kim et al. (2006) proposed an evolutionary algorithm to deal with both balancing and sequencing problems in mixed-model U-shaped lines. Kara et al. (2007) dealt with the balancing and sequencing problems of mixed-model U-lines simultaneously to minimize the number of workstations. Lian et al. (2012) dealt with the balancing and sequencing problems of mixed-model U-lines simultaneously to minimize the ADW. Hamzadayi and Yildiz (2012) minimized the number of workstations and smoothing the workload between-within workstations at the end of all cycles for MMUL.

The paper intents to solve the balancing problem and the sequencing problem of MMUL with the objective of minimizing ADW. Due to the balancing and sequencing problem is more complicated for a U-line, meta-heuristic based on genetic algorithm is designed. The remainder of this paper is organized as follows. Sect. 2 describes the mathematical programming model to formulate the balancing and sequencing problem of MMUL. Sect. 3 presents meta-heuristic based on GA solution procedures. A comprehensive computational study in Sect. 4 investigates algorithmic performance. Finally, Sect. 5 concludes the paper.

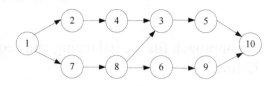

Figure 1. Combined precedence diagram.

2 PROBLEM FORMULATION

First, take an example to explain how the objective of ADW changes under the interaction of balancing and sequencing problem in MMUL. In the sequencing of products, the Minimal Production Set (MPS) concept is used. Therefore, only optimizing the sequence for the products in one MPS is considered. Products in one MPS are denoted with $(d_1, d_2, \ldots, d_M)$, where M is the number of models, d_m is the number of products in the mth model that needs to be assembled in one MPS. Assumed that the MMUL which is comprised with 4 workstations assembles 3 models, the demand of each model in one MPS is 1, 3 and 2, respectively. Then, there are 6 products to be assembled in one MPS. The tasks constituting these 3 models and the processing time for each task is listed in Table 1. The combined precedence diagram is shown in Fig. 1. Two candidate balancing patterns are shown in Fig. 2, which both meet the precedence relations of the example problem. Table 2 lists the ADW under different balancing and sequencing pattern. From Table 2 it is found that both the balancing pattern and sequencing pattern have an effect on ADW. Therefore, the effect of balancing and sequencing on ADW in MMUL should be considered simultaneously.

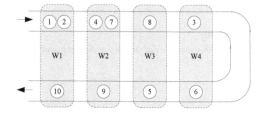

(a) Balancing pattern 1

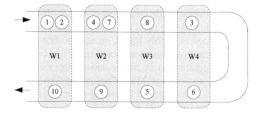

(b) Balancing pattern 2

Figure 2. Two balancing pattern for the MMUL.

Table 2. ADW under different balancing and sequencing pattern.

	Balancing pattern 1	Balancing pattern 2
Sequencing pattern 1: ABCBCB	85.67	83.67
Sequencing pattern 2: BBABCC	91.5	85.83

Table 1. Processing time of each task for each model.

	T1	T2	T3	T4	T5
A	2	0	4	5	3
B	10	3	0	2	6
C	6	1	7	0	9

	T6	T7	T8	T9	T10
A	0	2	8	1	5
B	7	2	0	1	5
C	5	0	2	4	4

The problem in the paper is: given a MPS and the precedence constraints, and the number of stations is limited, how to find an optimal solution of task assignment and model production sequence with the objective of minimizing ADW.

Notation:

ST	Set of all tasks;
N	Number of all tasks;
SP	Set of all precedence relations of tasks for each model;
K	Number of workstations;
SM	Set of all model;
M	Number of different models produced in MMUL;
t_{im}	Processing time for task i of model m;
d_m	Demand for model m in one MPS;

S Number of products in on MPS;

STF_k Set of tasks in workstation k located on the front of the U-line;

STB_k Set of tasks in workstation k located on the back of the U-line;

f_{kc} Model produced on the front of workstation k at production cycle c;

b_{kc} Model produced on the back of workstation k at production cycle c;

T_{ave} Average workload of workstation;

T_{kc} Workload of workstation k at production cycle c;

x_{mt} 1, if model m is i^{th} launch MMUL; 0, otherwise.

Average workload of workstation T_{ave} is defined below.

$$T_{ave} = \frac{1}{K \times S} \sum_{i=1}^{N} \sum_{m=1}^{M} d_m t_{im} \tag{1}$$

Workload of workstation k at production cycle c T_{kc} is defined below.

$$T_{kc} = \sum_{i \in STF_k} t_{if_{kc}} + \sum_{i \in STB_k} t_{ib_{kc}} \tag{2}$$

The model for the integrated problem of the balancing and the sequencing problem of MMUL can be formulated as follows:

$$\text{Min } ADW = \sum_{k=1}^{K} \sum_{c=1}^{S} | T_{kc} - T_{ave} | \tag{3}$$

$$\bigcup_{k=1}^{K} STF_k \cup STB_k = N \tag{4}$$

$$STF_k \cap STB_k = \varnothing, k = 1, ..., K \tag{5}$$

$$STF_k \cap STB_{k'} = \varnothing, k \neq k' \tag{6}$$

$$\forall (i, j) \in SP, if\ i \in STF_k, j \in STF_{k'}, then k \leq k';$$
$$if\ i \in STB_k, j \in STB_{k'}, then k \geq k'; \tag{7}$$

$$\sum_{m \in SM} x_{mt} = 1 \tag{8}$$

$$\sum_{t=1}^{S} x_{mt} = d_m \tag{9}$$

Objective function is denoted in Equation (3). Constraint (4) assures that each task must be allocated to at least one workstation. Constraint (5-6) guarantees that each task can only be assigned to the front or back of at most one workstation. Constraint (7) assures that the precedence relations must be satisfied. Constraint (8) ensures that exactly one model is assembled per production cycle. Constraint (9) enforces the products to be produced in the demanded quantities.

3 META-HEURISTIC FOR MMUL

3.1 Flow framework

In this paper, we propose a two-stage multi-population genetic algorithm (MPGA) to solve the balancing and sequencing integrated problem in the mixed-model U-line. In the first stage, two populations evolve for BP and SP problems respectively. Considering the influence on ADW due to the interaction between the BP and SP problems, the individuals in the two populations of the first stage are combined together and are arranged into several sub-populations, which become the initial populations of the second stage. Each sub-population then evolves separately while an elitist strategy preserves the best individuals for the integrated BP and SP problems. Flow chart is illustrated in Fig. 3.

3.2 Solution representation

Due to balancing problem and sequencing problem are considered separately in the first stage, two chromosomes of two populations which are shown in Fig. 4 (a) and (b) represent the encoded solution for balancing problem and sequencing problem, respectively. Fig. 4 (a) illustrates the chromosome encodes for the balancing pattern 1 and Fig. 4 (b) shows the chromosome encodes for the sequencing pattern 1 for the example given in Sec. 2. Note that task sequence

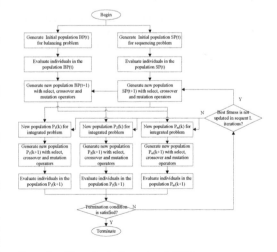

Figure 3. Flow chart of two-stage MPGA.

in the Fig. 4(a) must satisfy the precedence constraints among tasks and station sequence is a non-decreasing sequence with the same length of task sequence. The value of each element in station sequence belongs to [1, 2*k*], where [1, *k*] and [*k*+1,2*k*] represent the front and back of the corresponding station, respectively.

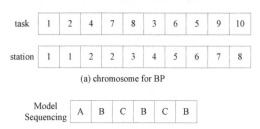

(a) chromosome for BP

| Model Sequencing | A | B | C | B | C | B |

(b) chromosome for SP

Figure 4.   Solution representation scheme.

3.3   *Initial population construction*

In order to ensure the diversity of the initial population in the first stage, the chromosomes in the population for BP and SP problem are generated randomly according to the encoding scheme described above, respectively. However, the task sequence and station sequence in the chromosomes for the BP problem may be infeasible in the initial generation procedure, crossover and mutation operation because the precedence constraints among tasks may be violated and the non-decreasing limitation of station sequence may be not satisfied. Therefore, a repairing procedure which adjusts an infeasible task or station sequence into a feasible one is necessary which is omitted because of the space limitation.

3.4   *Crossover operator*

Fig. 5 (a) and (b) illustrate the crossover procedure for BP and SP problem, respectively. First, taking arbitrary two chromosomes (Parent1 and Parent2) from

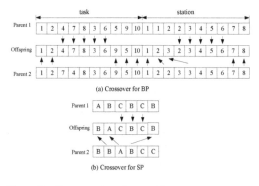

(a) Crossover for BP

(b) Crossover for SP

Figure 5.   Crossover operator procedure.

mating pool that are generated by selection operator; Randomly selecting two crossover point in task part and station part, respectively; Dividing task part and station part of the two parents into three parts by the two crossover points; Taking the middle part of Parent1 directly into offspring, deleting the corresponding genes in the middle part from the Parent2, and inserting the rest the genes in Parent2 into offspring consecutively. Note the offspring is infeasible because the station sequence does not meet the non-decreasing limitation and it must be adjusted with the repairing procedure mentioned above.

4   EXPERIMENTAL RESULTS

We randomly generated different numbers of MPS (see Table 3) for problems in order to evaluate the performance of the proposed algorithm. The number of tasks generated randomly with an average value of 10. Task processing times of the tasks are randomly generated from [1, 20] for being able to better represent general characteristics of the proposed approach. The proposed algorithm is coded in VC6.0 and run on a 3.00 GHz Pentium 4 computer. The parameters that are used in the test problems are set as below. Population is 100, maximum iteration is 100, probability of crossover and mutation is 0.7 and 0.03, respectively. The experiment is repeated five times for every test problem and the minimum, mean and maximum value of the solutions are shown in Table 4, respectively.

From Table 4, the algorithm exhibits excellent convergence on different size problems. And it is also found that the value of ADW may be varying according to different MPSs even the number of models and jobs are the same. This situation indicates that the sequences in which different models are produced have an effect on the balancing and sequencing integrated problem. In fact, this is because of the difference of the combination of models assigned to workstations according to varying task assignments.

Table 3.   Test problems.

Problem	Number of models	Number of jobs	MPS
1	3	3	{1,1,1}
2	3	5	{1,2,2}
3	3	5	{3,1,1}
4	4	4	{1,1,1,1}
5	4	8	{1,2,2,3}
6	4	8	{2,2,2,2}
7	5	5	{1,1,1,1,1}
8	5	11	{1,2,2,3,3}
9	5	11	{1,4,1,3,2}

Table 4. Results of experiments.

Problem	Min.	Average	Max.
1	1.1263	1.1263	1.1263
2	2.5317	2.5317	2.5317
3	2.4572	2.4572	2.4572
4	2.4714	2.4714	2.4714
5	3.4261	3.4471	3.4923
6	3.5459	3.5736	3.6219
7	2.9852	3.0117	3.1987
8	4.7362	4.7592	4.7952
9	5.1298	5.3834	5.5012

5 CONCLUSION

This paper proposes a novel genetic algorithm MPGA that solves the mixed-model U-line balancing and sequencing (MMUL/BS) problem. The innovations of this paper are: We combine the balancing problem and the sequencing problem of the U-type assembly line when products are to be put into production and consider them as a whole. The objective function is beneficial for improving workers' working efficiency and reducing the bottlenecks of the production line. To effectively solve the discussed problem, MPGA solves the integrated problem in a two-stage multi-population evolutionary pattern. A new representation scheme for MMUL/BS problem is suggested. This representation scheme is simple to implement and efficient in solving the MMUL/BS problem.

Computational experiments are conducted to validate the performance of the proposed algorithms.

ACKNOWLEDGEMENTS

This research is supported by the Funding Projects for Teachers in Universities of Shanghai under Grant No. ZZegd12026, the Innovation Program of Shanghai Municipal Education Commission under Grant No. 14YZ155 and the Key Cultivating Academic Discipline Projects of Industrial Engineering of Shanghai Second Polytechnic University under Grant No. XXKYS1403, No. XXKPY1311.

REFERENCES

Hamzadayi, A. & Yildiz, G. 2012. A genetic algorithm based approach for simultaneously balancing and sequencing of mixed-model U-lines with parallel worksta-tions and zoning constraints. *Computers & Industrial Engineering* 62: 206–215.

Kara, Y.; Ozcan, U. & Peker, A. 2007. An approach for bal-ancing and sequencing mixed-model jit u-lines. *The International Journal of Advanced Manufacturing Technology* 32: 1218–1231.

Kim, Y.K.; Kim, J.Y. & Kim, Y. 2006. An endosymbiotic volutionary algorithm for the integration of balanc-ing and sequencing in mixedmodel u-lines. *European Journal of Operational Research* 168: 838–852.

Lian, K. & Zhang, C. 2012. A modified colonial compet-itive algorithm for the mixed-model U-line balancing and sequencing problem. *International Journal of Production Research* 18(50): 5117-5131.

McMullen, P.R. & Frazier, G.V. 2000. A simulated anneal-ing approach to mixed-model sequencing with multiple objectives on a jit line. *IIE Transactions* 32(8): 656–679.

Computational Intelligence in Industrial Application – Ling (ed.)
© 2015 Taylor & Francis Group, London, ISBN: 978-1-138-02818-0

Study on evaluation system of industrial clusters innovation ability

MengJi Yin

School of Business Administration, Zhengzhou Institute of Aeronautical Industry Management, Zhengzhou, China

ABSTRACT: This article built an evaluation index system for the innovation ability of industrial clusters, and optimized the index system and determined weight of each index by factor analysis. We evaluated the innovation ability of industrial clusters by fuzzy comprehensive evaluation.

KEYWORDS: Industrial clusters, Innovation ability, Evaluation system, Fuzzy comprehensive evaluation.

1   INTRODUCTION

As a common economic phenomenon, industry cluster has attracted the attention of many economists back in the early twentieth Century. As a kind of industrial space organization form, industry cluster is a kind of institutional arrangements conducive to innovation and constructing innovation advantage, having group competition and benefit of scale economies of agglomeration. Industrial cluster is an effective way to enhance regional innovation capability, and is the important carrier of regional innovation system. It plays an important role in the construction of regional innovation system and regional economy development. Today the regional economic competition is becoming increasingly fierce, market competition changes from the enterprise layer to industrial clusters layer. Cluster development has become the trend of the world economic development, and industrial cluster has become an important part of regional economy. Therefore, all countries and regions pay attention to the development of industrial clusters, implement active cluster initiative, and formulate regional economic development policy based on industrial cluster.

At present, in many economy areas of our country, especially the coastal area and Jiangzhe area appear a large number of industrial clusters. They have effectively promoted technological progress and diffusion and promoted the development of local industry and economy. However, the development of industrial clusters in our country is very unbalanced, and most of the industrial clusters consist of small and medium-sized enterprises, mainly engaged in labor-intensive industries. The technology level is low and the product grade is not high, industrial clusters lack innovation ability [1-2,4].

Therefore, the analysis and comprehensive evaluation of the innovation ability of industrial clusters will help enterprises realize their own development environment, understand their own innovation and development situation scientifically, and formulate reasonable strategy of innovation, strengthening the competition advantage. It is also conducive to understand industrial clusters development situation under the jurisdiction objectively for government departments and formulating reasonable policy to stimulate the development of industrial clusters [5-6].

2   CONSTRUCTION OF EVALUATION INDEX SYSTEM OF THE INNOVATION ABILITY OF INDUSTRIAL CLUSTERS

The innovation ability of industrial clusters is defined as the comprehensive ability to transform innovative ideas into new products, new processes and new service in a range of industrial cluster, through professional division of labor and cooperation, giving full play to the enthusiasm of innovation main body of industrial clusters, allocating innovation resources of industrial clusters efficiently. It is the organic integration ability of industrial clusters to knowledge accumulation, learning ability, competition and cooperation ability, innovation ability and development ability [3,5-6].

Evaluation index system of innovation ability of industrial clusters will take the innovation theory as guidance, combined with the characteristics of industrial clusters, according to the principles of system comprehensive, scientific and practical, operational, flexibility, combination of qualitative and quantitative to build a multi-level evaluation index system. The construction mentality of the index system, first through the methods of literature review, in-depth interview, preliminarily explore the factors of influencing industrial clusters innovation ability evaluation and its terms; then pre test in a small range and remove part of the unnecessary or unimportant terms; finally, in large range of formal tests, carry out questionnaire survey, and conduct correlation coefficient analysis and factor analysis of the data, the index system optimized and getting the factor load of each index. See Table 1.

This paper constructs evaluation index system of industrial clusters innovation ability from four aspects.

A. Technology level

Technology level reflects the overall technical strength of industrial clusters and is the main factor to decide the innovation ability of industrial clusters. Through its effect on levels and depth of technology innovation activity, it restricts the development of industrial cluster innovation activities. Technology level is mainly reflected in the following six aspects: the number of new technology use, technology innovation diffusion degree, technical personnel proportion, degree of information, the number of new product development and enterprise equipment updated speed.

B. Innovation performance

The innovation ability of industrial clusters must ultimately reflect on the innovation performance. The innovation performance of industrial clusters is measured by four indicators, the number of high-tech enterprise in the industrial cluster, patent application and grant amount, economic benefit of industry cluster and the contribution to regional economy.

Table 1. Factor analysis results of each term.

ORDER	Index	Main factor1 Technology level	Main factor2 Innovation performance	Main factor3 Innovation input	Main factor4 Innovation environment
1	The number of new technology use	0.8157			
2	Technology innovation diffusion degree	0.8233			
3	Technical personnel proportion	0.7028			
4	Degree of information	0.5943			
5	The number of new product development	0.7115			
6	Enterprise equipment updated speed	0.6627			
7	The number of high-tech Enterprises in the industrial cluster		0.7644		
8	Patent application and grant amount		0.8594		
9	Economic benefit of industry cluster		0.6452		
10	The contribution to regional economy		0.5769		
11	R&D funding			0.7422	
12	R&D funding proportion of sales			0.8026	
13	Training expenditure per capita of the enterprise in cluster			0.6564	
14	The number of intermediary service institutions				0.6297
15	The number of universities and research institutions				0.6318
16	government support				0.7065
17	Innovation culture in industrial cluster				0.5853
	variance contribution	40.8872	21.2565	13.4427	8.0693

C. Innovation input

Innovation investment fund has great influence on the innovation ability of industrial clusters. R&D funds are necessary to guarantee the enterprises in the cluster to have continuous innovation ability. R&D funding, R&D funding proportion of sales, training expenditure per capita of the enterprise in industrial cluster, the three indicators reflect the intensity of industrial clusters innovation investment.

D. Innovation environment

Innovation environment is the external forces of effective operation of industry cluster innovation activities. Environment has a decisive influence on the innovation ability of industrial clusters at the premise of the hardware set. Innovation capability of the whole industry cluster can be upgraded only with a good innovative environment and atmosphere. From the process of innovation, technical strength decided by enterprise, government, intermediary service institutions, universities and research institutions, local culture and values, has an important influence on the innovation ability of industrial cluster. Therefore, the number of intermediary service institutions, the number of universities and research institutions, government support, innovation culture in industrial cluster, the four indicators are the basis of industrial cluster innovation activities and decide the quality of innovation environment [5-7].

3 EVALUATION OF INNOVATION ABILITY OF INDUSTRIAL CLUSTERS BY FUZZY COMPREHENSIVE EVALUATION METHOD

Assume now we need to evaluate the innovation capability of a certain industry cluster to provide reference for investors or the relevant government departments to formulate policies.

A. Establish evaluation index set.

According to the evaluation index system of innovation ability of industrial clusters established above, we can get factor set

$$E = \left\{ \begin{array}{l} technology\ level\ I_1, innovation\ performance I_2, innovation\ input I_3, \\ innovation\ environment I_4 \end{array} \right\}$$

B. Determine the fuzzy weight vectors for factor set:

$$A = (a_1, a_2, a_3 \ldots a_m),\ A_i = (a_{i1}, a_{i2}, a_{i3} \ldots a_{im}).$$

In the second part we have already mentioned, the index weight is determined by factor analysis method, which has larger objectivity and superiority compared to the traditional analytic hierarchy process.

On the basis of getting each index factor load, the secondary indexes weights are determined according to each main factor's variance contribution ratio of factor analysis, the relative weights of the level 3 indexes are determined by its factor load size, taking "the number of new technology use" for example, its weight is 0.8157/(0.8157+0.8233+0.7028+0.5943+0.7115+0.6627). The specific calculation process no longer lists because of the length of this article. The eventual weight of index for evaluation of innovation ability of industrial clusters is showed in Table 2.

C. Determine comment set:

$V = \{v_1, v_2, v_3 \ldots v_n\}$, v_j (j=1,2,...,n) represents the comments from high to low. Fuzzy comprehensive evaluation is mainly to take all the factors that have influence on the evaluation object into comprehensive consideration, so it can get the optimal evaluation result from comment set V. To get more accurate evaluation result, the comment level should be better, then the corresponding evaluation process will be complex, uneasy to grasp, so comment level should be appropriate related to evaluation problem. Considering the technical side, simple and feasible, this paper adopts comment set at five levels, $V = \{v_1, v_2, v_3, v_4, v_5\}$ = {best, better, common, worse, worst}.

D. Establish the fuzzy relation matrix R

Due to the complexity of evaluation index of innovation ability of industrial clusters and the existence of a lot of qualitative or quantitative index, we take the way of expert judging and scoring to determine. Set the experts number as X, x_{ij} is the number of experts who judge factor I_i as comment level v_j, so:

$$r_{ij} = \frac{x_{ij}}{X} \tag{1}$$

We can get R from the formula $R = (r_{ij})_{m \times n}$

We invite relevant authority staffs, scholars, enterprise managers to evaluate the testing industrial cluster and screened 10 effective questionnaires, after finishing showed in table 2.

According to the fuzzy comprehensive evaluation model,

$$R_1 = \begin{bmatrix} 0 & 0.2 & 0.6 & 0.2 & 0 \\ 0 & 0.1 & 0.6 & 0.2 & 0.1 \\ 0.2 & 0.5 & 0.3 & 0 & 0 \\ 0.1 & 0.4 & 0.3 & 0.2 & 0 \\ 0.2 & 0.3 & 0.4 & 0.1 & 0 \\ 0 & 0.2 & 0.5 & 0.2 & 0.1 \end{bmatrix}$$

$$A_1 = (0.1892, 0.1910, 0.1631, 0.1379, 0.1651, 0.1537)$$

$$B_1 = A_1 \circ R_1 = (0.07943, 0.27392, 0.46131, 0.15087, 0.03447)$$

Table 2. Index weight and evaluation summary of industrial clusters innovation ability.

Secondary index and weight	Third-class index	Evaluation opinions					Third-class index weight
		Best	Better	Common	Worse	Worst	
Technology level (0.4888)	The number of new technology use	0	2	6	2	0	0.1892
	technology innovation diffusion degree	0	1	6	2	1	0.1910
	technical personnel proportion	2	5	3	0	0	0.1631
	degree of information	1	4	3	2	0	0.1379
	The number of new product development	2	3	4	1	0	0.1651
	enterprise equipment updated speed	0	2	5	2	1	0.1537
Innovation performance (0.2541)	the number of high-tech enterprises in cluster	0	2	6	1	1	0.2686
	patent application and grant amount	0	0	3	5	2	0.3020
	economic benefit of industry cluster	1	3	6	0	0	0.2267
	the contribution to regional economy	1	2	5	2	0	0.2027
Innovation input (0.1607)	R&D funding	1	5	3	1	0	0.3372
	R&D funding proportion of sales	1	3	3	2	1	0.3646
	training expenditure per capita of the enterprise in cluster	2	4	3	1	0	0.2982
Innovation environment (0.0964)	the number of intermediary service institutions	2	5	3	0	0	0.2466
	the number of universities and research institutions	1	3	3	2	1	0.2475
	government support	2	6	2	0	0	0.2767
	innovation culture in industrial cluster	2	5	1	1	1	0.2292

$B_2 = A_2 \circ R_2 = (0.04294, 0.16227, 0.48913, 0.21840, 0.08726)$

$B_3 = A_3 \circ R_3 = (0.12982, 0.39726, 0.3, 0.13646, 0.03646)$

$B_4 = A_4 \circ R_4 = (0.17525, 0.47817, 0.22649, 0.07242, 0.04767)$

E. The secondary fuzzy comprehensive evaluation. Making compound operation, we get the comprehensive evaluation result:

$$B = A \circ R \qquad (2)$$

Let the vectors B_1, B_2, B_3, B_4 above be the fuzzy relation matrix of secondary fuzzy evaluation, then

$$R = \begin{bmatrix} 0.07943 & 0.27392 & 0.46131 & 0.15087 & 0.03447 \\ 0.04294 & 0.16227 & 0.48913 & 0.21840 & 0.08726 \\ 0.12982 & 0.39726 & 0.3 & 0.13646 & 0.03646 \\ 0.17525 & 0.47817 & 0.22649 & 0.07242 & 0.04767 \end{bmatrix}$$

while $A = (0.4888, 0.2541, 0.1607, 0.0964)$

then $B = A \circ R = (0.08749, 0.28506, 0.41982, 0.15815, 0.04948)$

F. Analysis of fuzzy comprehensive evaluation results

For a variety of different evaluation results, there are different corresponding processing methods. Processing methods commonly used in practice are the simple average method, the maximum membership degree method and weighted average method. Simple average method fits the condition that the evaluation results are constants, so we choose the weighted average method for the need to quantify the result of fuzzy comprehensive evaluation. Because the problem we want to solve is an evaluation of innovation capability of industry cluster, not choosing one from multiple objects, we use the maximum membership degree method. This method is relatively simple in practical application, the rationality relatively strong.

Set the fuzzy comprehensive evaluation results as $B = \left(b_1, b_2, b_3, ..., b_n \right)$, $b_s = \max \left\{ b_j \right\}$, Then fuzzy evaluation result of innovation ability of the industrial cluster is b_s, belonging to the comment level s. From secondary fuzzy comprehensive evaluation results above, $\max \left\{ b_j \right\} = 0.41982$, $s = 3$, we can judge that the innovation ability of the industrial cluster belongs to "common".

4 CONCLUSIONS

There is a very important practical significance to evaluate the innovation ability of industrial clusters. Combining with the meaning of innovation ability of industrial clusters and the development status of Chinese industry clusters, based on reviewing literature and interviewing, this article constructed evaluation index system of innovation ability of industrial clusters, optimized index system and determined the index weight with factor analysis method. Then evaluated the innovation ability of industrial clusters by fuzzy comprehensive evaluation method and introduced the whole evaluation process. The evaluation system is simple, strongly rational and operational, and has certain application value in reality.

REFERENCES

Chen Linsheng. Study on the construction of regional innovation system promoted by industry clusters. *Exploration of Economic Problems*. (4): 108–110, 2005.

Zhao Zhonghua. The innovative industrial clusters network structure and Performance Research. Harbin Institute of Technology Press. pp1–72, 2009.

Meng Fang. Construction of evaluation model of characteristic industrial cluster innovation ability. *Commercial Times*. 2011, (7):116–117.

Yang Dongmei, Zhao Liming, Chen Liuqin. Construction of regional innovation system based on industrial clusters. *Science of Science and Management of S, T*, 2005 (10): 79–81.

Hu Bei, Gu Jiajun. Empirical study on industry cluster innovation ability evaluation based on back propagation neural networks. *Science & Technology Progress and Policy*, 2008 (7): 144–147.

Gu Jiajun, Xie Fenghua. Empirical study on industry cluster innovation ability evaluation based on AHP. *Science and Technology Management Research*, 2008 (6): 480–481.

Liu Feng, Lin Tao, Gong Lufang. Study of fuzzy comprehensive evaluation of industrial clusters innovation ability. *Jiangxi Science*, 2007, (6): 334–337.

Computational Intelligence in Industrial Application – Ling (ed.)
© 2015 Taylor & Francis Group, London, ISBN: 978-1-138-02818-0

New construction building epidermis digitized

Ying Li
Jilin Architectural University, Jilin, China

ABSTRACT: With the economic development, China has become a powerhouse in the development of increasing awareness of the importance of environmental protection, under the high-tech industrial society and the ecological environment worsening conflict, environmental protection reforms in various industries are a long way. Development of the construction industry is a high-energy-intensive industry, in construction decoration materials, construction, renovation process will consume a lot of energy. This paper describes the green interior design connotation and denotation, and interior designers from all angles argument should establish green interior design concept is based.

KEYWORDS: Green Design; Indoor Environment; Sustainable Development.

In recent years, "green design" concept swept up in our businesses, but it comes with almost no human-related industries inevitably began to promote the principles of a "sustainable development". Popular these ideas will undoubtedly expresses the relationship between the human self and the external environment philosophy.

1 ON THE "GREEN INTERIOR DESIGN" CONCEPT ANALYSIS

In the interior decoration industry, "green" design has a vague understanding of who are the minority, in order to "green interior design," the concept of a more precise definition, the following a few common misconceptions will be explained and corrected.

2 GREEN INTERIOR DESIGN IS NOT EQUAL TO "HIGH-TECH"

As we all know, the use of high-tech green building is an important means to achieve the goal of green building technology, when construction is complete the enclosed indoor space began to produce. "Green design" has its material standards: standards for indoor air, the designer may not use cheap and hazardous materials on the human body; indoor lighting to rationality and structure, designers are not free to change the size of the window opening to facilitate appropriate room temperature, designers must consider the heating, air conditioning settings and so on. These mandatory targets or principles for the use of materials, construction techniques proposed

mandatory requirements that green interior must meet the "flexible and efficient" and "health and comfort", "save energy", "environmental protection" of these four basic elements. This paper argues that meeting these four contents does not necessarily have to be "sophisticated" techniques. Because the combination of these four elements, the green interior decoration must be "suitable" and serve humanity. One pair of air-conditioning duct allergies, living in borrowed by high-tech means to achieve green building ventilation system running inside, obviously not as good as living in the natural ventilation of the room comfortable, low-skilled and often better than the high-tech nature of effective coordination. So in a "humane principles" dialectical designers should treat high-tech and low-tech, while conducting interior design, to both harmony and unity, in order to improve the quality of people's living environment that goal.

3 GREEN INTERIOR DESIGN IS NOT NECESSARILY THE ECONOMY

In China, green interior design is often a compromise. Because we cannot afford not to consider those energy-efficient materials and equipment to bring down the cost, but cannot because of the price factor and the reduction of the value of the works. In fact, under different conditions, the designer of the "green" design is always looking for the highest price the critical point. Since the owner is subjected to the wishes of cost estimates and a variety of factors, designers often cannot do everything. For example, designers will meet the standards of air quality index in the premise of design objects controllable adjustment is necessary, the use of non-toxic green materials certainly increases the material cost, but can be used to reduce other unnecessary decoration "throttling. "

You can say that the real green interior design is the most cost-effective work. Although the use of renewable raw materials may increase the difficulty of technology and construction, design and production time will subsequently be extended, but in the long run, this temporary pay can be rewarded after the operation, and its indirect the social effects and economic benefits completely incalculable. So, here designers proposed a requirement that the interior space to create a process must not only consider the immediate, but the vision must be liberalized, considering the long-term perspective.

Green interior decoration or a process of sustainable development involves many factors of economic, regulatory, social and cultural. Designers on the formation of human life, the environment cannot shirk responsibility, so designers should have a good work ethic norms judgment and behavior, responsible for themselves and for the customer is responsible, is responsible for optimizing the living environment of mankind.

4 GREEN INTERIOR DESIGN PRINCIPLES OF SUSTAINABLE DEVELOPMENT

4.1 The principle of people-oriented energy conservation

Green interior design services for the owners is an activity, the object is human must meet the physical and psychological needs of people. Green interior design is in harmony with the environment not only in material terms, but also in the spirit of the human dimension of aesthetic. Designers should not only ensure the creation of an environment can bring convenience for people living and working environment, but also in the construction of the production process focused on changes in the environment caused by a series of consequences, but also pay attention to whether they can create space for the occasion owners bring a ray of consolation and comfort in mind, from this point of view, green interior design to create a place of green principles in the gas field, and more likely to be recognized and accepted.

Energy conservation is a core principle of physical materials savings and fully utilized. Redesign process will be environmental factors and pollution prevention measures into planning, to reduce material and energy consumption, reduce emissions of harmful substances. How to arrange the construction process of construction, the use of construction materials is part of what green design requires consideration. The interior environment is the biggest energy equipment demand from the physical environment, in order to conserve energy; the designer should choose the right equipment to match the space.

4.2 The principle of less is more

Visual effects designer in order to pursue a variety of designs so that the blind will be stacked vocabulary, not only brings beauty, but will form a visual pollution, people feel cluttered. Indeed, the needs of the owners are the first one, like the luxurious style of people in today's society is still not confused minority, but whether because of the requirements of owners, designers should give up the pursuit of truth it? If the owners in order to pursue luxurious atmosphere, requiring the demolition of a wall, which communicated with the two spaces, but this wall is load-bearing wall building body, in accordance with the structure of rationality, it should not be removed, if the split then after the wall reinforcement and repair, not only does not help, but will increase the burden on other structures, causing a huge waste, are more likely to cause structural problems. Then designers should play the role of values correctly guide the owners, to implement the correct and environmentally friendly design concepts, creating efforts in space rather than on the real decoration efforts. Great architect Mies once said, "less is more" when "less" when we will be reduced to a reasonable space environment, which contains the meaning is often "more". This design philosophy, it should be the criteria of green interior design aesthetic, but also the construction guidelines.

5 CONCLUSION

Green design in our atmosphere is getting into a very bad situation, consider green design is not just technical level, it is more important to change the concepts, the designer cannot simply consider the space unique, must fully consider the entire life cycle of space science and technology in the economy and the rapid development of today, green design ideas will be fresh blood design industry for sustainable development.

REFERENCES

SOUTHERN. On the green decoration design errors Chinese residential facilities. 2006.
Gui Lai modern interior design ecological values and its application] Hulunbeier College. 2006.

Computational Intelligence in Industrial Application – Ling (ed.)
© *2015 Taylor & Francis Group, London, ISBN: 978-1-138-02818-0*

Discovering scheduling rules with a machine learning approach based on GEP and PSO for dynamic scheduling problems in shop floor

X.G. Wang, L. Nie & Y.W. Bai
School of Mechanical & Electronic Engineering, Shanghai Second Polytechnic University, People's Republic of China

ABSTRACT: Scheduling problems that exist widely on the practical shop floor are dynamic scheduling problems because there are always all kinds of random and unpredictable events that occur. To solve these problems, developing customized scheduling rules is an effective way. In the paper, a new approach based on GEP (Gene Expression Programming, GEP) and PSO (Particle Swarm Optimizing algorithm, PSO) is proposed to discover effective customized scheduling rules according to the feature of shop configurations, operating conditions and performance objectives. First, a new chromosome structure named DSS (Double String Structure) is proposed. Then, a hybrid algorithm is designed which incorporates PSO into GEP's framework based on the new chromosome structure DSS. The results of experiments demonstrate that the efficiency of evolution is improved markedly compared to the previous approach.

1 INTRODUCTION

Scheduling plays an important role in a shop floor control system, which has a significant effect on the performance of the shop floor. Scheduling problems that exist widely in practical shop floor are dynamic scheduling problems because there are always occurrence of all kinds of random and unpredictable events. For example, jobs arrive over time, machines break down randomly and the due dates of jobs are changed during processing. Although many static scheduling approaches are available in literatures, they are not suitable for dynamic scheduling problems because the assumptions that jobs arrive simultaneously, machines are available all the time and the attributes of jobs or machines are unchanged are always made.

Scheduling rules are one of the most prevalent approaches in dynamic production environment due to the easiness of implementation and the speed of application. The most popular classical scheduling rules included first in first out (FIFO) rule, shortest processing time (SPT) rule, earliest due date (EDD) rule and minimum slack time rule (MST), etc. Many other human-made rules were also developed based on expertise and experience (Jayamohan and Rajendran, 2000). Investigations found that no rule performed consistently better than all other rules under a variety of shop configurations, operating conditions and performance objectives because the rules had all been developed to address a specific class of system configurations relative to a particular set of performance criterion and generally had not performed well in another environment or for other criteria.

Therefore, it became one of the hot topics to develop customized scheduling rules with artificial intelligent approaches based on the feature of shop configurations, operating conditions and performance objectives. Geiger et al. (2006), Jakobovic and Budin (2006), Tay and Ho (2008), Nguyen et al. (2013), Pickardt et al. (2013) and Pan et al. (2013) implemented the machine learning approaches based on GP (Koza, 2007) for static scheduling problems. The approaches construct scheduling rules with fundamental elements rather than synthesize existing rules so that they can identify innovative and potentially rules. For dynamic scheduling problem, Nie et al. (2010) proposed the approaches with machine learning in which evolutionary algorithm GEP (gene expression programming, GEP) (Ferreira, 2001) served as the computation engines. It was demonstrated that scheduling rules obtained through evolution of the machine learning algorithms were markedly better than conventional scheduling rules. And it was also proved that the approaches performed better than the approaches based on GP in the aspect of solution quality and computation time (Nie et al., 2010). The approaches provide inspiration for further development of advanced scheduling rules and make a contribution to the solution of the dynamic scheduling problems in practical shop floor. It is encouraging to improve further the performance of the approaches and the intelligibility of the scheduling rules evolved by them. However, one of the bottlenecks is that original GEP has not the ability to express appropriately the weights for each bit in the chromosome so that the intelligibility of the evolved scheduling rules is not obvious enough. Although GEP has been applied successfully in various fields, such as symbolic regression, function finding, classification, clustering analysis, time series analysis and neural networks design, its ability is limited when the weight of each term in the target function is complex, and few literatures tackle this problem well. It is necessary to extend the ability of GEP to adjust bit weights in the chromosome meticulously in its evolving process.

PSO (particle swarm optimizing algorithm, PSO) is a kind of evolutionary computing technology based on swarm intelligence. It was designed by Kennedy and Eberhart in 1995 (Poli, 2007). Since its introduction in 1995, PSO is widely applied in various research area, including the field of scheduling problem (Qi 2014). It has performed as a powerful optimization tool. Considering PSO as a meta-heuristic based on swarm intelligence and it has been proved that it has strong ability to search optimal solution in real number space, it is promising to combine the advantages of GEP with that of PSO to discover more effective and intelligible scheduling rules.

In this paper, a new chromosome structure named DSS (Double String Structure, DSS) is designed. It makes each bit in the chromosome correspond with a weight so that scheduling rules could be expressed more precisely. Based on the new chromosome structure DSS, a hybrid algorithm which incorporates PSO into GEP is proposed to evolve the appropriate mathematical expression and term weights for scheduling rules simultaneously. The paper is organized as follows: Section 2 describes the scheduling problem considered. In Section 3, the hybrid algorithm based on GEP and PSO techniques is described in detail. Section 4 provides the experiments and results by presenting a comparative study between the hybrid algorithm and the original approach. Finally, Section 5 concludes the paper and gives future search.

2 DYNAMIC SCHEDULING PROBLEM

Scheduling problem on a single machine is a typical scheduling problem in the shop floor and also the foundation of solving other more complex scheduling problems. In the paper, single machine scheduling problem is examined. The problem is described as below:

The shop floor consists of one machine and n jobs, which are released over time and processed once on the machine without preemption. Each job i can be identified with several attributes, such as processing time p_i, release date r_i and due date d_i. The attributes of a job are unknown in advance unless the job is currently available at the machine or arrives in the immediate future. It is also assumed that the machine cannot process more than one job simultaneously.

The scheduling objective is to determine a sequence of jobs on the machine in order to minimize one or more optimization criteria, in our case, flow time.

Since the hybrid algorithm aims to evolve customized scheduling rules, the scheduling rules are developed and evaluated on training sets and test sets which consisted of problem instances that represent the different operating conditions relative to different performance criteria, respectively. In order the training sets or test sets to have a similar influence on the overall performance estimation of a scheduling rule, average criterion value over the training set or test set of problem instances is defined as below.

$$|F| = \frac{1}{k} \sum_{j=1}^{k} \frac{F_j}{n_j \cdot \overline{p}_j} = \frac{1}{k} \sum_{j=1}^{k} \frac{\sum_{i=1}^{n_j} (c_{ij} - r_{ij})}{n_j \cdot \overline{p}_j} \qquad (1)$$

where F_j denotes the total flow time of problem instance j; n_j denotes the number of job in problem instance j; $\overline{p}_j$ denotes the mean processing time of all jobs in problem instance j; c_{ij} and r_{ij} denote completion time and release date of job i in problem instance j; k denotes the number of instances in a training set or test set; and $|F|$ represents the average value of flow time over the training set or test set of problem instances. It is obvious that scheduling rules with less objective values of $|F|$ are better.

3 HYBRID ALGORITHM BASED ON GEP AND PSO

3.1 Framework of the hybrid algorithm

The hybrid algorithm consists of two levels of iteration, evolutionary iteration and weight-adjusting iteration. The outer level of iteration (i.e. evolutionary iteration) is based on the framework of GEP which is responsible for the reproduction and variation of chromosomes in the population, while inner iteration (i.e. weight-adjusting iteration) is based on the framework of PSO which is responsible for the adjustment of the bit weights of the chromosomes. Through incorporating PSO into GEP framework, the ability of GEP to evolve more exquisite chromosomes is extended. The framework of the hybrid algorithm is shown in Fig. 1.

The evolutionary iteration starts with an initial population which consists of n_{gep} chromosomes that are randomly generated (denoted with expression (2)). Through the weight-adjusting iteration, the bit weights of each chromosome are optimized. Then, these chromosomes in the population are assessed using the performance measures of the average flow time on a set of problem instances (calculated with expression (1)). The values of the performance measures for all individuals are shown in expression (3). Fitness of each chromosome is calculated with expression (4). A variety of genetic operations (refer to Nie (2010)) are executed on the current population and new offspring which inherit the excellent genes from the current high-fitness individuals are generated. Weight-adjusting iteration operates again so that each chromosome in the newborn population is endowed with optimal bit weights. This cycle is repeated until the termination condition is satisfied and the best chromosome is obtained.

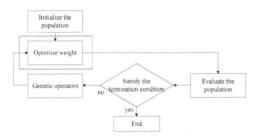

Figure 1.   Framework of the hybrid algorithm.

$$p_{gep}^{(t)} = (ch_1^{(t)}, ch_2^{(t)}, ..., ch_i^{(t)}, ..., ch_{n_{gep}}^{(t)}) \quad (2)$$

$$O_{gep}^{(t)} = (O_1^{(t)}, O_2^{(t)}, ..., O_i^{(t)}, ..., O_{n_{gep}}^{(t)}) \quad (3)$$

$$f_i^{(t)} = \frac{O_{max}^{(t)} - O_i^{(t)}}{O_{max}^{(t)} - O_{min}^{(t)}} \quad (4)$$

$P_{gep}^{(t)}$ denotes the GEP population in tth iteration, $ch_i^{(t)}$ denotes chromosome i in population $P_{gep}^{(t)}$. $O_i^{(t)}$ denotes values of the performance measures for chromosome i. $f_i^{(t)}$ denotes the fitness of the chromosome i. $O_{max}^{(t)}$ and $O_{min}^{(t)}$ denote the maximal and minimal criterion value over all the chromosomes in the population $P_{gep}^{(t)}$, respectively. Since the scheduling objection is minimization, the better chromosome is assigned the bigger fitness. It is noticed that the heuristic used in the evaluation of chromosomes refer to Nie (2010).

3.2   Representation of chromosome

3.2.1   Designing of FS and TS

Each chromosome of GEP is generated randomly at the beginning of the search and modified during evolutionary progress with the elements from Function Set (FS) and Terminal Set (TS). The FS and TS are defined as follows:

Function Set: including functions such as "+", "–", "*", which express the corresponding arithmetic functions, respectively, and "/" which expresses the protected division which returns 1 when the denominator is equal to 0.

Terminal Set: including elements that denote the current status and attributes of the candidate jobs for scheduling, such as:

p job's processing time;
r job's release date;
d job's due date;
sl job's positive slack, max $\{d - p - \max\{ct, r\}, 0\}$, where ct denotes the idle time of the machine;
st job's stay time, max $\{ct - r, 0\}$, where ct is defined as above;
wt job's wait time, max $\{r - ct, 0\}$, where ct is defined as above.

3.2.2   Chromosome structure DSS

In the paper, a new chromosome structure named DSS is designed. It makes each bit in the chromosome correspond with a weight so that chromosomes express scheduling rules more precisely.

Each chromosome consists of a symbol string and a numerical string. Each number in the numerical string is corresponding with a bit in the symbol string, representing the weight of the bit.

The symbol string of a chromosome is composed of one or more genes. A gene is a fixed length symbolic string with a head and a tail. Each symbol is selected from FS or TS. The symbols which come from FS mean to perform a certain operation on arguments. For example, "+" returns the weighted sum of two arguments. The symbols come from TS have no arguments. For example, "a" directly returns the product of variable a and its weight.

It is stipulated that the head of a gene may contain symbols from both the FS and the TS, whereas the tail consists only of symbols come from TS. Suppose the symbolic string has hl symbols in the head, and tl symbols in the tail, then the length of the tail is determined by the expression $tl=hl * (n_a-1)+1$, where n_a is the maximum number of arguments for all operations in FS, which ensure the correctness of gene. Therefore, the length of a gene $gl=hl+tl=hl*n_a+1$. The length of a chromosome $m=gl*n_g$, where n_g denotes the number of the genes in the symbol string of a chromosome.

Assume we use $hl=6$ and $n_a=2$ for arithmetic operations. Thus, the tail length must be $tl=7$. So the total gene length gl is 13. According to the FS and TS in the above section, a randomly generated chromosome with 3 genes (each gene's size is 13) is illustrated in Fig. 2. The tail is underlined. x_i, y_i and z_i ($i=1,...,13$) are bit weights of the chromosome. They are real numbers.

367

Gene 0

$$*.\ *:\ +.\ p.\ sl.\ /.\ d.wt.\ r.\ st.\ d.\ p.\ r$$
$$x_1.\ x_2.\ x_3.\ x_4.\ x_5.\ x_6.\ x_7.\ x_8.\ x_9.\ x_{10}.\ x_{11}.\ x_{12}.\ x_{13}$$

Gene 1

$$+.\ +.\ st.\ p.\ *.\ /.\ sl.\ d.\ r.\ wt.\ d.\ r.\ sl$$
$$y_1.\ y_2.\ y_3.\ y_4.\ y_5.\ y_6.\ y_7.\ y_8.\ y_9.\ y_{10}.\ y_{11}.\ y_{12}.\ y_{13}$$

Gene 2

$$/.\ wt.\ *.\ p.\ sl.\ -.\ st.\ r.\ d.\ p.\ wt.\ r.\ d$$
$$z_1.\ z_2.\ z_3.\ z_4.\ z_5.\ z_6.\ z_7.\ z_8.\ z_9.\ z_{10}.\ z_{11}.\ z_{12}.\ z_{13}$$

Figure 2. Illustration of chromosome.

3.2.3 *Mapping mechanism between chromosomes and scheduling rules*

Each gene can be mapped into an ET following a depth-first fashion. Specifically, first element in gene corresponds to the root of the ET. Then, below each function is attached as many branches as there are arguments to that function. A branch of the ET stops growing when the last node in this branch is a terminal. Fig. 3 shows the ET corresponding with the chromosome in Fig. 2. Each gene codes for a sub-ET and the sub-ETs interact with each other in a way of addition to form a more complex multi-subunit ET. The multi-subunit ET showed in Fig. 3 represents a scheduling rule represented with expression (5).

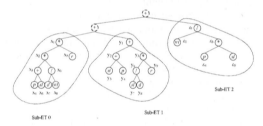

Figure 3. Illustration of ET.

$$x_1\{x_2[x_3(x_4p + x_5sl) * x_6(x_7d / x_8wt)] * x_9r\} +$$
$$y_1\{y_2(y_3st + y_4p) + y_5[y_6(y_7sl / y_8d) * y_9r]\} + \qquad (5)$$
$$z_1[z_2wt / z_3(z_4p * z_5sl)]$$

3.2.4 *Optimizing bit weights of chromosomes*

In the hybrid algorithm, PSO is employed to optimize bit weights of chromosomes and implement the weight-adjusting iteration. The step is described below:

Step1. Define the parameters of PSO and the objective function (expression (1)).

Step2. Set counter $t = 0$. Initialize a population $P_{pso}^{(t)}$ which consists of n_{pso} particles (denoted with expression (6)). Each particle $p_i^{(t)}$ is composed of three m-dimensional vectors, where m is the dimensionality of the search space (denoted with expression (7)). The three vectors represent its current position

$x_i^{(t)}$, its current best position $pb_i^{(t)}$ and its velocity $v_i^{(t)}$, respectively. Each initial position and initial velocity of each particle is generated by randomly selecting a value with uniform probability over each dimension in a predefined interval.

$$P_{pso}^{(t)} = (p_1^{(t)}, p_2^{(t)}, ..., p_i^{(t)}, ..., p_{n_{pso}}^{(t)}) \qquad (6)$$

$$p_i^{(t)} = \{x_i^{(t)}, pb_i^{(t)}, v_i^{(t)}\} \qquad (7)$$

Where

$$x_i^{(t)} = (x_{i1}^{(t)}, x_{i2}^{(t)}, ..., x_{ij}^{(t)}, ..., x_{im}^{(t)})$$

$$pb_i^{(t)} = (pb_{i1}^{(t)}, pb_{i2}^{(t)}, ..., pb_{ij}^{(t)}, ..., pb_{im}^{(t)})$$

$$v_i^{(t)} = (v_{i1}^{(t)}, v_{i2}^{(t)}, ..., v_{ij}^{(t)}, ..., v_{im}^{(t)})$$

Step3. Evaluate the fitness of each particle according to the objective function. In this situation, the lower objective function value is, the better the corresponding particle performs.

Step4. Update the global best positions $gb^{(t)}$ and local best positions $pb_i^{(t)}$ for each particle $p_i^{(t)}$, $i=1,...,n$.

Step5. If the stopping criterion is not satisfied go to step 6. Otherwise, terminate the iteration and obtain the best weight set with the global best solution $gb^{(t)}$.

Step6. Update counter $t=t+1$ and update the velocity and position of particle $p_i^{(t)}$, $i= 1,..., n$ in the kth dimension ($k=1,...,m$) according to expressions (8) and (9).

$$v_{ik}^{(t)} = \omega^{(t)} \times v_{ik}^{(t-1)} + l_1 \times r_1 \times (pb_{ik}^{(t-1)} - x_{ik}^{(t-1)})$$
$$+ l_2 \times r_2 \times (gb_k^{(t-1)} - x_{ik}^{(t-1)}) \qquad (8)$$

$$x_{ik}^{(t)} = x_{ik}^{(t-1)} + v_{ik}^{(t)} \qquad (9)$$

where l_1 and l_2 are called cognitive and social learning rates and aim to balance between the global and local search abilities. Empirically these two parameters are kept constant at 2. r_1 and r_2 are independent random numbers uniformly distributed in the range of [0, 1]. $\omega^{(t)}$ is called the inertia factor. Usually it is set according to the expression (10).

$$\omega^{(t)} = \omega_{max} - k \times (\omega_{max} - \omega_{min}) / iter_{max} \qquad (10)$$

where ω_{max} is the initial weight, ω_{min} is the final weight, $iter_{max}$ is the iteration number.

4 EXPERIMENTS AND RESULTS

To validate the effectiveness of the new proposed algorithm, partial problem instance sets generated in Nie (2010) are employed. In the experiments, totally 9 problem instance sets as follows are considered to be training sets. (1) Number of jobs n = 10, 50 or 100; (2) Processing time of jobs p is drown from U[1, 100], U[100, 200] or U[200, 300] (U denotes uniform distribution). In each set, due date tightness (T) and due date range (R) are assumed the values of 0.1, 0.5, and 0.9 in various combinations (3×3=9). The number of instances generated for each of 9 combinations of parameter in each set is 3. Hence, each problem instance set consists of 27 problem instances. In addition, 9 different test sets of the similar composition using the same parameters are used for validating purposes. Five runs are conducted in total for each training set.

In the hybrid algorithm, parameters relevant to GEP are configured as below. Population size n_{gep} is 10. Termination condition is that the best solution has not been improved for consecutive 100 evolutionary iterations. The length of head hl is 5 and gene number n_g is 1. Therefore, the total length of the symbol string m is 11. Mutation probability is 0.03. Probabilities of IS, RIS and Gene transposition are 0.3, 0.1 and 0.1, respectively. Probabilities of one-point, two-point and gene recombination are 0.2, 0.5 and 0.1, respectively. The parameters relevant to PSO are set as follows. Population size n_{pso}=5, the maximum iteration number $iter_{max}$=10. The inertial weight decreases from ω_{max}=0.9 to ω_{min}=0.4 linearly and the accelerated parameters l_1=l_2=2.

To validate its efficiency, the hybrid algorithm is compared with the approach based on original GEP in Nie (2010). The former is denoted with HGEP-PSO and the latter is denoted with GEP. Tables 1 and 2 show the flow time results of the best scheduling rules evolved by the hybrid algorithm and the original algorithm on different training sets and test sets. "AVE" is the average performance of the evolved rules over the training sets or test sets from the five runs. The "BEST" columns summarize the performance of the best rules over the five runs, respectively. From the result, it is easy to conclude that the efficiency of evolution of the hybrid algorithm HGEP-PSO is better than the original approach in Nie (2010).

5 SUMMARY AND FUTURE WORK

As for single-machine dynamic scheduling problem which exists widely in practical shop floor, scheduling rules are one of the most prevalent approaches because of its easiness of implementation and speed of application. However, it is a hard work to develop

Table 1. Average flow time on training sets.

Training Set	GEP		HGEP-PSO	
	AVE	BEST	AVE	BEST
1	2.7128	2.7118	2.7105	2.7101
2	3.2673	3.2661	3.2675	3.2673
3	3.3692	3.3691	3.3691	3.3691
4	8.9827	8.9820	8.9795	8.9761
5	11.9560	11.9549	11.9550	11.9547
6	12.7377	12.7375	12.7375	12.7375
7	16.3969	16.3958	16.3941	16.3941
8	23.3597	23.3593	23.3589	23.3578
9	25.7711	25.7711	25.7713	25.7710

Table 2. Average flow time on test sets.

Test Set	GEP		HGEP-PSO	
	AVE	BEST	AVE	BEST
1	2.7922	2.7909	2.7900	2.7884
2	3.2416	3.2377	3.2416	3.2392
3	3.3136	3.3129	3.3129	3.3129
4	8.6307	8.6291	8.6304	8.6289
5	11.8170	11.8151	11.8184	11.8151
6	12.7607	12.7604	12.7608	12.7608
7	16.3186	16.3135	16.3178	16.3178
8	23.1457	23.1438	23.1474	23.1443
9	25.8164	25.8164	25.8140	25.8128

advanced scheduling rules. In the paper, the approach that evolves customized and effective scheduling rules automatically according to the feature of the shop configurations, operating conditions and performance objectives is investigated. A new chromosome structure named DSS is designed so that chromosomes may express scheduling rules nicely because each bit in the chromosome takes along a weight that could be tuned meticulously during the process of the construction of scheduling rules. Based on DSS, a hybrid algorithm based on GEP and PSO is proposed. In the hybrid algorithm, GEP and PSO constitute a two-level-iteration framework which makes it possible to represent effective scheduling rules with appropriate elements and their weights in reasonable mathematical expressions.

Although only the single machine dynamic scheduling problem concerning the performance objective of flow time is considered in the paper, the work may be extended for many other scheduling problems. The main advantage of the proposed approach is that it could be applied easily on the practical large-size manufacturing system because it may provide advanced and practical scheduling rules according to the feature of the system without profound expertise and complicated reasoning process.

In our future work, the improvement of the current approach base on GEP is continued. Meanwhile, we also try to explore other approaches to construct advanced scheduling rules. The possible direction includes the combination of multi agent system and machine learning algorithm such GEP.

ACKNOWLEDGEMENTS

This research is supported by the Funding Projects for Teachers in Universities of Shanghai under Grant No. ZZegd12026, the Innovation Program of Shanghai Municipal Education Commission under Grant No. 14YZ155 and the Key Cultivating Academic Discipline Projects of Industrial Engineering of Shanghai Second Polytechnic University under Grant No. XXKYS1403, No. XXKPY1311.

REFERENCES

Ferreira, C. 2001. Gene expression programming: a new adaptive algorithm for solving problems. *Complex Systems*. 13(2):87–129.

Geiger, C., Uzsoy, R., & Aytug, H. 2006. Rapid modeling and discovery of priority dispatching rules: an autonomous learning approach. *Journal of Scheduling*. 9(1):7–34.

Jakobovic, D., & Budin, L. 2006. Dynamic scheduling with genetic programming. *Lecture Notes in Computer Science*, 3905, 73–84.

Jayamohan, M., & Rajendran, C. 2000. New dispatching rules for shop scheduling: A step forward. *International Journal of Production Research*. 38(3): 563–586.

Koza, J. 2007. Introduction to genetic programming. In: Lipson H (ed) *Proceedings of GECCO 2007: Genetic and Evolutionary Computation Conference*, 7–11 July 2007 London. ACM, London, pp. 3323–3365.

Nguyen, S., Zhang, M., & Johnston, M. 2013. A Computational Study of Representations in Genetic Programming to Evolve Dispatching Rules for the Job Shop Scheduling Problem. *IEEE Transactions on Evolutionary Computation*. 17(5): 621–639.

Nie, L., Shao, X., Gao, L., & Li, W. 2010. Evolving scheduling rules with gene expression programming for dynamic single-machine scheduling problems. *International Journal of Advanced Manufacturing Technology*. 150(5–8):729–747.

Pan, Y., Xue, D., Gao, T., Zhou, L., & Xie, X. 2013. Research on Strategy of Dynamic Flexible Job-Shop Scheduling. *Applied Mechanics and Materials*. 423–426, 2237–2243.

Pickardt, C., Hildebrandt, T., & Branke, J. 2013. Evolutionary generation of dispatching rule sets for complex dynamic scheduling problems. *International Journal of Production Economics*. 145(1): 67–77.

Poli, R., Kennedy, J., & Blackwell, T. 2007. Particle Swarm Optimization. *Swarm Intelligence*. 1(1): 33–57.

Qi, J., Liu, Y., Jiang, P., & Guo, B. 2014. Schedule generation scheme for solving multi-mode resource availability cost problem by modified particle swarm optimization. *Journal of Scheduling*. 10.1007/s10951-014-0374-0.

Tay, J., & Ho, N. 2008. Evolving dispatching rules using genetic programming for solving multi-objective flexible job-shop problem. *Computers & Industrial Engineering*. 54(3): 453–473.

Computational Intelligence in Industrial Application – Ling (ed.)
© 2015 Taylor & Francis Group, London, ISBN: 978-1-138-02818-0

A new processing technology of solidified soybean yogurt

XingRong Liu, HaiLe Ma & ChuanJun Li
School of Food and Biological Engineering, Jiangsu University, Zhenjiang, Jiangsu, China

ABSTRACT: In order to overcome the shortcomings of current soybean yogurt with unstable types and bad activity of diversity, heavy soybean odor and adding thickening agent, a new processing technology was developed to produce natural solidified soybean yogurt with multi strain domesticating method. Through the experimental optimization, the effect of different factors on solidified soybean yogurt quality was investigated, including the different pretreatments of soybean grinding, different fermentation strains ratio, beans–water ratio, bacteria amount, sugar amount, homogenization pressure, fermentation temperature, fermentation time, ripening time and so on. And the optimal conditions of fermentation process were obtained. The results show that in the optimal conditions, the number of bacteria is 6.7 times higher than that of before the strain domestication, the acidity is 70% higher than that of the before, the thickening agent is not needed, the flavor has been obviously improved.

KEYWORDS: Soybean yogurt, Domestication, Sensory evaluation, Processing technology.

1 INTRODUCTION

In the recent years, soybean yogurt with its unique low price and the health care effect, whose consumption in the USA increases with the speed of 20% [1–5]. Because China's production of soybean yogurt has bacteria species and their activity is unstable, with the existence of nutritional defects that trypsin inhibitor, allergic factors, flatulence factor, the beany flavor and so on. Meanwhile its flavor needs to be improved, so its share of the world market share is small [6–12]. In this paper, taking this as the breakthrough point, study the effect of many factors on the quality of solidified soybean yogurt, including, different soybean grinding pretreatment, different fermentation strain selection and its adding amount, bean–water ratio, sugar amount and homogeneous pressure, fermentation and ripening, etc. On the one hand, to ensure the nutritional content of milk, while inhibitors of trypsin, lipoxygenase, blood omentin and other anti-nutritional factors need to be completely passivated, solving the flatulence, beany and bitter taste side effects, on the other hand, it can be better guaranteed sensory quality of yoghurt (color, taste, odor, texture state, etc.), Jiernieer degree and the number of Lactobacillus bacteria, providing a theoretical basis for the industrial fermentation of soybean products.

2 MATERIALS AND METHODS

2.1 Reagent and instrument

Reagent: *Lactobacillus bulgaricus*, *Streptococcus thermophilus*, lactobacillus, skim milk powder, soybeans, sugar, etc.

Instrument: Kjeldahl nitrogen instrument, soya bean milk machine, homogenizer, constant temperature incubator, acidity meter, electronic balance.

2.2 Test methods

2.2.1 Establishment of sensory evaluation methods

At present, the quality of most domestic yoghurt varies greatly, domestic standards are not uniform. According to the Yoghurt standards of GB 19302-2010 by the assessment team with 5 trained people, they assess soybean yogurt according to 5 sensory, including, the color, flavor and smell, taste, viscosity of tissue state, etc. itemized scoring fill in the prescribed form, and calculate the average total score, average score is the final score of 5. The indicators of sensory evaluation are color, texture and odor, state of organization, taste, viscosity.

2.2.2 Determination of acidity

According to GB 5413.34-2010 determination of acidity in milk and milk products. Weigh 5 g sample in a 150 mL conical flask, and add 40 mL boiling water placed to about 40 °C, mix, then add 2–3 drops of phenolphthalein indicator solution, with a concentration of 0.1 mol/L sodium hydroxide is titrated to a reddish, it is not disappear within 0.5 min as the endpoint, the consumption of sodium hydroxide standard solution of ML number multiplied by 20, this value is the Jill Nel acidity acid soya bean milk(°T).

2.2.3 Count of lactic acid bacteria

According to GB 478935-478935, the national food safety standards, food microbiology inspection and the count of lactic acid bacteria inspection.

2.2.4 Determination of soybean yogurt processing technology route

With scores and acidity of sensory evaluation as major test indicators, comparing study of acid pretreatment quality effects of three kinds of soybean grinding pretreatment technology: boiled after grinding the green beans, grinding in hot water after soaking the beans, grinding the cooked beans.

The bean paste with the skin, in the premise of not affecting the taste of soybean yogurt, as far as possible, keep the dietary fiber nutrition of soybean, increase the solids content of soymilk, and keep a certain viscosity, filter can use the way of only the bottom of the soybean milk over a layer of gauze, then add skim milk powder, sugar and mix thoroughly, 75–80 bar homogeneous, rapidly cooling to 45°C, after inoculation, Fermentation about 4 h in cultivating box with constant temperature of 43°C, under the temperature of 4°C refrigerated and ripening for about 12 h, finished product inspection.

2.2.5 Domestication of strain

Lactobacillus bulgaricus, Streptococcus thermophilus, Lactobacillus plantarum three kinds of bacteria to the quality ratio of 0.75% according to the technical route 2 respectively made of soybean yogurt, analysis of the sensory evaluation score and acidity values after fermentation, select a set of optimal to train as the first generation of products of soybean yogurt, then take 1% mass fraction of the first generation of product samples, gradually reduce the content of skim milk powder of soymilk (7.5%, 6.5%, 5.5%, 4.5%, 4%, 3%), domestication and cultivation by generation, until obtain the optimum amount of milk powder adding quantity and domestication times.

2.2.6 Optimization of soybean yogurt processing parameters

Through the single factor experiment research of bean–water ratio (1:5, 1:6, 1:7, 1:8, 1:9), the amount of bacteria added,(0.2%, 0.6%,1%, 1.4%, 1.8%), the amount of sugar(5%, 6%, 7%, 8%, 9%), fermentation time, (3h, 4h, 5h, 6h, 7h), fermentation temperature(40°C, 42°C, 44°C, 46°C) ,homogenization pressure (50bar, 60 bar, 70 bar, 80 bar, 90 bar), ripening time (10h, 11h, 12h, 13h, 14h)influence the quality and acidity of soybean yogurt, determine the significant influence factors and the optimal process parameters.

The fermentation of soybean yogurt is a complex process of interaction of many facts, For visual analysis of this process, the test will be on the basis of single factor experiment, focusing on the effect of four factors on the quality of yogurt, namely: the inoculum size, the amount of sugar, the fermentation time and temperature, According to the processing conditions of orthogonal test design to make, and in accordance with the optimal process parameters obtained for verification testing.

3 RESULTS AND ANALYSIS

3.1 Impact of different soybean refining pretreatment on yogurt quality

According to the method for the determination above, studied the influence of three kinds of soybean grinding pretreatment on the quality of yogurt was studied, such as boiled after grinding the green beans, grinding in hot water after soaking the beans, and grinding the cooked beans. The results are shown in tab.1.

Table 1. Comparison of three techniques route.

Grinding pretreatment	Acidity (°T)	Sensory value	Evaluation of description
Boiling after grinding the green beans	80–90	75–83	soy yogurt with whitish color, heavier beany, delicate tissue, but tastes astringent feeling
Grinding in hot water after soaking the beans	73–85	85–92	uniform color, lighter beany, with fragrant yogurt flavor, delicate tissue
Grinding the cooked beans	65–78	70–77	whitish color, with the whey separation, poor viscosity, more bean dregs

See from Tab.1, in case of grinding in hot water after soaking the beans relative to boiling after grinding the green beans and grinding the cooked beans, the score of less beany flavor in the sensory evaluation is higher. Because when raw soybean is crushing the grain refined, linoleic or linolenic acid produces hydrogen peroxide in the lipoxygenase action, and finally degraded into aldehydes, ketones, epoxides, etc. wherein hexanal related beany flavor, in this way the beany flavor is heavy, but grinding the cooked beans lead to heat denatured and solubility of soy protein become worse, a slurry is not high, more bean dregs, the solids content reduces after filtration, which appears less viscosity, whey separation phenomenon. It suggests that the bean hot water slurry can effectively to fishy, can get sweet and delicious (80–85°T), high protein content (5.5–5.7%), rich in dietary fiber products.

3.2 Analysis of the results of species domestication

When it was elected with Streptococcus thermophilus and Lactobacillus bulgaria lactic acid bacteria fermentation, the finished product viscosity is high, easy to hang a wall, and join a large number of plant lactobacillus, there are more whey separation, bean smell is aggravating, and the three strains in appropriate proportion mix, finished products curd state is very good, moderate acidity, and a faint sweet bean fragrance. By reducing the proportion of soybean milk, skimmed milk powder, according to the strains of domestication method above, the results show that lactic acid bacteria can breed produce acid in soybean milk, but the measured pH value significantly lower than the same conditions of acid milk acidity, when the quality of skim milk powder less than 3%, significantly lower acidity, may be due to less nutrients required for bacterial growth, results in the decrease of strain energy. Considering strain energy and economic cost, using *Lactobacillus bulgaria*: thermophilic streptococcus: plant lactobacillus to 5:5:1 mixing ratio, 5 strains of domestication, skimmed milk powder mass fraction of 4.0%. See from Table 2, when the soybean yogurt was produced with the fermentation of the strain after domestication, its acidity was 85 °T, which was increased by 70%. Its bacterial count was 1.6×108 cfu/mL, increased by 6.7 times. And its flavor had been obviously improved.

3.3.1 Effect of the adding strains on the quantity and acidity of soybean yogurt

Bacteria amount is one of the important factors affecting the quality of yogurt. If the amount is too large, can make yogurt too fast solidification, texture coarse, and if quantity is too small, can make yogurt solidification speed too slow, the fermentation time is too long. The test result is shown in Figs. 1 and 2.

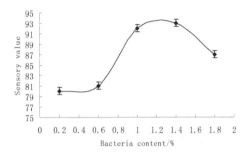

Figure 1. Effects of bacteria content on quality.

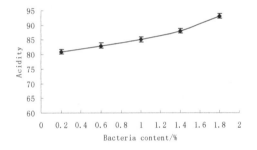

Figure 2. Effects of bacteria content on acidity.

Strains can be seen from Figs. 1 and 2, the adding amount of sour soybean milk quality has a great influence, and it has a certain influence on the acidity. When the bacteria content is 1.0%, sour soybean milk quality is better, acidity is appropriate. Strains of bacteria amount is lower than 0.8, with the increase of adding amount of quality raise and the acidity gradually ascending to sour milk normal levels, that is due to the strains of adding quantity is little, the same fermentation condition, bacteria content, the more the more fermented soymilk, evaluation and the better the taste, Strains in more than 1.4%, fermentation curd formation too fast, uneven, and whey separation, the product acidity is too high, a drop in quality. Strains can choose to add the amount of 1.0–1.4%.

Table 2. Characters of fermented soybean milk before and after strain domesticating.

	Strain before domesticating	Strain after domesticating
Time of fermentation / lagering (h)	6	6
Acidity in lagering(°T)	50	85
Bacterial count (x 107cfu/mL)	3.8	16
Evaluation of description	soy yogurt with whitish color, heavier beany, delicate tissue, but tastes astringent feeling	uniform color, lighter beany, with fragrant yogurt flavor, delicate tissue

3.3 Optimization of sour soybean milk process parameters of single factor and orthogonal experiment

When the concentration of soybean milk is too thin, the finished product is not easy to flocculation precipitation and whey, and consistency of excessive concentration of finished product is too high, ease straws, absorption phenomenon of wall hanging. Tests show that bean water was no better than 1:8.

3.3.2 Effect of sugar on the soybean yogurt quality and acidity

Added sugar to acid beany flavor, which not only provides the carbon sources for lactic acid bacteria growth, promote the ability to produce acid, but can also increase the dry matter content, makes sour soymilk coagulation, and also can improve product quality and flavor, but instead of sugar is too high. The test results are shown in Figs. 3 and 4.

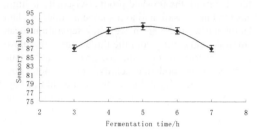

Figure 5. Effects of fermentation time on quality.

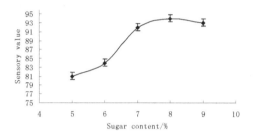

Figure 3. Effects of sugar content on quality.

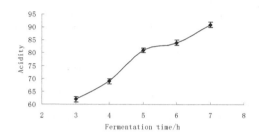

Figure 6. Effects of fermentation time on acidity.

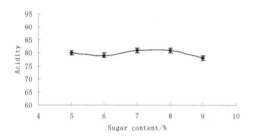

Figure 4. Effects of sugar content on acidity.

In Figs. 3 and 4, the amount of sugar to acid soybean milk quality has a great influence and less obvious influence on acidity. This is because less than 5% sugar fermentation was due to the lack of sugar which can lead to less growth of lactic acid bacteria, also it does not produce enough lactic acid. Add 7% of the soft sugar can meet the needs of the growth of lactic acid bacteria, and generate enough acid, flavor of fermented milk have very good effect. *Lactobacillus bulgaria* cannott use sucrose decomposition, produce less acid, thermophilic streptococcus has the ability to ferment sucrose, but weaker capacity of producing acid and sour soybean milk acidity mainly comes from the decomposition of lactose in milk and lactic acid bacteria produce lactic acid. Comprehensive flavor, it can choose sugar of 7–9%.

3.3.3 Effect of acidity fermentation time on the quality of soybean yogurt

The experimental results are in Figs. 5 and 6.

In fig. 5 and fig. 6, the fermentation time of the qualityof soybean yogurt has a certain impact, but it has significant effects on the acidity. Fermentation time of 5 h when product quality is good, sweet and sour moderate. During the early stage of the fermentation, fermentation speed faster, thermophilic streptococcus decompose protein, to provide a certain amount of nitrogen source for Bulgaria lactobacillus, with soybean milk fermentation degree gradually deepen, produce acid increased, the curds gradually getting better, taste also improved, When the fermentation time more than 5 h, acidity with the increase of fermentation time, but at a slower pace, and finished product viscosity is too high, too produce acid amount and affect the taste, and quality decline. Can choose the fermentation time is 4 h to 6 h.

3.3.4 Effects of fermentation temperature on the quality and the acidity of soybean yogurt

The incubation temperature of lactic acid bacteria fermentation of milk is usually between 38°C–45°C, different strains of the optimum fermentation temperature is not the same. The experimental results are in Fig. 7 and fig. 8.

From Figs. 7 and 8 it can be seen that fermentation temperature has large effects on the quality and the acidity of soybean yogurt. When the fermentation temperature is lower than 42 °C, The curd is slow, it required longer fermentation time, to achieve the same desired acidity condition, with the increase of fermentation temperature, fermentation

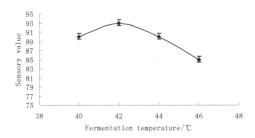

Figure 7. Effects of fermentation temperature on quality.

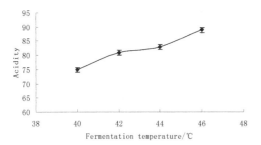

Figure 8. Effects of fermentation temperature on acidity.

time is greatly shortened, taste becomes rough, flavor is insufficient, the quality of soybean yogurt decreased. It can choose the fermentation temperature of 42–44 °C.

According to the four factors and three levels orthogonal experiment, the optimized parameters were obtained. Order of the influences of various factors on the quality of soybean yoghurt is: bacteria amount >fermentation time >fermentation temperature >sugar. In this experiment, the optimized parameters were: bacteria amount 1%, fermentation time 6 h, fermentation temperature 44 °C,sugar 9%. In the verification experiment, the average score of sensory evaluation was 95 °T, and its acidity was 85 °T.

4 CONCLUSION

The factors of the different pretreatments of soybean grinding, different fermentation strains ratio, beans-water ratio, bacteria amount, sugar amount, homogenization pressure, fermentation temperature, fermentation time, ripening time, etc. determine the quality of the solidified soybean yogurt.

The process parameters of optimized soybean yogurt: grinding in hot water after soaking the beans, beans–water ratio of 1:8, *Lactobacillus plantarum* 5:5:1, 5 strains of domestication, after acclimation (reduce the mass fraction of skim milk powder in the soybean milk is 7.5%, 6.5%, 5.5%, 4.5%, 4%, 3%), the inoculums size 1%, sugar 9%, homogenization pressure 75–80 bar,fermentation temperature 44 °C, fermentation time 6 h, ripening about 12 h after the storage of temperature of 4 °C.

Under the optimized process parameters, the soybean yogurt has high quality with no additive, moderate acidity (80–85 °T), high protein content (5.5–5.7%).Compared with no domesticating production, its acidity increases 70%, its bacterial count increases 6.7 times. And its flavor has been obviously improved.

ACKNOWLEDGEMENTS

This work was financially supported by Emulsion Biotechnology National Key Laboratory Open Foundation (SKLDB2011003) and Academic Program Development of Jiangsu Higher Education Institutions (PAPD).

REFERENCES

Yang Qian, Wu Zuxing, Han Ming. The research on production process of mixing soybean probiotics composite yogurt. *Food Science and Technology*, 2008, 34(3):95–98.

Zhang Bingwen, Song Yongsheng, Hao Zhenghong, etc. The effect of fermentation processing on the isoflavone content and components in soybean products. *Food and Fermentation Industry*, 2002, 28(7):6–9.

Ana N. Rinaldoni, Mercedes E. Campderrós, etc. Physico-chemical and sensory properties of yogurt from ultrafiltreted soy milk concentrate added with inulin. *LWT - Food Science and Technology*, 2012, 45(2):P142–147.

E.R. Farnworth, I. Mainville, M.-P. Desjardins, etc. Growth of probiotic bacteria and bifidobacteria in a soy yogurt formulationOriginal Research Article. *International Journal of Food Microbiology*, 2007, 116(1):P174–P181.

V. Ferragut, N.S. Cruz, A. Trujillo, etc. Physical characteristics during storage of soy yogurt made from ultra-high pressure homogenized soymilkOriginal Research Article. *Journal of Food Engineering*, 2009, 92(1):P63–P69.

Tang Yumeng, Chen Jie, Fu Xuebiao, etc. Research on the domestication of yogurt fermentation strains. *Science and Technology of Food Industry*, 2012, 33(2):185–188.

Cui Ruijing, Gao Haisheng, Liu Xiufeng. Study on no odor soybean processing the soybean milk and its stability. *Journal of the Chinese Cereals and Oils*, 2005, 20(3):54–57.

Zhu Dongsheng, Ma Kuikui, Li Qingqing, etc. Selection of protective agent and liquid nitrogen precool in the process of lactobacillus acidophilus freeze-drying. *Food Science*, 2010, 31(1):198–200.

Meng Xiangchen, Du Peng. Lactic acid bacteria and dairy starter cultures. Beijing: Science Press, 2009:323–325.

Meng Yuecheng, Lin Haizhi, Chen Jie. The effect of different thickener on solidifying fermented sour soybean milk quality and structure. *China's Dairy Industry*, 2010, 38(4):38–41.

Chen Tao, Ma Yingkun, Chen Fusheng. Suitable for soybean milk fermented by lactic acid bacteria screening and its application. *Food and Fermentation Industry*, 2014, 40 (3):76–82.

Wang Shuiquan, Bao Yan, Zhang Yanchao, etc. Fermentation Characteristics of lactobacillus fermentation with potential good properties in fermented soybean milk. *Journal of China's Dairy Industry*, 2010, 38(5):7–12.

Section 4: Automation and control, information technology and MEMS

Computational Intelligence in Industrial Application – Ling (ed.)
© *2015 Taylor & Francis Group, London, ISBN: 978-1-138-02818-0*

Reliability analysis of two dissimilar components parallel systems under Poisson shocks

YuTian Chen & XianYun Meng

Department of Applied Mathematics, Yanshan University, Qinhuangdao, China

ABSTRACT: This paper considers a degenerate parallel system with two dissimilar components and a repairman who may take a vacation. Assuming that the operating component can be targeted by the shocks and the arrival time of the shocks follow a Poisson process, with the intensity $\lambda > 0$. The system may fail whenever the amount of shocks exceeds the threshold of the operating component. Time on repairing the failed component follows the exponential distribution; the length of vacation time obeys general continuous distribution. With the Markov process theory, a supplementary variable method, we derive a number of reliability indices of the system. Lastly, a numerical example is given to simulate the effectiveness of the results.

KEYWORDS: Poisson shocks; Geometric process theory; Vacation; Reliability.

1 INTRODUCTION

The shock model is a familiar model in the reliability theory. It attracts the attention of many scholars [1–5]. There are three kinds of shock models that have been extensity studied: extreme shock model, cumulative shock model andδ-shock model. Wu [1–2] researches the extreme shock models: the system fails when an individual shock is too large for the component to endure. He obtains the reliability indices of this system and derives abivariate replacement policy for it. Recently most literatures [3] focus on one component under Poisson shocks with the assumption that the damage caused by a single shock may be neglected and the system fails only when the damage has accumulated to a certain level.

However, this assumption does not apply to all the problems. Wang and Zhang [4] research a repairable shock model with two-type failures. In their paper, they assume that two different kinds of the shocks in a sequence will cause the system failure. Ultimately, they also obtain the reliability indices of the system under this assumption. Furthermore, they also deliberate a $k/n(G)$system which under Poisson shocks, they assume every shock will attract the system independently, the system fails only when the number of the operating components is less than k. Under this assumption, the system reliability function and the system average operating time have been derived.

Lots of research studied parallel components' reliability and maintenance policy [6–7]. Wu [8] considers a two-dependent component and assume the failed components cannot be repaired "as good as new", by the geometric process they obtain a number

of reliability indices of the system. However, system failure may be caused by many reasons, like external shocks, components wear, or aging. Based on the above idea, a model of a pair able parallel system with stand by components has been presented [9]. This model includes two types of failure of the system: human error and common-cause failures. Under this assumption, Laplace transforms of state probabilities and the steady-state availability of the system are developed.

In the practical application, due to the ageing effect or the accumulative wear the systems are deteriorative. The system after repair cannot be "as good as new". This pattern of the operating and repair time can be described as a geometric process [10–11]. Lam [12–13] studies the maintenance policy for a geometric-process. Zhang [14] analysis a repairable cold standby system with priority in use and repair, they also obtain the optimal replacement for this system.

In the practical applications, the repairman need to procure parts for the system or go out to do other works, then he may not be able to repair the failed component in time. Yuan and Xu [15] study a deteriorating system with a repairman who can take multiple vacations. Ke and Wang [16] research a machine repair problem with a single vacation and multiple vacations, respectively.

The reliability indices of two components parallel degenerate system are important to the industries. Analyzing and deriving the indices for this system impacted by shocks seem to be very interesting for the shocks occur at any time in our real lives. Under this motivation, we analyze the reliability of a parallel system with two components under Poisson shocks and obtain the liability indices of this model. Finally, we give a numerical example to simulate the effectiveness of the results.

2 SYSTEM ASSUMPTIONS

A1. At the beginning time, both of the components are new and in the working state.

A2. After repair component is not "as good as new", component 2 is "as good as new".

A3. The distribution of the operating time and the repair time of component i in the n th cycle are, respectively,

$$F_n^{(1)}(t) = 1 - e^{-a^{n-1}\lambda_1 t}, F_n^{(2)}(t) = 1 - e^{-\lambda_2 t}$$

$$G_n^{(1)}(t) = 1 - e^{-b^{n-1}u_1 t}, G_n^{(2)}(t) = 1 - e^{-u_2 t}$$

where $t \geq 0, a > 0, 0 < b < 1, \lambda_1 > 0, \lambda_2 > 0, u_1 > 0, u_2 > 0; n = 1, 2...$

A4. The system may subject to the shocks. The advent of the shocks obeys a Poisson process $\{S(t), t \geq 0\}$ with the intensity of $\lambda > 0$. X, the magnitude of every shock is an independent variable.

A5. When a shock comes, it will only affect the operating component. The operating component fails if the magnitude of one shock exceeds the threshold, where the threshold of component i is a random variable τ_i with a distribution function of ϕ_i.

A6. When a component fails with the repairman in the system, it will be repaired at once. The repair rule follows that "first-in-first-out". If the nearly component fails when the other is being repaired, the newly one must wait for repair and the system is down. If there is no failed component in the system after repairman vacation, he needs to stay in the system and remain idle until the first failed component appears. Denote Z as the vacation time, its distribution is:

$$H(t) = \int_0^t h(x)\mathrm{d}x = 1 - e^{-\int_0^t v(x)\mathrm{d}x}, E(H) = \frac{1}{c}$$

A7. All of the random variables mentioned above are independent.

3 MODEL DEVELOPMENT

Based on the above model assumption A4 and A5, we can obtain the probability of one shock leads to the operating component i loss of efficiency in the nth cycle is

$$p_i = P\{\hat{X}_n > \tau_n\} = \int_0^\infty P\{\tau_n < x | \hat{X}_n = x\}\mathrm{d}P\{\hat{X}_n \leq x\}$$

Lemma 1(see[1]).The distribution function of ξ_n^i is

$$V_n(x) = 1 - e^{-\lambda p_i x}, x > 0, (i = 1, 2; n = 1, 2, 3...)$$

In the same way, we can get the common working time of both units that it fails caused by the shocks in the nth cycle, its distribution can be denoted as

$$W_n(x) = 1 - e^{-\lambda p_1 p_2 x}, x > 0, (n = 1, 2, 3...)$$

Let $N(t)$ be the state of the system at time t, then we have:

1: component 1 and component 2 are both in the working states; the repairman is idle.

2: component1 and component 2 are both in the working states; the repairman is taking a vacation.

3: component1 is waiting for repair while component 2 is working; the repairman is taking a vacation.

4: component1 is being repaired while component 2 is working.

5: component1 is working while component 2 is waiting for repair; the repairman is taking a vacation.

6: component1 is working while component 2 is being repaired.

7: component1 and component 2 are both waiting for repair; the repairman is taking a vacation.

8: component1 is being repaired while component 2 is waiting for repair.

9: component1 is waiting for repair while component 2 is being repaired.

Obviously, the operating state is $W = \{1, 2, 3, 4, 5, 6\}$ and the failure state is $\{7, 8, 9\}$. With the assumption we know that $\{N(t), t > 0\}$ is not a Markov process. Hence, we introduce the supplementary variables as follows

$Y(t)$: Elapsed vacation time when the repairman is taking vacation;

$I(t)$: Number of cycles of component 1 at time t.

Then $\{N(t), Y(t), I(t), t \geq 0\}$ is a continuous vector Markov process with the following state space:

$$P_{i,k}(t) = P\{N(t) = i, I(t) = k\}, i = 1, 4, 6, 8, 9;$$

$$P_{j,k}(t, y)\mathrm{d}y = P\begin{cases} N(t) = j, I(t) = k, \\ y < Y(t) \leq y + \mathrm{d}y \end{cases} j = 2, 3, 5, 7$$

By system analysis, we have the following equations:

$$\left(\frac{\mathrm{d}}{\mathrm{d}t} + \alpha + \beta + r_k\right)P_{1,k}(t) = \int_0^\infty v(y)P_{2,k}(t, y)\mathrm{d}y \quad (1)$$

$$\left(\frac{\partial}{\partial t}+\frac{\partial}{\partial y}+\alpha+\beta+r_k+v(y)\right)P_{2,k}(t,y)=0 \quad (2)$$

$$\left(\frac{\partial}{\partial t}+\frac{\partial}{\partial y}+\beta+v(y)\right)P_{3,k}(t,y)=r_kP_{2,k}(t,y) \quad (3)$$

$$\left(\frac{d}{dt}+b^{k-1}u_1+\beta\right)P_{4,k}(t)=r_kP_{1,k}(t)$$
$$+\int_0^\infty v(y)P_{3,k}(t,y)dy+u_2P_{9,k}(t) \quad (4)$$

$$\left(\frac{\partial}{\partial t}+\frac{\partial}{\partial y}+r_k+v(y)\right)P_{5,k}(t,y)=\beta P_{2,k}(t,y) \quad (5)$$

$$\left(\frac{d}{dt}+u_2+r_k\right)P_{6,k}(t)=\beta P_{1,k}(t)$$
$$+\int_0^\infty v(y)P_{5,k}(t,y)dy+b^{k-2}u_1P_{8,k-1}(t) \quad (6)$$

$$\left(\frac{\partial}{\partial t}+\frac{\partial}{\partial y}+v(y)\right)P_{7,k}(t,y)=\beta P_{3,k}(t,y)$$
$$+\alpha P_{2,k}(t,y)+r_kP_{5,k}(t,y) \quad (7)$$

$$\left(\frac{d}{dt}+b^{k-1}u_1\right)P_{8,k}(t)=\alpha P_{1,k}(t)+\beta P_{4,k}(t)$$
$$+\int_0^\infty v(y)P_{7,k}(t,y)dy \quad (8)$$

$$\left(\frac{d}{dt}+u_2\right)P_{9,k}(t)=r_kP_{6,k}(t) \quad (9)$$

$$P_{2,k}(t,0)\overset{.}{=}b^{k-2}u_1P_{4,k-1}(t)+u_2P_{6,k}(t) \quad (10)$$

$$P_{2,1}(t,0)=P_{3,1}(t,0)=P_{5,1}(t,0)=P_{7,1}(t,0)$$
$$=P_{6,1}(t)=P_{9,1}(t)=0 \quad (11)$$

$$\left(\frac{d}{dt}+\alpha+\beta+r_1\right)P_{1,1}(t)=1 \quad (12)$$

$$\left(\frac{d}{dt}+u_1+\beta\right)P_{4,1}(t)=r_1P_{1,1}(t) \quad (13)$$

$$\left(\frac{d}{dt}+u_1\right)P_{8,1}(t)=\alpha P_{1,1}(t)+\beta P_{4,1}(t) \quad (14)$$

The initial condition is $P_{1,1}(0)=1$, otherwise is 0.
The Laplace transform of the above equations are, respectively, given by:

$$P_{1,k}^*(s)=\frac{h^*(\Delta_k)}{\Delta_k}P_{2,k}^*(s,0) \quad (15)$$

$$P_{2,k}^*(s,y)=e^{-\Delta_ky}\bar{H}(y)P_{2,k}^*(s,0) \quad (16)$$

$$P_{3,k}^*(s,y)=\frac{\left[e^{-(s+\beta)y}-e^{-\Delta_ky}\right]r_k}{\alpha+r_k}\bar{H}(y)P_{2,k}^*(s,0) \quad (17)$$

$$P_{4,k}^*(s)=\frac{C_kP_{2,k}^*(s,0)+u_2P_{9,k}^*(s)}{\Pi_k} \quad (18)$$

$$P_{5,k}^*(s,y)=\frac{\beta}{\alpha+\beta}\left(e^{-(s+r_k)y}-e^{-\Delta_ky}\right)\bar{H}(y)P_{2,k}^*(s,0) \quad (19)$$

$$P_{6,k}^*(s)=\frac{B_kP_{2,k}^*(s,0)+b^{k-2}u_1P_{8,k-1}^*(s)}{s+u_2+r_k} \quad (20)$$

$$P_{7,k}^*(s,y)=\bar{H}(y)P_{2,k}^*(s,0)\times$$
$$\left[\frac{r_k}{\alpha+r_k}e^{-(s+\beta)y}+\frac{\beta}{\alpha+\beta}e^{-(s+r_k)y}\right]+$$
$$\bar{H}(y)P_{2,k}^*(s,0)\times\left[e^{-sy}-\frac{\alpha^2-\beta r_k}{(\alpha+r_k)(\alpha+\beta)}e^{-\Delta_ky}\right] \quad (21)$$

$$P_{8,k}^*(s)=\frac{\alpha h^*(\Delta_k)\Pi_k+\beta\Delta_kC_k}{(\Pi_k-\beta)\Pi_k\Delta_k}P_{2,k}^*(s,0)$$
$$+\frac{\beta u_2}{(\Pi_k-\beta)\Pi_k}P_{9,k}^*(s) \quad (22)$$

$$P_{9,k}^*(s)=\frac{r_k\left[B_kP_{2,k}^*(s,0)+b^{k-2}u_1P_{8,k-1}^*(s)\right]}{(s+u_2)(s+u_2+r_k)} \quad (23)$$

$$P_{1,1}^*(s)=\frac{1}{\Delta_1},\quad P_{4,1}^*(s)=\frac{r_1}{\Pi_1\Delta_1} \quad (24)$$

$$P_{8,1}^*(s)=\frac{\beta(\alpha+r_1)+\alpha(s+u_1)}{(s+u_1)\Delta_1\Pi_1} \quad (25)$$

Where $\Delta_k=s+\alpha+\beta+r_k,\Pi_k=s+\beta+b^{k-1}u_1$

$$B_k=\frac{\beta h^*(\Delta_k)}{\Delta_k}+\frac{\beta\left[h^*(s+r_k)-h^*(\Delta_k)\right]}{\alpha+\beta}$$

$$C_k=\frac{r_kh^*(\Delta_k)}{\Delta_k}+\frac{r_k\left[h^*(s+\beta)-h^*(\Delta_k)\right]}{\alpha+r_k}$$

4 RELIABILITY INDICES

4.1 System availability

Based on the definition of system availability of $A(t)$, we have

$$A(t) = \sum_{k=1}^{\infty} \{\sum_i P_{i,k}(t) + \sum_j \int_0^{\infty} P_{j,k}(t,y)dy\}$$

$$i = 1,4,6; j = 2,3,5$$

The Laplace transform of $A(t)$ is

$$A^*(s) = \sum_{k=1}^{\infty} \{\sum_i P_{i,k}^*(s) + \sum_j \int_0^{\infty} P_{j,k}^*(s,y)dy\}$$

$$i = 1,4,6; j = 2,3,5$$

With Eqs. (15–25) we obtain the explicit Laplace transform representation system availability

$$A^*(s) = \sum_{k=2}^{\infty} \{\sum_i P_{i,k}^*(s) + \sum_j \int_0^{\infty} P_{j,k}^*(s,y)dy\}$$

$$+ \frac{\Pi_1 + \gamma_1}{\Delta_1 \Pi_1}, i = 1,4,6; j = 2,3,5$$

4.2 System reliability

In order to analyse the reliability of the system, we let above three failure states {7,8,9} be the absorbing states, in this way we can get another vector Markov process $\{\tilde{N}(t), I(t), Y(t), t \geq 0\}$. With the similar method as Section 2, we get the following differential equations:

$$\left(\frac{d}{dt} + \alpha + \beta + r_k\right)Q_{1,k}(t) = \int_0^{\infty} v(y)Q_{2,k}(t,y)dy$$

$$\left(\frac{\partial}{\partial t} + \frac{\partial}{\partial y} + \alpha + \beta + r_k + v(y)\right)Q_{2,k}(t,y) = 0$$

$$\left(\frac{\partial}{\partial t} + \frac{\partial}{\partial y} + \beta + v(y)\right)Q_{3,k}(t,y) = r_k Q_{2,k}(t,y)$$

$$\left(\frac{d}{dt} + b^{k-1}u_1 + \beta\right)Q_{4,k}(t) = r_k Q_{1,k}(t)$$
$$+ \int_0^{\infty} v(y)Q_{3,k}(t,y)dy$$

$$\left(\frac{\partial}{\partial t} + \frac{\partial}{\partial y} + r_k + v(y)\right)Q_{5,k}(t,y)$$
$$= \beta Q_{2,k}(t,y)$$

$$\left(\frac{d}{dt} + u_2 + r_k\right)Q_{6,k}(t) = \beta Q_{1,k}(t)$$
$$+ \int_0^{\infty} v(y)Q_{5,k}(t,y)dy$$

The boundary conditions are:

$$Q_{2,k}(t,0) = b^{k-2}u_1 Q_{4,k-1}(t) + u_2 Q_{6,k}(t)$$

$$\left(\frac{d}{dt} + \alpha + \beta + r_1\right)Q_{1,1}(t) = 1$$

$$\left(\frac{d}{dt} + u_1 + \beta\right)Q_{4,1}(t) = r_1 Q_{1,1}(t)$$

$$Q_{2,1}(t,0) = Q_{3,1}(t,0) = Q_{5,1}(t,0) = Q_{6,1}(t) = 0$$

The initial condition is $Q_{1,1}(0) = 1$, otherwise is 0. The solutions of the Laplace transform for the above differential equations are respectively, given by

$$Q_{1,k}^*(s) = \frac{h^*(\Delta_k)}{\Delta_k}Q_{2,k}^*(s,0) \tag{26}$$

$$Q_{2,k}^*(s,y) = e^{-\Delta_k y}\bar{H}(y)Q_{2,k}^*(s,0) \tag{27}$$

$$Q_{3,k}^*(s,y) = \frac{\left[e^{-(s+\beta)y} - e^{-\Delta_k y}\right]r_k}{\alpha + r_k}\bar{H}(y)Q_{2,k}^*(s,0) \tag{28}$$

$$Q_{4,k}^*(s) = \frac{C_k}{\Pi_k}Q_{2,k}^*(s,0) \tag{29}$$

$$Q_{5,k}^*(s,y) = \frac{\beta}{\alpha + \beta}\left[e^{-(s+r_k)y} - e^{-\Delta_k y}\right]\bar{H}(y)Q_{2,k}^*(s,0) \tag{30}$$

$$Q_{6,k}^*(s) = \frac{B_k}{s + u_2 + r_k}Q_{2,k}^*(s,0) \tag{31}$$

With the boundary condition and initial conditions, we can get:

$$Q_{1,1}^*(s) = \frac{1}{\Delta_1}, Q_{4,1}^*(s) = \frac{r_1}{\Delta_1 \Pi_1} \tag{32}$$

$$Q_{2,2}^*(s,0) = \frac{u_1 r_1(s + r_2 + u_2)}{(s + r_2 + u_2 - \beta u_2)\Delta_1 \Pi_1} \tag{33}$$

According to the definition of the system reliability $R(t)$, we have:

$$R(t) = \sum_{k=1}^{\infty} \{\sum_i Q_{i,k}(t) + \sum_j \int_0^{\infty} Q_{j,k}(t,y)dy\} \tag{34}$$

$$i = 1,4,6; j = 2,3,5$$

With the above equations (26-33), the Laplace transform of $R(t)$ is given by:

$$R^*(s) = \frac{\Pi_1 + \gamma_1}{\Delta_1 \Pi_1} + \sum_{k=2}^{\infty} \{\sum_i Q_{i,k}^*(s) + \sum_j \int_0^{\infty} Q_{j,k}^*(s,y)dy\} \tag{35}$$

$$i = 1,4,6; j = 2,3,5$$

4.3 The mean time to the first failure of the system (MTTFF)

$$MTTFF = \sum_{k=2}^{\infty} \{ \sum_i Q_{i,k}^*(0) + \sum_j \int_0^{\infty} Q_{j,k}^*(0,y)dy \}$$

$$+ \frac{\Pi_1 + \gamma_1}{\Delta_1 \Pi_1}, i = 1,4,6; j = 2,3,5$$

With $MTTFF = \int_0^{\infty} R(t)dt = \lim_{s \to 0} R^*(s)$ and the result of 3.2 imply the result.

4.4 The idle probability of repairman

According to the system analysis, easily get that the repairman is idle only when the vacation time is over and both of the components are operating. Thus, the idle probability of the repairman at time t is:

$$I(t) = P\{N(t) = 1\} = \sum_{k=1}^{\infty} P_{1,k}(t)$$

The Laplace transform of $I(t)$ is

$$I^*(s) = \sum_{k=1}^{\infty} P_{1,k}^*(s)$$

With the above result $P_{1,1}^*(s) = \Delta_1^{-1}$, the solution Laplace Transform of $I(t)$ is:

$$I^*(s) = \frac{1}{\Delta_1} + \sum_{k=2}^{\infty} P_{1,k}^*(s)$$

4.5 The vacation probability of the repairman

Let $V(t)$ be the probability of the repairman vacation by the analysis of the system, only when the system at states $\{2, 3, 5, 7\}$. We can get the equations

$$V(t) = \sum_{k=1}^{\infty} \int_0^{\infty} \sum_j P_{j,k}(t,y)dy, j = 2,3,5,7$$

The Laplace transform of $V(t)$ is

$$V^*(s) = \sum_{k=1}^{\infty} \int_0^{\infty} \sum_j P_{j,k}(s,y)dy, j = 2,3,5,7$$

With the result of Section 2, we can obtain

$$V^*(s) = \sum_{k=2}^{\infty} \int_0^{\infty} \sum_j P_{j,k}(s,y)dy, j = 2,3,5,7$$

The above is the Laplace transform of the vacation probability of repairman.

5 NUMBER EXAMPLES

Based on these results above, we can find that it is difficult to obtain the transient results of the reliability indices for the system proposed in this paper. We can only obtain the steady-state results of some reliability of the system. To validate the above derivation,

We conduct the following numerical experiments. Here, we assume

$$F_1(x) = \begin{cases} 1 - e^{-\theta x}, x > 0; \\ 0, x \le 0. \end{cases} \quad \Phi_1(x) = \begin{cases} 1 - e^{-\theta x}, x > 0; \\ 0, x < 0. \end{cases}$$

$$\Phi_2(x) = \begin{cases} 1 - e^{-\frac{\theta}{3}x}, x > 0; \\ 0, x < 0. \end{cases}$$

We obtain $p_1 = 0.5, p_2 = 0.25$. For Eq. (35) we make inverse Laplace transform. Let $\lambda_1 = 0.1, \lambda_2 = 0.4, \mu_1 = 0.4, \eta = 0.01, \lambda = 0.8$. When $k = 1, t = 1, 3, 5, 7, 9$. by Matlab, we can obtain these solutions of $Q_{1,1}(t), Q_{4,1}(t)$ (see Table 1). We can obtain the solution of $R(t) \approx Q_{1,1}(t) + Q_{4,1}(t)$. When $t = 1, 3, 5$ 7, 9, the results are $R(1) = 0.5940$, $R(3) = 0.1942$, $R(5) = 0.0605$, $R(7) = 0.0185$, $R(9) = 0.0056$. Clearly, we can see that the value of the system reliability $R(t)$ is decreasing when t increases (see Fig. 1). For a deteriorating system, the observation is consistent with our intuition.

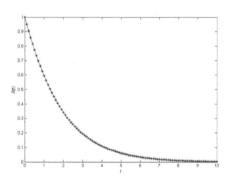

Figure 1. The curve of the reliability $R(t)$ against t.

6 CONCLUSION

In this paper, we consider the reliability indices of a parallel system with two different components and a repairman who may take vacation. We assume the operating component may subject to shocks. Such a system is one of important repaired systems in the reliability engineering and is also difficult to analyze as it is a degenerate system which has many random variables with general distributions included. Lastly, a numerical data experiment shows the relationship between the derived reliability indices.

REFERENCES

Ma, S, 2011.The parallel system with repair of non-new and repairman vacation. *Mathematics in Practice and Theory* 41(16):111–117.

Liu, H&Meng, X, 2012. A geometric process model for parallel system with two dependent components. *Journal of natural science of Hei Long Jiang University* 29(2):169–173.

Liang,X, 2010. Study on a cold standby repairable deteriorating system with repairman vacation. *Journal of Systems Engineering*25(3):426–432.

Wu, S, 1995.Reliability analysis of two-dependent unit system without being repaired "as possible as new".*ACTA Automatica Sinica* 2(12):232–237.

Ke, J C& Wang, K H, 2007. Vacation policies form machine repair problem with two type spares. *Applied Mathematical Modeling* 31(5):880–894.

Liu H&MengX, 2012. A warm standby with repair of non-new and repairman vacation. *Acta Automatica Sinica* 38(4):639–645.

Meng X, 2006. Reliability analysis of warm standby repairable system of two components with continuous life-time switch and priority. *Journal of Yanshan University* 30(1): 51–56.

Shanthilumar, J, 1983. General shock models associated with correlated renewal sequences. *J. of Appl. Prob* 20:600–614.

Shanthilumar J, 1984. Distribution properties of the system failure time in a general shock model. *Adv. Appl. Prob,* 16:363–377.

Zhang, Y, 1998.Reliabilityof consecutive-k-out-of-$k/n(G)$ repairable system. *International Journal of Systems Science* 29(12):1375–1379.

WangG & Zhang.Y, 2009. Reliability analysis for$k/n(G)$ system under Poisson shocks. *Chinese Journal of Applied Probability* 29(25):1–11.

Liang.X,2010. Reliability analysis for$k/n(G)$system with possible different components under Poisson shocks. *Journal of Nanjing Audi University* 7(2):80–83.

WuQ.&WuS, 2011. Reliability analysis of two-unit cold standby repairable systems under Poisson shocks. *Applied Mathematics and Computation*218(1): 171–182.

ZhangY.&Wang G, 2009. A geometry process repair model for a repairable cold standby system with priority in use and repair. *Reliability Engineering and System Safety*94(5): 1782–1787.

Yuan L, 2011.A deteriorating system with its repairman having multiple vacations. *Applied Mathematical and Computation* 217(10):4980–4989.

KenJ. & WangK, 2007. Vacation policies for machine repair problem with two type spares. *Applied Mathematical Modelling* 31(5):880–894.

Computational Intelligence in Industrial Application – Ling (ed.)
© 2015 Taylor & Francis Group, London, ISBN: 978-1-138-02818-0

Research and optimization of micro-oil ignition system relates to the protection logic for 350MW opposed fire boiler

B. Lin
State Grid Shandong Electric Power Research Institute, Jinan, China

W.Y. Sui
Shandong University, Jinan, China

ABSTRACT: This paper mainly introduced the application of micro-oil gun ignition system on 2×350MW opposed fire boiler in Xinfa Power Plant. According to the design of micro-oil ignition technology characteristics and operation scheme, we compared the big oil ignition control scheme, researched and optimized the boiler main protection logic and the protection logic of the original bottom pulverizer. In the mean time, we improved the micro-oil ignition system self-protection and realized the safe starting and operation of the boiler.

KEYWORDS: Opposed Fire Boiler; Micro-oil Ignition System; FSSS; The Original Bottom Pulverizer; Protection Logic.

1 INTRODUCTION

With the development of our national economy, the total power installed capacity and unit capacity have been increasing sharply. The light diesel oil that used only for boiler start-stop, low-load combustion and stable combustion in the power plant has reached several tens of millions of tons annually. In order to effectively reduce the loss of boiler's start-stop, low load stable combustion to save cost, technology of micro-oil ignition has been widely used. This has proposed new requirements for the protection related to thermal control.

Xinfa Power Plant adopts the supercritical pressure 350MW boiler which is developed, designed and manufactured by Harbin boiler Co., ltd with independent intellectual property rights. Its furnace is HG-1150 / 25.4-PM1, an once-through boiler with opposed wall burning, single reheat and supercritical pressure swing operation. It also has a single furnace, balanced ventilation, solid slag, all steel frame, full suspension structure, π-type layout, and adopts air start expansion system without a recirculation pump. By studying micro-oil ignition system's successful application of the furnace, this paper analyzed system design and problems that occurred during the application process. This paper studied micro-oil ignition system's successful application of the furnace and analyzed system design and problems that occurred during the application process. At the same time it is based on micro-oil ignition technology. It optimized FSSS system, pulverizer system and micro-oil

ignition system's own protection logic. The operational reliability of the micro-oil ignition system can be further improved.

2 MICRO-OIL IGNITION PRINCIPLE

Micro-oil ignition uses the fuel's atomizing process which can atomize fuel into ultra-fine droplets. Then the fuel can be burned. And at the same time using the heat generated by the burning of fuel for initial heating, expansion, and post-heating. This process can make oil droplets evaporation completed within a very short time. This can let oil gun directly burn gas state fuel thereby greatly improving the efficiency of fuel combustion and the flame temperature. The flame propagates at supersonic speed, extremely rigid, completely transparent and the center temperature up to 1500°C. The flame with high combustion temperature formatted by small gasification oil gun is exposed directly with coal particles, changing the characteristics of coal particles, which make volatile and precipitation with high heating rates so much faster. With the sharp rise of the temperature, coal particles make itself violently crush and burn, and then ignite the remaining pulverized coal particles.

Micro-oil ignition system achieves staged combustion of coal with energy progressively larger. It can reach to the combustion of pulverized coal accelerate purpose and greatly reduce the ignition energy required for the other pulverized coal to meet the needs of boiler's start-stop and steady combustion of low load.

3 PROTECTION LOGIC AND OPTIMIZATION STUDIES

3.1 *FSSS system*

FSSS system on the boiler is an important thermal protection system. It is composed by BCS and FSS. The function is that if meeting the serious security issues, it can timely trip to protect boiler equipments and do the correct operation on the related systems equipment to ensure the safety. FSSS system completes the protection of the main fuel trip (MFT), OFT, boiler ignition, furnace purge and fuel system leaks trials and other.

Micro-oil Ignition system is mainly involved in the whole furnace extinguishment and ignition failure protection of FSSS. The Whole furnace extinguishment protection uses the same logic of big oil gun protection. Ignition failure protection uses the large oil gun crew. Since the large amount of oil in, using oil inlet close as the sign, detect the pulse signal. And if 3 times ignition unsuccessful, it will trigger MFT, to prevent excessive amount of fuel flowing into the furnace, causing deflagration. The specific logic is as follows:

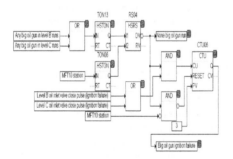

Figure 1. Big oil gun ignition failure 3 times.

The micro-oil Ignition oil gun ignition fails every time, the amount of fuel leaking into the furnace is significantly lower than big oil gun. So we adopt the method that big oil gun is inappropriate. We have to judge micro-oil gun separately. The micro-oil gun is modified to 6 times the ignition failure triggered. The specific logic is as follows:

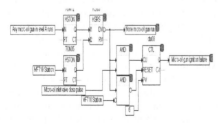

Figure 2. Micro-oil ignition failure 6 times.

This system can be used in boiler starting ignition and stable combustion at low load. In extreme conditions, the micro-oil gun ignition failure occurs, resulting in unnecessary MFT. So we could join any burner to lock and avoid triggering MFT. The specific logic is as follows:

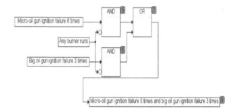

Figure 3. Ignition failure to MFT.

3.2 *Pulverizer system*

The pulverizer system is BBD mill with primary air of positive pressure direct injection. The boiler totally fixes up 6 layers burners (front 3, back 3), each layer for 4 burners, and altogether 24 low NOX axial-flow swirl burners. The micro-oil ignition system enters into the single-ended tripping protection of the pulverizer system. The front and back wall on the same layer of the opposed fire boiler can be operated separately, therefore the unified pattern of the micro-oil ignition based on the tangential fire boiler isn't suitable any more. In order to make the judgment of the micro-oil ignition more accurate, the ignition pattern is subdivided into front wall and back wall.

BBD mill costs much more time to place coal taking the coal feeder runs within 180s, judging the micro-oil flame detection 2/4 fireless, unable to properly trigger the mill single-ended tripping. Primary air mix at the entrance of the mill. The instruction of pneumatic control valves is over ≥4%, that can be thought the powder has been in the boiler, put the coal mill into actual operation, through the signal and judgment can solve the problem well. The logic is as follows:

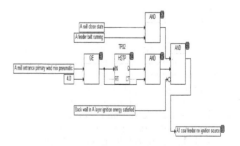

Figure 4. Single-ended tripping.

3.3 Micro-oil Ignition System Protection

Micro-oil ignition system lays out on the bottom of the boiler, corresponding to A mill 8 burners. Mainly by the composition of oil gun, air register (steady burner), high-energy ignition device etc. This system can be used in boiler starting ignition and stable combustion at low load. Under the control of DCS, it can implement protection, automatic ignition, fuel injection ignition and send the ignition signal out by virtue of the flame inspection device.

In the micro-oil ignition system self-protection logic, MFT, OFT, A mill tripping, primary air fan are long signals. To avoid the abnormal operation of the micro-oil ignition, all the mill outlet doors closing, the temperature of the burner center cylinder>600°C, oil inlet valve separating the off- stop 25S and flame inspection fireless is the pulse tripping signal. Take the high temperature of burner center cylinder as an example, if tripping out of the oil gun, the operating crew have not checked the reason of tripping seriously and put in the oil gun at once. When the tripping does not interlock, then it will cause the higher temperature to break down the oil gun. So during the starting of the oil gun, the control logic of RS trigger is permitted with the no tripping condition interlocking.

4 CONCLUSION

After the study and optimization, the logic sufficiently considered the influence of the micro-oil ignition system devoted working condition to the thermal protection. The improved scheme is feasible, implementing protecting the entire input, so that ensures the normal starting and operating while the boiler adopts the micro-oil ignition system, further improving reliability and security.

ACKNOWLEDGMENT

The research work of this thesis was conducted in Xinfa Power Plant, received enthusiastic help from the staff of the power plant, Zhongshiyitong Group and Hollysys, I really want to express my sincere thanks!

REFERENCES

Jia, Q.Y. & Liu, C.Y. 2012. Application of micro-oil ignition technique in Dagang Power Plant's 328.5MW unit. *North China Electric Power* 1: 36–45.

Huang, S.F. 2009. Problems of micro-oil ignition of boiler in Shajiang C Power Station. *Guangdong Electric Power* 22(4): 56–58.

HG-1150 / 25.4-PM1 Supercritical Once-through Boiler Specifications. 2012. Harbin boiler Co., ltd Press.

Liu, W.W. & Liang, W.Z. 2013. Successful experiences about the application of micro oil ignition technique on super-critical boiler. *Shanxi Electric Power* 6(183): 56–58.

Wang, Y.J. 2009. *The principle and application of thermal protection of thermal power plant*. Beijing: China Power Press.

Xie, W.G. 2009. Influence of making use of oil-saving ignition on protective logic and its improvement schemes. *Journal of Electric Power* 24(4): 331–333.

Xun, G.Y. & Zhao, J. 2004. The control strategies for the direct-fired system of ball pulverizer in DCS. *SCI / TECH Information Development & Economy* 14(10): 212–214.

Liu, H.L. 2010. Practice and thinking of domestic micro-oil ignition transformation of 600MW boiler. *Energy Conservation Technology* 28(159): 87–90.

Tang, Y.H. & Liu, H.B. 2012. Calculating as-fired coal quantity for double-inlet and double-outlet mill based on soft sensing. *Instrument Technique and Sensor* 5: 27–31.

Wang, L.D. & Yu, G. 2010. Control logic implementation of tiny-oil ignition technology applied in opposed firing boiler. *Journal of Shanghai University of Electric Power* 26(3): 262–265.

Computational Intelligence in Industrial Application – Ling (ed.)
© *2015 Taylor & Francis Group, London, ISBN: 978-1-138-02818-0*

The intelligent control system of agricultural greenhouses based on Internet of Things

FuPing Wang, XingZe Li & ZhiXin Liu
Innovative-venturing Education Centre, Beifang University of Nationalities, Yinchuan, China

ABSTRACT: Because the traditional artificial cultivation management methods cannot achieve the scientific requirements, the intelligent control system in the greenhouse is a trend. But currently there are few effective intelligent control systems in China, and the one from foreign countries fails to accord with our national situations. Therefore, this paper presented a kind of intelligent control system of an agricultural greenhouse based on Internet of Things. In this system, the environmental data acquired by sensors of nodes could be transmitted to ZigBee center node by wireless ZigBee node and wired RS485 node ZigBee node, and then it will be passed on to the host computer using GPRS. Control center conducts analysis and judgment for the acquired data and issues the appropriate action command to make the slave computer execute actions such as opening the window, pulling the curtain, irrigation, alarm and so on.

KEYWORDS: Host computer; ZigBee; Intelligent control.

1   INTRODUCTION

In recent years, the agricultural greenhouse cultivation has brought great convenience for the improvement of people's living standard, so it has been rapidly popularized and applied. The temperature, humidity, light intensity, CO_2 concentration and other factors of planting environment have a great impact on crop production [1]. However, the greenhouse system is a nonlinear, time-varying and lagging system, so it is difficult to establish mathematical model for it. The conventional control methods cannot obtain satisfactory static and dynamic performance [2]. According to the characteristics of the greenhouse environment control, this paper presents a kind of intelligent control system of agricultural greenhouse based on Internet of Things. The system consists of three subsystems: host computer system, transmission system and slave computer system. The host computer system includes the control center, environmental data analysis, environmental data storage and historical data query. The transmission system includes the GPRS remote module, wireless transmission module of ZigBee node and RS485 cable transmission module; the data is mainly the environmental information data acquired by the transmission sensor [3]. The slave computer mainly includes environmental information collection module and action execution module, which could acquire environmental information by sensors, achieve real-time uploading and execute the appropriate actions according to instructions issued by the host computer. The three interrelated systems

interrelate and restrain each other, and they have common goals and core control strategies, aiming to complete the intelligent control task of agricultural greenhouse cultivation. This improves scientific cultivation and intelligent management of agricultural greenhouse.

2   OVERALL ARCHITECTURE OF THE SYSTEM

The intelligent control system of agricultural greenhouses includes control center module, data storage module, network transmission module, field data acquisition module and action execution module. Wherein the control center module is the core part of the system and it is responsible for the analysis, storage and processing of on-site data. The staff could conduct real-time monitoring for the on-site changes of environment, query historical data and make the appropriate control commands to adjust the field environmental factors to keep its scientificity. The data storage module is mainly in charge of storing the on-site video data, environmental factors, and some data analysis reports, which will then be used by management personnel. The network transmission module consists of wireless ZigBee and wired RS485 combined with GPRS components. Its function is to transmit the field data information acquired by child nodes to the control center for analysis. The field data acquisition module is to collect air temperature humidity, light intensity, carbon dioxide concentration, soil temperature and humidity as well as video data, etc. Actions execution module is composed by an automatic watering module and shutter control sub-module, which could adjust the environment in the greenhouse according to the commands of conrol conter. The overall structure of the system is as shown in Fig. 1.

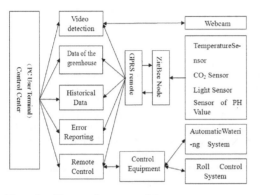

Figure 1. The overall structure of the system.

3 HARDWARE DESIGN OF THE SYSTEM

3.1 *Slave computer system*

The slave computer is in the control environment, mainly composed by environmental information collection module and action execution module, as shown in Fig. 2. It could achieve real-time data acquisition, processing and display of greenhouse environment, as well as the regulation for the greenhouse environment. The acquired parameters could be transmitted to the host PC via RS485, ZigBee and GPRS and then it could make action commands according to the control decisions of the host computer and control the actuator to carry out regulation; it could achieve offline operation and work independently when the host computer is power off. Management staff or specialists could preset and acquire environment parameters by the keyboard and then autonomously operate the decision-making process of the slave computer, aiming to achieve automatic control of the greenhouse [4]. The independent control flow chart of slave computer is as shown in Fig. 2.

3.2 *Data acquisition and control module*

3.2.1 *Sensor selection*
The system adopts the SM2801B soil moisture sensors from Sonbest Company. SM2801B, the self-developed products of SONBEST, has superior accuracy and long term stability due to its core elements with industrial-grade precision. The sensor has small volume and is easy to install and carry. It has

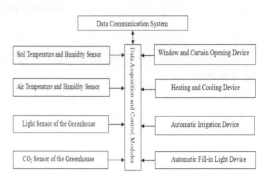

Figure 2. Slave machine structure.

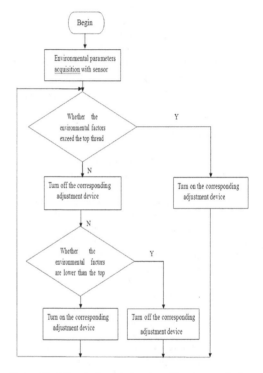

Figure 3. The independent control process of slave computer.

reasonable structure design and the sealing is good. The stainless steel probe could ensure applicability and breadth. The soil moisture detection sensors could be buried in the soil and dams for a long term and then be used in conjunction with data acquisition. It can be used as the sentinel surveillance or mobile measuring instrument of moisture and temperature.

The system adopts the WTA-100 soil temperature sensor, which is made by precious platinum resistance probe with less thermal response time and could reduce dynamic errors. It could

390

accurately measure the temperature of deep layer, shallow layer and surface of the soil and has a variety of signal output formats. WAH-C10 temperature and humidity sensor could block the dirt effectively, preciously and stably due to its stainless steel shell and the filter at the front mounted convection mouth. WLS-TH100 light sensors use silicon optical principles to measure all the radiation from sun and sky, the received signal has higher accuracy through the cosine correction; the sensor cable is also loaded with a resistance, converting the microampere level current of the sensor into millivolt voltage signal. WSD-C200 wind sensor uses hard corrosion-resistant and anti-aging thermoplastic plastics, stainless steel shell with electroplated aluminum enclosure, which makes the device simpler and lighter; the output signal has a standard format. Fig. 4 is the real product of soil temperature and humidity sensor node.

Figure 4. The real product of soil temperature and humidity sensor node.

3.2.2 Data acquisition and control module

The system uses DMG-1 data acquisition and control module and the core is 16-bit ultra-low power microcontroller. It uses MODBUS statute which is open, simple and easy to get text. It is a universal data acquisition and control module with RS485 interface and RE232 interface, so it has higher communication rates and longer communication distance when using the RS485 interface.

3.3 Actuator architecture

The actuators of this system include windows opening system, curtain pulling system, fans – wet curtain cooling system, greenhouse heating system, greenhouse irrigation system and video surveillance system. These systems could give orders to the data acquisition and control module by the host computer and utilize the corresponding electromagnetic valve controller switches, thus achieving the automation and intelligence.

4 SYSTEM SOFTWARE DESIGN

4.1 Development of ZigBee wireless network software

The system adopts the CC2530 wireless chip from TI Company as hardware support, IAR Embedded Workbench integrated compiler environment from IAR Company as basis and TIZ-Stack software as the operating system. On this basis, the sensor module parameters reading function used to monitor the greenhouse environment is utilized, including soil temperature and humidity sensors, air temperature and humidity sensors and carbon dioxide concentration sensor, etc. For the acquired sensor data, read the data according to the sensor datasheets and steps of install agreement.

4.1.1 IAR integrated development environment

IAR Embedded Workbench (EW for short) is a set of powerful software development tools for compilation and debugging of embedded applications written by C, C ++ i. This integrated development environment includes the IAR C/C ++ optimal compiler, assembler, linker, text editor, file manager, project manager and C-SPY Debugger [5]. This program could generate reliable and efficient FLASH/PROMable code for the processor hip through its built-in code optimizer for different chips.

4.1.2 Application of ZigBee protocol stack

The software design of the network nodes is based on the project template creation of Z-Stack-CC2530-2.2.2-1.3.0 protocol stack development package provided by TI Company. Add the programs and function modules in the application layer of the project template. The template itself also provides the underlying function library related to the CC2530 chip, such as ADC operating function library, UART operating function library, so users could achieve the desired functions by simple API interface function calls. This will greatly facilitate the project development and effectively shorten the project development cycle.

Z-Stack achieves programming in accordance with the hierarchy of IEEE 802.15.4 standard and ZigBee standard, including the following modules: APP application layer, HAL hardware abstraction layer, MAC media access layer, NWK network layer, OSAL operating system layer and Service, Security and ZDO and so on.

Z-Stack protocol stack uses the event polling mechanism with priority and the system goes to sleep mode after the initialization. If events occur, the system is waken up and then processes the events in priority order. Then the system will go to the sleep mode again and wait for the next wake up. This kind of mechanism could effectively reduce system power consumption.

4.2 Software design of data transmission

After each sensor node completing the data acquisition and conversion, it is necessary to send the data to the coordinator node for summarization process. This relates to data communication between wireless network nodes. In what format and order should the data be transmitted to the coordinator node and how does the coordinator node process the data. To solve these problems, it is necessary to design a data communication protocol, which means to determine the frame format of the data packet.

The system adopts the fixed-length frame format to constitute the data acquired by the sensor nodes to a data packet in a certain order, then add the header and trailer, and send the data to the coordinator. Then the coordinator will transmit the data to the control center by the serial port for processing. Define the frame format of data communication protocol, as shown in Table 1.

Table 1. The frame structure of the data communication protocol.

Byte	1	1	1	2	2	1	1	1	1	1
Implication	Frame Header	Node ID	Frame Count	Soil Temperature	Soil Temperature	Voltage	Electric Current	RSSI	Check Code	Frame Trailer

4.3 Design of control algorithm

We end up using the fuzzy control algorithm to design the intelligent control system for agricultural greenhouse based greenhouse control manner. Fuzzy control, also known as fuzzy logic, is a kind of control technology on the basis of fuzzy sets, fuzzy linguistic variables and fuzzy logic. It was proposed by L.A. Zadeh from the United States in 1965 who introduced fuzzy set theory and gave the definition and related theorems of fuzzy logic control in 1973. In 1974, EHMamdani from Britain successfully used a fuzzy controller comprised of vague statements to control the steam engine and boiler in a lab. Fig. 5 shows the diagram of fuzzy control system.

Soil temperature and humidity of the greenhouse are treated as the error variable E of the fuzzy

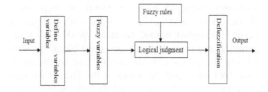

Figure 5. Fuzzy control system.

control system in this system. A fuzzy control table is designed through the operation of a series of fuzzy control rules, which is stored in the controller's memory. The system analyzes and processes the acquired data to obtain data of the fuzzy control table as output values to control the system.

5 CONCLUSION

The ever-changing science and technology has brought brighter prospects to agriculture greenhouses, enabling agriculture science to make optimal control strategy to achieve scientific cultivation in combination with the environment information obtained from the sensor. This paper has introduced intelligent control technology agricultural greenhouses based on Internet of Things, which uses a distributed control structure and centralized operations management with relatively independent design to achieve the intelligent control of a single agricultural greenhouse via the integration of computer network communications and fuzzy control technology.

Foundation project: The National Natural Science Foundation Project. (number 61261045).

The National College Students' innovative projects.(number 201311407023)

REFERENCES

Peng Qisheng. Liu Songling. *Monitoring System of Microcontroller Parameters of Greenhouse Cultivation*[N]. Journal of South-Central University for Nationalities: Natural Science, 2004.

Xie Yonghui. *Design of Intelligent Greenhouse Control System based on PLC* [D]. Shandong: Shandong University 2008.

Xu Lingling, Yang Jingchang, Hao Minggang. *The Community ECG Monitoring System based on ZigBee Network*. Microcomputers and applications, 2012.

Zhou Jianmin, Xu Dongdong, Zhou Qixian, Liu Yande, Zhang Hailiang. *Main Program of Modern Architecture and Development of Greenhouse Monitoring System*]. Anhui Agricultural Sciences, 2010.

Luan Xinying *Research and Application of Integrated Development Environment of Tsinghua Vehicles* Hunan: Hunan University, 2005.

Computational Intelligence in Industrial Application – Ling (ed.)
© 2015 Taylor & Francis Group, London, ISBN: 978-1-138-02818-0

Data warehouse quality control scheme for electricity market analysis

YanMin Guo, SuYan Long & GaoQin Wang
China Electric Power Research Institute, Nanjing, Jiangsu, China

ZhongYuan Chen & HaiChao Wang
State Grid Anhui Electric Power Company, Hefei, Anhui, China

WenZhe Zhang
State Grid Chongqing Electric Power Company, Chongqing, China

YiTing Zhao
State University of New York, Buffalo, NY, USA

ABSTRACT: In order to achieve electricity market analysis and business decision, data warehouse technology is used to process massive data. Data quality is a key factor affecting the data warehouse project. In order to define data check rules about the dataset and data items, the metadata are extended, expressions are defined based on the formula editor, and the dynamic database query script. Check tasks are executed regularly or manually. On the basis of the verification results, indicators about rationality, consistency and timeliness are calculated, and are analyzed through reports and curves. Abnormality information about data quality is sent to a data maintenance person automatically via SMS or e-mail, to promote that the problems are solved timely.

KEYWORDS: Electricity market; Data warehouse; Metadata; Data quality control; Check rule.

1 INTRODUCTION

With the development of smart grid construction and electricity market in China[1], the function of transaction operation system changes from information management to transaction control [2,3] and business decisions [4,5]. The system runs for several years, and accumulates massive data which are important resource for companies to make market strategy analysis [6]. In this paper, decision support system about electricity market, include transaction management and control, market decision support and market trend analysis [7], is constructed using data warehouse technology. High quality data is a prerequisite for an advanced decision support applications such as optimization [8, 9]. For the data warehouse project, complete data quality solution scheme is given.

In this paper, platform tools for data quality control are implemented based on extended metadata and the data check rules. Check rules configured by equation editor, and dynamic data query technology. Through task scheduling and timing services, data quality scans in the background cycle. For data rationality, consistency and timeliness, three control indicators are designed and analyzed through a flexible report tool. Abnormality information about data quality is sent to data maintenance person automatically via SMS or e-mail, so that the problems are timely solved finally. The solution protects for electricity market advanced analysis and application.

2 ANALYSIS OF CHECK DEMAND

2.1 Data warehouse model

In the electricity trading data warehouse, after ETL, data are stored in fact table and dimension tables. According to the data characteristics of electricity trading, the time dimension, physical body dimension and business applications dimension are used to describe business data. These dimensions are metadata describing the data relationships. The time dimensions include year, season, month, day, hour. The physical body dimensions include all properties about market members, business units, economic units, Tie-line, electricity meter, etc. The business applications dimensions include electricity caliber, transactions roles, transaction components, transaction sequence, contract sequence, etc. By the combination of three kinds of dimensions, a series of related data are described and then data sets are formed.

Electricity transactions data items is not an independent data, but data collection is associated with each other. In the settlement data case, data items include electricity quantity, price, fee with period flag, and have fifteen filed totally. In the meter data case, data items are electricity quantity with period flag, and have five filed. The period flags are total period, spikes period, peak period, flat period, bottom period in the system. The fact table supports maximum 20 data items in the database, which does not include the primary key. Chinese descriptions for data items are defined through the interface, and help us to use data and display data. The composition of data with the same item definition is a dataset. Due to different business, different dataset are formed. Based on these dataset, advanced market analysis applications are designed. The database model structure is shown in Fig. 1.

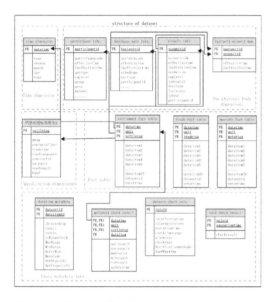

Figure 1.   Diagram of database model structure.

2.2 Data check demand

Data quality check scheme, not only for the data in warehouse, but also for the data in source system, gives data quality monitor method and control strategy from source to end. There are three kinds of data that need quality check.

1 Quality management for data from exchange interface. Electricity market analysis, closely related to grid operation, need to collect much data from other system in the same market and from the lower market.
2 Quality management for ETL process. After the data cleaning and conversion, data consistency must be checked between resource and warehouse to avoid the loss of data.
3 Quality management for warehouse itself. Based on logical relationships between datasets and data items, the horizontal and vertical comparison of data is needed to verify of data quality further.

The check nodes set is shown in Fig. 2.

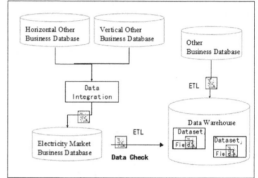

Figure 2.   Diagram of check nodes set.

3 DEFINITION OF CHECK RULE

In the data warehouse, metadata describes the data structures and utilize methods, provides use guidelines of data resources [10]. In the data warehouse, relay on metadata, loose data are composed orderly. Technical metadata is often used to describe the data model, data sources, extraction rules and conversion rules. In this paper, extend metadata are further used to describe quality requirements for data items, dimensions data and original business data. Deferent check rule is defined to meet complex validation logic.

3.1 Check definition based on metadata

Base on metadata, check message are defined which include null check, zero check, range check, max check, min check, etc. If the data item refers to menu, then menu code is defined. If the data item refer to other table, table message and field message are defined. Data is not qualified if it is not within the reference resource. Check task sequentially scan data to determine whether the data meets quality requirements based on metadata information.

If the check is for the dimension data, or is for the source system data, metadata about check is defined in the field information table. There is a lot of menus info in the dimension table, so for the dimension table, references check is particularly suitable.

Table 1. The metadata for check.

Item name	Chinese discription	Remark
IsNull	is null check	0 no,1 yes
IsZero	is zero check	0 no,1 yes
IsRangeCheck	is range check	0 no,1 yes
MaxRange	Maximum Range	value
MinRange	Minimum Range	value
ReferMode	Reference Mode	0 not refer, 1 refer to menu,2 refer to table.
MenuCode	Menu Code	Code marked menu group
Refobject id	Reference object id	Table message
Refproperty id	Reference property id	Field message

3.2 Check rules based on the expression

The data in the same dataset can check each other horizontally and vertically. Setting check logic between data items is called horizontal check. For example, in the settlement dataset, energy quantity for total period is equal to the sum of quantity for each period. Setting check logic between data collection, calls vertical check. In the dataset, data are marked by the combination of three dimensions. There are business logic relations between different dimensions of data. For example, for the same participant body and the same date, net energy should be less than gross energy. Net and gross is two values for the same dimension.

Rules expressions are recorded by the editor. In the editor, operator can select dataset, data item and can set filter conditions for data item. Expression is described by math formula, which includes conventional operator. Underlying services provide compile and calculate function.

$$D_i.EngT = D_i.EngS + D_i.EngP$$
$$+ D_i.EngF + D_i.EngB \tag{1}$$

$$D_j.EngT[\ D_j.appdimen = 'gross']$$
$$\geq D_j.EngT[\ D_j.Keyapp = 'net'] \tag{2}$$

In the formula (1): D_i is dataset, left is the total energy quantity, followed by the right side of the spikes, peak, flat, bottom four period energy quantity. It is hidden that the data have the same primary key.

In the formula (2): D_j is dataset, the left is the energy quantity for gross caliber, the right is the energy quantity for net caliber. It is hidden that the data have the same primary key except for energy caliber.

3.3 Check rules based on dynamic sql expression

In the same dataset, for a set of data that meet a certain criteria, we can check them through row number, the total value or average value. For different datasets, there are business relationships between them. Between data in the warehouse and data in the source system, there are consistency relationship. These relationships are more flexible, and not for single dataset only, so flexible define methods are need.

To meet these needs, dynamic sql expression is defined. By optional mathematical operator, The expression connect two sql statements, or connect sql statements with constant. Based on table info and field info, sql statements can be defined through inferface. Conventional database syntax keywords are provided.

$$HAVING(D_i.a)=96 \tag{3}$$

$$GROUP\ BY(D_i.Keyunit,D_i.Keyapp,trunc(D_i.Keytime,'yyyymmdd')) \tag{4}$$

The formula (3) is an expression. The formula (4) is constraints. Combination of them means that data are grouped by the conditions, each group should have 96 data. This case is uesd to check plan data with 15 minutes frequency.

3.4 Check rules for timeliness

Generally, in a dataset, data have the same business meaning and have the same generation time. According to this, data are checked for timeliness.

For example,for settlement dataset, the time primary key is 0:00 on June 1, 2014, which denotes data are for June settlement business. In check rule,

Table 2. Time parameters about timeliness check.

Parameter name	Parameter discription	Example
datatime	Actual business time for data, one primary key for fact table.	2014-06-01 00:00:00
demandtime	Time demand for data enter warehouse first, described by time offset to actual business time.	960h
facttime	Time stamp indicates that the data first enters the warehouse	2014-07-02 16:00:00
tasktime	Task trigger time, which include Frequency and time	Month/**/**/ 11H2M0S0

demand time for data enter warehouse is defined time offset, is 960h. So real demand time is 0:00 on July 1, 2014. Fact enter time is 16:00 on July 4, 2014. The check task is executed at 2:00 on July 10, 2014.

In check rule, we can check whether to delay demand time when holidays happen. If yes, when the demand time is in holiday, it is automatically delay to the first working day after the holidays.

4　CONTROL INDICATORS DESIGN

4.1　Data reasonable indicators

Reasonable indicators are used to judge whether the data is within a reasonable range based on metadata, which include null check, zero check, range check, max check, min check and reference range check. Any one is not pass, data is marked for being unreasonable. Reasonable indicators are statistics for all check result.

Reasonable rate for dataset is listed.

$$\alpha_i = \sum_{j=1}^{N}\sum_{k=1}^{M} \eta(i,j,k) \Big/ N \times M \qquad (5)$$

where $N \times M$ is the data number that in the range of check time, the product of rows and columns of data. The closer to 1 for αi, and the quality of the data is the better.

4.2　Data consistency indicators

Consistency indicators are used to judge whether the data is consistent with business logic relationship. Check rules are defined by the expression or dynamic sql statement. If data are checked by other data, and check result is bad, then both sides of data will be marked as bad. Through consistency check, we can find abnormal data, or missing data in cleaning process.

Consistency rate for dataset is listed.

$$\alpha_i = \sum_{j=1}^{N} C(i,j) \Big/ N \qquad (6)$$

where $C(i,j)$ is the data number in the dataset i which check result is bad. N is data number in the dataset i which are checked.

4.3　Data timeliness indicators

Timeliness indicators are used to judge whether the data exist in the database before demand time. After demand time, market analysis application will use the data. If the data is lost, analysis result is untrusted.

Timeliness rate for dataset is listed.

$$\alpha_i = \sum_{j=1}^{N} T(i,j) \Big/ N \qquad (7)$$

where $T(i,j)$ denotes that in the dataset i, whether data element with dimension j is exist. Yes is 1, no is 0. N is data row number checked.

5　SYSTEM FUNCTION DESIGN

The system is based on the Java2 platform, and uses the three-layer structure and design patterns based on model-view-controller (MVC). The technical structure is shown as Fig. 3.

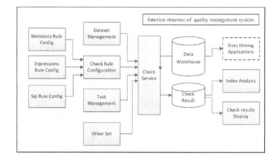

Figure 3.　Diagram of system function structure.

5.1　The function of dataset management

In this module, static information about dataset are defined, which include name, code, time frequency, stored fact table, check type, metadata check time, chinese discription for fields, etc. If the data comes from the power transactions business system, ETL extract views is need. If the data come from other system through data exchange interface, ETL views are not necessary, and interface program is responsible to fill data into dataset.

In addition, the module can query fact data for dataset through specify dimensions conditions. The module can query all related dataset check rules and data items check rules.

5.2　The function of check rule management

5.2.1　Set metadata for data item

In the module, we can define check metadata for data items. Except metadata described in Table 1, we can define warning level, which includes general, important, emergency. We can define indicators type, which include rationality, consistency and timeliness.

5.2.2 Set check rule expression for dataset

In this module, we can define check rule for dataset or data item. Check rule message include rule name, check type, warning level, rule expression, rule description, execute time, result notification method, is effective, etc.

The rule expression is important. We can select dataset, data item, and data collections. Common mathematical functions and operators symbol can be used. Check rules are stored via text form.

5.3 The function of task management

One rule corresponds to one task. Metadata check corresponds to a default system task. All check tasks are scanned timing by background permanent process. If check time is reached, then the task is triggered. Check task take much computer resources, so it is suggested that check time is set in the midnight, in order to avoid system use peak period.

After tasks running, scan results are stored into the quality results table and used to statistical analysis and query. Once the data quality flag is bad, red flashing warning light are showed in the lower right corner of all interfaces. Click it, then the detailed quality alarm query page are displayed.

In addition to automatic tasks, the system also supports to run check tasks manually.

5.4 The function of other parammeter management

In this module, we can set parameters about holidays. We can edit person information for dataset maintenance, which includes name, department, phone, e-mail, etc. Through platform interface, the system can send data quality scanning results to relevant personnel by phone or e-mail.

5.5 The function of check result query

Data check results are showed through data mining reveal tool which is designed based on data warehouse. Operator can query three indicators for dataset, which are reasonable indicators, consistency indicators, timeliness indicators. Click indicators, detail check results are displayed. By different background color of check result, different warning levels are distinguished.

6 CONCLUSIONS

In this paper, based on check metadata and check rule definition, the data quality control tool for electricity market data warehouse project is achieved. Through cycle tasks and manual tasks, system scans data quality of dataset, record scan results, and calculate rationality, consistency and timeliness control indicators. History trends for data quality are analyzed. Through tracking the problem solving process, data maintenance person is encouraged to solve data quality problem timely. Data quality is improved continuously, and the demand for transaction control and decision support is satisfied. The system has been applied in some grid companies of China.

The system is characterized by the following aspects:

1 Metadata are used to describe data quality verification requirements for data items, which include null check, zero check, range check, max check, min check and reference range check.
2 Check rule expressions are used to describe data check logic for data items, data elements, and different datasets.
3 Cycle tasks are used to trigger scan program, check result are sent to data maintenance personnel quickly. The trends for data quality are given to help make control measures.

ACKNOWLEDGMENTS

The author gratefully thanks the financial support of the Science Foundation of State Grid Corporation of China, Grant No. DZN17201300045 and Science Innovation Foundation of China Electric Power Research Institute, Grant No. 5342DZ130002.

REFERENCES

Lu Gang, Wei Bin, Ma Li. "Smart grid construction and development of the electricity market", *Automation of Electric Power System*, 2010 34(9):1–6.

Long Suyan, Yang Zhenglin, Geng Jian. "Research and application of smart grid cross regional trading control technology", *Automation of Electric Power System*, 2014, 13:108–114, July, 2014.

Long Suyan, Zhang Xian. "Research and application of flexible energy settlement methods based on component library", *Automation of Electric Power System*, .37:89–94 July 2013.

Zheng Yaxian, Zhang Xian, "Energy Resource Optimization Model In Large-scale Based On Aggregate Node". IPEC2012, pp. 279–283.

Geng Jian, Gao Zhonghe. "A preliminary Investigation on power market design considering social energy efficiency" [J], *Automation of Electric Power System*, 31:18–21, October, 2007.

Guo Yanmin, Shao ping, Guo Junhong. "Design of purchasing costs analysis system for grid companies". 2012, 10th International Power and Energy Conference, IPEC2012, pp.295–299.

Shaoping, Guo Yanmin. "Design and Implementation of Multi-caliber Power Purchase Cost Analysis System based on Data Warehouse" International Conference on Computer Science and Artificial Intelligence, ICCSAI 2013, pp. 315–321.

Zhang Yu, Pan Hongfang. "Application of data quality management platform in inner mongolia electric power company", *Electric Power Information and Communication Technology*, 12(3):104–107, 2014.

Chen Ping, Liu Songxian. "data warehouse model construction Based on SAP BW--data warehouse technology in the power of enterprise ERP". Journal of Liaoning Technical University(Social Science Edition), vol16(2):113–117, 2014.

Yang Hongbin, Song Ming. "Research on architecture of metadata management platform", *Computer Systems Applications*, (11):17–20, 2007.

Study on the structure and control modes of the microgrid

Qing Hai Zhang, Jia Wen Liang, Xing Hao Shi, Xin Tao Wang, He Xian Wang & An Hua Liu
Liaocheng Power Supply Company, State Grid Shandong Electric Power Company, Liaocheng, China

ABSTRACT: Based on the introduction of the microgrid and its basic structure, the structure characteristics of AC microgrid, DC microgrid and AC/DC microgrid are summarized. Its respective block diagrams are given. For the two kinds of normal operation mode of microgrid, it introduces the operation state of the microgrid and the mutual transformation relationship, analyzes the three kinds of control mode of the microgrid in independent operation mode detailedly. The traditional multi-agent in power electronic is applied in the microgrid control mode, the hierarchical control mode of microgrid based on multi-agent is developed.

KEYWORDS: Microgrid, Structure characteristics, Control mode.

1 INTRODUCTION

Microgrid, as the new-emerging cutting-edge technology, could improve the energy efficiency of distributed generation [1]. Microgrid is a supplement to bulk grid and more and more distributed microgrid will appear in the small and medium sized smart distribution grid in the future.

Microgrid plays an important role in improving the reliability of bulk grid as well as the power quality. Due to the increasingly prominent vulnerability of the bulk grid, the important loads with geographic proximity are used to constitute microgrid. Taking appropriate control methods to provide reliable power supply for them could save cost of the reliability and quality improvement of the grid and reduce the economic losses caused by the power outages.

According to the network structure and functional characteristics of microgrid, this paper summarizes the structural characteristics of AC microgrid, DC microgrid and AC / DC hybrid microgrid. Based on its two normal operation modes, this paper introduces various operation modes of microgrid and their mutual transformation relationship and analyzes three control methods in independent operation mode. The traditional multi-agent technology in the power electronic system has been applied in the microgird control mode to develop the hierarchical control mode of microgrid based on multi-agent, which could meet the characteristics of distributed microgrid.

2 MICROGRID AND ITS BASIC STRUCTURE

Figure 1 shows the typical configuration diagram of microgrid system, including several distributed power (such as photovoltaic arrays, wind power), the power electronic conversion device (such as an inverter) and the energy storage device (such as battery, super capacitor) and they are combined to supply power for the load. The distributed power is the renewable energy generation system and the loads contain conventional power loads and some cooling loads and thermal loads. Under normal circumstances, microgrid could work in two modes, grid-connected mode and off-grid mode, and then connect to the external power grid by the PCC (point of common coupling). When PCC is cut off from the main network, the system could also keep powered up to the important loads of microgrid in the off-grid mode [2].

The microgrid can be regarded as a small power system, and it could keep the optimal allocation and balance of energy due to its excellent energy management capabilities, thus ensuring the operation economy. In addition, the microgrid can be seen as a "virtual" power or load in the power distribution system. By the coordination control of output power of distributed power, it could carry out load shifting for the power grid and control the fixed value or given range of the power exchange capacity with the external grid, thus effectively reducing the difficulty of the operation scheduling and eliminating the influence on the external distribution network and the surrounding users caused by the fluctuations of renewable energy power.

Generally speaking, microgrid is connected to the bulk grid, including two operation modes, grid-connected mode and off-grid mode (islanding mode). Under grid-connected mode, the microgrid is connected to the mid&low-voltage distribution network and they support each other to achieve two-way energy flow. If the plan is islanded or there is external

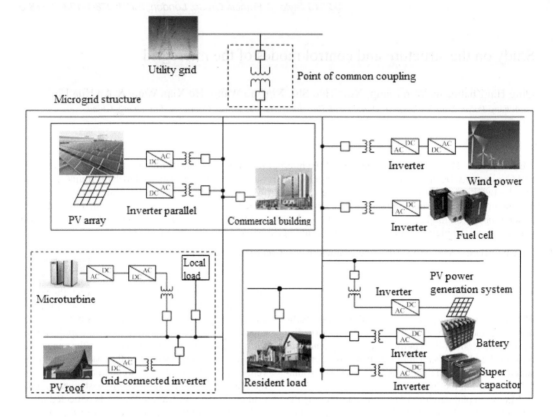

Figure 1. A typical microgrid system structure.

power failure, the microgrid will be converted to the off-grid operation, continuing to power for the critical loads and improving their reliability of power supply. Taking advanced control methods could ensure high-quality power supply and achieve the smooth switching between two operation modes [3].

3 STRUCTURAL FEATURES OF MICROGRID

Microgrid can be divided into AC microgrid, DC microgrid, and AC-DC hybrid microgrid according to the network structure and functional characteristics.

3.1 AC microgrid

Currently, AC microgrid is the main form of microgrids [4]. As shown in Figure 2, distributed power and the energy storage device in AC microgrid are both connected to the AC bus through the power electronic conversion apparatus. By controlling the grid-connected and off-grid toggle switch at PCC, we can achieve a smooth transition between grid-connected mode and islanding operation mode.

3.2 DC microgrid

In contrast to AC microgrid, DC microgrid has a characteristic that distributed power, energy storage devices, and load are all connected to the DC bus, which then connect the DC network with an external AC power grid through the power electronic inverter device [5]. Figure 3 shows the structure of DC microgrid. It is able to provide AC and DC load of different voltage levels with power through electronics power conversion devices, and fluctuations in distributed power and load are adjusted by the energy storage device at the DC side.

In view of users' requirements for different levels of power quality and characteristics of distributed power, two or more DC microgrid can form a double loop or multiple loop power supply [6]: a DC feeder connected to distributed power with obvious intermittent feature is for supplying power to ordinary load use; another DC feeder connected to a stably operating distributed power and energy storage devices is for supplying power to high load use. Compared with AC microgrid, there only exists first-class voltage conversion device between the distributed power and DC bus in DC microgrid, reducing the construction cost of

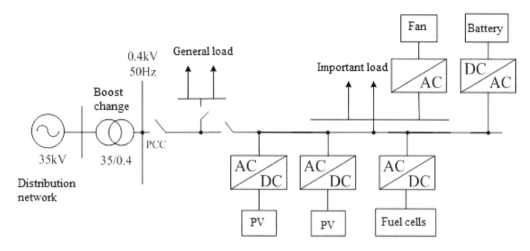

Figure 2.   Structure diagram of AC microgrid.

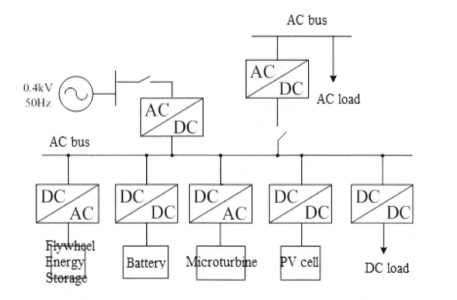

Figure 3.   Structure diagram of DC microgrid.

the system and making it easy to control; Meanwhile, without the consideration of synchronization problems between the various distributed power supplies, it is more competitive in terms of inhibiting the circulation between different distributed power supplies.

3.3   AC and DC hybrid microgrid

Figure 4 shows the structure of AC and DC hybrid microgrid. It contains both AC bus and DC bus, which is capable of supplying power to the AC load in direct current and directly powering DC load [7].

However, from the perspective of analysis of the overall structure, it is actually still seen as Ac microgrid when the DC microgrid power can be regarded as a power supply connected to AC bus through power electronic inverter.

4   CONTROL MODE OF MICROGRID

Microgrid often switches between two operation modes, the grid-connected and off-grid mode [8-10]. Figure 5 shows the operation states of microgrid

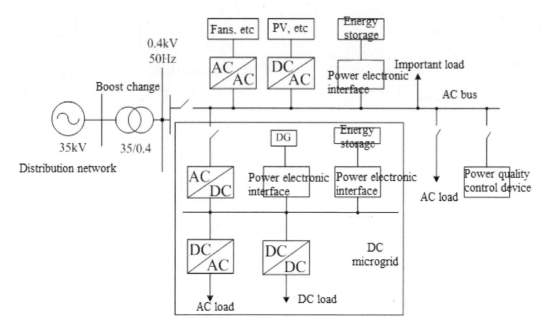

Figure 4. AC/DC microgrid structure.

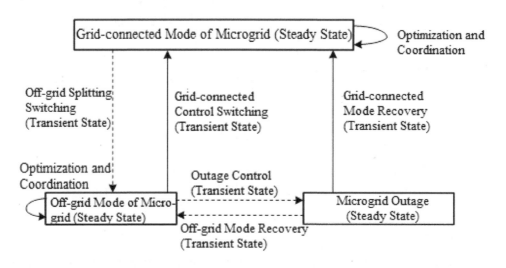

Figure 5. Running state transition of microgrid.

and the transformation relationship between them. As there are various energy input (wind, light, etc.), energy conversion (optical / electrical, thermal / power, wind / power, AC / DC / AC conversion), energy output (electricity, heat, cold) and operating states (grid-connected, off-grid) in the microgrid, its dynamic characteristics are more complex compared with the single distributed power. According to the different roles played by each distributed power

of microgrid in the stand-alone mode, the microgrid control mode can be divided into three types: master-slave control mode, point to point mode and hierarchical control.

4.1 Master-slave control mode of microgrid

Figure 6 shows the master-slave control pattern. It is a hierarchical control mode which means that a

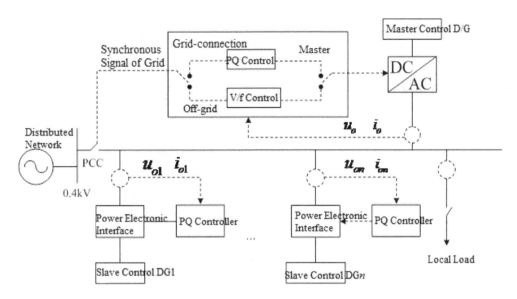

Figure 6. Master-slave control structure of microgrid.

distributed power or energy storage device of the microgrid could take droop control or constant voltage and fixed frequency control (V / f control) to provide voltage reference and frequency reference for other distributed power when the microgrid is in the off-grid operation mode. Meanwhile, other distributed power supply uses the constant power control (PQ control). The distributed power controllers or storage device with V / f control is called the master controller; other distributed power controller is called the slave controller. Due to the master-slave control mode, each unit can be effectively coordinated and the harmful effect of the switch between two operation modes could also be eliminated.

Master-slave control mainly includes centralized control and parallel control. To achieve centralized control by the control center of microgrid, there are three methods: optimal control, fuzzy control and neural network control. Parallel control requires control units with large volume which limit the flexibility of the microgrid operation.

When the microgrid is running off-grid, the distributed power, as the slave controller unit, generally takes the PQ control, the load changes should be tracked by the distributed power, the master control unit. Therefore, its power output should be controllable in a certain range and is capable of following the fluctuation of loads. In the microgrid with master-slave control, when the microgrid is in the grid-connected state, all the distributed power generally adopts PQ control; once in the islanding mode, the distributed power, as the master control unit, converts from PQ control mode to the V / f control mode, which

requires the master controller to meet the requirements of fast switch between these two operation modes. V / f control mode, similar to the second FM of conventional generators, adopts the double-loop control, outer voltage loop and inner current loop, to set the reference value of the voltage and frequency. Compared with the measured values, the output deviation goes through the PI regulator to adjust the output voltage and frequency of inverter. This control strategy could achieve power demand-supply balance, stabilize voltage and frequency, improve the dynamic response characteristics of the inverter and strengthen anti-disturbance capacity. But this requires spinning reserve capacities of the distributed power control to cover all power supply of bulk grid.

The main controller with master-slave control method includes the master control unit with the energy storage device, the master control unit with the distributed power, and the one with energy storage device and distributed power. With the storage system as the master control unit, it will not take too much time for the microgrid to operate in the islanding mode. The fluctuant renewable distributed power output by the photovoltaic and wind power is not suitable for the master control power. Therefore, it is necessary to combine it with the energy storage system to serve as a master control unit, for it could make full use of the rapid charge-discharge capability of the energy storage system to maintain the long-time independent operation of microgrid. Using this mode, the energy storage system can provide power support for the system when the microgrid converts to the independent operation. Compared with the mode with the storage system as the master control unit, this one could effectively reduce the capacity of the energy storage system and improve the economic efficiency of system.

4.2 Peer-to-peer control mode of microgrid based on plug and play

Peer-to-peer control mode means that there is no master-slave relationship between controllers, and the distributed power has equivalent control effect, and every distributed power could control the field information according to the voltage and frequency of the access system, as shown in Figure 7. The main method is the droop control method [11-13].

This method is based on the "plug and play" and "peer-to-peer" control mode of power electronics technology. According to the control target of microgrid, it adopts the drop character curve, similar to that of the conventional generators, as the control mode of the distributed power of microgrid, which means to dynamically distribute the power to each distributed power to ensure the power supply and demand balance. It is applicable to the microgrid system containing multiple inverters in parallel operation. When in the off-grid mode, the distributed power with the droop control plays an important role in the microgrid voltage and frequency regulation. When the microgrid loads take changes, this distributed power could automatically share the load power according to the droop

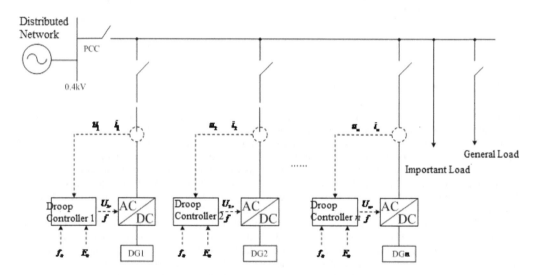

Figure 7.　Peer-to-peer control structure of microgrid based on the droop characteristics.

coefficient, which means to adjust the output voltage frequency and amplitude of the distributed power to make the microgrid achieve a new stable operating point, thereby realizing the reasonable and accurate distribution of output power. However, the steady-state voltage and frequency may change before and after the load varies, which means that this control mode is a kind of differential control and it fails to solve the problem of voltage and frequency restoration, namely, the secondary frequency modulation of traditional generator. Therefore, when the microgrid suffered severe damage or interference, it is difficult to ensure the system frequency quality.

In contrast with the master-slave control mode, distributed power using peer-to-peer control is easy to implement the plug and play and seamlessly switch operating modes. Among microgrids using peer control, some distributed power supplies can use PQ

control. However, for the multiple distributed power supplies that use droop control, they jointly undertake the control tasks of the main control unit in the master-slave controller. They can achieve a reasonable distribution of external power changes in distributed power supplies by setting a reasonable droop factor, thus meeting the needs of load changes and ensuring the stability of microgrid voltage and frequency.

4.3 Hierarchical control mode of microgrid based on multi-agent technology

Multi-Agent technology [14] is a kind of active entity possessing knowledge, goals and capabilities, which is able to reason and make decision independently or under the guidance of human beings. It is characterized by autonomy, spontaneity, intelligence and capability of communications, which can meet the needs of microgrid distributed characteristics. Hierarchical control mode of microgrid based on multi-agent technology is a control mode that applies the multiple agents in traditional power electronic systems to microgrid. Multi-agent system not only coordinates the steady-state operation of distributed power and the load of microgrid in grid-connected or off-grid mode, but also can provide black start to the transition process run in islanding operation.

In the three-level control structure of a typical Autonomous Electricity Networks (AEN), it is a multi-agent system consisting of DER agents, database agents, control agents and user agents. The first-level structure is comprised of a distributed power DG and load control, which are responsible for transient power balance and load management of microgrid and ensuring its reliable operation. The second-level structure is in charge of optimizing the power quality and reducing fluctuations in voltage and frequency. The third-level structure belongs to the top-level operation and management system for the distribution network, whose responsibility is to manage and schedule the multiple microgrids based on market and scheduling needs.

5 CONCLUSION

Microgrid technology has aroused widespread concern and attention around the world, which has an important role in improving the reliability of power supply and power quality of China. At present, China has established a number of cogeneration projects. Problems like how to choose the nearest suitable heat users and power users to form microgrid and a best combination of power generation have strategic significance for our country to improve energy efficiency, optimize energy structure, and reduce environmental pollution.

REFERENCES

Zhang Qinghai, Peng Chuwu, Chen Yandong, et al. *A Control Strategy for Parallel Operation of Multi-inverters in Microgrid* [J]. Proceedings of the CSEE, 2012,32 (25): 126–132.

Chen Yandong, Luo An, Long Jigen, et al. *Circulating Current Analysis and Robust Droop Multiple Loop Control Method for Parallel Inverters Using Resistive Output Impedance* [J]. Proceedings of the CSEE, 2013,33 (18): 18–28.

Chen Yandong, Luo An, Xie Sanjun, et al. *A Single-phase Photovoltaic Grid-connected Power Control Method Without Delay* [J]. Proceedings of the CSEE, 2012,32 (25): 118–125.

Lu Xiaonan, Sun Kai, Huang Lipei, et al. *Improved Droop Control Method in Distributed Energy Storage Systems for Autonomous Operation of AC Microgrid* [J]. Automation of Electric Power Systems, 2013,37 (1): 180–185.

Shi Jie, Zheng Zhanghua, Ai Qian, et al. *Modeling of DC micro-grid and stability analysis* [J]. Electric Power Automation Equipment, 2010,30 (2): 86–90.

Zhang Li, Sun Kai, Wu Tianjin, et al. *Energy Conversion and Management for DC Microgrid Based on Photovoltaic Generation* [J]. Transactions of China Electrotechnical Society, 2013,28 (2): 248–253.

Yin Xiaogang, Dai Dongyun, Han Yun, et al. *Discussion on Key Technologies of AC-DC Hybrid Microgrid* [J]. High Voltage Apparatus, 2012,48 (9): 43–46.

Zheng Jinghong, Wang Yanting, Li Xingwang, et al. *Control Methods and Strategies of Microgrid Smooth Switchover* [J]. Automation of Electric Power Systems, 2011,35 (18): 17–23.

Qiu Lin, Xu Lie, Zheng Zedong, et al. *Control Method of Microgrid Seamless Switching* [J]. Transactions of China Electrotechnical Society, 2014,29 (2): 171–175.

Liu Zhiwen, Xia Wenbo, Liu Mingbo. *Control Method and Strategy for Smooth Switching of Microgrid Operation Modes Based on Complex Energy Storage* [J]. Power System Technology, 2013,37 (4): 906–912.

Jing Long, Huang Xing, Wu Xuezhi. *Research on Improved Microsource Droop Control Method* [J]. Transactions of China Electrotechnical Society, 2014,29 (2): 145–152.

Zhang Qinghai, Luo An, Chen Yandong, et al. *Analysis of Output Impedance for Parallel Inverters and Voltage Control Strategy* [J]. Transactions of China Electrotechnical Society, 2014,29 (6): 98–105.

Zhang Ping, Shi Jianjiang, Li Rongguiet al. *A Control Strategy of 'Virtual Negative' Impedance for Inverters in Low-voltage Microgrid* [J]. Proceedings of the CSEE, 2014,34 (12): 1844–1852.

Zhang Mingguang, Lu Yunyun. *A Self-healing of Smart Distribution Control Based on Multi-agent System* [J]. Application of Electronic Technique, 2012,38 (11): 77–79.

Computational Intelligence in Industrial Application – Ling (ed.)
© *2015 Taylor & Francis Group, London, ISBN: 978-1-138-02818-0*

Research on flow control characteristics of LS and LUDV based on AMESim

JunXia Li & ZiMing Kou

College of Mechanical Engineering, Taiyuan University of Technology, Taiyuan, China
Mine Fluid Control Engineering Research Center (Laboratory), Shanxi Province, Taiyuan, China

ABSTRACT: In this paper, the working principles of two different control modes, load sensing control technology (LS) and load independent flow control technology (LUDV), are introduced. And their characteristics are analyzed. Meanwhile, mathematical models of the two systems are built and their working characteristics are simulated in different working conditions. Results show that the LS control system can meet the flow requirement of actuators in the case that the system has an adequate flow, even if the load fluctuation is higher. With an increasing flow requirement of them, the moving of actuators will be seriously uncoordinated. The actuators with the maximum load may stop running. However, the LUDV control system can not basically affect the moving coordination of the actuators no matter how the loads change. consequently, it verifies that this system is superior to the LS control system in flow distribution and provides a theoretical basis for the design of hydraulic system for engineering machinery.

1 INTRODUCTION

As a construction equipment, engineering machinery is widely used in construction sites of various projects. The development of engineering machinery is playing a significant role in adjusting the industrial structure, promoting the development of related industries and raising the level of machinery equipment industry. Control system is the most central element of engineering machinery and the hydraulic control system is used for driving control to the main actuators of engineering machinery. In the blossom of engineering machinery, its hydraulic control system also experienced a series of development, from the use of throttle governing and volume governing to the LS control technology adopted at present, from the main consideration of hydraulic control system's performance by now, it should take into account not only its control performance, more importantly, but also its energy-saving effect and moving coordination. The greater heat is produced due to the large throttling and overflowing loss in the throttle governing, causing varieties of problems for the system. So the control system is gradually developing to the load sensing control state(LS), which can make the output pressure and flow of pump adapt automatically to loading requirement and greatly improve the efficiency of hydraulic system. But one drawback of the hydraulic control system for LS control technology is that its speed will reduce and even stop moving when the loads of one actuator increase to a certain extent,

which can contribute to the fact that the movement of engineering machinery's actuators lose coordination. Since the LS control system exists the problem that when the flow demand of actuators is greater, the pressure oil will flow to the actuators with smaller loads according to the size of loading pressure, while the speed of actuators with greater loads will decline or even stop moving. In order to solve the problem, the Rexroth developed the load independent flow distribution control technology, namely, LUDV control technology. This paper make a comparative study on the characteristics of the LUDV control technology and the LS control technology in flow distribution.

2 CONTROL PRINCIPLE OF LUDV AND LS

2.1 *LS control principle*

By using the sensing valve to detect the change of loads, the load sensing technology can make relative adjustment to the system flow automatically to maintain the constant pressure differential before and after main valve, which can achieve the best match between flow and loads of the system . It is a hydraulic control technology that can provide the flow and pressure corresponding to the loads through automatically detecting the flow and pressure requirement of the system.

LS control system is also a hydraulic control system that based on load sensing technology and its hydraulic diagram is shown in Figure 1(a). The system

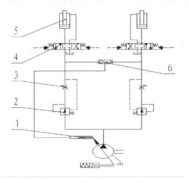

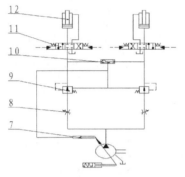

(a) LS control system

(b) LUDV control system

Figure 1. Schematic diagram of LS and LUDV control system1, 7–sensing valve 2, 9–pressure- compensated valve 3, 8–simulation operated valve 4, 11–direction valve 5, 12–simulated load 6, 10–shuttle valve.

consists of pressure-compensated valve, load-sensing variable pump, shuttle valve and operated valves. When the simulated loads are working simultaneously, the pressure-compensated 2 can detect the change of loads characteristic automatically and controls the operated valve 3 to adjust the pressure differential automatically to remain constant.

2.2 *LUDV control principle*

LUDV control technology is a load independent flow distribution technology in line with the principle that meeting the maximum load demand to compensate pressure. The hydraulic control system based on this technology can distribute the flow automatically to actuators according to their flow demands, rather than simply flow to the actuator with the lower loads. In the LUDV control system, the sensing valves can feedback the maximum load signal to the variable pump and pressure-compensated valves to limit the output pressures of multi-outlet control valve below the maximum load pressure and keep the pressure

differential constant before and after main valve under the action of pressure-compensated valves, and still remain the same pressure differential of multi-outlet control valve even when the flow load requiring is changing.

LUDV control system consists of operated valves, pressure-compensated valves, load, shuttle valves and sensing valves, whose schematic diagram is shown in Figure 1(b). Its working principle is that when the two simulated loads 1 and 2 are working simultaneously, the pressure differential before and after operated valves is kept consistent under the action of pressure-compensated valves, and the shuttle valve 10 can detect the maximum pressure loads requiring, the same input pressure of shuttle vale and pressure-compensated valve , the same pressure of operated valve's inlet and variable pump's outlet, the basically constant pressure before and after operated valve , which ensure the independence of flow distribution of the system.

3 SIMULATING AND MODELING

AMESim software includes model library of muti-disciplinary fields, such as mechanics, electronics, hydraulics, pneumatics, thermal fluid, control, magnetics and so on. By applying the hydraulic model library in AMESim software library ,it can be easy and direct to build the simulation model of the studied hydraulic system, providing a good reference for the researchers to design and study.

For the LUDV and LS control systems in this paper, most of the elements can be found in the model library. The pressure- compensated valve and load sensing valve need to be built by choosing appropriate hydraulic models based on its physical structure and working principle. Their hydraulic simulation models are shown in the following Figure 2 and Figure 3.

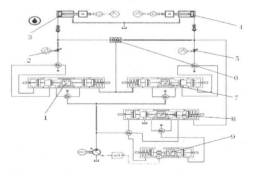

Figure 2. Simulating model of LS system.1–pressure-compensated valve 1, 2–simulation operated valve 1, 3–simulated load 1, 4–simulated load 2, 5–LS shuttle valves 6–simulation operated valve 2, 7–pressure-compensated valve 2, 8–sensing valve, 9–variable cylinder.

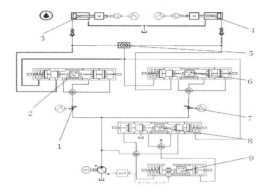

Figure 3. Simulating model of LUDV system.1–simulation operated valve 3,2–pressure-compensated valve 3,3–simulated load 3,4–simulated load 4,5–LS shuttle valve, 6–pressure-compensated valve 4,7–simulation operated valve 4,8–sensing valve ,9–variable cylinder.

Table 1. Model parameters table of LS and LUDV systems.

Motor	Rotational speed	1000r/min
Variable pump	Swept volume	100ml/r
	Spring stiffness	10N/mm
	Spring pre-compression force	95N
Load sensing valve	Piston diameter	7mm
	Piston rod diameter	4.5mm
	Spool displacement	–3~5mm
Variable cylinder	Spring stiffness	8N/mm
	Spring pre-compression force	210N
Pressure-compensated valve	Spring stiffness	6N/mm
	Spring pre-compression force	187N
	piston diameter	15mm
	piston rod diameter	7.8mm
simulated load (hydraulic cylinder)	piston diameter	40mm
	piston rod diameter	22mm
	spool stroke	1m

After building the simulating system, the parameters for the simulating models are setted. The common parameters of simulating system are shown in table 1.

4 SIMULATION AND ANALYSIS

For LUDV and LS control system, taking the simulated loads 2, 4 as the research object, the simulated load 1,3 are set of 10 KN, and the stroke signal of simulation operated valve 1,3 are set for 20; The pressures of the simulated load 2,4 are changing from 5KN-10KN-20KN with equal time, and the simulation time is 30s ,and the stroke signals of simulation operated valve 2,4 are 10, 20, 40. The simulation results are shown in the following figures.

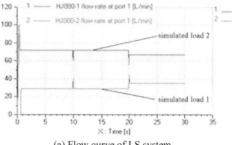

(a) Flow curve of LS system

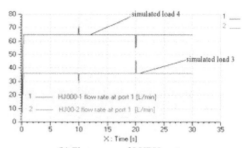

(b) Flow curve of LUDV system

Figure 4. Flow rate curve of input 10.

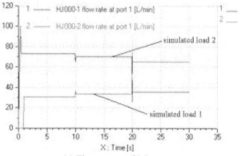

(a) Flow curve of LS system

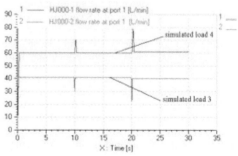

(b) Flow curve of LUDV system

Figure 5. Flow rate curve of input 20.

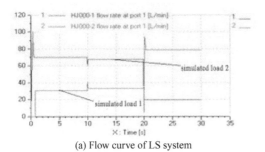

(a) Flow curve of LS system

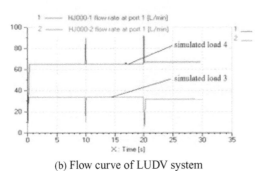

(b) Flow curve of LUDV system

Figure 6. Flow rate curve of input 40.

It can be seen from the Figure 4, 5 and 6:

1 When the input signal of the simulation operated valve is 10, the loads will start increasing at t=10s, t=20s, but the pressure-compensated valve can ensure the pressure differential basically unchanged before and after the operating valve ,so are the flows of two systems.
2 When the input signal of the simulation operated valve is 20, the loads will start increasing at t=20s, but the flows of two control models keep unchanged basically; When the load pressure changes at t=20s, the flow of simulated load 1 in LS control system will increase and the flow of simulated load 2 will decrease and the flow of LUDV control system keeps unchanged basically.
3 When the input signal of the simulation operated valve is 40 and the flow of LS control system is at t=20~30s, the simulated load 1 will almost consume most flow of the system while the simulated load 2 only take up very little, which may lead to the almost stop-moving of load 2 .

When the required flow of actuators is smaller, the LS control system can ensure the flow basically constant that flows to actuators even when the load changes greatly, which will ensure the unchanged moving speed and normal running of actuators basically. But with the increasing of the flow loads requiring, when it increases to some extant, instead, the flow provision for the actuators with large load will decrease, affecting the normal running of actuators. While for the LUDV control system, its flow can keep constant no matter how the loads change, which ensure the system can run normally and then strengthen the reliability of system.

5 CONCLUSION

The working principles of LS and LUDV hydraulic control technology are analyzed and their simulation models are established.

Through the simulation and comparative analysis of LS and LUDV control systems, LS control system can remain the moving coordination of actuators when the flow loads requiring is small. With the increasing of the required flow , the actuators of LS control system appear obviously uncoordinated motion while the LUDV control system can well control.

The LUDV control system is superior to the LS control system in flow distribution, which provide a theoretical basis for the design of hydraulic system for engineering machinery.

ACKNOWLEDGMENTS

The research was supported by the National Natural Science Fund for youth project (51105265), the New Century Excellent Talents under the project NCET-12-1038 and the Shanxi College of outstanding young academic leaders project. The first author gratefully acknowledged the helpful discussions with the research group and colleagues in the School of Mechanical Engineering at Taiyuan University of technology.

REFERENCES

Fu Yongling, Qi Xiaoye. 2005. Modeling and Simulation of AMEsim System. Being: Beihang University Press.
Li Huacong Li Ji. 2006. Modeling and Simulation Software AMESim for Mechanical/Hydraulic System. Computer Simulation 29(12):294–297.
Li Jianmin, Chen Fei, Qu Lihong. 2011. Simulation and Analysis of Load Sensing Hydraulic System. Coal Mine Machinery 32(7):53–56.
Wang Guanlei, Tong Zhixue, Zhang Zhanping, Han Yingfei. 2011. Simulation study on load sensing system of hydraulic manipulators .Mining&Processing Equipment 39(12): 105–107.
Wang Weshen, Ma Xiao, Wang Baoming. 2002. The Load Sensing Control Energy-Saving Technology of Hydraulic System. Modular Machine Tool & Automatic Manufacturing Technique (9):52–54.
Wu Xiaoguang, Song Zhentao, Yin Xinsheng, Fan Dong, Liu Qingxiu. 2008. Simulation Analysis of Drilling Load Sensing System Based on AMESim. Machine Tool & Hydraulics 36(13): 163–165.

Computational Intelligence in Industrial Application – Ling (ed.)
© 2015 Taylor & Francis Group, London, ISBN: 978-1-138-02818-0

"S" Shape walking path design of a new car without carbon

Yan Li

Zhengzhou University of Industry Technology, Zhengzhou, China

ABSTRACT: In the practical application of engineering, caring for mathematics modeling with respect to the question is the essential link. To study the methods and measures of improving the car's movement steadiness, motion commutation mechanism, drive and periodic steering to key issue and so on, which not only improve the accuracy of the car's movements, but also determine the car's structure parameters and the overall design scheme. In order to simplify the solving process established tracks, using the value repetitive process established based on the slanting plate organization new car without carbon mathematical model. In order to simplify the solving process of car walking path, using the value repetitive process establishment based on the slanting plate organization new car without carbon mathematical model. Considering the car exists inevitable setup error, using the micro-element method to solve the model analysis, and also through the MATLAB software to simulate walking path of the car.

KEYWORDS: Car without carbon; Numerical iterations; Swash plate mechanism; MATLAB; Walking path.

1 INTRODUCTION

Car without carbon, which is on the base of not using carbon energy, according to the principle of energy conversion, though a series of institutional movement conversion, convert the gravitational potential energy into the kinetic energy of the car, finally achieve the scheduled action in the process of car travel. The turning part controls the walking path of the car, how to realize the automatic steering, periodic rotation, complete the desired trajectory accurately, and also avoid obstacles as much as possible, that is the key. Although the CAM mechanism can realize the car's trajectory control, the CAM contour machining is difficult, the size cannot reversible change, precision is difficult to guarantee, heave weight, low efficiency and large energy loss. Take crank rocker for example, the structure is relatively simple, but there is a sliding friction pair, the transmission efficiency is low, and also the fast returning character decided that it is difficult to design a good agency[3].

In this thesis based on swash plate of car without carbon, simple and compact structure, easy to design, can complete the desired trajectory, which makes up for the inadequacy of the CAM and linkage mechanism, through the MATLAB programming can realize the visualization of the walking path.

2 GETTING STARTED

2.1 The car's overall structure design

In this thesis, it requires that the car can avoid the predefined cylindrical pin of equal distance in the process of travel. According to how much the car travel distance and avoid the obstacles to comprehensive evaluation of the car performance is good or bad. The car is made up of the car plate, the front wheel, rear wheel, gear set, inclined plate, direction guide rod, weight, thread and so on. Weight through the thin line drives rear wheel rotation, the real wheels through gear group transfer power to the swash plate, inclined plate agencies in connection with the front wheel steering shaft in the direction of the guide bar, ultimately drive the front wheel steering. To achieve flexible steering car, two rear wheels of the car design, take a turn round, round of follow-up, in order to reduce the rear wheels in the process of friction with the ground loss, specific requirements which are as follows:

1 Require the rear wheel drive without differential drive;
2 The front wheel angle must be satisfied;

$$r_1 = \frac{L}{\tan(\alpha)} \qquad (1)$$

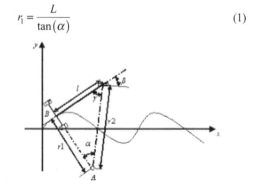

Figure 1. The car instantaneous walk path diagram.

The corresponding author Li Yan, email: lytangshao@163.com.

3 Curve circumference should be multiple of the rear wheel circumference in order to ensure that the cyclical nature of steering mechanism;

4 In order to ensure the car walking distance far enough in the process of marching. When designing, the structure should be lighter, and also the transmission chain should be short as far as possible in order to reduce the energy loss. In addition, the car still needs to maintain at a low speed in the process of marching, there are two reasons:

A. Suppose that the damping in the system is linear damping, a Rayleigh damping [3] function is:

$$D = \frac{1}{2}cv^2 \qquad (2)$$

B. The car is by the weight of gravitational potential energy into kinetic energy to drive the car, the faster the weight falling, the faster the car's speed;

5 The car's cyclical turning requirements

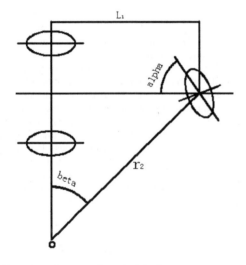

Figure 2.   The car turning principle diagram.

Suppose that the distance between a front axle and rear axle is L, according to the geometric relationship can be concluded that: when $\alpha = \beta$, the largest turning angle of the front wheel is:

$$\alpha_{\max} = \arcsin\frac{L}{r_{2\min}} \qquad (3)$$

2.2 Establish the car walking trajectory model

Because of the car's front wheel is swinging in a row, the geometric relationship between the front wheel and the rear wheel in the process of marching which is shown in Fig. 1. Suppose the walking trajectory

curve equation of the car is $y(x)$, the length of the car is L, the curvature of the front wheel turning radius is r_2, the curvature radius corresponding to the rear wheel is r_1, then the function relation is:

$$\begin{cases} r_1 = \dfrac{\left(1+\left(y(x)'\right)^2\right)^{\frac{3}{2}}}{\left|y(x)''\right|} \\[4mm] \tan(\alpha) = \dfrac{L}{r_1} \end{cases} \qquad (4)$$

1 The relationship analysis about inclined plate and guide role.

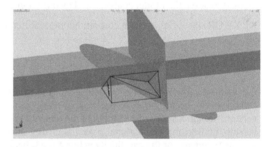

Figure 3.   The 3d graphic of the swash plate.

Suppose that angle between the inclined plate and the level of the shaft central line is γ, the inclined plate angle is α_1, the inclined plate radius is R_1, the rear wheel radius is R, the turning angle corresponding to rear wheel is φ, through the analysis of the Fig. 3 and structural theorem you can get:

$$R_1^2 = \left(x\tan\gamma\cot\alpha_1\right)^2 + \left(\frac{x}{\cos\gamma}\right)^2 \qquad (5)$$

So x is a function of α_1.

2 Using infinitesimal method analysis the car's rear wheel movement.

As is shown in Fig. 1, divide the inclined plate week corresponds to the motion of the car to walk the length of the track on the x axis, write it as:

$\Delta x = [\Delta x_1, \Delta x_2, \Delta x_3 \ldots \ldots \Delta x_n]^T$, the m paragraph of the corresponding horizontal axis is $x_m = \sum_{i=1}^{m} \Delta x_i$. As each turn, you can get a swash plate angle, at the same time, the rear wheel corresponding turns an angle, that is $\Delta s_0 = R_1 \phi_0$. In the period of m, the walking path of the car has the following relationship:

$$\begin{cases} \Delta y_1 = \tan(\varphi) \Delta x_i \\ \Delta s_i = \sqrt{\Delta x_i^2 + \Delta y_i^2} \end{cases} \quad (6)$$

In type (6), $\tan(\varphi)$ is $y(x_i)$, when Δx is close to zero, it can be approximately regarded as a straight line, that is $s_m = \sum_{i=1}^{m} s_i$. Assume that the car's rear wheel radius is R, when the car's walking trajectory is equal to s_m, then the rear wheel turning angle is equal to $\theta_m = \dfrac{s_m}{R}$, you can get:

$$\theta_{rearwheel} = \frac{1}{R}[s_1, s_2, s_3 \ldots \ldots s_N]^T \quad (7)$$

The inclined plate rotates a week, the corresponding car walks are length is:

$$s_n = \int_0^{x_n} \sqrt{1 + \left(y(x)'\right)^2}\, dx \quad (8)$$

At this time, the corresponding inclined plate rotates angle $\theta = \dfrac{s_n}{2\pi R}[s_1, s_2, s_3 \ldots \ldots s_n]^T$, in this way, you can get the car's desired trajectory.

2.3 The car's walking trajectory simulation and visualization

The car's side double check calculation.

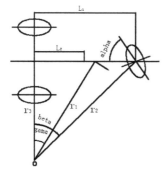

Figure 4. The turning principle diagram of the weight.

Through geometric relations can get:

$$\alpha = \beta, r_3 = r_1 \cos_\beta, \beta = \arcsin \frac{L}{r_1} \quad (9)$$

Then:

$$r_2 = \sqrt{r_3^2 + L_2^2} \quad (10)$$

Among them: the front wheel turning radius is r_1, a major piece of turning radius is r_2, and the turning radius axle center is r_3.

5 The car's walking trajectory simulation.

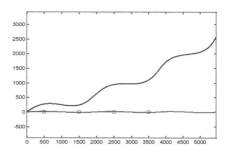

Figure 5. The car's walking path.

As can be seen from Fig. 5, in the procession of marching, the car deviates from the orbit gradually. The reason for this problem is that when the car is in the procession of marching, there exists installation error which is inevitable, the center of the car's front wheel and rear wheel is not in a line. In the process of travel, with the car's front wheel deflecting around constantly, like Fig. 5. In order to improve this situation, adopting the direction guide rod which connected bottom of the car is equipped with reset spring. Rear mount fine-turning mechanism, correct the error, ensure that the swash plate mechanism driven by a driving force when rotating, through the direction of the guide rod the front wheel and body into a certain angle between axis of swing, realize the car's desired trajectory [4].

At any time the car's movement is as follows:

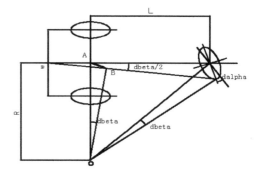

Figure 6. The car's movement at any time.

A car space between two rear wheels and space between front and rear wheel for, instantaneous center point of the rear wheel speed radius, front wheel and the car axis angle, initial coordinate can transmission ration. When the car is in any position, the rear wheel turning, the car front wheel deflection, the next moment, instantaneous center around car thought speed, the front wheel turning, arc AB for long, the arc length respectively projection to the shaft, shaft available:

$$
\begin{cases}
dy = -2 \times \rho \times \sin\left(\dfrac{d\beta_i}{2}\right) \times \sin\left(\beta_i + \dfrac{d\beta_i}{2}\right) \\[2ex]
dx = 2 \times \rho \times \sin\left(\dfrac{d\beta_i}{2}\right) \times \cos\left(\beta_i + \dfrac{d\beta_i}{2}\right)
\end{cases}
\tag{11}
$$

Using the iterative method for coordinates:

$$
\begin{cases}
x_i = x_{i-1} + dx_i = x_{i-1} + 2 \times \rho \times \sin\left(\dfrac{d\beta_i}{2}\right) \times \cos\left(\beta_i + \dfrac{d\beta_i}{2}\right) \\[2ex]
y_i = y_{i-1} + dy_i = y_{i-1} - 2 \times \rho \times \sin\left(\dfrac{d\beta_i}{2}\right) \times \sin\left(\beta_i + \dfrac{d\beta_i}{2}\right)
\end{cases}
\tag{12}
$$

Draw the curve (x, y), you can get the car's wheel center curve equation.

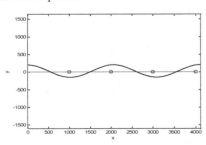

Figure 7.   "S" shape tracks of the car.

3   CONCLUSION

On the basis of accurately mathematical modeling which is based on the new car without carbon with the inclined plate institutions, use MATLAB powerful scientific analysis, calculation and visual function, after that use the appropriate method, analyze the trolley routes to realize the car's walking trajectory simulation in a fixed period, and further expounds the importance of mathematical modeling in actual engineering application.

REFERENCES

Sun Heng, Chen Zuomo, Ge Wenjie etc, The Mechanical Principle, Higher Education Press, Beijing, 2005.
Yin zeming, Ding Chunli etc, Proficient in MATLAB 6[M], Tsinghua Education Press, Beijing, 2002.
Chen Haiwei, Zhang Qiuju etc, Mathematical Modeling-an Important Part of the College Student's Engineering Training, Journal of Wuxi Vocational and Technical College, 2011.
Pu Lianggui, Ji Chunming etc, Mechanical Design 8[M], Higher Education Press, Beijing, 2005 [5] Wang Bin, Wang Yan, Li Runlian etc, "Car without Carbon" Innovative Design, Journal of Shanxi Datong University (National Science Education), 2012.

Computational Intelligence in Industrial Application – Ling (ed.)
© 2015 Taylor & Francis Group, London, ISBN: 978-1-138-02818-0

Research on the method of automatic recognition of DMI information of CTCS-3 onboard equipment

Yong Zhang, JianXiong Wu

School of Electronic and Information Engineering, Beijing Jiaotong University, Beijing, China

ABSTRACT: The method for automatic recognition of Driver-Machine Interface (DMI) information of CTCS-3 Onboard Equipment was proposed. First, the Hough Transform was used to detect the outline of the DMI display area, and the affine transformation was adopted for tilt correction. Secondly, according to the characteristics of DMI display, such as fixed position of display areas, specific size and font of characters, enumerable displaying information, normalization and feature extraction were conducted for icons, numeric numbers and letters, as well as Chinese characters respectively. Finally, the decision tree method and SVM method were adopted for classification and recognition of the extracted features. The experiment results proved the effectiveness of this method.

KEYWORD: CTCS-3; DMI; Image recognition; SVM; Automatic testing.

1 INTRODUCTION

The CTCS-3 train control system has been widely applied in high speed railways in China. To verify whether the system can meet relative standards[1, 2], it is necessary to carry out a series of testing, such as laboratory testing, field testing and interoperability testing.

For the laboratory testing of the onboard equipment, the black-box method based on data-driven mechanism is usually adopted[3]. Before the testing, a lot of test data shall be prepared. During the testing, the testers shall observe and record the test results at the visible interfaces such as DMI. Finally manual analysis is required for verification of the test results, by comparing the expected results of the test data with the test results from different channels such as DMI display and data recorder. Therefore, to obtain the display information of DMI is an important link in the testing.

A solution to the automatic recognition of DMI information is presented: during the testing of on-board a video camera is used to record the DMI display in real time, and the method based on image recognition and data mining is adopted to recognize the information automatically, thus laying a good foundation for the automatic testing of the on-board equipment.

2 ANALYSIS OF DMI INFORMATION

2.1 Characteristics of DMI display

According to *Specification of Driver Machine Interface information for CTCS-3 Onboard equipment*[4], the DMI display area is divided into six regions, named as A, B, C, D, E and F, which can be sub-divided, as shown in Fig.1.

The DMI display consists of such elements as icons, numeric numbers and letters, as well as Chinese characters. It is featured by:

1 Black background, with a distinct contrast with the displayed information.
2 All information is displayed in areas with fixed size and position, and each type of information is enumerable.
3 All Chinese characters are displayed in Youyuan font, in gray or white color, and in the size of 14, 16 or 18 pounds.
4 All the numeric numbers and letters are displayed in Arial font, in black or white color, and in the size between 10 and 22pounds.
5 The color type of DMI display is enumerable (17 types in total).
6 All text information is displayed in the form of single lines.
7 The number of Chinese characters and icons is limited, and not more than 300.

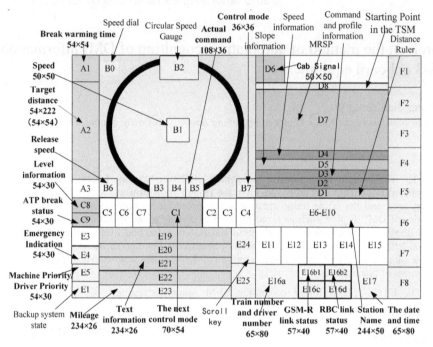

Figure 1. DMI information distribution.

2.2 The areas to be identified

In principle, only the information related to the testing will be extracted, which is displayed in the areas labelled with pixel size logos, as shown in Fig.1.

The icons can be classified into fixed ones and dynamic ones such as the brake warning time icon in A1 zone. For the former type, the decision tree method is adopted for its recognition. For the latter type, the color, size and other features are extracted for classification and recognition.

The SVM method is adopted for the classification and recognition of numeric numbers, letters and Chinese characters.

3 DMI INFORMATION RECOGNITION

The automatic recognition process of DMI information is shown as in Fig.2. Firstly frame images are extracted from the DMI video; secondly features are extracted from the image data; finally the image data are classified and identified according to the features. At the same time, the extracted features are also used to supplement the feature library, forming a close-looped control of the feature library to realize the automatic training of the feature library with the aim to improve the recognition rate.

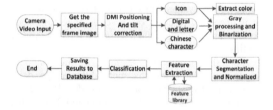

Figure 2. The process of DMI automation recognition.

3.1 DMI image acquisition

As the shooting angle, the orientation and the vision range of the video camera may be not fixed, the DMI video shall be processed by tilt correction and clipping, to obtain a rectangular image of the DMI display area.

3.1.1 Detection of outline of DMI display

Hough Transform[5] is used to detect the outline of DMI display which is enclosed by four straight line segments. OpenCV provides two versions of Hough Transform implementation, with the basic version as cv:: HoughLines. In order to eliminate the detection errors due to the simultaneously passing of a pixel by multiple lines, the Probabilistic Hough

Transform with the implementation function as cv:: HoughLinesP is adopted. In order to avoid the detection of excessive number of line segments, we can configure the minimum length and the slope of the line segments in view of the nearly rectangular feature of DMI display region. Thus, the four longest straight line segments are chosen from in the list of detected line segments, and the closed area formed by these four straight line segments is the outline of the DMI display region.

3.1.2 Tilt correction

In tilt correction, affine correction shall be carried out by affine matrix. An affine transform function: cvGetPerspective Transform is provided by Open CV for this purpose. A DMI image after tilt correction can be obtained by using the function cvWarpPerspective, which is further clipped to obtain a rectangular image of DMI display region. The process of tilt correction is shown as in Fig.3.

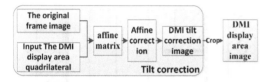

Figure 3.   Process of tilt correction.

3.2 Pre-procsessing

Pre-processing mainly involves several processes such as image clipping, greyscale transformation, thresholding, as well as character segmentation and normalization.

3.2.1 Image Clipping

As the locations of the interested areas are fixed，it is quite easy to get the images of them. Then the images are classified into icons, numeric numbers and letters, as well as Chinese characters, for further processing.

3.2.2 Grayscale transformation and thresholding

As the color information is normally not necessary for the recognition, we transform the color images into grayscale images, and then into binary images. By the way, the recognition of icon colors shall be conducted before the grayscale transformation.

Grayscale transformation: a gray image contains only brightness information, and each pixel has the value representing the brightness of its location, i.e. the gray value which is divided into 256 quantization levels in an 8-bit gray image.

Thresholding: a binary image has a smaller size, with the speed advantage for such operations as storing and retrieving. There are many ways to transform gray images into binary images.

3.2.3 Character segmentation and normalization

As stated in section 2.1, all the text information is displayed in the form of single lines. Therefore, for character segmentation, we firstly clip off the excessive parts above and below a single line, then segment the line to obtain the images of single characters.

In view that all the characters in a single line are aligned and with the same height, the level integral projection method is used in the horizontal direction [6], which can achieve good results

3.3 Feature extraction

As the icons, numeric numbers and letters, and Chinese characters are different from each other, they are handled separately in the feature extraction.

3.3.1 Icons

As the icons are displayed in fixed areas and are enumerable, the feature extraction based on color and shape is adopted. After a binary icon image is normalized, we calculate the number of pixels with the logic value of 1, to obtain the geometry feature of an icon.

3.3.2 Numeric numbers and letters

The feature of numeric numbers and letters includes the statistics feature and the structural feature. As many numeric numbers and letters have the same size of area after being normalized, it is difficult to distinguish them directly by the size of area. Therefore, we divide the whole area into 9 small sub-areas. Thus, we can obtain a 9-dimensional feature vector by counting the number of pixels with logic value of 1 in the sub-areas. The area division is shown as in Fig. 4.

Figure 4.   9-dimensional grid feature.

Regarding the structural feature of numeric numbers or letters, we use an improved four-line feature extraction method. As all the numeric numbers and letters are in Arial font, all of them are almost identical in shape and area after being normalized. Based on this, the pixels indicating the color change on the four straight lines are extracted. The first, third and fourth lines is counted from left to right, while the second line is counted in reverse direction, as shown in Fig.5.

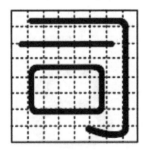

Figure 5. Improved four-line feature extraction method.

3.3.3 *Chinese characters*

As the Chinese characters used in DMI display amount to no more than 300 and are in the same font, a template matching method is adopted for their recognition. We firstly obtain the feature templates of the standard Chinese characters through the recognition channel, secondly feed a character to be recognized to the same channel to obtain certain features, and then calculate the similarity between the feature of the character to be recognized and that of the template, and finally get the result by choosing the template with the maximum similarity.

The rough grid feature of a Chinese character is extracted by uniformly dividing the character image into n × n grids, and counting the number of the pixels occupied by the character in each grid, to obtain a n×n feature vector. This feature vector reflects the distribution of the overall shape of the character and its complexity, whose accuracy is in positive correlation with the value of n. An example is given Fig.6, showing the grid feature of Chinese character '司', with the value of n as 8.

Figure 6. Features of 8×8 rough grid.

Meanwhile, the structural feature of the Chinese character is also considered by extracting the crossing line feature, including the 8-dimensional vertical and 8-dimentional horizontal crossing line feature.

The method of extracting the 8-dimensional vertical crossing line feature is as follows.

1 A normalized image of Chinese character is vertically and uniformly divided into 4 meshes.
2 For each mesh, count the number of non-character pixels from the top, until a character pixel is encountered, and calculate the size of the area containing the non-character pixels, to obtain the first 4 vectors.
3 For each mesh, likewise, continue to count the number of non-character pixels until the next character pixel, and calculate the size of area containing the non-character pixels, to obtain another 4 vectors..

Thus, an 8-dimensional vertical feature vector can be obtained; an 8-dimensional horizontal feature vector can also be obtained likewise. Hence, a 16-dimensional feature vector is obtained.

Figure 7 shows the process of extracting the vertical crossing line feature.

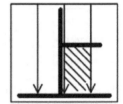

Figure 7. Vertical crossing line feature.

3.4 *SVM based recognition*

During the recognition progress, the extracted features are used to classify the images. In this paper, the decision tree method is used for the classification and recognition of icons, and the SVM[7] method is used for classification and recognition numeric characters, letters and Chinese characters.

Because the SVM is the two-pass classifier, several SVM classifiers shall be used in combination to identify multiple types of samples. Here, the SVM functions provided by LIBSVM library are utilized for training and recognition respectively. The LIBSVM based recognition process is shown as in Fig.8.

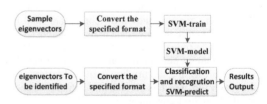

Figure 8. LIBSVM flowchart.

4 EXPERIMENTAL RESULTS

4.1 *Tilt correction*

Fig.9 (a) shows a frame image of the DMI video; the Hough Transform Straight Line Detection is used to obtain the outline of the DMI display, as shown in Fig.9 (b); then affine transformation is used to realize tilt correction; finally, the unwanted parts are clipped off, to obtain the final picture as shown in Fig.9 (c).

(a) A frame image   (b) DMI contour detection (Blue Line)   (c) Tilt correction cutting results

Figure 9. Skew correction result.

4.2 *Feature Extraction*

For numeric numbers, the extracted features shall be normalized, as shown in Fig.10.

Figure 10. Extracted feature vectors of numeric numbers.

In Fig.11, the process of extracting the features of Chinese character '完' is presented.

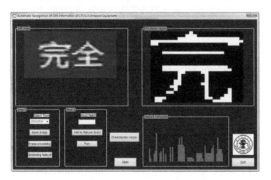

Figure 11. Extracting the features of Chinese characters.

4.3 *SVM based classification and recognition*

The SVM classifier is c_svc type with RBF kernel. The grid.py tool of python platform is used to select the optimal parameters of c and g, as shown in Fig.12.

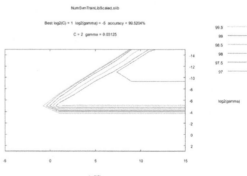

Figure 12. Selecting SVM optimal parameters.

For the feature libraries extracted from Chinese characters, numeric numbers and letters, as well as icons, the SVM optimal parameters were obtained, respectively, as shown in table 1.

Table 1. SVM parameter optimization.

Lib	C(penalty coefficient)	Gamma(rbf)	accuracy
CharSvmTrainLib.slib	512	0.0001220703125	95.2381
NumSvmTrainLib.slib	2.0	3.0517578125e-05	99.5204
IconSvmTrainLib.slib	8.0	0.0078125	88.0952

The optimized penalty coefficient and γ value are used to train the feature library, to obtain a model which can improve the accuracy when conducting regressive prediction based on unknown samples. Fig.13 shows the model parameters obtained by using SVM to train the feature library of numeric numbers and letters.

Figure 13. SVM model parameters.

The number of the support vectors is 276, reflecting the linear inseparability of the sample itself. In addition, after the parameter optimization, the number of the support vectors is reduced by nearly a half, demonstrating the necessity and effectiveness of conducting parameter optimization before training the feature library.

5 CONCLUSIONS

The method for automatic recognition of DMI information of CTCS-3 onboard equipment is proposed. Based on analyzing the characteristics of DMI displaying, the display elements are divided into three categories: icons, numeric numbers and letters, as well as Chinese characters. Specific methods are adopted to extract the feature of each type of element, and the SVM method is used for the classification and recognition of these three types of elements.

Regarding the recognition performance, following aspects shall be considered:

1 When obtaining frame image from the DMI video, with consideration of the slow change in the display contents, it is allowed to read three frames per second, so as to improve the recognition efficiency.

2 When determining the outline of DMI display area, it is suggested that the size of the image shall be scaled down before conducting straight line detection, to obtain the coordinates of the four vertexes of the DMI display area, and finally obtain the original coordinates of the four vertexes based on the scale ratio. In this way, the amount of computation can be greatly reduced.

ACKNOWLEDGEMENT

This work was supported in part by the key project of China Railway Corporation under Grant 2013X001-F.

REFERENCES

China Railway Corporation, "CTCS-3 train control system of technological innovation overall scheme" Beijing: China Railway Publishing House, 2008.58-59.

China Railway Corporation, "CTCS-3 train control system requirements specification (SRS)" Beijing: China Railway Publishing House, 2009.

Liu Y, Tang T. Research on the method of interoperability test for the onboard equipment of CTCS-3 train control system of Chinese railway[C]. Autonomous Decentralized Systems (ISADS), 2011 10th International Symposium on. IEEE, 2011: 415-419.

China Railway Corporation, "CTCS-3 train control board equipment DMI displays specifications V2.0" Beijing: China Railway Publishing House, 2010.

Laganière R. OpenCV 2 Computer Vision Application Programming Cookbook: Over 50 recipes to master this library of programming functions for real-time computer vision[M]. Packt Publishing Ltd, 2011.

Zhang Y, Qin Y, Li K, et al. The Method for Recognizing Driver Machine Interface Information of the Onboard Equipment in CTCS-3[J]. China Railway Science, 2010, 4: 020.

Zhang H, Berg A C, Maire M, et al. SVM-KNN: Discriminative nearest neighbor classification for visual category recognition[C],Computer Vision and Pattern Recognition, 2006 IEEE Computer Society Conference on. IEEE, 2006, 2: 2126-2136.

Computational Intelligence in Industrial Application – Ling (ed.)
© 2015 Taylor & Francis Group, London, ISBN: 978-1-138-02818-0

Research on the method and tool for automated generation of test data for CTCS-3 on-board equipment

Yong Zhang, Qian Huang & Yuntao Liang
School of Electronic & Information Engineering, Beijing Jiaotong University, Beijing, China

ABSTRACT: CTCS-3 train control system is the key equipment to ensure train operation safety and efficiency. In order to provide an environment for researching the key technical issues concerning the automatic testing of CTCS-3 on-board equipment, a testing platform which contains three modules such as automatic generation of test script, automatic test execution and automatic analysis of test results was designed. In this paper, we focus on the method and tool for automatic generation of test data. Firstly the source and contents of the test data were introduced; secondly the test data generation process, including some key technical issues such as RBC message generation was described; thirdly the design and implementation of the test data generation tool were introduced.

KEYWORDS: CTCS-3; On-board equipment; Test sequence; Test data; RBC message; XML script.

1 INTRODUCTION

As a unified technology platform for Chinese high-speed railway, the CTCS-3 train control system provides an important guarantee for the safe and efficient operation of the train. In order to verify whether a CTCS-3 system is consistent with system requirements specification, it is necessary to carried out a series of testing, including laboratory testing, pilot track testing, integrated testing and commissioning, and interoperability testing.

The laboratory testing of the on-board equipment is a kind of black-box test, which mainly involves three aspects: the preparation of test data, the execution of test script and the analysis of test results. The automation in these three aspects is critical for achieving the automatic testing of the on-board equipment.

At present the preparation of test data is conducted manually according to test sequence, which is cumbersome and time-consuming. This paper focuses on the method and tool for the automatic generation of test data.

It is organized as follows. In section 2 the on-board equipment testing method is described; in section 3 the test data generating method and the key technical issues on generating the RBC(Radio Block Center) radio messages are introduced; in section 4 the architecture and functions of the test data generation tool are introduced, together with the test data generation process and test data application; in section 5 some conclusions are drawn.

2 ON-BOARD EQUIPMENT TESTING

2.1 *Structure of on-board equipment*

The major function of the CTCS-3 on-board equipment is to supervise train operation based on train positioning, and train control related information received from the trackside equipment.

The on-board equipment consists of: Vital Computer(VC), Track Circuit Receiving module (TCR), Radio Transmission Module (RTM), Balise Transmission Module (BTM), Driver Machine Interface (DMI), Train Interface Unit (TIU), Juridical Recording Unit (JRU) and etc. The VC processes the data from other modules and calculates train speed profile for supervising train operation; the DMI presents various kinds of train control information to the driver; BTM, RTM and TCR receive and process trackside data and send relevant information to VC; JRU records all information related with train operation and equipment state[1] .

2.2 *Operating mode of on-board equipment*

The working process of CTCS-3 on-board equipment is achieved through the transition of its operating modes which include Full Supervision (FS), On Sight (OS), Calling On (CO), Shunting (SH), Isolation (IS), Stand By (SB), Sleeping (SL), Trip(TR), and Post Trip (PT)[1] .

2.3 The test method

The laboratory testing of the on-board equipment is a kind of black-box test based on data driven mechanism [2], which uses the test data contained in the test script to drive the test process.

2.4 Testing platform

The on-board equipment testing platform consists of four modules: scenario controller, on-board simulator which can be replaced by the real equipment, trackside simulator and data recorder.

The scenario controller reads the test data from the test script and distributes the data to the relevant sub-module of the trackside simulator. It is in charge of controlling the testing process.

The on-board simulator simulates the functions of the on-board equipment and train motion dynamics.

The trackside simulator consists of three submodules: the track circuit module for sending track circuit code, the balise module for sending balise telegrams, and the Radio Block Center (RBC) module for sending radio messages such as movement authority (MA). All the information is sent to the on-board simulator based on event triggering mechanism.

The data collector collects and stores all the data produced during the test process.

3 TEST DATA GENERATION METHOD

3.1 Sources of test data

The test data comes from two sources: the test sequence and track data.

A test sequence is generated by concatenating a series of test cases, which is used to test the functions of on-board equipment during a test trip [3-5].

A test sequence contains: (1) identity information of the test sequence, such as the number of the test sequence, the major functional points to be tested; (2) the test prerequisites, including the start position of the train, the initial state of the on-board equipment; (3) the track layout graph, which is used to illustrate the overview track layout of the test line, the train path, the test conditions (such as track occupation, Temporary Speed Restriction (TSR)) and some prompt boxes for the focus points during a test trip; (4) the test procedures, which describe the status of the train step by step, the mode transformation of the on-board equipment, the test conditions, the related cases and the expected test results.

The track data contains: (1) the position of and the relationship among trackside elements such as balises, signals, switches, track sections; the

gradient information; the start position and length of neutral zones; the range of shunting areas; the preannouncement and execution points for level transition and RBC/RBC handover; (2) route information, including the number and name of routes; the start and end position of each route [7].

3.2 Contents of test data

The test data contains: train position, RBC message, balise message, driver operation, test case ID and expected result. The test data is expressed in XML script, shown as follows:

```
<?xml version="1.0" encoding="UTF-8"?>
<datarootxmlns:od="urn:schemas-microsoft-com:
officedata" generated="2014-05-21T15:05:08">
<Multiple Vehicles Xml Num="1">
<Action Position>625280</ Action Position>
< RBC Message File>0xFF </ RBC Message File>
<Balise Message File >0xFF</ Balise Message File>
< Drivers Actions >0x20</ Drivers Actions >
< Test Cases ID >0xFF </ Test Cases ID>
<Expected Result>0xFF</ Expected Result >
</ Multiple Vehicles Xml >
……
</data root>
```

3.3 The general design thought

The process of generation of test data is shown as in Fig.1, with following steps:

Step 1 generate a test sequence based on the test cases, test line and train path selected by the user.

Step 2 generate the RBC messages according to the track data, TSR information.

Step 3 according to the test step of the test sequence, insert the balise telegrams retrieved from the test line database and the RBC message into the test data structure.

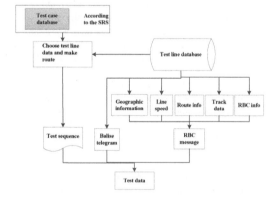

Figure 1. Test data generation process.

3.4 Key technical issues

3.4.1 RBC message generation

RBC message consists of two categories: (1)MA message, which can contain other optional packets; (2) other messages, which are used for the communication between RBC and the on-board equipment.

During the execution of the test sequence, each test case contained in the test sequence is checked, to verify whether it is necessary to generate radio messages needed by the test case.

Fig.2 shows the process of the messages exchanging between RBC and on-board equipment during the on-board registration process, in which three messages M32, M156, M39 shall be sent by RBC.

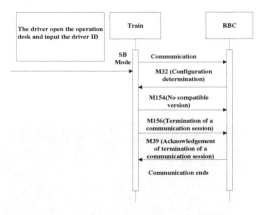

Figure 2. Message exchange in radio session establishment.

3.4.2 Choosing optional packets for MA

RBC message is composed of message header and optional data packets, as shown in Table1 and Table 2.

Table 1. Radio message structure.

	Variable	Note
Header	NID_MESSAGE	Number of Message
	L_MESSAGE	Length of Message
	T_TRAIN	Time related to RBC
	M_ACK	Whether the on-board equipment need to confirm the message.
	NID_LRBG	Balise number (LRBG)
...	...	......
...	Basic Packet	
	Optional Packet	

Table 2. Optional packets.

Message Name	Message ID	Optional packet number
MA	3	3,5,21,27,41,65,68,80
General Message	24	3,5,21,27,41,42,57,58, 65,66,72
MA adjusting the reference position	33	5,21,27,41,65,68,80

The process for choosing optional packets for MA message is as follows.

Step 1, packet 15, 5, 27, 21, 3, and 68 are default packets.

Step 2, if there are level transitions, packet 41 shall be added.

Step 3, if there are temporary speed restriction, packet 65 shall be added.

Step 4, if there are shunting operation or calling on routes , packet 80 shall be added.

4 TEST DATA GENERATION TOOL

4.1 Software architecture and functions

The software was developed under MS Studio 2010, with C# language and SQL Server 2005.

It consists of two modules, the test sequence management module and the test data automatic generation module, whose overall structure is shown as in Fig.3.

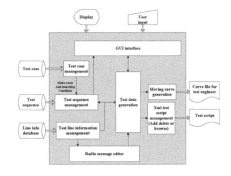

Figure 3. Software architecture.

Test sequence management module [7]

This module generates the test sequence according to the test cases selected by the user, as well as necessary test conditions, such as track occupation, temporary speed restriction (TSR).

Test data automatic generation module

This module generates the test data according to the test sequence and track data selected by the user. It also provides the test data management functions such as browsing, deleting and modifying the test data.

4.2 Test data generation process

The test data generation consists of following steps:

Step 1: the user choose some test cases through the graphic user interface.

Step 2: according to the test requirements and track condition, the user can set up some test conditions, such as track occupation, TSR, etc.

Step 3: some basic information are extracted from the test sequence structure, such as on-board operating mode, test case execution position.

Step 4: according to the route chosen by the user, the tool extracts information from the route table, which are used to generate the MA messages.

Step 5: save the test data as XML script.

The process is shown in Fig.4.

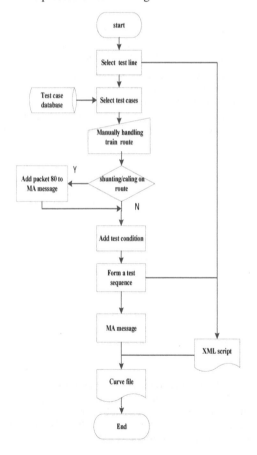

Figure 4.   Software process.

4.3 Test data application

The test data generated by the tool has been used in the on-board testing platform. Fig.5 shows the test script. Fig.6 shows the main window of the test execution software, in which blue highlight part at the lower part indicates the test data being executed.

Figure 5.   Test script.

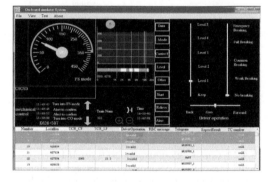

Figure 6.   Main window of test execution software.

5   CONCLUSIONS

In this paper the method and tool for generating test data for CTCS-3 on-board equipment was studied. The general process and some key technical issues were introduced; the test data generation tool was designed and implemented with C# language.

ACKKNOWLEDGEMENT

This work was supported in part by the key project of China Railway Corporation under Grant 2013X001-F.

REFERENCES

Transportation Bureau of MOR, System Requirements Specification for CTCS-3 Train Control System, Beijing: China Railway Publishing House, 2009.

Liu Y, Tang T. Research on the method of inter-operability test for the onboard equipment of CTCS-3 train control system of Chinese railway[C]. Autonomous Decentralized Systems (ISADS), 2011 10th International Symposium on. IEEE, 2011: 415–419.

Transportation Bureau of MOR, Test Cases for CTCS-3 Train Control System (v3.0), Beijing: China Railway Publishing House, 2009.

Zhang Y, Sha S, Wang S, An Expert System Approach for Generating Test Sequence for CTCS-3 Train Control System, 2013 Fourth International Conference on Intelligent Control and Information Processing (ICICIP), Beijing China, June 9–11, 2013:271–276.

Zhang Y, Tang H, Chen Y, et al. Research and Implementation of Test Case Management System for CTCS-3 Train Control System, Dec. 2011 : 230–234.

Zhang Y, Chen Y. Research on Data Management and Configuration in RBC Simulator, 2011 3rd International Conference on Information Electronic and Computer Science, (ICIECS2011), Tianjin, Dec.2011:414–418.

Zhang Y, Wang S, Zhang X, et al. A Computer-aided Tool for Generating Test Sequence for CTCS-3 Train Control System [J]. Advances in Information Sciences and Service Sciences, 2012, 4(11).

Computational Intelligence in Industrial Application – Ling (ed.)
© *2015 Taylor & Francis Group, London, ISBN: 978-1-138-02818-0*

Design and implementation of CTCS-3 on-board equipment testing software based on multithreading

Yong Zhang & Bin Zhang

School of Electronic & Information Engineering, Beijing Jiaotong University, Beijing, China

ABSTRACT: A testing platform was designed for the purpose of supporting the research on the key technical issues concerning the automatic testing of CTCS-3 on-board equipment, which consisted of 3 parts responsible for automatic generation of test script, automatic test execution and automatic analysis of test results respectively. This paper focused on the automatic test execution software. Firstly the architecture and working process of the software were introduced; secondly, the architecture and functions of its four functional modules including on-board simulator, trackside simulator, scenario controller and data collector were described; finally the software implementation based on C# language and multithreading technique was introduced.

KEYWORDS: CTCS-3; On-board equipment; Simulation; Testing; C#; Multithreading.

1 INTRODUCTION

The CTCS-3 train control system is the key equipment to ensure train operation safety and efficiency, which has been applied in nearly 10 high speed lines. In order to verify whether a CTCS-3 system is consistent with the system requirements specification, it is necessary to carry out a series of testing including laboratory testing, pilot track testing, integration testing and commissioning, as well as interoperability testing [1,2].

The architecture of the laboratory testing platform of the on-board equipment normally consists of three parts: the Equipment Under Test (EUT), the test environment providing the necessary information to the EUT, and the test tools for controlling the test execution and recording the test results, etc.[3]

The laboratory testing of the on-board equipment is a kind of black-box test based on data-driven mechanism [4], which mainly involves three aspects: the preparation of test data, the execution of test script and the analysis of test results. The automation in these three aspects is critical for achieving the automatic testing of the on-board equipment.

For the purpose of conducting the research on the key techniques concerning the automatic test of the on-board equipment, a testing platform is designed and implemented. The main purpose of this paper is to present the part concerning the automatic execution of the test script.

This paper is organized as follows : in Section 2 the architecture working process of the platform are introduced; in Section 3 to 6 the architecture and functions of the on-board simulator, the trackside simulator, the scenario controller and the data collector are described respectively; in Section 7 the implementation of the software based on C# and multithreading technique are introduced; in Section 8 some conclusions are given.

2 ON-BOARD EQUIPMENT TEST SOFTWARE

2.1 *Overall architecture*

As shown in Fig.1, the software consists of four functional modules: scenario controller, on-board simulator, trackside simulator, and data collector. The scenario controller is in charge of controlling the testing process; the on-board simulator is responsible for simulating train motion dynamics and the functions of the onboard equipment; the trackside simulator sends the necessary trackside information to the on-board simulator; the data collector gathers all the data produced during the testing process.

The input of the software is the test scripts generated by a test script generator.

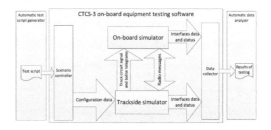

Figure 1. Architecture of the test software.

Figure 2.　Test script.

2.2　Main process

1　Before the testing, the test script generator produces test script and test data containing balise telegrams and radio messages based on the test sequence [5] and relevant track data, and stores the test script into the database.

2　During the testing, the scenario controller reads the data from the test script and distributes them to relevant sub-modules of the trackside simulator, which sends the data to the on-board simulator.

3　At the end of the testing, the data collector stores all relevant data into the database for further analysis.

3　ON-BOARD SIMULATOR

3.1　Architecture

The on-board simulator integrates the simulation of onboard equipment, train motion and driving operation.

The on-board equipment consists of Vital Computer (VC), Radio Transmission Module (RTM), Track Circuit Reader (TCR), Balise Transmission Module (BTM), Drive Machine Interface (DMI), Judicial Recording Unit (JRU) and Train Interface Unit (TIU), as shown in Fig.3.

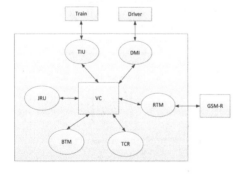

Figure 3.　Modules of on-board equipment.

3.2　Functions

The main functions of the on-board simulator are as follows:

1　Simulating the functionality of the on-board equipment, including receiving and processing track circuit information, balise telegram and radio message, realizing train positioning, sending position report and MA request to RBC, calculating speed control curve to supervise train operation, etc.

2　Calculating train speed and running distance based on train motion dynamics and the command from driver simulation.

3　Simulating driving operation, providing such means as controlling the traction/braking, and the forward/backward movement of the train.

4　TRACKSIDE SIMULATOR

4.1　Architecture

The trackside simulator consists of following sub-modules: track circuit, balise, and Radio Block Center (RBC), which send relevant data to the on-board equipment at an appropriate time.

4.2　Functions

The main functions of the trackside simulator are as follows:

1　The track circuit module sends the information indicating the number of free block sections ahead of the train, to the on-board equipment, and conducts track occupation detection.

2　The balise module sends the balise telegrams containing track data related information such as speed restriction and gradient, to the on-board simulation module.

3　The RBC module sends such messages as Movement Authority (MA), track data, balise link information and emergency stop to the onboard equipment according to the position reports or MA requests sent by the on-board module. [6]

The track circuit module and balise module communicate with the on-board simulation module uni-directionally. They send relevant data to the onboard module according to the current position of the train.

The RBC module communicates with the on-board simulation module bi-directionally. It generates and sends radio messages under following conditions:

1　The on-board module has sent a position report that contains special locations such as a neutral zone.

2 A specific time interval has expired, e.g. a train will send a position report to the RBC every 6s.
3 Upon receiving specific messages such as MA request from the train.
4 Certain special situations like unexpected occupation of track circuit have occurred.

The main process of the trackside simulator is shown in Fig. 4.

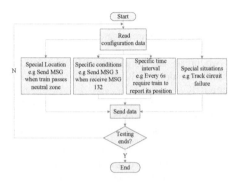

Figure 4. Main process of trackside simulator.

5 SCENARIO CONTROLLER

5.1 *Architecture*

The scenario controller communicates with other modules, to send configuration data to them and monitor their status, whose architecture is shown in Fig. 5.

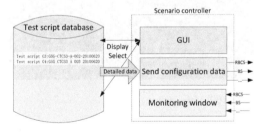

Figure 5 Architecture of scenario controller

5.2 *Functions*

The main functions of the scenario controller are as follows:

1 Retrieving a test script from the database and displaying it in a list, from which a tester can select a test script.

2 Sending the test data to each sub-module of the trackside simulator as configuration data.
3 Supervising the testing process.

6 DATA COLLECTOR

6.1 *Functions*

The data collector collects all the information exchanged among other modules as well as the internal state of the sub-modules, and stores the data into the database after the testing is finished.

The on-board and trackside simulator will send relevant data to the data collector by means of messages, if the internal state of a sub-module has changed.

The data received by data collector is shown in Fig. 6.

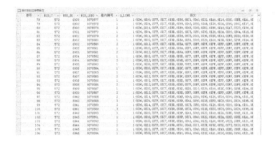

Figure 6. Test result data.

7 SOFTWARE IMPLEMENTATION

The software was developed under MS Studio 2010, with C# programming language and SQL Server 2005. The structure of the system is shown in Fig.7.

Figure 7. The structure of the system.

7.1 *Multithreading in Visual C#*

In Visual C#, the classes of multithreading are contained in the System. Threading name space, with the main methods as follows:

1 Thread. Start, starting thread.
2 Thread. Suspend, suspending a thread.
3 Thread. Resume, resuming a thread.
4 Thread. Abort, quiting a thread.
5 Thread. Sleep, suspending a thread for certain period of time and then resuming it.

Synchronization problems occur when more than one threads access the same data simultaneously. Synchronous data access technology is used to protect the data integrity.

7.2 Test result database design

The test result database stores the information about the train whose entity relationship is shown as in Fig.8, where the Field 'SendMSG'/'RecvMSG' indicate the radio message sent/received by the train respectively. The format of radio messages is shown in Fig.9.

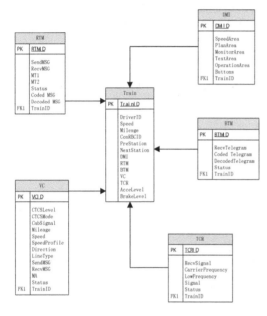

Figure 8.   Entity relationship of train.

MSG3: Movement Authority

Segment No.	Values/Packets	Comment
1	NID_MESSAGE	
2	L_MESSAGE	
3	T_TRAIN	
4	M_ACK	
5	NID_LRBG	
6	CTCS-3 MA	Packet 15
7	Optional Packets	

Figure 9.   Format of radio message.

7.3 Implementation of the software based on multithreading

All the 4 functional modules mentioned above are implemented in one single software by means of multithreading technique. There are totally 7 threads, which are divided as follows:

1   One thread is used by the scenario controller.
2   In the on-board simulator, in order to reduce the time delay in thread execution, two threads are used to simulate the onboard equipment function and train motion dynamics respectively. These two threads communicate with each other via messages.
3   Two threads are designed for RBC simulation, one for the handover RBC and the other for the receiving RBC.
4   One thread is shared by the track circuit sub-module and the balise sub-module.
5   One thread is used by the data collector.

The main window of the software is shown in Fig.10, which consists of the monitoring window and the operating window. The former contains a mimic DMI of the on-board equipment, showing the status of the train and the on-board equipment; the latter contains a mimic driver panel, allowing the tester to drive the train in manual mode.

Figure 10.   Main window.

8   CONCLUSIONS

The CTCS-3 on-board equipment testing software was designed and implemented. The software was divided into four sub-modules: scenario controller, on-board simulator, trackside simulator and data collector. It was developed based on C# and multithreading technique, which can provide a good environment for conducting research on the key technical issues concerning the automatic testing of on-board equipment. Future works include improving the on-board simulator and trackside simulator to support CTCS-2 functions and realizing semi-auto or automatic driving of the train.

ACKNOWLEDGEMENT

This work was supported in part by the key project of China Railway Corporation under Grant 2013X001-F.

REFERENCES

Transportation Bureau of MOR, Test Cases for CTCS-3 Train Control System (v3.0), Beijing: China Railway Publishing House, 2009.

Ji X, LI K, ZHANG Y, et al. Research on the Method of Generating Test Cases of CTCS Level 3 Train Control System [J]. Railway Signaling and Communication, 2009, 45(10): 1–5.

UNISIG, Subset-094 Functional Requirements for an on board Reference Test Facility, 2009.

Liu Y, Tang T. Research on the method of inter-operability test for the onboard equipment of CTCS-3 train control system of Chinese railway[C]. Autonomous Decentralized Systems (ISADS), 2011 10th International Symposium on. IEEE, 2011: 415–419.

Zhang Y, Wang S, Zhang X, et al. A Computer-aided Tool for Generating Test Sequence for CTCS-3 Train Control System [J]. Advances in Information Sciences and Service Sciences, 2012, 4(11).

Zhang Y, Chen Y. Research on Data Management and Configuration in RBC Simulator[C]. International Conference on Information, Electronic and Computer Science, (ICIECS2011), Tianjin, China, Dec. 2011: 414–418.

Computational Intelligence in Industrial Application – Ling (ed.)
© 2015 Taylor & Francis Group, London, ISBN: 978-1-138-02818-0

Fuzzy optimal selection of similar slope for evaluation of slope stability

XuHua Wang, DeShen Zhao & WenJing Liang
Civil and Architectural Engineering College, Dalian University, Dalian, China

ABSTRACT: Considering the randomness and fuzziness of rock and soil engineering, a method of fuzzy optimal selection of similar slope is put forward to analyze slope stability based on the theory of fuzzy optimization. Calculation shows that using this method to evaluate slope stability can achieve good results.

KEYWORDS: Slope stability; Fuzzy optimal selection of similar slope; Maximum relative membership degree.

1 INTRODUCTION

The evaluation of slope stability is often made by the limit equilibrium theory. But the theory fails to fully reflect the randomness and fuzziness of the factors influencing slope stability. Moreover, the slope usually needs to be sliced and numerous preparations are to be made before calculation, it seems to be a little troublesome. In fact, only a general recognition and an accurate and quick judgment on the slope stability are needed in most cases, it is necessary to adopt a method to analyze slope stability that only simple preparations are needed and the numerous influential factors can be reflected comprehensively. Many engineering cases that the slope stability states have been detected clearly provide abundant information for the slope stability analysis that leads to the application of the fuzzy set theory in the slope stability evaluation. Considering the randomness and fuzziness in slope stability evaluation is introduced and a method of fuzzy optimal selection of similar slopes is put forward based on the fuzzy optimization theory, and the practical calculation shows good results.

2 THE BASIC IDEA OF FUZZY OPTIMAL SELECTION OF SIMILAR SLOPES

The slope to be evaluated its stability is determined as the object sample F_a. The index characteristic value vector of the object sample is x_a. N representative stable and failure engineering cases are selected as the source samples that will be compared with the object sample. The index characteristic value vector of the source sample F_j. is x_j, $j=1,2,...,n$ The fuzzy optimization theory is used to calculate the relative membership degree vector $U=(U_1,U_2,...,U_n)$ of the source samples similar to the object sample. U_j. describes the similarity of the object sample to the source sample F_j. The source sample with the maximal relative membership degree is regarded as the best similar one to the object sample, and the stability of the object sample is same as that of the best similar source sample.

Suppose there are n source samples, $F_1,F_2,...,F_n$, and m indexes are used to describe the similarity between the source samples and the object sample, we have the vectors x_j of the m index characteristic values of the source samples and x_a the object sample:

$$\begin{cases} x_a = \left(x_{1a}, x_{2a}, \cdots, x_{ma}\right)^T \\ x_j = \left(x_{1j}, x_{2j}, \cdots, x_{mj}\right)^T \end{cases}$$

So far as the similarity concerned, the similar degree between the single index i of the source sample F_j and that of the object sample F_a is 1 if $x_{ij}=x_{ia}$, that is, the relative membership degree of F_j similar to F_a concerning the single index i is $r_{ji}=1$. According to this, there is the following equation:

$$r_{ij} = 1 - \frac{\left|x_{ij} - x_{ia}\right|}{\underset{i}{\max}\left|x_{ij} - x_{ia}\right|} \tag{1}$$

Eq. (1) can be used to turn the index characteristic value vector x_j of each source sample F_j into the corresponding index relative membership degree vector: $r_j=(r_{1j},r_{2j},...,r_{mj})^T$, which shows the similar degree between F_j and F_a concerning the m single indexes. Where $\underset{i}{\max}\left|x_{ij} - x_{ia}\right|$ in the eq. (1) is the maximum one of the differences between the characteristic value of the single index i of every source sample and that of the object sample F_a.

The eq. (2) is better in describing the differences among the source samples than the corresponding equation in the literature.

Suppose the relative membership degree of F_j similar to F_a is U_j, then U_j can be calculated by eq. (2) according to the fuzzy optimization theory, where w_{ij} is the weight of the m indexes of F_j, and p is the distance parameter.

$$U_j = \cfrac{1}{1 + \left[\cfrac{\sum\limits_{i=1}^{m} \left(w_{ij} \left| r_{ij} - 1 \right| \right)^p}{\sum\limits_{i=1}^{m} \left(w_{ij} r_{ij} \right)^p} \right]^{2/p}} \tag{2}$$

of the n source samples the one with the maximal relative membership degree is selected as the most similar sample to the object slope, and its stability state is regarded as that of the object slope.

3 CASE STUDY

This paper chooses the 18 slope engineering cases presented in the paper as the research object to expound and validate the method of fuzzy optimal selection of similar slopes for slope stability analysis. Influential factors and the stability states of the slopes are listed in table 1. There are 6 influential factors: bulk specific gravity of rock, cohesive force, internal friction angle, slope angle, slope height and pore water pressure ratio. There are two kinds of stability states: stable and failure. The first 12 slopes are selected as the source samples and the last 6 slopes as the object samples.

Here the object slope F_{13} is taken as the example to explain the appraisal process of the slope stability by the method of fuzzy optimal selection of similar slopes. From table 1 the characteristic value vector of the 6 indexes of the object slope F_{13} is obtained as follows:

$$x_a = \left(x_{1a}, x_{2a}, \cdots, x_{6a} \right)^T = (23.00, 0, 20, 20, 100, 0, 0.30)^T$$

The characteristic value matrix of the 6 indexes of the 12 source samples is $X_{6 \times 12}$:

The index characteristic value matrix of the 12 source samples can be changed to the corresponding matrix of the index relative membership degree by eq.(1).

Then the vector of the relative membership degree of the 12 source samples which are similar to the object sample F_{13} can be calculated. In general, the distance may be Haiming distance or O's distance when the parameter p equals to 1 or 2 respectively, and the results are consistent. The Haiming distance is representative and easier to calculate than the O's,

Table 1. Practical slope samples and their stability states.

Source sample	Object sample	Bulk specific gravity	Cohesive force	Internal friction angle	Slope angle	Slope height	Pore water pressure ratio	Slope stability state	Best similar source sample
				Influencing factors of slope stability					
F_1		20.41	20.90	13	22	10.67	0.35	Stable	
F_2		19.63	11.97	20	22	12.19	0.40	Failure	
F_3		20.82	8.62	32	28	12.80	0.49	Failure	
F_4		20.41	33.52	11	16	45.72	0.20	Failure	
F_5		18.84	15.32	30	25	10.67	0.38	Stable	
F_6		21.43	0	20	20	61.00	0.50	Failure	
F_7		19.06	11.71	28	35	21.00	0.11	Failure	
F_8		21.51	6.94	30	31	76.81	0.38	Failure	
F_9		22.40	100.0	45	45	15.00	0.25	Stable	
F_{10}		24.00	0	40	33	8.00	0.30	Stable	
F_{11}		18.00	5.00	30	20	8.00	0.30	Stable	
F_{12}		20.00	20.00	36	45	50.00	0.25	Failure	
	F_{13}	23.00	0	20	20	100.0	0.30	Failure	F_6
	F_{14}	20.00	0	25	20	8.00	0.30	Stable	F_{11}
	F_{15}	14.00	11.97	26	30	88.00	0.45	Failure	F_8
	F_{16}	22.00	0	40	33	8.00	0.35	Stable	F_{10}
	F_{17}	22.40	10.00	35	45	10.00	0.40	Failure	F_3
	F_{18}	18.84	14.36	25	20	30.50	0.45	Failure	F_2

Table 2. The evaluation results of the slope stability.

Object sample	Evaluation result	Best similar source sample	Stability state of best similar source sample	Maximal relative membership degree
F_{13}	Failure	F_6	Failure	0.8574
F_{14}	Stable	F_{11}	Stable	0.9769
F_{15}	Failure	F_8	Failure	0.9107
F_{16}	Stable	F_{10}	Stable	0.9824
F_{17}	Failure	F_3	Failure	0.9096
F_{18}	Failure	F_2	Failure	0.9566

so it is selected in this example. So far as the similarity of two slopes concerned, the 6 influential factors or indexes have approximately the same importance, that is to say, the weight of any one of the 6 indexes is $w_i=1/6$, i=1,2,...,6. we can obtain the vector of the relative membership degree of the 12 source samples which are similar to the object sample F_{13}.

$$U=(U_1,U_2,...,U_n)=(0.72\ 0.71\ 0.43\ 0.67\ 0.51\ 0.86$$
$$0.30\ 0.83\ 0.14\ 0.66\ 0.68\ 0.42)$$

Of the relative membership degrees of the 12 source samples that of F_6 is the maximal, $U_6=0.8574$. So the sample F_6 is the best similar one to F_{13}, and the stability state of F_{13} is accordingly thought to be the same as that of F_6, which is failure. In the same way, the stability states of the other object samples can be evaluated, from F_{14} to F_{18}. the corresponding results are listed in table 2.

We can see from the table 2 that all evaluation results conform to the practical states very well, proved that the method of fuzzy optimal selection of similar slopes presented by this paper is feasible for evaluation of the slope stability.

4 CONCLUSIONS

1 The slope stability is an uncertain problem to a great extent with fuzziness and randomness. The method of fuzzy optimal selection of similar slopes for slope stability evaluation can take the numerous uncertain influential factors into account, obtain characteristics and knowledge from the known engineering cases, and find the best slope sample that is similar to the object slope to be evaluated. The stability of the object slope is thereby evaluated with that of the best similar one.

The presented method offered another effective approach to evaluate the slope stability.

2 This paper only analyzed the 18 slope engineering cases presented in the literature [3], and obtained the satisfying results. Whether or not the method advanced in this paper is applicable and accurate completely to other slopes needs further research.

ACKNOWLEDGEMENT

The paper was supported by the National Natural Science Funds of China 51274051.

REFERENCES

Shen,Liangfeng & Gu,Sulin. 2001. Fuzzy Comprehensive Evaluation for the Slope Stability Analysis. *Journal of Harbin University of Commerce Natural Sciences Edition.* vol. 17(4):111–113.

Xu, Chuanhua, Zhu, Shengwu & Fang, Dingwang. 2000. ISODATA Fuzzy Clustering Analysis of Slope Stability, *Metal Mine.* 12:24–26.

Liu Muyu, Ruigeng Zhu.2002.Case-Based Reasoning Approach to Slope Stability Evaluation Based on Fuzzy Analogy Preferred Ratio. *Journal of Rock Mechanics & Engineering.* Vol. 21:1188–1193.

Dodagoudar, G.R. & Venkatachalam, G. 2000. Reliability Analysis of Slopes Using Fuzzy Sets Theory. *Computers and Geotechnics.* Vol. 27:101–115.

Juang C.H. Jhi Yuin-Yao & Lee Der-Her.1998. Stability Analysis of Existing Slopes Considering Uncertainty. *Engineering Geology.* Vol. 49:111–122.

Chen, Shouyu. 2002. *Fuzzy Pattern Recognition Theory and Application to Complex Water Resource System Optimization.* Jilin: Publishing House of Jilin University.

Chen, Shouyu. 1998.*The Fuzzy Sets Theory and Practice for Engineering Hydrology and Water Resources System.* Dalian:Dalian University of Technology Press.

Computational Intelligence in Industrial Application – Ling (ed.)
© 2015 Taylor & Francis Group, London, ISBN: 978-1-138-02818-0

Research on the automatic analysis and validation methodology of the CTCS-3 onboard equipment testing results

Yong Zhang & WenTing Zhou

School of Electronic and Information Engineering, Beijing Jiaotong University, Beijing, China

ABSTRACT: The CTCS-3 on-board equipment is an important part of the train control system. It is necessary to make sure that on-board equipment meets the requirement specification of the train control system. Since the testing contents and the testing steps are very complicated in the whole process, the work of evaluation of the testing results is very hard. Therefore, the research of the automatic analysis and validation method makes sense. By maintaining the testing data from DMI, JRU and MER, and combining the testing script documents, the automatic analysis and validation method can be achieved.

KEYWORDS: On-board equipment; Test case; Testing results; Automatic analysis and validation.

1 INTRODUCTION

With the maturity of the technology in CTCS, CTCS has been put into the practical application in the modern railway system. CTCS-3 system has two parts. They are on-board equipment and ground equipment. On-board equipment is responsible for receiving commands from ground equipment, generating the speed pattern, monitoring the train's operation and ensuring the safety of train's operation. Therefore, it is necessary to test on-board equipment if it meets the SRS & FRS of CTCS-3 before the formal operation. The analysis and validation of the testing results has an important significance for the train's safe monitor, the accident analysis, the state diagnostic and the statistical analysis.

European has released Subset-094, the functional requirements and specifications to test each module in on-board equipment. Multitel, a company in Belgian, has provided a complete set of solutions for automated testing on-board equipment of ERTMS. It includes DEM (DMI event manager), SA-TAV (semi-automatic trip and analysis validation) and LIT (log importation tool). It can achieve the automatic identification of the information from DMI and semi-automatic analysis and validation of the testing results.

Due to the undisclosed key technologies and the different design goals between China and European countries, the tools from abroad cannot be applied in our country. We have only to study the automatic test technologies and the testing tools.

In the current testing environment, testers need to manually record data for the comparison and analysis after the test. This work is time-consuming and labor-intensive. Therefore, the research of the automatic analysis and validation method makes sense.

This paper introduces the test platform of the on-board equipment, the details of the test solutions and the sources and the types of the testing results. And then, it indicates the method of the automatic analysis and validation of the CTCS-3 onboard equipment testing results.

2 THE INTRODUCTION OF THE TEST PLATFORM OF ON-BOARD EQUIPMENT

The test platform of the on-board equipment includes three parts: the test script import module, the simulation module of the on-board equipment and the analysis and validation module. Its overall structure is shown as figure 1.

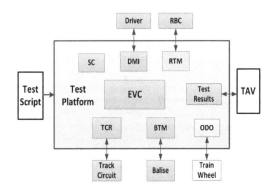

Figure 1.   The structure of the test platform of the on-board equipment.

The first step of the test process is that SC (the scene controller) queries the database to choose one test sequence. Then, SC sends these edited data in the test sequence to each module of the on-board equipment separately. All the modules provide a real environment to the simulation platform and transmit the corresponding data under the specified conditions in the script. In the whole test process, we need to record all information into the database to further analysis and validation.

In order to fully know the working conditions of the on-board equipment, we should focus on the following types of data which are producing in the test. They are the operating level, controlling modes and state information of on-board equipment, track circuit information, speed information, balise information and the information from the wireless communication module.

The above data can be obtained from DMI, MER and JRU module. MER is used to record useful information from other tools and to download data from JRU. It allows offline analysis and validation. There are many modules in the simulation platform which can exchange data with MER. These modules are mainly related to send information and records to MER in order to provide data for the offline analysis. JRU is the judicial record unit. It is used to record information about the operation of the train.

3 AUTOMATIC ANALYSIS AND VALIDATION OF THE TESTING RESULTS

3.1 The system overall structure

The automatic analysis and validation of the testing results is the process to analyze and validate all the data producing in the test. We collect the data from DMI, MER and JRU, and compare the data with the contents in test cases. Then we can get the validation conclusion that the result is right or not in this test and some useful information to further analyze this testing process. The structure of the automatic analysis and validation is shown in figure 2.

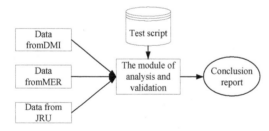

Figure 2. The structure of the automatic analysis and validation module.

3.2 The introduction of the test cases

The test cases are edited with SRS of CTCS-3. A feature is a group of SRS requirements, which can be tested at the available standardized interfaces.

A test case is one test of a feature. In order to test a feature completely, one or more test cases may be necessary.

A test case contains following items.

Basic information: describing the number and name of a test case, the equipment to be tested (onboard or trackside), the test target, as well as SRS requirements related to the test case, etc.

Test method: describing the method and the constraints related to the test case.

Initial conditions (internal state): specifying the mode and level of the onboard equipment, as well as other internal state before executing the test case.

Initial conditions (interface state): specifying the initial state of the available standardized interfaces before executing the test case.

Test procedures: describing step by step the actions necessary for executing the test case. For each step, the step number, I/O event on the related interface and expected result shall be described.

End condition (internal state): specifying the mode and level of the train, as well as other internal state after the test case is executed.

End condition (interface state): specifying the state of the available standardized interfaces after the test case is executed.

A test sequence is a concatenation of test cases. And the all information of the test sequence is included in the test script. The testing results are checked against the expected output of the test cases to obtain the test reports.

3.3 The design and implementation of the automatic analysis and validation method

In one test of the on-board equipment, we first need to collect the testing data from DMI, MER and JRU modules, and put the dada into their own tables in the database.

For the different types of the data, the time we collect it is also different. To the information from BTM and RTM, we can trigger the function which is used to collecting data from the corresponding module when the content is not empty in each row and column of the test script. And in the other hand, considering that the data displayed on the DMI changes all the time in the test, we get the data from DMI module in fixed time, such as 1 second or 5 seconds.

After obtaining the testing data, we can display the data in the playback screen. In the same way, we design different screens about DMI information, BTM information and RTM information, and use the combination of text and graphics to display the testing data. The whole process of analysis is shown as figure 3.

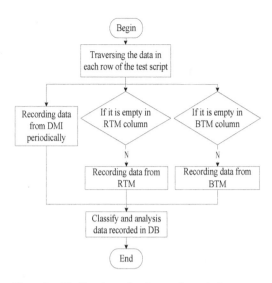

Figure 3. The flowchart of testing results analysis.

When one test of the on-board equipment is finished, we must need to verify the testing results of key steps.

After analyzing the testing results and the ending conditions of the internal state and the interfaces in the test cases, we find that the information about the expected results can be divided into four kinds: the information from DMI, JRU, BTM and RTM. So we only need to select the key information from the database.

For example, when we verify the message information about the BTM or RTM, we can extract the number of the message packet, and then check the contents of the packet according to the number. To the information from DMI, we can check the text information, the speed information, or the braking information if they are consistent with the data in the test script.

In one word, we need to classify the testing results and digitize every kind of the results data to make them have their own number. Then, we program the function of validation for every kind of the testing results. So when verifying the testing results, we just need to searcher the number of the expected results in the test script and use the corresponding method functions to verify the result.

3.4 *The example*

When one test script is involved the test case 98.2, which feature is about receiving a conditional emergency stop message packet and shortening the MA, the automatic analysis and validation module should obtain the numbers of the message packet from RBC to RTM and the message packet from RTM to RBC and the status of wireless connection between the on-board equipment and RBC from the table in the testing results database.

Meanwhile, from the DMI database, the module will get the text information about the emergency message and the latest target distance and speed information.

After obtaining the test results, the module will compare the data with the contents in the test case 98.2 that the numbers of the message packet from RBC to RTM and the message packet from RTM to RBC should be 15 (a conditional emergency stop message) and 147 (confirm the emergency stop message), and the status of wireless connection between the on-board equipment and RBC should be linked.

If the testing results are consistent with the contents in the test case, the module will get a correct conclusion of this step.

4 CONCLUSION

This paper mainly discusses the automatic analysis and validation methodology of the CTCS-3 onboard equipment testing results. We establish a database for storing different types testing data by classifying the data, design the method of automatic analysis the testing results and achieve the testing results playback. Meanwhile, after the analysis we further filter the testing results and use the different methods to validate the different types of data. At last, we get the result of this testing process and generate the reports for the testers to verify the results.

REFERENCES

The science and technology division of railway ministry. *The technology innovation scheme of CTCS-3.*Beijing.
The science and technology division of railway ministry. *The system requirement specification of CTCS-3.*Beijing.
The science and technology division of railway ministry. *The functional requirement specification of CTCS-3.*Beijing.
Liu. Y, Tang. T, Li. K. CH, Yuan. L. 2011 The interoperability testing methods of the on-board equipment of CTCS-3 train control system. *Railway communication and signal.* 47(12).4–7.
Tang. T, Gao.CH. H. 2004. The analysis of ETCS and the research of CTCS. *Electric drive for locomotives.* (6).
Gao. CH. H, Tang. T, Zhang. J. M. 1999.The software design of on-board equipment in high speed train control system. *Beijing jiaotong university joural.* 23.(5).
Xu.L. 2007. *The simulation test platform of CTCS-3—the research of the simulation subsystem of on-board equipment.* Beijing: Beijing jiaotong university.
Chen. J. Q. 2009. *The research of the automatic test method of on-board equipment in CTCS-3.* Beijing: Beijing jiaotong university.
Zhang.W. W, Zhang. Y. 2007. The research of the simulation test platform of on-board equipmentin CTCS-3. Beijing. *Railway computer application.*16(1).4–7.
ERTMS/ETCS-Class 1 UNISIG Functional Requirements for an on board Reference Test Facility. SUBSET-094-0. 04-November-2005.

Design and optimization of a new remotely controllable structure health monitoring system using FBG sensors

W. J. Sheng

College of Automation Engineering, Shanghai University of Electric Power, Shanghai, P R China
School of Electrical Engineering and Telecommunications, University of New South Wales, Australia

N. Yang

College of Automation Engineering, Shanghai University of Electric Power, Shanghai, P R China

ABSTRACT: We designed a new remotely controllable structure health monitoring system using Fiber Bragg Gating (FBG) sensors. Reference FBG's working point can be remotely controlled during measuring process. Parallel processing is employed to improve real-time data acquisition rate in field monitoring system. Self-adaptive sampling is proposed to optimize data acquisition performance of remotely controllable structure health monitoring system based on parallel processing. The experimental results show that the remotely controllable system can achieve the highest possible data acquisition rate without affecting the system operating performance on both local data acquisition and remote process control.

1 INTRODUCTION

Many major civil infrastructures involve a high capital cost and are generally designed for a very long service life. Engineers have long been seeking ways to obtain information about how a structure is behaving in service by incorporating, at the time of construction or subsequently, sensing devices (such as structurally integrated Fiber Optic Sensors (FOSs)) which can provide structural information or conditions such as strain, temperature, and humidity[1, 2]. The development of such structurally integrated fiber optic sensors has led to the concept of smart structures [3-8]. In recent years, remote monitoring is being developed for sending signals from smart structures over Internet to a central monitoring station where the signal are interpreted.

Recently, real-time structural monitoring has received strong interests from both academics and industries. Some research focuses on parallel processing to deal with real-time monitoring [9-11]. These promising results show parallel processing is an efficient multitask method for real-time structural monitoring. On the other hand, the advantages of remote control have also been utilized by many demonstrators [12-14]. For structural monitoring systems, both real-time high-efficient monitoring with parallel processing and remote control of data acquisition process are entirely necessary. However, little research has been focused on the combination of these two parts.

In this paper we proposed a novel scheme with combination of parallel processing and remote control, and incorporation of self-adaptive sampling to overcome the data acquisition rate deficiency due to continual listening of commands. With the combination of parallel processing and self-adaptive sampling, real-time acquisition rate is optimized under remote real-time monitoring and control circumstance.

2 SYSTEM OPTIMIZATION

2.1 System structure

A schematic FBG sensor based structural monitoring system with remote monitoring and controlling capability is shown in figure 1. This system can be divided into three major sub-systems, i.e. local data acquisition sub-system (DAS), local data acquiring PC sub-system and remote monitoring & control station. The DAS collects and transfers data to the local data acquiring PC, which is located in a secure facility adjacent to the structure. The data acquiring PC acquires raw data and performs preliminary processing of the FBG reflection data, which is then sent to remote monitoring & control station via Ethernet. Static or dynamic strain and temperature applied to a FBG sensor change the index modulation, and cause spectral shifts in the FBG sensor's reflectivity spectrum. The whole control process including monitoring spectral shifts and setting working point are implemented in the remote monitoring & control station.

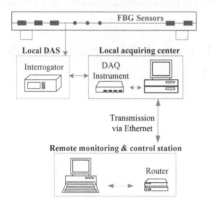

Figure 1. Schematic diagram of structure health monitoring system.

2.2 Principle of parallel processing

The traditional method of configuring data acquiring and processing is in serial operating mode, which is shown in figure 2. In each cycle, the sampled data of the DAQ card is processed and sent to remote PC. However, frequent start and stop executions in each cycle of the loop not only cost excessive time delay but also result in the I/O tasks out of synchronization. The scanning spectrum input and FFP setting output may have phase difference in each cycle that cause signal instability. Furthermore, the low data processing efficiency greatly limits data acquisition rate.

Figure 2. Serial operation of data acquiring and processing.

Data acquiring and processing are configured to work simultaneously without waiting using parallel processing shown in Figure 3. The DAQ starts and stops only once, and the signal input and output will execute in constant phase difference consequently eliminates signal instability problem.

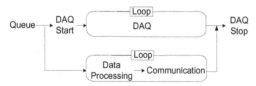

Figure 3. Parallel operation of data acquiring and processing.

2.3 Self-adaptive sampling

Parallel processing is efficient in increase real-time data acquisition rate, but in remotely controllable

structure health monitoring systems, after a part of data processing, local PC has to listen to commands of setting reference FBG working point. Frequent command listening slows down the data processing speed, and causes buffer overflow at the local data acquiring PC. The system will be required to work under much lower rate to avoid data stocked up in the queue buffer. So we introduce a mechanism of self-adaptive sampling on the basis of parallel processing. The workflow employing self-adaptive sampling at local data acquiring PC is shown in figure 4.

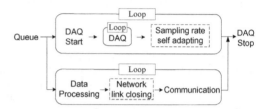

Figure 4. Combination of parallel processing and self-adaptive sampling.

The implementation of self-adaptive sampling includes closing network link and self adapting sampling rate, which are executed in the data acquiring loop and data processing loop separately. Both of these two parts will be executed only once, and the trigger event is the command of setting working point.

3 EXPERIMENT RESULTS

An experimental system for demodulating strain was built, as shown in Fig 5. The ASE light source outputs a broadband light. It illuminates 4 FBGs with different Bragg wavelengths through a 3dB coupler. The reflected light signal is propagated to a photo detector, which transform the intensity of optical signal into the amplitude of voltage and sends this voltage signal out to Labview DAQ card. There is a piezo-electric ceramics in fiber Fabry-Perot (FFP) filter,

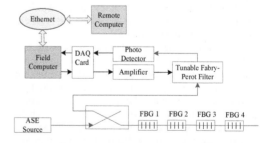

Figure 5. Schematic diagram of demodulation system based on tunable FFP filter.

and the ceramics is driven by an amplified voltage signal generated by DAQ card. The DAQ card sends out control voltage and receives sensing voltage from photo detector simultaneously, meanwhile, it communicates with a personal computer.

All four FBGs are immersed into water and kept under strain free condition to eliminate Bragg wavelength shift. The FFP filter collected 18000 voltage data point in every scanning spectrum. Each experiment is finished in 5 minutes, and the ambient temperature of the water tank during the experiment can be regarded as a constant. Dynamic strain measurement was performed in serial and parallel modes separately, and actual acquisition rates was observed as shown in Fig 6.

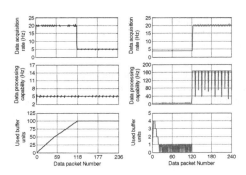

Figure 7. Real-time acquisition performance of parallel processing (a) without, and (b) with self-adaptive sampling.

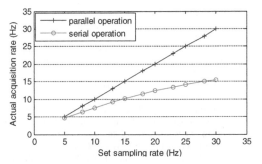

Figure 6. Comparision of actual acquisition rate in serial and parallel operation mode.

The actual data acquisition rate does not always equal to the set sampling frequency. In fact, Fig. 6 demonstrates that the data acquisition rate of the serial processing is much slower than the sampling rate setting initially, while in the parallel processing mode it equals to the sampling rate setting. As the sampling rate is increased, the data acquisition rate with serial operation cannot increase proportionally.

In the field monitoring system, 18000 sample data points can be continually collected, and data acquisition rate can achieve around 30Hz with stable measuring results. After that, we test the performance of parallel processing in the remotely controllable monitoring system, and introduce self-adaptive sampling into the parallel processing. The experiment results is shown as figure 7.

For the remotely controllable monitoring system only with parallel processing, the data acquisition rate is around 20Hz at first, while the data processing capability is about only 5Hz. Due to the speed difference between data acquisition and processing, more and more data packet are stocked up in the buffer. As figure 10(a) shows, the used buffer units continuously grows during the data acquisition process. When the 118th data packet comes, all 100 buffer units have

been filled full. And then the acquisition system has to force acquisition rate switching to a lower level (5Hz) to avoid buffer overflow. Since the data acquisition rate and data processing capability are both around 5Hz, the pushing and popping of the buffer runs at the same speed, and the buffer maintains full with the continually coming data samples.

The self-adaptive sampling is proposed to resolve the conflict between remote control and high acquisition rate. With the data processing capability is around 5 Hz at first, the initial sampling rate is lowered to 4Hz. After data acquisition starts, data samples are processed timely, and the usages of buffer units decrease very fast and steadily fluctuate between 0 and 1. After 120 data packets are processed, the remote command of setting working point is received. When local working point is set successfully on the local acquiring PC, the network link for the command of working point setting is closed, and data processing capability increases by a large margin. Subsequently, the sampling rate is self adapting to 20Hz and the coming data samples can still be processed timely and the buffer maintains empty.

The experiments results of data acquisition rate that compare performance of a system without and with self-adaptive sampling are shown in figure 8.

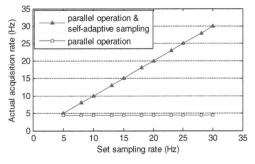

Figure 8. Comparison of before and after using the self-adaptive sampling on the basis of parallel processing.

443

The data acquisition rate can only maintain at about 4 Hz when the initial sampling rate is set at different values. However, by employing self-adaptive sampling, the data acquisition rate can always follow the self adjusted sampling rate. From the above experiments results, it is obvious that self-adaptive sampling is a necessary and effective complement to parallel processing in remote monitoring and control environment.

4 CONCLUSIONS

Remote structural monitoring systems have extremely large potential application areas. Higher acquisition rate is important to provide high frequency structural monitoring, while remote control is an essential extending to field structural monitoring. The mechanism introduced in this paper can be applied into remote structural monitoring systems to realize remote control without significant penalty of data acquisition rate. The mechanism is implemented without much coding, and is easy to be transplanted to the other structural monitoring and control systems. Alternatively, the technique can also be used to dynamically adjust data acquisition rate to decrease the transfer bandwidth in remote structural monitoring systems.

REFERENCES

Q. Zhang, T. Zhu, J. Zhang, K. Chiang, Photonics Technology Letters 25 (2013) 1751.

S. Isaacs, F. Placido, I. Abdulhalim, Applied Optics 53 (2014) H91.

S. Niu, Y. Liao, Q. Yao, Y. Hu, Optics Communications 285 (2012) 2826.

Y. Ma, C. J. Wang, Y. H. Yang, S. B. Yan, J. M. Li, Optics&Laser Technology 50 (2013) 107.

M. Han, T. Q. Liu, L. L. Hu, Q. Zhang, Optics Express 21(2013) 29269.

Y. C. Ma, H. Y. Liu, S. B. Yan, Y. H. Yang, Measurement Science and Technology 24 (2013) 055201.

H. Ding, J. Liang, J. Cui, X. Wu, Sensors and Actuators B: Chemical 138 (2009) 154.

K. Liu, T. Liu, J. Jiang, G. D. Peng, H. Zhang, Lightwave Technology 29 (2011) 15.

K. Liu, W. C. Jing, G. D. Peng, Y. M. Zhang, Optics communications 281(2008) 3286.

H. Qi, S Wei, C Wei, Optoelectronics Letters 9 (2013) 101.

H. Ding, X. N. Wu, J. Q. Liang, X. L. Li, Measurement 42 (2009) 1059.

B. Q. Jiang, J. L. Zhao, C. Qin, Z. Huang, F. Fan, Optics and Laser in Engineering 49 (2011) 415.

A. G. Kallapur, Low pass filter model-based offline estimation of ring-down time for an experimental Fabry-Perot optical cavity, in: IEEE International Conference on Control Applications (CCA), 2012, pp: 75–79.

S. Schultz, W. Kunzler, Z. Zhu, M. Wirthlin, R. Selfridge, Smart Materials and Structures 18 (2009) 115015.

Computational Intelligence in Industrial Application – Ling (ed.)
© *2015 Taylor & Francis Group, London, ISBN: 978-1-138-02818-0*

Author index

T - #0336 - 101024 - C0 - 246/174/62 [64] - CB - 9781138028180 - Gloss Lamination